Diana Bowers

Methods in Enzymology

Volume 230
GUIDE TO TECHNIQUES IN GLYCOBIOLOGY

METHODS IN ENZYMOLOGY

EDITORS-IN-CHIEF

John N. Abelson   Melvin I. Simon

DIVISION OF BIOLOGY
CALIFORNIA INSTITUTE OF TECHNOLOGY
PASADENA, CALIFORNIA

FOUNDING EDITORS

Sidney P. Colowick and Nathan O. Kaplan

Methods in Enzymology

Volume 230
Guide to Techniques in Glycobiology

EDITED BY

William J. Lennarz

DEPARTMENT OF BIOCHEMISTRY AND CELL BIOLOGY
STATE UNIVERSITY OF NEW YORK AT STONY BROOK
STONY BROOK, NEW YORK

Gerald W. Hart

DEPARTMENT OF BIOCHEMISTRY AND MOLECULAR GENETICS
UNIVERSITY OF ALABAMA AT BIRMINGHAM
BIRMINGHAM, ALABAMA

ACADEMIC PRESS, INC.
A Division of Harcourt Brace & Company
San Diego   New York   Boston   London   Sydney   Tokyo   Toronto

Front cover illustration (paperback edition only) : Cartoon of a Fast Atom Bombardment (FAB) ion source together with a partial negative FAB mass spectrum of sulfated
glycans derived from the 16K fragment of bovine proopiomelanocortin. Courtesy of
Professor Anne Dell, Department of Biochemistry, Imperial College of Science,
Technology and Medicine, London, United Kingdom.

Academic Press, Inc.
525 B Street, Suite 1900, San Diego, California 92101-4495

United Kingdom Edition published by
Academic Press Limited
24–28 Oval Road, London NW1 7DX

International Standard Serial Number: 0076-6879

International Standard Book Number: 0-12-182131-5 (Hardcover)
International Standard Book Number: 0-12-443665-X (Paperback)

PRINTED IN THE UNITED STATES OF AMERICA
94 95 96 97 98 99   EB   9 8 7 6 5 4 3 2 1

Table of Contents

v

Contributors to Volume 230

Article numbers are in parentheses following the names of contributors.
Affiliations listed are current.

JACQUES U. BAENZIGER (14), *Departments of Pathology and Cell Biology and Physiology, Washington University School of Medicine, St. Louis, Missouri 63110*

ANTHONY CALABRO (24), *Bone Research Branch, National Institute of Dental Research, National Institutes of Health, Bethesda, Maryland 20892*

WENGANG CHAI (28), *Clinical Mass Spectrometry Section, MRC Clinical Research Centre, Harrow, Middlesex HA1 3UJ, United Kingdom*

TEH-YING CHOU (26), *Department of Biochemistry and Molecular Genetics, Schools of Medicine and Dentistry, University of Alabama at Birmingham, Birmingham, Alabama 35294*

RICHARD D. CUMMINGS (6), *Department of Biochemistry and Molecular Biology, University of Oklahoma Health Sciences Center, Oklahoma City, Oklahoma 73190*

ANNE DELL (8), *Department of Biochemistry, Imperial College of Science, Technology, and Medicine, London SW7 2AZ, United Kingdom*

RICHARD R. DRAKE (20), *Department of Biochemistry and Molecular Biology, University of Arkansas for Medical Sciences, Little Rock, Arkansas 72205*

ALAN D. ELBEIN (19), *Department of Biochemistry and Molecular Biology, University of Arkansas for Medical Sciences, Little Rock, Arkansas 72205*

TEN FEIZI (28), *Glycoconjugates Section, MRC Clinical Research Centre, Harrow, Middlesex HA1 3UJ, United Kingdom*

CARL G. GAHMBERG (3), *Department of Biochemistry, University of Helsinki, Helsinki, Finland*

HILDEGARD GEYER (7), *Biochemisches Institut am Klinikum der Universität Giessen, University of Giessen, D-35385 Giessen, Germany*

RUDOLF GEYER (7), *Biochemisches Institut am Klinikum der Universität Giessen, D-35385 Giessen, Germany*

MARK R. HARDY (11), *Immunogen Inc., Cambridge, Massachusetts 02139*

GERALD W. HART (26), *Department of Biochemistry and Molecular Genetics, Schools of Medicine and Dentistry, University of Alabama at Birmingham, Birmingham, Alabama 35294*

VINCENT C. HASCALL (24), *Bone Research Branch, National Institute of Dental Research, National Institutes of Health, Bethesda, Maryland 20892*

SUMIHIRO HASE (13), *Department of Chemistry, Osaka University College of Science, Toyonaka, 560 Osaka, Japan*

SADAKO INOUE (27), *Department of Biophysics and Biochemistry, Faculty of Science, University of Tokyo, Tokyo 113, Japan*

YASUO INOUE (27), *Department of Biophysics and Biochemistry, Faculty of Science, University of Tokyo, Tokyo 113, Japan*

PETER JACKSON (15), *Biomethod Consultants (Fulbourn), Cambridge CB1 5DS, United Kingdom*

GARY S. JACOB (17), *Glycosciences Group, Department of Protein Biochemistry, Monsanto Corporate Research, St. Louis, Missouri 63167*

GUR P. KAUSHAL (19), *Department of Biochemistry and Molecular Biology, University of Arkansas for Medical Sciences, Little Rock, Arkansas 72205*

KAY-HOOI KHOO (8), *Department of Biochemistry, Imperial College of Science, Technology, and Medicine, London SW7 2AZ, United Kingdom*

vii

KEN KITAJIMA (27), *Department of Biophysics and Biochemistry, Faculty of Science, University of Tokyo, Tokyo 113, Japan*

AKIRA KOBATA (11), *Tokyo Metropolitan Institute of Gerontology, Tokyo 173, Japan*

ALEXANDER M. LAWSON (28), *Clinical Mass Spectrometry Section, MRC Clinical Research Centre, Harrow, Middlesex HA1 3UJ, United Kingdom*

ROBERT J. LINHARDT (16), *Division of Medicinal and Natural Products Chemistry, College of Pharmacy, The University of Iowa, Iowa City, Iowa 52242*

ROY A. MCDOWELL (8), *Department of Biochemistry, Imperial College of Science, Technology, and Medicine, London SW7 2AZ, United Kingdom*

ANANT K. MENON (25), *Department of Biochemistry, University of Wisconsin–Madison, Madison, Wisconsin 53706*

ROBERTA K. MERKLE (1), *Complex Carbohydrate Research Center, University of Georgia, Athens, Georgia 30602*

RONALD J. MIDURA (24), *Department of Orthopedic Surgery, The University of Iowa, Iowa City, Iowa 52242*

HOWARD R. MORRIS (8), *Department of Biochemistry, Imperial College of Science, Technology, and Medicine, London SW7 2AZ, United Kingdom*

LEILA K. NEEDHAM (23), *Department of Experimental Pathology, Guy's Hospital, London Bridge, London SE1 9RT, United Kingdom*

MONICA M. PALCIC (18), *Department of Chemistry, University of Alberta, Edmonton, Alberta, Canada T6G 2G2*

MARIA PANICO (8), *Department of Biochemistry, Imperial College of Science, Technology, and Medicine, London SW7 2AZ, United Kingdom*

R.B. PAREKH (5, 21), *Oxford GlycoSystems Ltd., Abingdon, Oxon OX14 1RG, United Kingdom*

T.P. PATEL (5), *Oxford GlycoSystems Ltd., Abingdon, Oxon OX14 1RG, United Kingdom*

THOMAS H. PLUMMER, JR. (4), *Department of Biochemistry, New York State Department of Health, Wadsworth Center for Laboratories and Research, Albany, New York 12201*

IZABELLA POPPE (1), *Complex Carbohydrate Research Center, University of Georgia, Athens, Georgia 30602*

ANNA RADOMINSKA (20), *Department of Medicine, University of Arkansas for Medical Sciences, Little Rock, Arkansas 72205*

ANDREW J. REASON (8), *Department of Biochemistry, Imperial College of Science, Technology, and Medicine, London SW7 2AZ, United Kingdom*

GERD REUTER (10), *Biochemisches Institut, Christian-Albrechts-Universität, D-24098 Kiel, Germany*

ELIZABETH P. ROQUEMORE (26), *Department of Biochemistry and Molecular Genetics, Schools of Medicine and Dentistry, University of Alabama at Birmingham, Birmingham, Alabama 35294*

ROLAND SCHAUER (10), *Biochemisches Institut, Christian-Albrechts-Universität, D-24098 Kiel, Germany*

RONALD L. SCHNAAR (22, 23), *Department of Pharmacology and Neuroscience, Johns Hopkins School of Medicine, Baltimore, Maryland 21205*

PETER SCUDDER (17), *Glycosciences Group, Department of Protein Biochemistry, Monsanto Corporate Research, St. Louis, Missouri 63167*

MARK S. STOLL (28), *Glycoconjugates Section, MRC Clinical Research Centre, Harrow, Middlesex HA1 3UJ, United Kingdom*

ANTHONY L. TARENTINO (4), *Department of Biochemistry, New York State Department of Health, Wadsworth Center for Laboratories and Research, Albany, New York 12201*

MARTTI TOLVANEN (3), *Department of Biochemistry, University of Helsinki, Helsinki, Finland*

R. REID TOWNSEND (12), *Department of Pharmaceutical Chemistry, University of California, San Francisco, San Francisco, California 94143*

FREDERIC A. TROY II (27), *Department of Biological Chemistry, University of California School of Medicine, Davis, California 95616*

HERMAN VAN HALBEEK (9), *Complex Carbohydrate Research Center, and Department of Biochemistry, The University of Georgia, Athens, Georgia 30602*

AJIT VARKI (2), *Cancer Center, School of Medicine, University of California, San Diego, La Jolla, California 92093*

MASAKI YANAGISHITA (24), *Bone Research Branch, National Institute of Dental Research, National Institutes of Health, Bethesda, Maryland 20892*

JEAN YE (27), *Department of Biological Chemistry, University of California School of Medicine, Davis, California 95616*

CHUN-TING YUEN (28), *Glycoconjugates Section, MRC Clinical Research Centre, Harrow, Middlesex HA1 3UJ, United Kingdom*

Preface

Two developments, one in academic research and another in industry, are clear-cut indicators of the maturing of a distinct field, glycobiology. The academic development is the publication of a journal devoted entirely to the subject of glycobiology. The industrial one is the commercial production of a variety of glycoproteins for use in medical applications. These developments, coupled with continuously improving methods to express recombinant proteins in a variety of eukaryotic hosts, and clear-cut evidence that the carbohydrate chain of glycoproteins is important for the stability of proteins and, in some cases, their biological function have created the need for easy access to a variety of methods for studying the structures of carbohydrate chains. Because of the large number of potential permutations and combinations in sugar types, linkage position, and anomeric configuration, the methods for carbohydrate chain analysis are far more diverse than those for protein or DNA sequence analysis.

The purpose of this *Guide* is to collate in one source and format this wide variety of methods. Each contribution provides background material explaining the basis of the methods for the less knowledgeable scientist, followed by detailed procedures. For individuals with more expertise, the background section can be skipped, and the researcher can proceed directly to the protocols. We hope this *Guide* will be useful to a wide group of researchers in industry and the academic world.

WILLIAM J. LENNARZ
GERALD W. HART

[1] Carbohydrate Composition Analysis of Glycoconjugates by Gas-Liquid Chromatography/Mass Spectrometry

By ROBERTA K. MERKLE and IZABELLA POPPE

Introduction

To determine the monosaccharide composition of a glycoconjugate it is necessary to cleave the oligosaccharide into its components before analysis. Three useful methods of cleavage are acid hydrolysis, methanolysis, and solvolysis by hydrogen fluoride. Following cleavage, the nonvolatile monosaccharide residues are often derivatized to volatile forms, and subjected to analysis by gas-liquid chromatography and combined gasliquid chromatography/mass spectrometry. This chapter is not intended to be an exhaustive review; instead, we discuss several methods of glycosidic bond cleavage as well as two major derivatization methods, formation of alditol acetates and trimethylsilylation, used for glycosyl composition analysis. Basic procedures for cleavage of oligosaccharides and for the derivatization methods used for composition analysis are described in detail.

Glycosidic Bond Cleavage

Hydrolysis

Glycosidic bonds differ in their susceptibility to acid hydrolysis. The rate of release is dependent on the position and anomeric configuration of the glycosidic linkage as well as on the identity of the monosaccharide.[1–4] Additionally, because on hydrolysis the released monosaccharides differ in their susceptibility to destruction by acid, it is necessary to choose conditions for hydrolysis that preserve the free sugars. This often necessitates a compromise between allowing complete hydrolysis and preserving the released residues. Depending on the nature of the monosaccharide composition of the glycoconjugate in question, it may be necessary to

[1] J. G. Beeley, *Lab. Tech. Biochem. Mol. Biol.* **16**, 100 (1985).

[2] J. H. Pazur, *in* "Carbohydrate Analysis: A Practical Approach" (M. F. Chaplin and J. F. Kennedy, eds.), p. 55. IRL Press, Oxford, 1986.

[3] J. Montreuil, S. Bouquelet, H. Debray, B. Fournet, G. Spik, and G. Strecker, *in* "Carbohydrate Analysis: A Practical Approach" (M. F. Chaplin and J. F. Kennedy, eds.), p. 143. IRL Press, Oxford, 1986.

[4] J. Conchie, *Methods Carbohydr. Chem.* **7**, 195 (1976).

METHODS IN ENZYMOLOGY, VOL. 230

determine the kinetics of hydrolysis and/or to employ differing acidic conditions, analyzing the composition at varying time points. For example, hexosaminyl linkages are often not labile under conditions sufficient for the hydrolysis of neutral hexosyl linkages. Additionally, the acidic conditions used for cleavage results in loss of N-acetyl groups from the amino sugars, and subsequent re-N-acetylation is necessary before derivatization for composition analysis. Alternatively, more than one set of hydrolysis and derivatization conditions may be necessary for a complete analysis. More detailed discussions of the chemical parameters of lability of the glycosidic linkage and lability of the released monosaccharides are found in several reviews.[4–6]

Methanolysis

An alternate method of cleavage is methanolysis. This method overcomes the problems of hydrolysis described above, resulting in high yields of the carbohydrates from glycoproteins as well as from other glycoconjugates because there is less destruction of carbohydrate than with aqueous acids.[7,8] An exception to this statement is that the GlcNAc attached to asparagine is not efficiently cleaved, and when it is, glucosamine is released as a free monosaccharide rather than as a methylglycoside.[3] The glycoconjugate is treated with methanolic HCl, and the result of the reaction is a mixture of methylglycosides (or methyl ester methylglycosides in the case of glycosyluronic acids).[9] We use methanolysis in conjunction with derivatization by trimethylsilylation of the resultant methylglycosides for composition determination by gas-liquid chromatography-mass spectrometry.[4]

Solvolysis

Another method, using anhydrous hydrogen fluoride, has been used for glycosidic bond cleavage. This solvolytic procedure cleaves most glycosidic linkages without affecting the N-acyl moieties of the acylamido sugars, the N-glycosidic linkage between asparagine and N-acetyl-

[5] C. J. Biermann, in "Analysis of Carbohydrates by GLC and MS" (C. J. Biermann and G. D. McGinnis, eds.), p. 247. CRC Press, Boca Raton, FL, 1989.
[6] C. J. Biermann, Adv. Carbohydr. Chem. Biochem. **46,** 251 (1988).
[7] R. E. Chambers and J. R. Clamp, Biochem. J. **125,** 1009 (1971).
[8] M. F. Chaplin, Anal. Biochem. **123,** 336 (1982).
[9] W. S. York, A. G. Darvill, M. McNeil, T. T. Stevenson, and P. Albersheim, this series, Vol. 118, p. 3.

glucosamine, or the peptide bonds.[10,11] As there is little, if any, degradation of the released monosaccharides, they can be derivatized for subsequent composition analysis by the procedures detailed below. However, because this procedure entails the use of a special apparatus to handle the highly toxic, corrosive HF, it may not be practical to set up for infrequent use.[12] For detailed procedures of the HF solvolytic method, the reader is referred to Mort et al.[10,11] and Rorrer et al.[12]

Derivatization

Because gas-liquid chromatographic separation depends on the differential distillation of the components of the mixture, a prerequisite for the analysis is the formation of volatile derivatives of the carbohydrates.[13] We generally use either of two methods to generate volatile derivatives for the glycosyl composition analysis of glycoconjugates: production of alditol acetates or trimethylsilylation (TMS). The former method involves a series of derivatization steps, but results in the formation of a single derivative per monosaccharide. Although the TMS method requires fewer manipulations, it results in a somewhat more complex chromatographic pattern. Because the monosaccharides remain in the ring form on methanolysis, a mixture of the pyranose and furanose forms, as well as the α and β anomers of each monosaccharide, are generated on derivatization. However, the multiple peaks derived from each monosaccharide can produce a characteristic pattern that further aids in identification by retention time.[13,14] Furthermore, the TMS method allows the simultaneous identification of neutral sugars and amino sugars as well as of acidic sugars (e.g., sialic acid and uronic acid).

Composition Analysis by Gas-Liquid Chromatography and Mass Spectrometry

Gas-liquid chromatography (GC) is a sensitive analytical technique for monosaccharide composition analysis, allowing detection of subnanomole

[10] A. J. Mort and D. T. A. Lamport, *Anal. Biochem.* **82**, 289 (1977).

[11] A. J. Mort, P. Komalavilas, G. L. Rorrer, and D. T. A. Lamport, *in* "Modern Methods of Plant Analysis" (H.-F. Linskens and J. F. Jackson, eds.), New Ser., Vol. 10, p. 37. Springer-Verlag, Berlin, 1989.

[12] G. L. Rorrer, M. C. Hawley, S. M. Selke, D. T. A. Lamport, and P. M. Dey, *Methods Plant Biochem.* **2**, 585 (1990).

[13] M. F. Chaplin, *in* "Carbohydrate Analysis: A Practical Approach" (M. F. Chaplin and J. F. Kennedy, eds.), p. 1. IRL Press, Oxford, 1986.

[14] T. Bhatti, R. E. Chambers, and J. R. Clamp, *Biochim. Biophys. Acta* **222**, 339 (1970).

amounts of carbohydrates.[13] The method provides information on both identity and quantity of the component monosaccharides.[15] It is useful for glycoproteins, as the protein generally does not interfere with the analysis. For glycolipids, the fatty acid components can also be analyzed separately from the carbohydrate components by the TMS derivatization method. A method for TMS derivatization of fatty acids and analysis by GC-MS has been described.[16]

The most frequently used GC columns are fused silica, wall-coated open tubular columns that allow high resolution of peaks even when derivatization results in more than one chromatographic peak per monosaccharide, as in the TMS procedure (see below). Although identification of the monosaccharide derivatives, many of which are isomers of one another, is made on the basis of GC retention time, combined gas-liquid chromatography-mass spectrometry (GC-MS) provides a further refinement in identification. The mass spectra from these derivatives give a fragmentation pattern that can be compared to known spectra of standards and/or by deduction from the known cleavage characteristics. Thus, although isomers will give identical mass fragments, the carbohydrate nature of a peak identified by GC retention can be confirmed, providing another level of identification. In combined GC-MS, the mass spectrometer serves as the detector for the GC. In our laboratory, we generally perform GC by using a chromatograph with a flame-ionization detector in addition to combined GC-MS.

Procedures

Alditol Acetate Derivatization of Neutral Sugars

The formation of alditol acetate derivatives involves four steps subsequent to hydrolysis[17]:

1. reduction to alditol by use of sodium borohydride (or borodeuteride)
2. removal of excess borate
3. acetylation of hydroxyl and amino groups
4. analysis by gas chromatography and gas chromatography-mass spectrometry

A detailed discussion of the chemistry of this derivatization method can be found in Fox *et al.*[17] (and references therein). The procedure for

[15] R. A. Laine, W. J. Esselman, and C. C. Sweeley, this series, Vol. 28, p. 159.
[16] W. R. Mayberry, *J. Bacteriol.* **143,** 582 (1980).
[17] A. Fox, S. L. Morgan, and J. Gilbart, *in* "Analysis of Carbohydrates by GLC and MS" (C. J. Biermann and G. D. McGinnis, eds.), p. 87. CRC Press, Boca Raton, FL, 1989.
[18] J. Lönngren and S. Svensson, *Adv. Carbohydr. Chem. Biochem.* **29,** 4 (1974).

derivatization is essentially as has been described previously for analysis of plant cell wall polysaccharides.[9]

Sample Preparation. The glycoconjugate sample, containing 5–500 μg of carbohydrate, free of salt and detergents, should be prepared by drying (e.g., lyophilization) in a 13 × 100 mm screw-cap tube. When analyzing samples containing small amounts of carbohydrates, special methods may be helpful (see below for a discussion of such methods). A standard sugar mixture should be prepared concurrently (see below for discussion of standard preparation and quantitation). Before drying, 2–20 μg of *myo*-inositol is added to each sample and to the standard mixture as an internal standard (less inositol than the amount of carbohydrate in the sample should be added; e.g., if the sample contains 5 μg of carbohydrate, add 2 μg of *myo*-inositol). The standard sugar mixture is carried through the entire procedure in parallel with the sample(s) to be analyzed.

Hydrolysis. For samples containing neutral sugars, hydrolysis is carried out with 2 *M* trifluoroacetic acid (TFA). Add 0.5 ml of 2 *M* TFA to the dry samples, cover with a Teflon-lined screw cap, and incubate in a temperature block at 121° for 2 hr.

The acid is evaporated by drying under a stream of nitrogen while incubating the samples in a water bath at 40–50°. We use an N-Evap analytical evaporator (Organomation Associates, Inc., Berlin, MA). Alternatively, a homemade apparatus, such as that described previously, may be used.[9] Remaining traces of acid are removed by addition of 250 μl of 2-propanol (or *tert*-butanol) to the dried sample, drying under a stream of nitrogen, and repeating the alcohol addition and drying once more.

Reduction. The dried, hydrolyzed sample is dissolved in 0.25 ml of 1 *M* ammonium hydroxide containing sodium borohydride (10 mg/ml) and incubated at room temperature for 2–18 hr. Alternatively, sodium borodeuteride can be employed in place of sodium borohydride in order to produce asymmetric mass fragments on mass spectral analysis.[18] This is more important when previously uncharacterized monosaccharide structures are present.

Borate Removal. The solution is acidified and excess borohydride destroyed by adding glacial acetic acid dropwise until bubbling (release of hydrogen gas) ceases. Evaporate to dryness at room temperature. Borate ions are removed by adding 0.5 ml of a mixture of methanol–acetic acid (9 : 1, v/v), and drying under a stream of nitrogen at room temperature. Repeat the addition of methanol–acetic acid and evaporation three more times. Finally, add 0.5 ml of methanol, and evaporate to dryness under nitrogen.

Acetylation. The alditols are O-acetylated by the addition of 0.1 ml each of acetic anhydride and pyridine. The tube is capped and heated for

20 min at 121° in a dry bath. The samples are dried under a stream of nitrogen (or air) at room temperature (or up to 40°). Complete removal of the reagents is facilitated by adding 50–100 μl of the toluene to the solution before evaporation.

The dried residue is dissolved in 0.5 ml of dichloromethane (or chloroform). Water (0.5 ml) is added and the sample is vortexed, then centrifuged, at 1000 g for 2 min at room temperature to separate the organic and aqueous phases. The organic (bottom) layer is transferred to a clean test tube with a Pasteur pipette, without transferring any of the aqueous phase. (*Note:* If some of the aqueous phase is inadvertently transferred, drying can be effected by adding a pinch of sodium sulfate to the organic phase, mixing by vortexing, and then filtering it over glass wool that has been packed into a Pasteur pipette.) Dry the organic phase under a stream of nitrogen.

The sample is now ready for analysis by GC and GC-MS. Dissolve the dried sample in 50–100 μl of acetone. This is a general guideline for a sample containing 100 μg of carbohydrate. If the starting sample contained less than this amount the final solution can be dissolved in a smaller volume of acetone. If, on GC injection, the sample is too concentrated it can be diluted; if the sample is below the limits of detection, the solvent can be evaporated and the sample dissolved in a smaller volume. The sensitivity of these procedures and of parameters affecting sample preparation are discussed in the last section of this article.

Analysis of Alditol Acetate Derivatives of Neutral Sugars by Gas-Liquid Chromatography/Mass Spectrometry

The sample (1 μl) is injected onto a gas chromatograph, using a Supelco (Bellefonte, PA) SP-2330 15-m fused silica capillary column (0.25-mm i.d.), and the alditol acetate derivatives are detected by a flame-ionization detector. An injection splitter is used at a ratio of 1:50. The oven temperature program includes an initial temperature of 190° for 2 min, then an increase to 240° at a rate of 10°/min, holding at 240° for 10 min. Identification of the component sugars is based on comparison of the retention times with those of standard sugars. The method for quantitation of the component monosaccharides is described below (Preparation of Standards and Quantitation of Monosaccharide Components of Analyzed Sample).

As discussed, further confirmation of the sugar peaks identified by GC retention times is accomplished by mass spectral analysis. For combined GC-MS analysis of alditol acetate derivatives of the neutral sugars we use a Hewlett-Packard (Palo Alto, CA) 5890 GC coupled to a 5970 mass

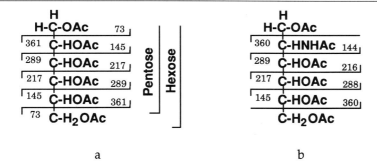

a   b

FIG. 1. Mass spectral fragmentation pattern of alditol acetate derivatives of (a) neutral pentose or hexose and (b) amino sugars GlcNAc and GalNAc.

selective detector (MSD) with a Supelco SP-2330 fused silica capillary column (30 m). On electron ionization (EI), characteristic fragments are produced from the alditol acetate derivatives.[18] For example, Fig. 1 shows the fragmentation patterns for the hexoses (e.g., galactose, mannose, and glucose) and pentoses (e.g., ribose, arabinose, and xylose). A number of references include detailed discussion of the parameters influencing fragmentation.[13,17,19]

The temperature program used for combined GC-MS is slightly different than that for GC, because splitless injection is employed to allow a larger sample application (necessary for good mass spectra) to the column. The initial temperature is 80° held for 2 min, with an increase at a rate of 30°/min to a final temperature of 235°, which is then held for 35 min. A gas chromatographic profile of the alditol acetate derivatives of the component neutral monosaccharides of the glycoprotein fetuin is shown in Fig. 2A. Also illustrated (Fig. 2B) is a sample mass spectrum of one of the peaks.

Alditol Acetate Derivatization of Amino Sugars

Although the preparation of the sample is the same as described above for neutral sugars, the hydrolysis of the glycosidic bonds of amino sugars requires stronger acid conditions. Additionally, to prevent formation of borate complexes with the free amino groups that result on hydrolysis, the amino groups are N-acetylated before the reduction step.

[19] D. Patouraux-Promé and J.-C. Promé, in "Gas Chromatography/Mass Spectrometry Applications in Microbiology" (G. Ohdam, L. Larsson, and P.-A. Moardh, eds.), p. 105. Plenum, New York, 1984.

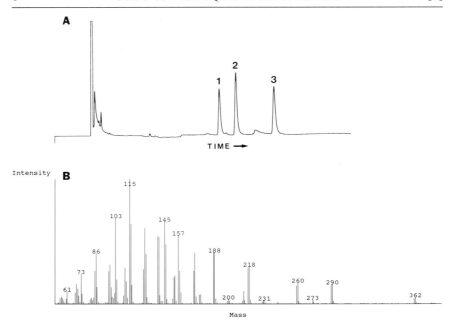

FIG. 2. Gas chromatography of the alditol acetate derivatives of neutral sugar components of bovine fetuin and mass spectrum of a neutral hexose. Fetuin (1 mg of protein, containing approximately 240 μg of carbohydrate) was hydrolyzed along with 2 μg of inositol, using 2 M TFA, and derivatized by the alditol acetate method for neutral sugars as described in text. The acetylated alditols were dissolved in 100 μl of acetone and 1 μl was subjected to gas-liquid chromatography (A). The peaks were identified by comparison of retention times with known standards: 1, mannose; 2, galactose; 3, inositol. (B) A portion of the sample was also subjected to combined GC-MS, and the mass spectrum for mannose (the mass spectrum for galactose is identical) is shown. Note that because the derivatization was carried out with sodium borodeuteride as the reductant, the mass spectrum includes the asymmetric fragments produced by the presence of deuterium on the anomeric carbon (C-1).

Sample Preparation. The sample is prepared as above for the neutral sugars, in parallel with a mixture of the standard amino sugars, *N*-acetylgalactosamine and *N*-acetylglucosamine. The sample and standard mixture should contain 2–20 μg of *myo*-inositol as an internal standard.

Hydrolysis. Hydrolysis is accomplished by adding 500 μl of 4 N HCl to the dried sample, sealing the tube with a Teflon-lined screw cap, and heating at 100° for 18 hr. Remove the HCl by evaporation under a stream of air or nitrogen. Add 500 μl of H_2O, and evaporate under air or nitrogen at room temperature or up to 40°. Repeat the addition of H_2O and evaporation once more.

Re-N-Acetylation. To the dried sample, add the following reagents:

Deionized H$_2$O, 100 μl
Acetic anhydride [5% (v/v) in H$_2$O, prepared fresh just prior to use],
   25 μl
Saturated sodium bicarbonate, 25 μl

Incubate the solution at room temperature for 3 min. Repeat the addition of the acetic anhydride and saturated sodium bicarbonate (25 μl each) twice more (for a total of three times), incubating for 3 min at room temperature after the second addition, and for 20 min at room temperature after the third (final) addition. Stop the reaction by heating for 4 min at 100°. After cooling to room temperature, evaporate the solution under a stream of nitrogen or air. The remainder of the procedure (reduction, borate removal, and acetylation) is identical to that for the neutral sugars (above).

Analysis of Alditol Acetate Derivatives of Amino Sugars by Gas-Liquid Chromatography/Mass Spectrometry

The sample (1 μl) is injected onto a gas chromatograph, using a Supelco DB-1 fused silica capillary column (20 m). The oven temperature program includes an initial temperature of 200° held for 3 min, with an increase at a rate of 3°/min to a final temperature of 260°. Identification of the component sugars is based on comparison of the retention times with those of standard sugars. The method for quantitation of the component monosaccharides is described below (Preparation of Standards and Quantitation of Monosaccharide Components of Analyzed Sample).

For combined GC-MS analysis of the amino sugars we use a Hewlett-Packard 5890 GC coupled to a 5970 MSD with a Supelco DB-1 fused silica capillary column (30 m). The less polar DB-1 column allows elution of the amino sugar derivatives at a lower temperature than from the more polar SP-2330 column used for the neutral sugar derivatives. However, the DB-1 column does not allow resolution of the neutral derivatives. The temperature program includes an initial temperature of 80° held for 2 min, with an increase at a rate of 20°/min to a final temperature of 180°. The fragmentation pattern for the amino sugars GlcNAc and GalNAc is shown in Fig. 1. For further information on the parameters of fragmentation, refer to Fox *et al.*,[17] Lönngren and Svensson,[18] and Patouraux-Promé and Promé.[19] A gas chromatogram and mass spectrum are shown in Fig. 3 for the alditol acetate derivatives of the amino sugars GlcNAc and GalNAc of fetuin.

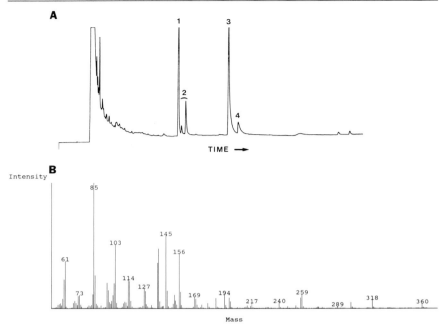

FIG. 3. Gas chromatography of the alditol acetate derivatives of the amino sugar components of bovine fetuin and mass spectrum of an amino sugar. Fetuin (1 mg of protein, containing approximately 240 μg of carbohydrate) was hydrolyzed along with 20 μg of inositol, using 4 N HCl, re-N-acetylated, and derivatized by the alditol acetate method for amino sugars as described in text. The derivatives were dissolved in 100 μl of acetone and 1 μl was subjected to gas-liquid chromatography (A). The peaks were identified by comparison of retention times with known standards: 1, inositol; 2, neutral hexoses; 3, N-acetyl-glucosamine; 4, N-acetylgalactosamine. (B) A portion of the sample was also subjected to combined GC-MS, and the mass spectrum for GlcNAc is shown (the mass spectrum for GalNAc is identical). Note that because the derivatization was carried out with sodium borodeuteride as the reductant, the mass spectrum includes the asymmetric fragments produced by the presence of deuterium on the anomeric carbon (C-1).

Derivatization by Trimethylsilylation

In the TMS derivatization reaction the monosaccharide hydroxyls are converted to silyl ethers.[20] This modification occurs quickly under mild conditions, converting the monosaccharides to volatile derivatives suitable for gas-liquid chromatography. Kakehi and Honda[20] provide further details on this procedure, including the use of alternative silylating agents, the

[20] K. Kakehi and S. Honda, in "Analysis of Carbohydrates by GLC and MS" (C. J. Biermann and G. D. McGinnis, Eds.), p. 43. CRC Press, Boca Raton, FL, 1989.

features of the resultant derivatives, and references on the variety of carbohydrates analyzed by this procedure.

Preparation of Methanolic HCl (1 M). We use reagents from the instant methanolic HCl kit (Alltech, Deerfield, IL). Prepare a 100-ml three-necked flask by heating in a drying oven. The flask, with a magnetic stir bar added, is placed in an ice bath on a stir plate in a fume hood. The two outside necks are used as an inlet and an outlet for a stream of nitrogen; an addition funnel with an equalizing line and stopcock is attached to the central neck. Add 20 ml of anhydrous methanol through the funnel, and then add 2.8 ml of acetyl chloride dropwise while stirring. After approximately 5 min, transfer aliquots of the reagent to dry screw-cap tubes and store at $-20°$ over desiccant. Alternatively, methanolic HCl can be purchased (Aldrich, Milwaukee, WI) and diluted to the desired concentration with anhydrous methanol.

Sample Preparation. The sample, containing 20–1000 μg of carbohydrate, and the mixture of standard sugars are prepared exactly as described above under Alditol Acetate Derivatization of Amino Sugars, adding 20 μg of *myo*-inositol as an internal standard.

Methanolysis. To the dried sample and standard mixture, add 500 μl of 1 M methanolic HCl. Incubate for 16 hr (overnight) at 80°. Ten minutes after placing the tubes in the heating block, double-check that the caps are tightly sealed, as it is important to avoid leakage of methanol or introduction of water into the reaction vessel.

Evaporate the methanolic HCl by drying at 40° under a stream of nitrogen or air. Add 250 μl of methanol and evaporate; repeat this addition and evaporation step once more.

N-Acetylation. (If the sample does not contain amino sugars, this step is not necessary.) N-Acetylation is effected by adding 200 μl of methanol, 40 μl of pyridine, and 40 μl of acetic anhydride. Seal the tube(s) with a Teflon-lined screw cap and incubate at room temperature for 6–18 hr (overnight). Evaporate the solvent and excess acetylating reagents under nitrogen or air at 40°.

Silylation. To silyate the methylglycosides (or methyl ester methylglycosides resulting from glycosyluronic acids) produced by methanolysis, the silylating reagents hexamethyldisilazane and trimethylchlorosilane are added in the solvent pyridine.[21] We add 200 μl to each dried sample from a 1-ml ampoule of the commercially available Tri-Sil reagent, which contains both the silylating agents and the solvent (Pierce, Rockford, IL).

[21] C. C. Sweeley, R. Bentley, M. Makita, and W. W. Wells, *J. Am. Chem. Soc.* **85**, 2497 (1963).

The tube is sealed with the screw cap and incubated at 80° for 20 min. After cooling to room temperature, evaporate the reagents under a gentle stream of nitrogen or air at room temperature. This evaporation should be carried out for a limited time (just until the sample is dry), as they are volatile after silylation. Add 1 ml of hexane, mix by vortexing, centrifuge at 1000 g for 2 min at room temperature to facilitate settling of insoluble salts, and then transfer the supernatant hexane to a fresh tube. Evaporate under nitrogen or air at room temperature just until dry. The sample is now ready for analysis by GC-MS. Resuspend the dried sample in 100 μl of hexane.

Analysis of Trimethylsilylated Derivatives by Gas-Liquid Chromatography-Mass Spectrometry

The sample (1 μl) containing the TMS methylglycosides is injected (split at 1 : 50) onto a gas chromatograph, using a Supelco DB-1 fused silica capillary column (20 m). The program used consists of an initial temperature of 160° held for 3 min, with a subsequent increase at a rate of 3°/min to a final temperature of 260°, which is held for 15 min. Identification of the component sugars is based on comparison of the retention times with those of standard sugars. Multiple peaks will be produced for the component sugars. Quantitation is carried out as described below, except that the peak areas of all the major derivatives of each sugar are added together for calculation of response factors and for calculation of monosaccharide components of the sample.

GC-MS analysis is performed with a Hewlett-Packard 5890 gas chromatograph coupled to a 5970 MSD, using a Supelco DB-1 fused silica capillary column (30 m), with splitless injection. The program used includes an initial temperature of 80° held for 2 min, with an increase at a rate of 20°/min to a final temperature of 160°. On electron ionization mass spectral fragmentation $m/z = 73$ for the characteristic base fragment of all TMS methylglycosides. For neutral sugars $m/z = 204$ is characteristic, for amino sugars $m/z = 173$ is characteristic, and $m/z = 298$ is characteristic of the sialic acids. An example of GC/MS analysis is shown in Fig. 4. For a detailed discussion of mass spectrometry of silyl ethers, see Kakehi and Honda.[20]

Preparation of Standards and Quantitation of Monosaccharide Components of Analyzed Sample

Depending on the nature of the glycoconjugate being analyzed, the standard sugars derivatized in parallel with the sample can include arabinose, ribose, rhamnose, fucose, xylose, mannose, galactose, glucose, N-acetylglucosamine, N-acetylgalactosamine, N-acetylneuraminic acid, galacturonic acid, glucuronic acid, and glucoheptose. The monosaccharide standards are stored dry in a desiccator after purchase. Stock solutions of each sugar are prepared at a concentration of 1 mg/ml in water. To prepare a standard mixture, 1-ml aliquots of each sugar relevant to the particular glycoconjugate being analyzed are combined in a single tube. Fivefold less *myo*-inositol (200 μl) is added to the mixture. This standard mixture may be frozen. When ready to analyze a sample, 500 μl of the standard mixture is lyophilized, and subjected to parallel derivatization with the sample. The amount (in milligrams) of each sugar in the standard mixture is a function of the number of sugars that were mixed together.

Detector response factors (the ratio of the peak area of the internal standard to that of each standard sugar) are determined for each standard and are used to calculate the amount of each component of the sample being analyzed.[22] The response factor for each component of the standard mixture is calculated according to the following formula, using mannose as an example:

Response factor

$$= \frac{\text{total peak area for mannose/peak area of inositol internal standard}}{\text{weight of mannose (mg)/weight of internal standard (mg)}}$$

After a peak is identified in the analyzed sample ("unknown") as mannose by comparison of retention time with those of known standards, it can then be quantified by weight, using the response factor calculated above:

Weight of mannose (mg) in unknown

$$= \frac{\text{peak area of mannose in unknown} \times \text{weight of inositol in unknown (mg)}}{\text{peak area of inositol in unknown} \times \text{response factor for mannose}}$$

Although the response factors are similar from run to run, a standard mixture should always be prepared and derivatized in parallel with the sample(s) being analyzed.

Parameters Affecting Sample Analysis and Considerations of Sensitivity

The capillary columns of small inner diameter used for these analyses provide high resolution. However, because of their small capacity a split injection technique is used, and consequently only a fraction of the injected sample is actually applied to the column.[23] Theoretically, with combined

[22] K. M. Brobst, *Methods Carbohydr. Chem.* **6**, 3 (1972).
[23] C. J. Biermann, *in* "Analysis of Carbohydrates by GLC and MS" (C. J. Biermann and G. D. McGinnis, eds.), p. 1. CRC Press, Boca Raton, FL, 1989.

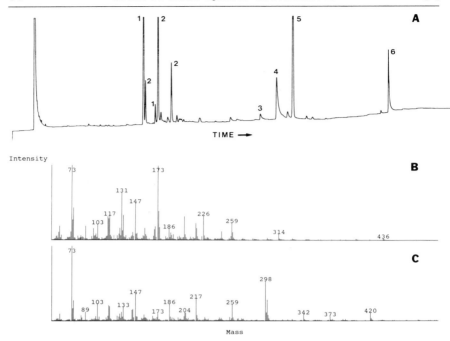

FIG. 4. Gas chromatography of the TMS derivatives of the sugar components of bovine fetuin and mass spectra of an amino sugar and sialic acid. Fetuin (1 mg of protein, containing approximately 240 μg of carbohydrate) was subjected to methanolysis, along with 20 μg of inositol, and derivatized by the TMS method as described in text. The derivatives were dissolved in 100 μl of hexane and 1 μl was subjected to gas-liquid chromatography (A). The peaks were identified by comparison of retention times with known standards: 1, mannose; 2, galactose; 3, N-acetylgalactosamine; 4, N-acetylglucosamine; 5, inositol; 6, sialic acid. A portion of the sample was also subjected to combined GC/MS, and the mass spectra are shown for GlcNAc (the mass spectrum for GalNAc is identical) (B) and for sialic acid (C).

GC-MS, in which the mass spectrometer is the detector, the limit of detection is 10 ng for a single monosaccharide. Practically speaking, however, the sensitivity is lower. For example, we carried out derivatization of differing amounts (50–100 μg) of the calf serum glycoprotein fetuin by the three methods described above. Fetuin has three N- and three O-linked oligosaccharides, with a sialylated triantennary oligosaccharide as the predominant structure of the N-linked chains and a variety of structures comprising the O-linked oligosaccharides, including the trisaccharide NeuAcα2-3Galβ1-3GalNAc.[24–30] The monosaccharide present in the low-

[24] R. G. Spiro, J. Biol. Chem. 239, 567 (1964).
[25] R. G. Spiro and V. D. Bhoyroo, J. Biol. Chem. 249, 5704 (1974).

est abundance is GalNAc, with 3 mol of GalNAc/mol glycoprotein, in an approximate ratio of 1 : 5 with GlcNAc. Samples containing 100 and 150 µg of fetuin would be expected to contain approximately 1.4 and 2.0 µg, respectively, of GalNAc. With the TMS derivatization method, we were able to detect GalNAc only in the larger sample. However, we have previously analyzed samples as small as 5 µg that are predominantly carbohydrate (e.g., 80% by weight) and detected 100 ng of a monosaccharide by this derivatization method.

To improve yields when analyzing samples containing small amounts of carbohydrate, micromethods can be employed.[31] Such modifications include using conical microvials as reaction vessels, silanizing the glassware, and using proportionally less reagent during the derivatization. Another means for detecting the sugars present in low abundance is to use multiple selected ion chromatography when doing GC/MS; that is, after chromatography, scanning for the combination of ions expected for the different monosaccharides.

Acknowledgments

We thank Dr. Russell Carlson and William York for helpful comments on the manuscript. Research was supported by the following grants to the Complex Carbohydrate Research Center, University of Georgia: NIH Biomedical Resource Center Program, Grant 1-P41-RR05351 (R.K.M. and I.P.) and DOE Center for Plant and Microbial Complex Carbohydrates, Grant DE-FG09-87ER13810 (I.P.).

[26] B. Nilsson, N. E. Nordén, and S. Svensson, *J. Biol. Chem.* **254**, 4545 (1979).

[27] R. R. Townsend, M. R. Hardy, T. C. Wong, and Y. C. Lee, *Biochemistry* **25**, 5716 (1986).

[28] E. D. Green, G. Adelt, J. U. Baenziger, S. Wilson, and H. van Halbeek, *J. Biol. Chem.* **263**, 18253 (1988).

[29] B. Bendiak, M. Harris-Brandts, S. W. Michnick, J. P. Carver, and D. A. Cumming, *Biochemistry* **28**, 6491 (1989).

[30] D. A. Cumming, C. G. Hellerqvist, M. Harris-Brandts, S. Michnick, J. P. Carver, and B. Bendiak, *Biochemistry* **28**, 6500 (1989).

[31] T. J. Waeghe, A. G. Darvill, M. McNeil, and P. Albersheim, *Carbohydr. Res.* **123**, 281 (1983).

[2] Metabolic Radiolabeling of Glycoconjugates

By AJIT VARKI

Introduction

If sufficient quantities of pure molecules are available, the complete and definitive structural analysis of glycoconjugates can be done with a combination of physical and chemical methods that are described elsewhere in this volume. However, isolating adequate quantities of a glycoconjugate may not be practical (e.g., in the analysis of biosynthetic intermediates, or of rare molecules). Alternatively, the biological question at hand might be adequately answered by partial structural analyses. In both these situations, metabolic labeling with radioactive sugars (or with other donors that label sugar chains) can be performed, and substantial structural information about the labeled oligosaccharide chains can be obtained.[1] The advantages of this approach include simplicity, ease of use, and the lack of need for sophisticated instrumentation (except for a scintillation counter). Furthermore, the purification of the glycoconjugate to "radiometric homogeneity" is sufficient.[2] The practical considerations in taking this approach include selection of the labeled precursor, understanding the specificity of labeling, and maximizing uptake and incorporation of the label. Most of the research using this approach has been carried out on animal cells, particularly mammalian cells. Thus, this chapter tends to focus more on metabolic labeling of glycoconjugates in such cells. It is assumed that the reader is familar with conventional protocols for handling, monitoring, shielding, and disposal of radioactive materials, and with the principles of tissue culture of cells and sterile handling of media. Following the choice of a specific radiolabeled precursor and the optimization of labeling conditions, preparative labeling is performed. The labeled glycoconjugate of interest is then isolated, and identification and separation of individual glycosylation sites may be necessary. Structural analyses can then be carried out on the labeled oligosaccharides.

Although metabolic labeling can provide much useful information regarding glycoprotein oligosaccharides, there are significant limitations in its use. First, it is often difficult to determine when true "equilibrium labeling" of a cell is reached (three to four doublings are usually assumed to be sufficient, see later). Thus, unless the specific activity of individual

---

[1] R. D. Cummings, R. K. Merkle, and N. L. Stults, *Methods Cell Biol.* **32**, 141 (1989).
[2] A. Varki, *FASEB J.* **5**, 226 (1991).

monosaccharide pools is carefully determined, the numerical ratio between various labeled glycoconjugates can be misleading. Second, individual precursors have greatly differing uptake and incorporation in different types of cells. Third, almost all available labeled precursors are only partially specific for certain monosaccharides. The degree of this specificity can be variable, depending on the cell type, and with prolonged labeling some radioactivity can even enter molecules other than oligosaccharides. Fourth, although monosaccharides that compete with glucose for uptake show improved incorporation when glucose concentrations are lowered, the low glucose supply can have direct effects on the oligosaccharide precursors in some cell types.

Background and General Principles

Sugar nucleotides are the immediate donors for glycosylation reactions. Because these negatively charged molecules cannot be taken up by cells in culture, metabolic labeling of the oligosaccharides is accomplished with radiolabeled monosaccharide precursors.[2-4] These commercially available molecules can be taken up by cells and activated to sugar nucleotides, which are directly utilized for cytosolic and nuclear forms of glycosylation[5] or transported into the Golgi apparatus,[6,7] where lumenally oriented transferases can add the monosaccharides to lumenally oriented acceptors (Fig. 1). Modifications of oligosaccharides such as sulfate esters, acetate esters, and phosphate esters can also be metabolically labeled with [^{35}S]sulfate, [^{3}H]acetate, or ^{32}Pi.

The labeling protocol to follow will depend on the objectives of the investigator. If labeling is being performed only to obtain labeled oligosaccharides for structural characterization, maximum yield may be the objective, regardless of the actual labeling protocol. If the objective is to establish precursor–product relationships, a pulse–chase protocol must be used. If there is a need to compare the masses of two molecules in the same sample, or to quantitate the mass of the same molecule in two different samples, it is necessary to label to constant radiospecific activity. As discussed later, this may not always be achievable.

[3] P. D. Yurchenco, C. Ceccarini, and P. H. Atkinson, in "Methods in Enzymology" (V. Ginsburg, ed.), Vol. 50, p. 175. Academic Press, New York, 1978.

[4] M. Yanagishita, A. Salustri, and V. C. Hascall, in "Methods in Enzymology" (V. Ginsburg, ed.), Vol. 179, p. 435. Academic Press, Orlando, FL, 1989.

[5] G. W. Hart, R. S. Haltiwanger, G. D. Holt, and W. G. Kelly, Annu. Rev. Biochem. **58**, 841 (1989).

[6] C. B. Hirschberg and M. D. Snider, Annu. Rev. Biochem. **56**, 63 (1987).

[7] B. Fleischer, in "Methods in Enzymology" (S. Fleischer and B. Fleischer, eds.), Vol. 174, p. 173. Academic Press, Orlando, FL, 1989.

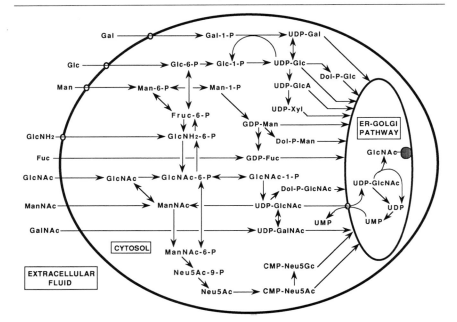

FIG. 1. Uptake of exogenously added monosaccharide precursors, and cytosolic pathways for their interconversion and activation to nucleotide sugar forms—a composite of the commonly known pathways. Individual pathways may be more or less prominent in different cell types. The monosaccharide pools can also be diluted by a contribution from degradation and recycling of endogenous glycoconjugates. Most of the sugar nucleotides and/or their dolicholphosphoryl derivatives are shown as being used for reactions occurring in the lumen of the endoplasmic reticulum (ER)–Golgi pathway. However, some can also be used for glycosylation reactions occurring in the cytosol (e.g., UDP-GlcNAc as a donor of O-linked GlcNAc residues). The transport and fate of one sugar nucleotide (UDP-GlcNAc) within the Golgi apparatus is shown.

Selection of a labeled monosaccharide precursor is based mainly on the efficiency of uptake (see later), and the type of glycoconjugate to be labeled (the distribution of monosaccharides among different types of vertebrate oligosaccharides is not random, as shown in Table I). The labeling can either be carried out for a long time (an attempt at reaching "equilibrium" distribution of the label), or for a short time (for a pulse–chase analysis). The type of monosaccharide decided on, and the period of labeling, will determine whether normal or reduced glucose in the medium should be used (see later). The type of monosaccharide precursor used and the position of the isotope can also affect specificity and final distribution of the label. The metabolic pathways for the biosynthesis, activation, utilization, and interconversion of various monosaccharides

TABLE I
MONOSACCHARIDES AND MODIFICATIONS: OCCURRENCE IN GLYCOCONJUGATES
OF MAMMALIAN CELLS[a]

Mono-saccharide	Type of glycoconjugate					
	N-GlcNAc-linked glycoprotein	O-GalNAc-linked glycoprotein	O-Xyl-linked proteoglycan	Glyco-sphingolipid	GPI[b] anchor	O-Linked GlcNAc
Man	+ + +	−	−	−	+ +	−
Fuc	+ +	+ +	+ / −	+	−	−
Gal	+ + +	+ + +	+ (Core)	+ + +	+ / − (Side chain)	−
Glc	+ (Precursor)	−	−	+ (Core)	−	−
GlcNAc	+ + +	+ + +	+ +	+ +	+ (Free amine)	+ +
GalNAc	+ / −	+ + +	+ +	+ +	+ / − (Side chain)	−
Sia	+ + +	+ + +	−	+ + +	+ (Side chain)	−
GlcA	+ / −	−	+ + +	+	−	−
SO₄	+	+	+ + + +	+	−	−
Pi esters	Man-6-P	−	Xyl-P	−	M-6-P in core	−
O-Acetyl	Sia-OAc	Sia-OAc	−	Sia-OAc	−	−
O-Acyl	−	−	−	+ + + +	+ / − (On inositol)	−

[a] The relative distribution of the different monosaccharides and modifications in the most commonly occurring glycoconjugates are indicated by + to + + +, and is valid for many cell types. However, an uncommon monosaccharide or modification may be commonly found in a given cell type or type of glycoconjugate. Also, some of the negative listings may well turn out to be incorrect in the future.

[b] GPI, Glycosylphosphatidylinositol.

and their nucleotide sugars have been studied extensively in the past (see Fig. 1 for the common pathways in animal cells, and the points at which exogenously added sugars can enter the endogenous pathways). The final distribution and specific activity of label can be significantly affected by dilution from endogenous pathways, the conditions of labeling, and the particular cell type under study.[3,4,8,9] The interconversion between monosaccharides (see Fig. 1 and Table II) includes several epimerization reactions that can result in loss of label from a precursor. Thus, the position of the label within the monosaccharide used can affect the ultimate fate

[8] A. Varki and S. Kornfeld, J. Biol. Chem. **258**, 2808 (1983).
[9] J. J. Kim and E. H. Conrad, J. Biol. Chem. **251**, 6210 (1976).

TABLE II

COMMONLY USED RADIOLABELED PRECURSORS: INCORPORATION INTO MONOSACCHARIDES OF MAMMALIAN CELLS[a]

Precursor	Entry of label into:							
	[³H]Man	[³H]Fuc	[³H]Gal	[³H]Glc	[³H]GlcA	[³H]GlcNAc	[³H]GalNAc	[³H]Sia
[2-³H]Man	+++	+++	−	−	−	−	−	−
[6-³H]Gal	+/−	+/−	++++	+++	−	+/−	+/−	+/−
[1-³H]Gal	+/−	+/−	++++	+++	+	+/−	+/−	+/−
[1-³H]Glc	++	++	+++	++++	++	++	++	++
[6-³H]GlcNH₂	+/−	+/−	+/−	+/−	−	++++	+++	++
[6-³H]ManNAc	+/−	+/−	+/−	+/−	−	++	+	++++
[6-³H]GalNAc	+/−	+/−	+/−	+/−	+/−	++	+++	+

[a] Extent of conversion from the original monosaccharide into minor pathways will vary considerably, depending on the cell type studied, and the length of the labeling (longer time allows more conversion into other monosaccharides).

of the radioactivity. For example, label originating from the 6-position of [6-^{3}H]GlcNH$_2$, [6-^{3}H]GlcNAc, or [6-^{3}H]ManNAc is eventually found at either the 6-position of GlcNAc or GalNAc residues or at the 9-position of terminal sialic residues.[10] However, eventual interconversion with glucose can be expected for most labeled monosaccharides. Thus, the label can spread into other monosaccharides and even into the nonoligosaccharide components of the cell. An exception is the extreme specificity of labeling obtained with [2-^{3}H]mannose, which should remain confined to mannose and fucose residues, regardless of how long the labeling is done. However, exceptions to this specificity have been reported.[11,12]

Many other factors can affect the uptake and incorporation of radioactive precursors into oligosaccharides. With radioactive amino acids, high specific activity labeling can be obtained simply by omitting the unlabeled molecule from the medium. Labeling with radioactive sugars is usually not as efficient, making it necessary to optimize incorporation into the glycoconjugate of interest. Many factors can affect the uptake and incorporation of radioactive monosaccharide precursors into macromolecules: The general factors are relatively self-evident, and include the amount of label, the concentration of label in the media, the number of cells, the duration of labeling, and the number of cell doublings that occur during the labeling. Some of these factors are obviously at odds with one another, and the correct balance between them must be individualized to the particular cell type and the experimental question at hand.

Whether a radioactive monosaccharide precursor competes with glucose for uptake is also an important factor affecting uptake and incorporation. Because the concentration of glucose in typical tissue culture media is about 5 mM, those monosaccharides that compete for active transport cannot be taken up any better than glucose itself. In these cases, lowering the concentration of glucose in the medium can improve the uptake of such precursors. The sugars that do not compete with glucose seem to be taken up only by inefficient mechanisms that are noncompetitive and passive. In these cases, manipulation of the glucose concentration has little effect on their relatively poor uptake. A comprehensive study of these issues has not been performed. Past published and unpublished experience indicates that glucosamine, galactosamine, galactose, and mannose compete with glucose for uptake into most cells, whereas N-acetylglucosamine, N-acetylmannosamine, mannosamine, fucose, and xylose do not.[3] However, it is now recognized that the glucose transporters

[10] S. Diaz and A. Varki, *Anal. Biochem.* **150,** 32 (1985).
[11] K. E. Creek, S. Shankar, and L. M. De Luca, *Arch. Biochem. Biophys.* **254,** 482 (1987).
[12] I. R. Rodriguez and S. J. Fliesler, *Exp. Eye Res.* **51,** 71 (1990).

are actually a family of gene products that are tissue specific in their expression.[13] Thus, a given cell may express more than one of these transporters, each of which may have distinctive kinetic and stereospecific uptake properties. Monosaccharides should be tested with the specific cell type under study, to determine which ones compete with glucose. When glucose-free medium is used in some cell types, synthesis of the lipid-linked oligosaccharide precursor of N-linked oligosaccharides can become altered.[14]

Other factors can affect the final specific activity and distribution of the cellular label, including the pool size of individual monosaccharides and nucleotides, the dilution of the label by endogenously synthesized monosaccharides, and the flux rates between interconverting pathways, as mentioned above. Also, when certain cell types are cultured in labeled medium for what appear to be reasonable periods of time, the glucose level in the medium can fall substantially.[2,8,9] In this case, the specific activity of labeled sugar nucleotides can actually rise during the latter part of the labeling, as the medium glucose concentration falls below the K_m of the glucose transporter(s) in the particular cell type. Although all of the factors discussed above must be considered in designing the final labeling protocol, some may not be easy to control. The ultimate goal is to obtain a sufficient amount of radioactivity in the glycoprotein of interest to permit the studies planned. However, this must be done without substantially altering the metabolic state of the cell.

Modifications of oligosaccharide chains such as acetylation, sulfation, acylation, and phosphorylation can have significant effects on the behavior of oligosaccharides in biological systems and during analysis. In these cases, the oligosaccharides can also be metabolically labeled with appropriate precursors such as [^{3}H]acetate, [^{35}S]sulfate, [^{3}H]palmitate, and ^{32}Pi. Naturally, such precursors are expected to enter into a wide variety of other cellular macromolecules, and the release and isolation of the labeled oligosaccharides is required before further analysis. In obtaining release, specific endoglycosidases are preferred, because chemical methods such as hydrazinolysis and alkaline β elimination can partially or completely destroy the modifications of interest. In many cases, double-labeling of the modification and the underlying sugar chain (e.g., with a ^{14}C-labeled monosaccharide precursor and ^{3}H-labeled acetate) can be useful in monitoring the purification and in carrying out the subsequent structural analysis.[15]

[13] G. W. Gould and G. I. Bell, *Trends Biochem. Sci.* **15,** 18 (1990).
[14] J. I. Rearick, A. Chapman, and S. Kornfeld, *J. Biol. Chem.* **256,** 6255 (1981).
[15] A. E. Manzi, E. R. Sjöberg, S. Diaz, and A. Varki, *J. Biol. Chem.* **265,** 13091 (1990).

Sulfate and phosphate ions are components of normal tissue culture media. Although their selective removal can improve labeling with the respective precursor, the biological effects of depletion of these ions can also occur. In the case of labeling with [^{35}S]sulfate, the efficiency of labeling varies widely with different cell types. In some cells, the endogenous pool of the sulfate donor 3' phosphoadenosine 5' phosphosulfate (PAPS) appears to be small, and the specific activity of the [^{35}S]sulfate is practically the same inside the cell as in the exogenously added label.[4] However, in other cell types, the endogenous sulfate pool is constantly diluted by the endogenous breakdown of the sulfur-containing amino acids methionine and cysteine.[16,17] In the latter situation, altering the cysteine and methionine concentrations in the medium can improve the incorporation of label into macromolecules.[16] Certain antibiotics, for example, gentamicin, are provided as sulfate salts, which can contribute significantly to the final sulfate concentration of the medium. Likewise, undialyzed serum contains sulfate ions.

In many cases, the precise specific activity of each monosaccharide pool does not need to be determined to allow interpretation of the experiment. If "equilibrium" labeling (see the next section) is attempted, a plateau in the rate of incorporation of label per milligram of cell protein can be taken as an approximate indication that a steady state has been reached. However, in some cases, the precise specific activity of a given monosaccharide may be of interest. Detailed discussions of these matters can be found elsewhere,[2–4,8,9] but certain points are worth noting. Endogenous glucose is the normal precursor for hexoses and hexosamines (see Fig. 1). Experimental manipulations that alter the concentration of glucose can thus alter the concentration of the internal pool of other hexoses and hexosamines. Also, cells frequently recycle their monosaccharides after degradation in the lysosomes. Thus, exogenously added labeled hexosamines become diluted in the cell, making specific radioactivity in the cell generally lower than that of the starting material. Yanagishita et al.[4] describe in detail a protocol to determine the specific activity of hexosamines in metabolically radiolabeled molecules. This approach takes advantage of the fact that both [^{3}H]glucosamine and [^{35}S]sulfate are incorporated into chondroitin sulfate and that defined disaccharides from this sugar chain can be isolated. It assumes that the [^{35}S]sulfate label is not significantly diluted from cellular sources (however, see discussion above). With the advent of high-pH anion-exchange chromatography and pulsed

[16] L. Roux, S. Holoyda, G. Sundblad, H. H. Freeze, and A. Varki, *J. Biol. Chem.* **236**, 8879 (1988).
[17] J. D. Esko, A. Elgavish, T. Prasthofer, W. H. Taylor, and J. L. Weinke, *J. Biol. Chem.* **261**, 15725 (1986).

amperometric detection (HPAE-PAD) analysis of monosaccharides it is now possible to measure monosaccharides accurately in the low picomole range (see [12] in this volume). The specific activity of a labeled monosaccharide component of a purified oligosaccharide can be obtained by acid hydrolysis of a portion of the sample, followed by HPAE-PAD analysis. Other high-performance liquid chromatography (HPLC)-based methods for the accurate measurement of low quantities of sulfate and sialic acids are also now available.

Ideally, the radiochemical purity of the precursor to be used should be checked before starting the experiment. The radioactive precursor may have become chemically degraded during storage, and even established commercial sources have been known to provide precursors contaminated by unknown compounds.

"Equilibrium" Labeling of Cells with Radioactive Precursors

When studying glycoconjugates in established tissue culture cell lines, it is often desirable to label the molecules to allow structural characterization. Whenever feasible, such molecules should be labeled as close as possible to "equilibrium," under conditions of normal growth of the cells. True equilibrium labeling is defined as reaching a constant level of radioactivity per mass unit of a given monosaccharide in all the glycoconjugates in the cell. In practice, this is difficult to achieve with most cell types because some macromolecules are more long lived than others. Regardless, the goal is to expose the maximum number of cells possible to the maximum amount of label feasible in the minimum volume possible for the longest possible period of time, in complete growth medium. Before labeling is attempted, the following growth characteristics should be determined for the cell type of interest:

1. doubling time under standard conditions of growth
2. maximum degree of dilution on splitting that is compatible with proper regrowth of cells
3. maximum cell density compatible with healthy growth and metabolism ("confluence")
4. minimum volume of medium in which cells can sustain full regrowth of cells from a full split [see point (2)] to the maximum density [see point (3)]

If these parameters are properly defined, growth from point (2) to point (3) should permit several doublings (at least three) for most cell lines. There may be cases (e.g., slowly growing cell lines, lines requiring frequent medium changes, or cells that cannot be split far back) in which this might not be entirely feasible.

Materials

Complete tissue culture medium (with supplements) suitable for long-term growth of the cell line

Sterile tissue culture supplies (e.g., sterile pipettes, 15-ml conical tubes with caps, flasks, or dishes)

Disposable 0.22-μm pore size filters suitable for sterile filtration of media, with Luer-lock fittings

Sterile plastic syringes with Luer-lock fittings; alternatively, flasks fitted with 0.22-μm pore size filters for sterile filtration of media

Tissue culture hood

Humidified incubator, equilibrated at 37° and 5% CO_2 (or whatever conditions are appropriate)

Radioactive precursor (e.g., ^{3}H- or ^{14}C-labeled monosaccharide, [^{3}H]acetate, or [^{35}S]sulfate)

Separate containers for liquid and solid radioactive waste

Phosphate-buffered saline (PBS), pH 7.2, chilled on ice

Labeling Procedure

1. Dry the radioactive precursor if necessary [e.g., to remove organic solvents such as 70% (v/v) ethanol that are used for storage], add the label to a small volume of complete medium, and filter sterilize. If the final volume of medium is 10 ml or greater, it is practical to use a disposable 50-ml flask fitted with a 0.22-μm pore size filter and vacuum suction device. Transfer the label with ~5 ml of medium onto the filter, wash the container with the remaining volume, and pass the wash through the same filter. Smaller volumes of media can be sterilized with a 0.22-μm pore size filter fitted directly onto the tip of a disposable sterile syringe. Remove the plunger from the syringe, maintain its sterility (e.g., by keeping it upright on a low-profile rack), screw the filter onto the barrel of the syringe (save the filter wrapping), and transfer the labeled medium directly into the barrel of the syringe. The plunger can then be placed back in the barrel, and the fluid and some air pushed through the filter into an appropriate sterile receptacle. Cover the filter with its original packaging (still sterile), and partly unscrew the filter from the barrel of the syringe. Carefully remove the plunger of the syringe, retighten the filter to syringe, wash the original container with an additional volume of medium, and pass this through the same filter. Dispose of the filter and/or syringe in the

appropriate container for radioactive waste. Add additional medium if needed to bring the volume up to the amount planned for the experiment.

As an alternative, the precursor may be available for purchase in a sterile aqueous medium, ready for use. Such preparations tend to be more expensive, and after repeated opening of the packages, filtering may be necessary in any case, to ensure sterility.

2. Place the sterile labeling medium in the incubator or water bath to warm to labeling temperature. Placing it in the incubator with a loose cap has the advantage that equilibration with CO_2 can begin immediately.

3. Prepare the cells for splitting by whatever standard protocol is appropriate for the cell type (e.g., trypsin for monolayer cultures, direct dilution or concentration of cells by centrifugation for suspension cultures). The number of cells needed will be determined by the growth characteristics of the particular cell of interest.

4. With a sterile pipette tip, remove a small aliquot of the radioactive medium for counting (it may be necessary to dilute it to obtain an accurate count). Mix the radioactive medium and cells, and incubate under standard conditions for the cell line.

5. At the time of maximum cell growth ("confluence"), chill the cells and medium on ice and harvest (a rubber policeman is necessary for scraping monolayer cultures). Pellet the cells out of the medium with a table-top centrifuge at 4°. Remove the medium and count an aliquot. To study the glycoconjugates in the medium, the medium should preferably be refiltered immediately, to eliminate any broken cell debris that was not removed with the centrifugation. The medium may contain labeled macromolecules synthesized by the cells. In some cases, the cells may have taken up the label, processed it into other low molecular weight products, and excreted these into the medium. Checking the spent medium for labeled macromolecules and for intactness of the original precursor may be worthwhile.

6. Wash the cells twice in a >50-fold excess volume of ice-cold PBS, pH 7.2. Discard the washings appropriately. Process the labeled cell pellet immediately or store it frozen for future analysis.

7. Process the cell pellet in a manner appropriate for extracting the glycoconjugates of interest (e.g., extraction into detergent). Determine the efficiency of label incorporation by measuring the amount of radioactivity in cells compared to that left in the medium.

Pulse Labeling with Radioactive Precursors

In some situations, it is desirable to label the glycoconjugate of interest briefly, for a pulse–chase study, usually meant to establish precursor–product relationships. This approach may also be useful if the equilibrium labeling described does not give sufficient incorporation of label into the glycoconjugate of interest. In both cases, a pulse-labeling protocol can be tried. This approach is most useful when using monosaccharides that compete with glucose for uptake (see discussion earlier). It is of limited value for monosaccharides that are not taken up efficiently by the cells (unless large quantities of such labeled molecules can be used).

Materials

Many of the same materials used for equilibrium labeling can be used here.

Dialyzed fetal calf serum (FCS)
Complete unmodified medium (if a chase is to be performed)
Multiply deficient medium (MDM) (see below and Table III), or appropriate commercial substitute
Stock solutions of the compounds missing from MDM (see Table IV)

Preparation of Multiply Deficient Medium

Different types of labeling require different types of selectively deficient media. Some deficient media are available commercially at reasonable prices, and some companies will custom prepare selectively deficient media on request. For a laboratory that frequently does experiments using a variety of different deficient media, it is convenient to prepare a "multiply deficient medium" (MDM), which is completely lacking in many commonly studied components. This stock medium can be used to make up different kinds of selectively deficient media as needed. The MDM formula presented in Table III is based on α-MEM medium, which supports the growth of most kinds of tissue culture cells. It can be reconstituted for labeling with ^{3}H- or ^{14}C-labeled monosaccharides, [^{35}S]sulfate, ^{35}S-labeled methionine or cysteine, or [^{3}H]serine.

All reagents used for the preparation of this medium should be of tissue culture grade. Care should be taken to either "pour out" or use disposable spatulas to transfer reagents out of bottles. This will ensure that reagents are not contaminated by ubiquitous compounds, such as endotoxin, that may cause aberrant cell behavior or cell death. For the same reason, care must also be exercised to use clean, autoclaved glassware. Preferably, a set of glassware should be put aside just for this purpose. The components to be added are listed in Table III. The $100\times$ stocks of many of these can be purchased commercially, in sterile form. If not, each should be made up and filter sterilized separately. The salts and phenol red are first dis-

TABLE III
COMPONENTS OF MULTIPLY DEFICIENT MEDIUM

Component	Final concentration (mg/liter)	Stock
CaCl$_2$	200	Shelf, reagent grade
KCl	400	Shelf, reagent grade
MgCl$_2$	75	Shelf, reagent grade
NaCl	6800	Shelf, reagent grade
NaH$_2$PO$_4 \cdot$ H$_2$O	140	Shelf, reagent grade
Phenol red	10	Solution
Sodium pyruvate	110	100×
L-Alanine	25	100×
L-Arginine	126	100×
L-Asparagine	50	100×
L-Aspartic acid	30	100×
L-Cysteine	(100 or none)[a]	100×
L-Glutamic acid	75	100×
L-Glutamine	None	100×
L-Glycine	50	100×
L-Histidine	42	100×
L-Isoleucine	52	100×
L-Leucine	52	100×
L-Lysine	72	100×
L-Methionine	(15 or none)[a]	100×
L-Phenylalanine	32	100×
L-Proline	40	100×
L-Serine	(25 or none)[a]	100×
L-Threonine	48	100×
L-Tryptophan	10	100×
L-Tyrosine	36	100×
L-Valine	46	100×
Vitamins	1×	100×

[a] Components left out; add back as needed for specific experiments.

solved one by one in ~800 ml of tissue culture-grade water. Pyruvate, amino acids, and vitamins are then added sufficient for a 1× final concentration in 1000 ml. The medium is then made up to 1000 ml with water and filter sterilized. The solution should be yellow and the pH should not be adjusted. Aliquots of 25 ml are placed in sterile 50-ml tubes and stored at −20°.

Checking for Optimal Labeling Conditions

1. Reconstitute MDM medium to 100% levels with all components except the one being presented as a radioactive precursor. For radioactive

TABLE IV
COMPONENTS REQUIRED FOR RECONSTITUTION OF
MULTIPLY DEFICIENT MEDIUM

Component[a]	Final concentration ($\times$) (mg/liter)	Stock
NaHCO$_3$	1$\times$ (2200 mg/liter)	100$\times$
or		
HEPES-HCl[b]	20 mM	2 M, pH 7.3
Na$_2$SO$_4$	0.81 mM	100 mM, sterile
D-Glucose	1$\times$ (1000 mg/liter)	100$\times$
L-Cysteine	1$\times$ (100 mg/liter)	100$\times$
L-Glutamine	1$\times$ (292 mg/liter)	100$\times$
L-Methionine	1$\times$ (15 mg/liter)	100$\times$
L-Serine	1$\times$ (25 mg/liter)	100$\times$

[a] Individual components are added back at full strength, at lowered concentration, or left out altogether, depending on the experiment planned.
[b] The pH of the final medium is controlled by bicarbonate/CO$_2$ or by HEPES-HCl.

monosaccharides, initially leave out the glucose (see discussion for rationale). Add dialyzed serum at the required final concentration.

2. Make a stock of the deficient medium containing the radioactive precursor, divide into aliquots, and add back the missing component in increasing concentrations from a 100$\times$ stock solution (e.g., 0, 5, 10, 20, 50, and 100% of the concentration in normal medium). For monosaccharides that compete with glucose, the missing component added back is glucose. Use these aliquots to do small-scale pilot labelings with the cells of interest, for fixed or varying periods of time. Because pulse labelings are usually done for a short time (e.g., minutes or hours), the cells are usually used in a near-confluent state, to maximize uptake and incorporation. Determine the efficiency of incorporation of radioactivity into whole-cell glycoconjugates, and if necessary into the specific macromolecules of interest. Plot the percentage incorporation versus total concentration of precursor (radioactive plus unlabeled). Determine at what point the curve breaks (i.e., the percentage of label incorporated is markedly decreased by further addition of unlabeled compound).

3. On the basis of these pilot labelings, choose the optimal concentration of unlabeled precursor to be used, that is, just above the breakpoint in the curve, where unlabeled precursor is not limiting, but where label incorporation is still good. Using these conditions, check for linearity of uptake and incorporation of label over time.

4. Before proceeding further, determine if reducing the concentration of the compound into this useful range has other detrimental effects on the cells, that is, grow cells under similar conditions and obtain independent measures of cell growth, viability, and general biosynthetic capability such as cell counts, trypan blue exclusion, and radioactive amino acid incorporation. It may be necessary to reach a compromise between the lowered concentration, the time of labeling, and condition of the cells. The goal of these pilot experiments is to determine the lowest concentration of the unlabeled precursor that can be used for the desired period of time, without affecting cell growth, viability, and biosynthesis of macromolecules in general. However, the specific glycoconjugate in question could be selectively affected by using partially deficient medium (e.g., lowering sulfate concentration too much can result in undersulfation of glycosaminoglycans in some cell types).

Labeling in Partially Deficient Medium

1. Depending on the type of labeling planned, the MDM is reconstituted with the deficient components at the final concentrations needed (see Table IV). For example, optimal labeling of endothelial cells with [^{35}S]sulfate and [6-^{3}H]glucosamine requires a medium with no cysteine, low glucose, and low concentrations of methionine and sulfate (A. Varki, L. Roux, and K. Norgard, unpublished observations, 1993). To reconstitute 50 ml of MDM for this experiment, the following components are added back: NaHCO$_3$ (1 × final concentration), glutamine and serine (500 μl each of 100× stocks), 0.1× (200-mg/liter final concentration) glucose (100 μl of 50× stock), 0.1× methionine (50 μl of 100× stock), and a 20 μM final concentration of inorganic sulfate (10 μl of 100 mM stock). The final concentrations of sulfate and methionine were based on the pilot experiments of the type described above. At this point, the medium should have an orange-red appearance, indicating the correct pH range. The radioactive labels [^{35}S]sulfate and [6-^{3}H]glucosamine are then added. The total volume of reconstituted medium will be slightly more than 50 ml. This minor discrepancy is ignored for practical purposes.

2. Add dialyzed serum if needed, filter sterilize, adjust the pH if necessary, and carry out pulse labelings of near-confluent cells for the desired period of time.

3. If it is appropriate to the experiment, remove the labeling medium and chase the label for varying periods of time in complete unmodified medium.

4. Harvest and wash the cells as described in the first protocol.

Sequential Pulse Labeling with Reutilization of Radioactive Media

Even under optimally defined conditions with selectively deficient medium, incorporation of label into the glycoconjugate of interest may be inadequate. On the other hand, prolonged exposure to deficient medium may not be feasible. In these situations, an alternate approach is to expose a series of plates or flasks of cells sequentially to a small volume of medium containing a high concentration of label, that is, reuse the medium for several labelings.[16,18] The assumption is that a small fraction of the radiolabeled precursor is consumed from the medium during each short labeling cycle, and that the medium itself is not substantially changed in other respects. This approach can also be adapted to pulse–chase studies.[19,20] For expensive radiolabeled precursors, the medium can be frozen after one use. For reutilization, the medium is then thawed, and the pH adjusted before reuse. In all such instances of reutilization, one must be careful in interpreting the experimental results, especially with regard to the specific activity of the labeled products.

Procedure

1. If necessary, determine the maximum time for which the cells can be safely exposed to deficient medium (see above). Set up several identical flasks or plates of cells to be labeled sequentially.

2. Label one batch of cells for the appropriate period of time. The actual time will vary from minutes to hours, depending on the cell type and the objective of the labeling.

3. Remove the growth medium from the second plate of cells, and transfer the labeled medium into this plate. For suspension cultures where it is possible to use less than 1 ml of labeling medium, 1.5-ml microcentrifuge tubes can be used for the incubation. The tubes are capped tightly and incubated at 37° in a warm room, on a rotating end-over-end apparatus. The cells can be briefly centrifuged in a microfuge (Brinkmann), the labeling medium transferred to the next tube, and the cell pellet harvested.

4. Add regular medium to plate 1, and continue the incubation.

5. Repeat such transfers until all plates have been labeled.

6. Harvest all plates. If a pulse–chase experiment is being performed, the individual cell pellets are kept separate. If the objective is simply

---

[18] I. Tabas and S. Kornfeld, *J. Biol. Chem.* **255,** 6633 (1980).
[19] D. E. Goldberg and S. Kornfeld, *J. Biol. Chem.* **256,** 13060 (1981).
[20] E. A. Muchmore, M. Milewski, A. Varki, and S. Diaz, *J. Biol. Chem.* **264,** 20216 (1989).

to obtain high incorporation, pool all the pellets. The combined pellets represent molecules synthesized over the entire period of time. Alternatively, each plate can be processed as soon as the labeling medium is removed, and the pellet stored frozen until all the pellets can be processed together.

[3] Nonmetabolic Radiolabeling and Tagging of Glycoconjugates

By CARL G. GAHMBERG and MARTTI TOLVANEN

Introduction

We have been involved in developing methods to (1) label cell surface glycoproteins and glycolipids radioactively, (2) estimate the degree of exposure of individual oligosaccharides, and (3) modify preexisting oligosaccharides by introducing new terminal carbohydrates.[1-3] Other groups have coupled noncarbohydrate labels into cell surface glycoconjugates or have used glycosyltransferases to add monosaccharides to terminal sugars. The identification and characterization of cytoplasmic *N*-acetylglucosamine-substituted proteins, using nonmetabolic radioactive labeling, has been a novel concept.[4]

In this chapter we describe the various techniques from a practical point of view and discuss potential applications, focusing primarily on the labeling of cell surface glycoconjugates in intact cells. However, these methods are also applicable to other membrane glycoconjugates and can be used for labeling soluble glycoproteins. Ashwell and co-workers first radioactively labeled soluble glycoproteins in their classic studies on the turnover and recognition of serum glycoproteins by the hepatic lectin.[5,6]

[1] C. G. Gahmberg, *J. Biol. Chem.* **251**, 510 (1976).

[2] C. G. Gahmberg, *EMBO J.* **2**, 223 (1983).

[3] C. G. Gahmberg and S. Hakomori, *J. Biol. Chem.* **248**, 4311 (1973).

[4] G. W. Hart, R. S. Haltiwanger, G. D. Holt, and W. G. Kelly, *Annu. Rev. Biochem.* **58**, 841 (1989).

[5] A. G. Morell and G. Ashwell, *in* "Methods in Enzymology" (V. Ginsburg, ed.), Vol. 28, p. 205. Academic Press, New York, 1972.

[6] L. van Lenten and G. Ashwell, *in* "Methods in Enzymology" (V. Ginsburg, ed.), Vol. 28, p. 209. Academic Press, New York, 1972.

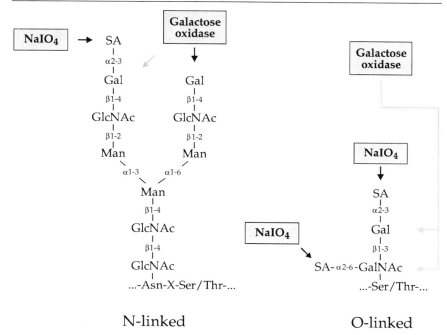

N-linked   O-linked

FIG. 1. Structures of common N-glycosidic (lactosamine type) and O-glycosidic oligosaccharides. The potential oxidation sites of periodate and galactose oxidase are indicated. Lighter arrows indicate residues that could be oxidized after glycosidase treatments.

Radioactive Labeling of Cell Surface Glycoconjugates with Galactose Oxidase and Tritiated Sodium Borohydride

Galactose oxidase from *Dactylium dendroides* oxidizes nonreducing terminal galactosyl and *N*-acetylgalactosaminyl residues in glycoproteins and glycolipids to their corresponding C-6 aldehydes.[7] Because of its high molecular weight, galactose oxidase cannot easily penetrate a cell membrane, and therefore only surface-exposed glycoconjugates will be oxidized.[3] Common nonreducing terminal carbohydrate structures in mammalian N- or O-linked glycoproteins are shown in Fig. 1. Penultimate galactosyl residues are not efficient substrates for the enzyme. Sialic acids must therefore be removed with sialidase (Fig. 2). Some oligosaccharides of these types are not substituted with sialic acid, and in such instances oxidation may occur without sialidase treatment. In fact, by comparing

---

[7] G. Avigad, D. Amaral, C. Asensio, and B. L. Horecker, *J. Biol. Chem.* **237**, 2736 (1962).

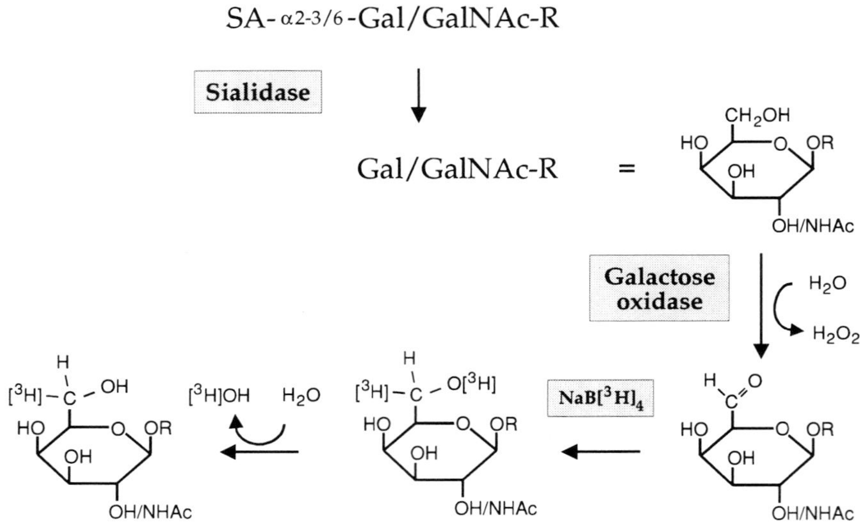

$$SA\text{-}\alpha2\text{-}3/6\text{-}Gal/GalNAc\text{-}R$$

Sialidase

Gal/GalNAc-R   =

Galactose oxidase

H_2O

H_2O_2

$NaB[^3H]_4$

FIG. 2. Labeling of glycoconjugates, using sialidase/galactose oxidase and NaB^3H_4. Terminal sialic acids are usually first removed wtih sialidase. Galactose oxidase then generates a C-6 aldehyde in galactose/N-acetylgalactosamine that can be reduced with NaB^3H_4.

the labeling with or without sialidase treatment, it is possible to obtain information on the degree of sialylation.

After washing away the enzyme(s), the cells are reduced with tritiated sodium borohydride. In this step the galactose/N-acetylgalactosamine aldehydes are reduced back to the original sugars, which are now radioactive (Fig. 2). An alternative to sodium borotritiate is sodium borodeuteride. In this case the degree of oxidation (and exposure) of individual oligosaccharides may be determined by mass spectrometry, as was done for the major glycolipids in human red cells.[8,9]

After labeling, the radioactive cells may be solubilized in detergent, and the extracts used for further analysis. Sodium dodecyl sulfate (SDS)-polyacrylamide gel electrophoresis is commonly used, often combined with immunoprecipitation. We often use labeled extracts or purified labeled glycoproteins as tracers in large-scale isolations of specific glycoproteins. Glycolipids are conveniently studied by thin-layer chromatography or high-performance liquid chromatography (HPLC) followed by radioactive analysis.

[8] A. Lampio, J. Finne, D. Homer, and C. G. Gahmberg, Eur. J. Biochem. **145**, 77 (1984).
[9] A. Lampio, H. Rauvala, and C. G. Gahmberg, Eur. J. Biochem. **157**, 611 (1986).

Procedure

Specific Reagents

Galactose oxidase (*Dactylium dendroides*)
Sialidase (*Vibrio cholerae*)
Tritiated sodium borohydride (6–15 Ci/mmol)

Galactose oxidase (*D. dendroides*) can be purchased from several commercial sources. It is usually supplied as a lyophilized powder. It is dissolved in 0.15 *M* NaCl–0.01 *M* sodium phosphate, pH 7.4 (PBS), and divided into aliquots at 1 unit/μl. Tubes containing 50–100 μl of enzyme are kept frozen at −20°. Alternatively, the tubes may be lyophilized and stored at 4°. The enzyme remains stable enough for at least 1 year.

Sialidase (*V. cholerae*) is commercially available and can be stored as such at 4°.

Sodium borotritiate is supplied as a dry powder sealed in glass vials. A specific activity of 6–15 Ci/mmol is recommended for most labeling experiments. Higher specific activities result in a more unstable reagent and preparations with lower specific activities result in weak labeling. The handling of the isotope needs some precautions. We usually use 100- or 250-mCi aliquots. The compound is rapidly dissolved at 1 mCi/μl in 0.01 *N* NaOH, which has been frozen on dry ice–ethanol, and then warmed until just melting. Eppendorf tubes are kept ready on dry ice–ethanol in a well-ventilated hood. The isotope is then quickly aliquoted and frozen at 50 mCi (50 μl)/tube. The tubes are stored at −70°, and in this way the sodium borotritiate remains stable for months.

For labeling, 20–100 × 10^6 cells in suspension, or one 250-cm² plate of growing cells, are washed by centrifuging or rinsing, respectively, three times in PBS. Then 1 ml of Dulbecco's PBS is added, along with 25 mU of neuraminidase and 10 U of galactose oxidase. The cells are incubated with the enzymes for 15–30 min at 37°, and washed twice in PBS.

After oxidation, monolayer cells are removed from the plates by ethylenediaminetetraacetic acid (EDTA) treatment, or by using a rubber policeman. The cells are then suspended in 0.5 ml of PBS, pH 8.0 (NaB³H₄ is more stable at higher pH), and 5–10 μl of sodium borotritiate solution is added in a well-ventilated hood. After incubation for 30 min at room temperature, the cells are washed three times by centrifugation, and lysed in PBS–1% Triton X-100 at 0°. Protease inhibitors may also be used. After centrifugation at 3000 *g* for 10 min the supernatants are recovered for further analysis.

If red cells are analyzed, the labeled cells are broken by hypotonic lysis, the membranes prepared, and then solubilized. It is sometimes useful

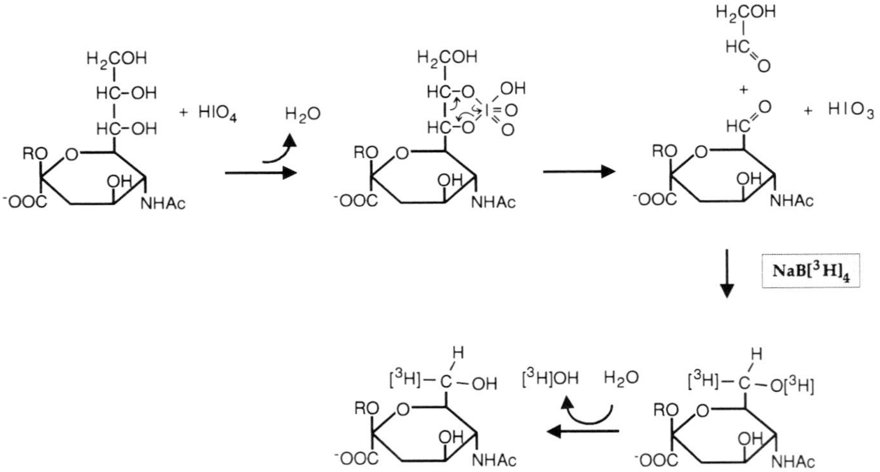

FIG. 3. Labeling of sialoglycoconjugates, using periodate and NaB^3H_4. Periodate generates sialic acid derivatives with aldehyde groups. The formation of a C-7 aldehyde is shown here. The aldehyde is then reduced with NaB^3H_4.

to omit the neuraminidase treatment for analysis of glycoproteins. When glycolipids are analyzed, sialidase treatment is usually not employed, because their structure would then be extensively changed. Furthermore, it is advisable to include a control without enzyme treatment. This is especially important if the primary goal is to ascertain which labeled proteins are glycoproteins. If isolated glycoconjugates are labeled, catalase can be added to dispose of the H_2O_2 formed in the reaction. H_2O_2 inhibits galactose oxidase at higher concentrations.

Radioactive Labeling of Cell Surface Sialoglycoconjugates with
    Periodate and Tritiated Sodium Borohydride

As shown in Fig. 3, mild periodate treatment specifically oxidizes vicinal OH groups in sialic acids, and C-7 or C-8 aldehyde derivatives are formed.[6] When the oxidation of living cells is performed at low temperature (0–4°), no transport of periodate into the cell takes place, and only externally located sialic acids are oxidized.[10] After washing, the cells are reduced with sodium borotritiate, and handled as described above.

[10] C. G. Gahmberg and L. C. Andersson, J. Biol. Chem. 252, 5888 (1977).

Procedure

Specific Reagents

Sodium metaperiodate
Tritiated sodium borohydride

The cells are washed with PBS, and suspended in or covered with 1 ml of PBS at 0°. Then 20 μl of 0.1 M sodium metaperiodate (freshly dissolved at 21 mg/ml in PBS) is added, and the cells are left on ice in the dark for 10 min. They are then washed twice in PBS, and suspended in 0.5 ml of PBS, pH 8.0 for reduction with NaB^{3}H$_4$. The subsequent handling is identical to that described above for galactose oxidase/ NaB^{3}H$_4$ labeling.

Radioactive Labeling of Glycoconjugates with Nucleotide Sugars and Glycosyltransferases

In principle, several nucleotide sugars and glycosyltransferases could be used to label glycoconjugates in living cells and isolated membranes.[10a] However, only UDP[^{14}C/^{3}H] galactose/galactosyltransferase has been used extensively for several reasons. First, bovine milk galactosyltransferase is commercially available and relatively inexpensive. Second, acceptor N-acetylglucosamines may be found in various types of oligosaccharides. Furthermore, both UDP[^{3}H]galactose and UDP[^{14}C]galactose are commercially available.

The labeling principle is shown in Fig. 4. Glycoconjugates containing nonreducing terminal N-acetylglucosaminyl residues act as acceptors of the sugar transferred from the nucleotide derivative and, when radioactive reagents are used, radioactively labeled glycoproteins are obtained. However, relatively few glycoproteins or glycolipids located at the cell surface contain terminal N-acetylglucosamines; these residues can be exposed by the use of glycosidases. In some cases such structures can be generated by treatment of cells with endo-β-galactosidase. This enzyme cleaves linear polylactosaminoglycan structures, leaving nonreducing terminal N-acetylglucosamine. The residues can then be tagged with galactose. In this way polylactosamine-type oligosaccharides can be identified.[11]

[10a] S. W. Whitehart, A. Passaniti, J. S. Reichner, G. D. Holt, R. S. Haltiwanger, and G. W. Hart, *in* "Methods in Enzymology" (V. Ginsburg, ed.), Vol. 179. Academic Press, San Diego, 1989.

[11] J. Viitala and J. Finne, *Eur. J. Biochem.* **138,** 393 (1984).

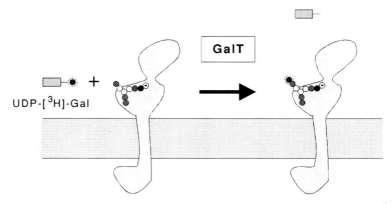

FIG. 4. Labeling of glycoconjugates, using galactosyltransferase (GalT) and UDP[³H]galactose. The transferase adds the radioactive galactose to terminal *N*-acetylglucosamine residues. Residues (circles): dark gray, GlcNAc; white, Man; black, Gal; white with enclosed minus, sialic acid.

An interesting discovery was made by Hart and co-workers, when they found that intracellular, cytoplasmically exposed proteins contain O-linked *N*-acetylglucosamines ([26] in this volume).[4,12]

Procedure

Specific Reagents

Bovine galactosyltransferase
UDP[¹⁴C]galactose or UDP[³H]galactose

Cells or membranes are suspended in 10 mM *N*-2-hydroxyethylpiperazine-*N*'-2-ethanesulfonic acid (HEPES) (pH 7.3)–NaHCO₃ (2 g/liter)–NaCl (8 g/liter)–sodium pyruvate (0.55 g/liter), and 0.2 U of bovine milk galactosyltransferase and 10 μCi of UDP[³H]galactose are added. After incubation at 37° for 30 min with gentle shaking, the cells or membranes are washed by centrifugation, dissolved in detergent, and further analyzed. If surface-located residues are specifically analyzed it may be advisable to include 10 mM galactose in the incubation medium to compete with nonspecifically hydrolyzed radioactive galactose, which could label metabolically intracellular proteins.

---

[12] C.-R. Torres and G. W. Hart, *J. Biol. Chem.* **259**, 3308 (1984).

Covalent Attachment of Mono- and Oligosaccharides to Surface
  Glycoconjugates of Living Cells

Few specific functions for cell surface carbohydrates are known. Blood
cell and endothelial cell selectins are carbohydrate-binding proteins (lectins) that specifically recognize oligosaccharides on other cells. Endothelial cells may contain the P- and E-selectins,[13-15] and leukocytes may
contain L-selectin.[16] These human selectins all seem to bind to sialyl-Le^X
or sialyl-Le^a epitopes on their ligands.[17-19] Another interesting finding is
that sperm cells bind to α-galactosyl-containing glycoconjugates on the
zona pellucida, which surrounds mouse egg cells;[15,20] but these examples
are exceptions. For the vast majority of cell surface carbohydrate, no real
function is known.

To elucidate possible carbohydrate-associated functions, different approaches can be taken. One way is to block N-glycosylation by use of
antibiotics, such as tunicamycin. Another possibility is to remove
N-glycosylation sites by site-directed mutagenesis. Alternatively, one can
use inhibitors of trimming glycosidases (see [19] in this volume). Such
treatments result in the accumulation of various intermediates in the biosynthesis of N-linked oligosaccharides.

We have taken another approach, recognizing that it could be advantageous to transfer "new" carbohydrates into living cells, and preferably
to specific sites in cell surface glycoproteins. In Fig. 5, we describe a
technique in which cell surface galactose/N-acetylgalactosamine or sialic
acid residues are first oxidized as described above, using galactose oxidase
or periodate treatment, respectively. Then the oxidized cells are incubated
with glycosylhydrazines, which react with the generated cell surface aldehydes.[21] In this way preexisting cell surface oligosaccharides, either N-
or O-linked, are elongated with new carbohydrates. The effect of these

[13] E. Larsen, A. Celi, G. E. Gilbert, B. C. Furie, J. K. Erban, R. Bonfanti, D. D. Wagner,
  and B. Furie, *Cell (Cambridge, Mass.)* **59,** 305 (1989).
[14] M. P. Bevilacqua, J. S. Pober, D. L. Mendrick, R. S. Cotran, and M. A. Gimbrone, Jr.,
  Proc. Natl. Acad. Sci. U.S.A. **84,** 9238 (1987).
[15] C. G. Gahmberg, P. Kotovuori, and E. Tontti, *Acta Pathol. Microbiol. Immunol. Scand.*
  100, 39 (1992).
[16] W. M. Gallatin, I. L. Weissman, and E. C. Butcher, *Nature (London)* **303,** 30 (1983).
[17] M. L. Phillips, E. Nudelman, F. C. A. Gaeta, M. Perez, A. K. Singhal, S. Hakomori,
  and J. C. Paulson, *Science* **250,** 1130 (1990).
[18] G. Walz, A. Aruffo, W. Kolanus, M. Bevilacqua, and B. Seed, *Science* **250,** 1132 (1990).
[19] M. Tiemeyer, S. J. Swiedler, M. Ishihara, M. Moreland, H. Schweingruber, P. Hirtzer,
  and B. K. Brandley, *Proc. Natl. Acad. Sci. U.S.A.* **88,** 1138 (1991).
[20] P. M. Wasserman, *Annu. Rev. Biochem.* **57,** 415 (1988).
[21] M. Tolvanen and C. G. Gahmberg, *J. Biol. Chem.* **261,** 9546 (1986).

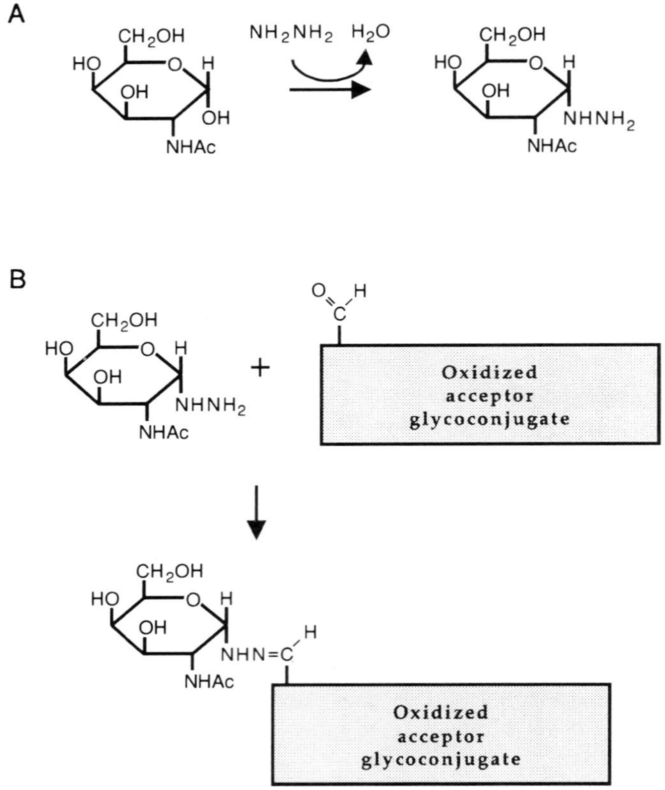

FIG. 5. Coupling of oligosaccharides to cell surface glycoconjugates, using hydrazine derivatives. The reducing oligosaccharide is first treated with hydrazine (A) and the derivative is attached to oxidized cell surface oligosaccharides (B).

modifications on the behavior of the cells and glycoconjugates may then be studied.

Procedure

Specific Reagents

Reducing mono- or oligosaccharides
Anhydrous hydrazine
Sodium metaperiodate, or sialidase and galactose oxidase

Preparation of Sugar Hydrazines. Desired mono- or oligosaccharides (up to 5 mg) containing reducing terminal groups are lyophilized in screw-cap tubes, and dried carefully *in vacuo*. Then 100–500 μl of anhydrous hydrazine is added, the tubes are filled with nitrogen, and the samples are left at 0° for 1 hr or longer. Care should be taken that the hydrazine used is really anhydrous. After incubation the samples are lyophilized, a small volume of toluene is added, and the samples are lyophilized again.

Labeling of Cells. Cells are oxidized, using sialidase plus galactose oxidase, galactose oxidase only, or periodate as described above. After washing, the oxidized cells are incubated with the glycosylhydrazines at 37° for 30–60 min. To avoid endocytosis of the attached structure, the incubation can be carried out at 0° for 1–3 hr. After washing, the cells are ready for further experiments. Sometimes it may be informative to use radioactive oligo- or monosaccharides. If not commercially available, they can be made, for example, by galactose oxidase or periodate/NaB^3H_4 labeling of intact glycoproteins or glycopeptides, followed by liberation of the labeled oligosaccharides by hydrazinolysis (see [5] in this volume) or by endoglycosidases such as endo-*N*-acetylglucosaminidase F. Isolated oligosaccharides can also be labeled radioactively by hydrazinolysis, followed by reacetylation of liberated carbohydrate amino groups with radioactive acetic anhydride.[22]

We have also made fluorescent oligosaccharides and attached them to living cells.[23] In this case the oligosaccharides were subjected to hydrazinolysis, followed by labeling of the free amino groups with fluorescein isothiocyanate.

Comments on Methods

Metabolic labeling of the carbohydrate portions of glycoconjugates in living cells and tissues, using monosaccharide precursors, is certainly feasible (see [2] in this volume) and useful for a number of purposes. It is, however, expensive and not always practical. In addition, the incorporation of carbohydrates into glycoconjugates of living cells is often slow, and it is hard to achieve high specific activities. Furthermore, such techniques do not reveal anything about the cellular localization of the labeled macromolecules. Some cells are not metabolically active, and cannot be labeled by using radioactive precursors. The most obvious example is the mature mammalian red cell, which does not synthesize any glycoproteins or glycolipids. Granulocytes are difficult to label metabolically with sugars, and there may well be other examples of cells behaving in a similar way. Glycoconjugates of isolated viruses cannot be labeled metabolically, but

[22] D. M. Carlson, *J. Biol. Chem.* **241,** 2984 (1966).
[23] C. G. Gahmberg and M. Tolvanen, *Anal. Biochem.* **170,** 520 (1988).

are easily labeled by oxidation of carbohydrates, followed by treatment with NaB^3H_4.[24]

There are numerous methods available for cell surface labeling of the polypeptide portions of glycoproteins. Most popular is lactoperoxidase-catalyzed iodination,[25,26] because it is convenient and can be used for identification of cell surface proteins. However, it has some disadvantages. [125]I, a strong γ emitter, requires special handling, and it has a relatively short half-life (60 days) compared to ^3H (13 years). The resolution of ^3H on polyacrylamide gels is superior, but tritium-labeled glycoproteins separated on gels need to be treated with scintillators for visualization.[27] There may also be cell surface proteins that cannot be ^{125}I-labeled with lactoperoxidase because they lack exposed tyrosines. In contrast, cell surface carbohydrates are usually readily available to external reagents.

It may also be necessary to study specifically the oligosaccharide portions of cell surface glycoconjugates. After labeling, specific glycoproteins can be isolated; then the labeled oligosaccharides are released and studied by a variety of techniques. Another application is to study the degree of exposure of individual oligosaccharides. We have determined the degree of oxidation (exposure) of red cell glycolipids by galactose oxidase-catalyzed oxidation, followed by reduction with sodium borodeuteride. After isolation of the glycolipids, the fractions available to the enzyme are determined by mass spectrometry of the isolated glycolipids.[8,9] These studies show interesting variations in the exposure rates between the red cells from different individuals. A substantial fraction of the major human red cell glycolipid, globoside, could not be oxidized even after long incubation periods with galactose oxidase, indicating that it is somehow hidden in the membrane.

For the labeling of soluble glycoproteins the oxidation steps are similar, but when using periodate oxidation the oxidized glycoproteins are reisolated by gel filtration on Sephadex G-25 in PBS, pH 8.0. After concentrating, if necesssary, the glycoproteins are reduced with NaB^3H_4, and then treated in a hood with dilute acetic acid to lower the pH to 3–5. This treatment decomposes any remaining NaB^3H_4. The labeled glycoproteins are then passed through Sephadex G-25 and the peak containing the labeled glycoprotein is recovered.

When comparing the galactose oxidase ($\pm$sialidase) and periodate/NaB^3H_4 labeling methods, we prefer periodate for most purposes, because it is inexpensive and stable in a dry form. Furthermore, the modification of the glycoconjugates is less when compared to samples treated with

[24] A. Luukkonen, C. G. Gahmberg, and O. Renkonen, *Virology* **76,** 55 (1977).
[25] D. R. Phillips and M. Morrison, *Biochemistry* **10,** 1766 (1971).
[26] A. L. Hubbard and Z. A. Cohn, *J. Cell Biol.* **55,** 390 (1972).
[27] W. M. Bonner and R. A. Laskey, *Eur. J. Biochem.* **46,** 83 (1974).

sialidase. The choice of technique, however, depends on the purpose and the structure of the substrate.

The method of attaching mono- and oligosaccharylhydrazines to oxidized cell surface glycoproteins and glycolipids is easy to apply. Although not widely used yet, it has several potential applications. Presently it is not possible to modify specific cell surface glycoconjugates; those containing suitable oxidizable residues, however, will be labeled. In many instances the terminal carbohydrate epitopes in various glycoproteins in a given cell are similar, and therefore a relatively uniform modification may not be a disadvantage. In other cases it may be desirable to introduce new carbohydrates in a large number of glycoconjugates in order to increase the density of the added epitope.

The selectins recognize sialyl-Le[x] and sialyl-Le[a] epitopes on glycoconjugates in blood cells and endothelial cells. In blood cells, the major carriers of these structures are cells of the myelomonocytoid lineage.[28] The selectins induced on endothelial cells interact with circulating cells, which results in the "rolling" phenomenon.[29-31] When a stronger interaction occurs, the granulocytes stick firmly to the endothelial cells, and later penetrate into the tissues. With the carbohydrate attachment method it may be possible to modify blood cells that do not contain sialyl-Le[x] or sialyl-Le[a] specifically, and in this way to make it possible for them to bind to endothelial cells. This could be important in combating infections or treating malignancies.

The glycosylhydrazines attached to cell surface molecules remain linked for hours, but are gradually lost into the surrounding medium.[21] This could be a disadvantage; however, it may be preferable that they do not stay permanently on cells *in vivo,* which could produce unwanted side effects.

Oxidized cell surface glycoconjugates can also be modified with other specific labels that are based on hydrazine chemistry. These methods have been reviewed previously in this series,[32] and they allow specific labeling of glycoconjugates with almost any type of label. Especially versatile is biotin hydrazide,[33] which can be used for coupling of any avidin-conjugated

[28] M. Fukuda, B. Bothner, P. Ramsamooj, A. Dell, P. R. Tiller, A. Varki, and J. C. Klock, *J. Biol. Chem.* **260,** 12957 (1985).

[29] K. Ley, P. Gaethgens, C. Fennie, M. S. Singer, L. A. Lasky, and S. D. Rosen, *Blood* **77,** 2553 (1991).

[30] U. H. von Andrian, J. D. Chambers, L. M. McEvoy, R. F. Bargatze, K.-D. Arfors, and E. C. Butcher, *Proc. Natl. Acad. Sci. U.S.A.* **88,** 7538 (1991).

[31] M. B. Lawrence and T. A. Springer, *Cell (Cambridge, Mass.)* **65,** 859 (1991).

[32] M. Wilchek and E. H. Bayer, *in* "Methods in Enzymology" (V. Ginsburg, ed.), Vol. 138, p. 429. Academic Press, Orlando, FL, 1987.

[33] H. Heitzmann and F. M. Richards, *Proc. Natl. Acad. Sci. U.S.A.* **71,** 3537 (1974).

probe or for affinity isolation of glycoconjugates on avidin or streptavidin columns.

Obviously there are many potential applications. The realization of the importance of oligosaccharide chains in many vital biological processes[15] has created an enormous interest in them among researchers in various areas. No doubt we can expect substantial progress to be made in the elucidation of carbohydrate structure and function, and in this process the methodologies outlined here should be of value.

Acknowledgments

The original work reported herein was supported by the Academy of Finland, the Sigrid Jusélius Foundation and the Finnish Cancer Society, the Magnus Ehrnrooth Foundation, and the Emil Aaltonen Foundation. We thank Yvonne Heinilä for expert secretarial assistance.

[4] Enzymatic Deglycosylation of Asparagine-Linked Glycans: Purification, Properties, and Specificity of Oligosaccharide-Cleaving Enzymes from *Flavobacterium meningosepticum*

By ANTHONY L. TARENTINO and THOMAS H. PLUMMER, JR.

Introduction

Reviews[1–3] on the oligosaccharide chain-cleaving enzymes of *Flavobacterium meningosepticum* describe the purification and properties of two distinct hydrolases with specificity directed at the invariant pentasaccharide core region of asparagine-linked glycans. One of these enzymes, peptide-N^4-(N-acetyl-β-glucosaminyl)asparagine amidase (PNGase F; EC 3.5.1.52), is actually an amidase (amidohydrolase) that hydrolyzes the glycosylamine linkage of a wide variety of glycoprotein/glycopeptide substrates (Fig. 1), generating an aspartic acid residue at the site of hydrolysis, and liberating a 1-amino oligosaccharide. The latter slowly hydrolyzes nonenzymatically to ammonia and an oligosaccharide with a di-N-acetyl-

[1] A. L. Tarentino and T. H. Plummer, Jr., this series, Vol. 138, p. 770.
[2] F. Maley, R. B. Trimble, A. L. Tarentino, and T. H. Plummer, Jr., *Anal. Biochem.* **180,** 195 (1989).
[3] A. L. Tarentino, R. B. Trimble, and T. H. Plummer, Jr., *Methods Cell Biol.* **32,** Part B, 111 (1989).

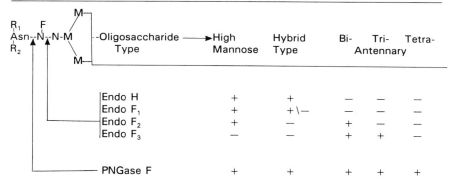

	High	Hybrid	Bi-	Tri-	Tetra-
	Mannose	Type	Antennary		
Endo H	+	+	−	−	−
Endo F$_1$	+	+\−	−	−	−
Endo F$_2$	+	−	+	−	−
Endo F$_3$	−	−	+	+	−
PNGase F	+	+	+	+	+

FIG. 1. Specificity of various endoglycosidases. Endoglycosidase refers to any hydrolytic cleavage of the asparagine-linked oligosaccharide core, including hydrolysis of the glycosyl-amine linkage by the amidase PNGase F, or the glycosidic cleavage of the di-N-acetylchitobi-ose moiety by the endo-β-N-acetylglucosaminidases Endo H, Endo F$_1$, Endo F$_2$, or Endo F$_3$. The invariant pentasaccharide core of a generic asparagine-linked glycan is shown to the left of the oligosaccharide type, and the type of attached oligosaccharide chain to the right. R$_1$, R$_2$, peptide bonds or H; N, N-acetylglucosamine; M, mannose; F, fucose; Asn, asparagine.

chitobiose unit on the reducing end. The second hydrolase, endo-β-N-acetylglucosaminidase F (Endo F; EC 3.2.1.96), is a true endoglycosidase[4] that cleaves the $\beta(1 \rightarrow 4)$-linked di-N-acetylchitobiose core in a manner analogous to Endo H from *Streptomyces plicatus* (Fig. 1): One N-acetylglucosamine residue remains attached to the asparagine moiety, and the other becomes the reducing terminus of the liberated oligosaccharide.

The structural determinants for PNGase F involve primarily the polypeptide chain and the di-N-acetylchitobiosyl region of the glycan chain.[1-3] These features account for the broad specificity of PNGase F because the peripheral oligosaccharide structure is largely unrecognized by the enzyme. In contrast, determinants for Endo F involve binding sites on the oligosaccharide backbone; its specificity was thought previously to be restricted to high mannose and biantennary oligosaccharide chains,[4-6] but new developments[5,6] have completely changed this picture.

The point of departure from our previous work began with a modification of the original purification procedure that incorporated fast protein liquid chromatography (FPLC)-controlled, high-resolution hydrophobic interaction chromatography[6] for the separation of Endo F and PNGase F.

[4] S. Alexander and J. H. Elder, this series, Vol. 179, p. 505.
[5] R. B. Trimble and A. L. Tarentino, *J. Biol. Chem.* **266**, 1646 (1991).
[6] T. H. Plummer, Jr. and A. L. Tarentino, *Glycobiology* **1**, 257 (1991).

This change provided essentially a one-step purification of PNGase F to near homogeneity, and led to the serendipitous discovery that Endo F is actually a mixture of three distinct endoglycosidases designated Endo F_1, Endo F_2, and Endo F_3.[5,6] A summary of the specificity of each of these enzymes is presented in Fig. 1 and details are presented in a later section (Properties of Enzymes) in conjunction with specific applications.

Assay Methods

All oligosaccharide-cleaving enzymes can be assayed, using the fluorescent dansyl (Dns) derivatives of their preferred substrates. Qualitative assays can be followed by simple paper chromatography.[1] Quantitative determination is more readily accomplished by chromatographic separation of deglycosylated product from original substrate in a high-performance liquid chromatographic (HPLC) assay. The methods will be outlined briefly as tabulated below.

Enzyme	Preferred substrate/source, oligosaccharide type	Suggested buffer
PNGase F	Dns-L-A-N(oligo)-AeC-S/ fetuin, triantennary	Sodium phosphate, 0.25 M, pH 8.6
Endo F_1	Dns-N-(GlcNAc)$_2$(Man)$_5$/ ovalbumin, high mannose	Sodium acetate, 0.25 M, pH 5.8
Endo F_2	Dns-L-M-G-E-N(oligo)-R/ fibrinogen, biantennary	Sodium acetate, 0.25 M, pH 4.5
Endo F_3	(Dns)$_2$-Y-ε-DnsK-N(oligo)-N-S-D-I-S-S-T-R/ IgM, biantennary with core 1 $\rightarrow$ 6 fucose	Sodium acetate, 0.25 M, pH 4.5

Generic Method

All reactions are conducted at 37°. Enzyme dilutions in appropriate buffers containing 0.1% bovine serum albumin (BSA) are chosen such that hydrolysis does not exceed 15–20% of input substrate during the incubation period. In the tip of a 0.5-ml microtube is added 2 μl (1.5–2.0 nmol) of dansylated substrate and the reaction is initiated by the addition of 2 μl of appropriately diluted enzyme. Reactions are scaled up (i.e., fivefold) if three or four time points are desired. At some time (usually 5–15 min), the reaction is terminated with 16 μl of 0.4 M phosphoric acid and 30 μl (7.5 nmol) of an internal marker (Dns-glycine) and 50 μl of water are added. Samples of 80 μl are automatically injected with a Waters Associates (Milford, NJ) WISP model 710B and separations are achieved with a linear gradient produced by a Waters Model 721 automatic gradient system (Waters curve 6). The flow rate is 2.0 ml/min and a gradient of 10–60% acetonitrile is developed from initial to 22 min, using 0.1 M sodium acetate, at pH 6.5, as buffer A and 100% (v/v) acetonitrile containing 0.01% trifluoroacetic acid as buffer B. Peak areas are automatically calculated with a Waters data module.

Enzyme assays using the synthetic [oligo-NH]-S-AAEANS substrates (see Materials) are identical to those described above for the Dns-glyco-peptide substrates except that only 0.2 nmol of substrate is used, and detection and quantitation are accomplished by in-line fluorimetry. In addition, because these compounds are much more polar than the Dns-glycopeptides, the gradient is modified accordingly, as follows: 5–30% acetonitrile from initial conditions to 10 min, using the same buffers.

Alternative Available Assays

Highly purified bovine ribonuclease B has been used to detect and estimate both PNGase F and Endo F_1 activities.[4] The resorufin derivative of ovalbumin glycopeptide is commercially available (Boehringer-Mannheim, Indianapolis, IN) for the analysis of PNGase F and Endo F_1, and it should also detect Endo F_2.

Materials

Flavobacterium meningosepticum is obtained from the American Type Culture Collection (Rockville, MD; ATCC 33958). Inoculum is prepared in 1% tryptone–0.5% (w/v) NaCl–0.5% (w/v) yeast extract and stored at $-70°$ in 12.5% (v/v) glycerol as outlined in Ref. 1. The triantennary glycopeptide L-A-N(oligo)-AeC-S is from a thermolytic digest of reduced, aminoethylated fetuin,[7] and the asparagine-oligosaccharide, N(GlcNAc)$_2$ (Man)$_5$, is from a pronase digest of ovalbumin.[8] The biantennary glycopeptides L-M-G-E-N(oligo)-R and V-E-N(oligo)-K are obtained from reduced carboxymethylated human fibrinogen by thermolytic and tryptic digestion, and purified by HPLC. The biantennary glycopeptide Y-K-N(oligo)-N-S-D-I-S-S-T-R, with a core fucose residue linked $\alpha(1 \rightarrow 6)$ to N-acetylglucosamine, is from a thermolytic digest of a Waldenström's IgM(patient Ga),[9] and corresponds to the C-1 oligosaccharide at position Asn-171. All glycopeptides are dansylated as described previously.[7] Biantennary and triantennary 1-aminooligosaccharides are produced by

[7] T. H. Plummer, Jr., A. Phelan, and A. L. Tarentino, *Eur. J. Biochem.* **163**, 167 (1987).
[8] C-G. Huang, H. E. Mayer, Jr., and R. Montgomery, *Carbohydr. Res.* **13**, 127 (1970).
[9] T. H. Plummer, Jr. and A. L. Tarentino, *J. Biol. Chem.* **256**, 10243 (1981).

PNGase F digestion of dansylated fibrinogen and dansylated fetuin glyco-peptides, respectively, and are amidinated with 2-iminothiolane hydro-chloride (Pierce, Rockford, IL). The newly incorporated sulfhydryl group is alkylated with 5-({[(2-iodoacetyl)amino]ethyl}amino)naphthalene-1-sul-fonic acid(1,5-IAEDANS; Molecular Probes, Eugene, OR) to produce new substrates for endoglycosidases (oligo-NH-S-AAEANS).

Purification Procedure

All chromatography is accomplished at room temperature. All interme-diate steps are held at 4°.

Step 1: Ammonium Sulfate Precipitation. Twelve liters of culture fil-trate[1] is concentrated to less than 1 liter by an Amicon (Danvers, MA) CH2PRS spiral cartridge concentrator and adjusted to 90% saturation by the addition of solid ammonium sulfate. The mixture is allowed to stand overnight at 4° and the precipitate is isolated by centrifugation at 8000 g for 30 min. The precipitate is extracted with 50 ml of 0.1 M sodium phos-phate (pH 7.0) containing 1 M ammonium sulfate and 10 mM ethylenedi-aminetetraacetic acid (EDTA). This step results in a 15-fold purification, based on the original absorbance.

Step 2: TSK-Butyl Chromatography. The extract (55 ml) from the ammonium sulfate precipitation is applied to a TSK-butyl-Toyopearl 650M (Supelco, Bellefonte, PA) column (2.0 × 39 cm) equilibrated in 100 mM sodium phosphate, pH 7.0, containing 1 M ammonium sulfate and 10 mM EDTA. The column is flushed with equilibration buffer for 1 hr and then developed in a linear gradient to 50 mM sodium phosphate (pH 7.0), containing 10 mM EDTA, over a period of 17 hr. The flow rate is 60 ml/hr and fractions of 5 ml are collected. The elution profile (Fig. 2), in order of increasing hydrophobicity, indicates the presence of several well-sepa-rated components, designated as pools 1–8, in addition to a nonretarded peak. Sodium dodecyl sulfate–polyacrylamide gel electrophoresis (SDS–PAGE) of the peak tubes (Fig. 2, inset) indicates that most compo-nents are highly enriched over the initial extract. Enzyme assays and SDS–PAGE identify pool 1 (Fig. 2) as 90 + % Endo F_1, pool 3 as a mixture of Endo F_2 plus Endo F_3, and pool 6 as 90 + % PNGase F.

Step 3: TSK-Phenyl Chromatography. Ammonium sulfate equivalent to 1 M salt is added to pools 1, 3, and 6 and each is individually reapplied to a TSK-phenyl-Toyopearl 650S (Supelco) column (1.0 × 30 cm) devel-oped at a flow rate of 15 ml/hr (not shown).[6] This step removes minor impurities from pool 1 (Endo F_1), more significant impurities from pool 3 (Endo F_2 and Endo F_3), and only trace impurities from pool 6 (PNGase F). Pertinent enzymes can be monitored by their physical parameters in the absence of available Dns-substrates (see Table I). Desorption patterns

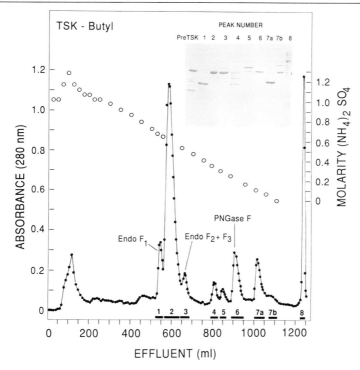

FIG. 2. Separation of *F. meningosepticum*-secreted proteins on TSK-butyl-Toyopearl 650 M. Enzymes were located by appropriate assays of Dns-substrates (see text). (●) absorbance at 280 nm; (○) molarity of ammonium sulfate. The bars represent fractions combined for further study. Inset, SDS–PAGE of 1-μg samples of individual peak tube of each pool.

TABLE I

PARAMETERS FOR MONITORING ENDOGLYCOSIDASES DURING HYDROPHOBIC
INTERACTION CHROMATOGRAPHY

Endoglycosidase	Apparent molecular weight (SDS–PAGE)	Molarity of ammonium sulfate required for desorption from:	
		TSK-butyl	TSK-phenyl
PNGase F	35,000	0.22	0.40
Endo F_1	31,000	0.65	0.90
Endo F_2	39,500	0.50	0.72
Endo F_3	31,000	0.50	0.72

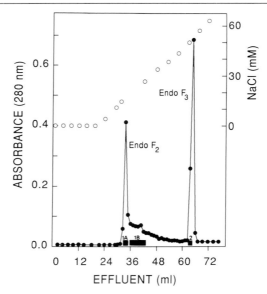

FIG. 3. Chromatography of the Endo F_2 plus Endo F_3 pool from TSK-phenyl-Toyopearl 650S on Protein-Pak 8P 8HR. (●) absorbance at 280 nm; (○) millimolarity of NaCl. The bars represent fractions combined for further study.

on the hydrophobic interaction TSK-butyl and TSK-phenyl columns have been remarkably consistent over a period of 2 years and, combined with SDS–PAGE of isolated peaks, can be used to follow the appropriate enzymes during purification.

Step 4: Sulfopropyl Ion-Exchange Chromatography. Pooled, dialyzed, enzyme-containing fractions from the TSK-phenyl chromatography are individually applied to a high-resolution sulfopropyl (Protein-Pak SP 8HR; Waters Chromatography Division) column (1.0 × 10 cm) equilibrated in 10 mM sodium acetate at pH 5.8 for Endo F_1 and PNGase F, and at pH 4.5 for Endo F_2 plus Endo F_3. The column is developed at a flow rate of 90 ml/hr and 1.5-ml fractions are collected. Figure 3 illustrates the separation of Endo F_2 (pools 1A and 1B) from Endo F_3 (pool 2), using an 80-min gradient to 100 mM NaCl. Endo F_2 pool 1A and pool 1B (Fig. 3) have the same specific activity, and are identical in apparent molecular weight on SDS–PAGE. It should be noted that ion-exchange chromatography at pH 4.5 removes the trace Dns-N-(GlcNAc)$_2$(Man)$_5$-cleaving activity reported as being inherent to Endo F_3 by our previous method of chromatography.[6] Only minor contaminants are removed from PNGase F and from Endo F_1 by similar chromatography at pH 5.8[6] (not shown).

General Comments

From 12 liters of cultural filtrate one can obtain approximately 5 mg each of PNGase F and Endo F_1, and about 0.28 mg of Endo F_2 and 0.47 mg of Endo F_3. Considering that commercially available PNGase F is rather expensive and roughly 150,000 milliunits (mU) of PNGase F is available from 12 liters of medium, there is a rather large incentive to undertake in-house purification of these enzymes, especially if more than occasional use is anticipated. Moreover, PNGase F and Endo F_1 can be obtained at 90 + % purity and free of cross-contamination, using only the first step of the protocol, namely TSK-butyl chromatography. We have found that the enzymes can be used at this stage for most purposes without further purification if metalloprotease inhibitors[1] are included in enzyme digests. PNGase F and Endo F_1 are easily purified to homogeneity by using two additional columns. Because Endo F_2 and Endo F_3 are not available commercially, they must be purified through the sulfopropyl chromatography stage for complete resolution and purity. Even at yields of 0.28 mg of Endo F_2 and 0.47 mg of Endo F_3 from 12 liters of medium, one finds considerable catalytic potential because the turnover of these enzymes is high (see Table II).

TABLE II
EFFECT OF CORE-SUBSTITUTED $\alpha(1 \rightarrow 6)$-FUCOSE ON HYDROLYSIS OF BIANTENNARY GLYCOPEPTIDES BY ENDO F_2 AND ENDO F_3

| | % Hydrolysis/60 min | | | |
| | IgM glycopeptide[a] + fucose | | Fibrinogen glycopeptide − fucose | |
Enzyme dilution[b]	Endo F_2	Endo F_3	Endo F_2	Endo F_3
1/10	83	82	100	40
1/100	24	68	40	1.8
1/500	6	25	12	—
1/1000	3	12	—	—

[a] Endo F_2 and Endo F_3 were assayed with a fucose-containing biantennary IgM glycopeptide and a biantennary (fucose-free) fibrinogen glycopeptide by HPLC (see Assay Methods).

[b] Endo F_2 and Endo F_3 were both adjusted to A_{280}/ml = 0.23. Serial dilutions were made from these enzyme stocks, using a 0.2 M sodium acetate (pH 4.75) buffer.

Properties of Enzymes

PNGase F

PNGase F is homogeneous following sulfopropyl chromatography, with a final specific activity of 25,000–30,000 nmol of a didansylfetuin glycopeptide hydrolyzed per minute per milligram enzyme protein (37°). The PNGase F gene has been cloned and sequenced[10] and codes for a mature protein of 314 amino acids with a molecular mass of 34,779 Da. A 1% solution of PNGase F gives an absorbance at 280 nm of 18, reflecting in part its high tryptophan content (nine residues). To convert from absorbance to enzyme activity, we use the factor A_{280}/ml of 1.0 = 15,000 milliunits.

The pH optimum for PNGase F activity is 8.6, but the enzyme is at least 80% active between pH 7.5 and 9.5. PNGase F is compatible with a wide variety of inorganic and organic buffers (0.1 M), including sodium phosphate, lithium carbonate, ammonium bicarbonate, Tris–HCl, glycylglycine, N-2-hydroxyethylpiperazine-N'-2-ethanesulfonic acid (HEPES), and triethylamine acetate, but sodium borate is inhibitory.

The specificity of PNGase F is well established[1–3] and, as summarized in Fig. 1, asparagine-linked glycans representing all major oligosaccharide classes are hydrolyzed by the enzyme. Asparagine-linked oligosaccharides containing an $\alpha(1 \rightarrow 6)$-fucose substituted on the asparagine-proximal N-acetylglucosamine residue are easily hydrolyzed by PNGase F, but the corresponding $\alpha(1 \rightarrow 3)$-fucose substituent,[11] found in the plant glycoproteins bromelin and horseradish peroxidase, completely blocks deglycosylation. To the best of our knowledge, this is the only known structural feature of an oligosaccharide moiety that confers resistance to PNGase F.

Endo F_1

Endo F_1 is homogeneous following sulfopropyl chromatography at pH 5.8 and at this pH it is stable at 4° for over 1 year. Endo F_1 shows maximum activity between pH 5 and 6; appreciable activity remains even at pH 8.5, but below pH 5 the Endo F_1 activity drops off rapidly. The Endo F_1 gene has been cloned and sequenced,[12] and codes for a mature protein of 289 amino acids with a molecular mass of 31,667 Da.

Endo F_1 and Endo H are similar in substrate specificity[5]; both enzymes are nearly identical in their ability to hydrolyze high-mannose oligosaccha-

[10] A. L. Tarentino, G. Quinones, A. Trumble, L.-M. Changchien, B. Duceman, F. Maley, and T. H. Plummer, Jr., J. Biol. Chem. 265, 6961 (1990).
[11] V. Tretter, F. Altmann, and L. März, Eur. J. Biochem. 199, 647 (1991).
[12] A. L. Tarentino, G. Quinones, W. P. Schrader, L.-M. Changchien, and T. H. Plummer, Jr., J. Biol. Chem. 267, 3868 (1992).

rides (Fig. 1). They differ only in their specificity for core-substituted fucose; Endo H hydrolyzes fucose-containing swainsonine-type hybrids (Fig. 1, +) at nearly the same rate as high-mannose glycans, but core-linked fucose reduces the rate of hydrolysis by Endo F_1 over 50-fold (Fig. 1, +/−) relative to high-mannose structures. Note that Endo F_1, like Endo H, does not hydrolyze any of the complex asparagine-linked oligosaccharides. Commercial preparations designated Endo F are mainly Endo F_1, with variably small amounts of Endo F_2, the component that cleaves biantennary oligosaccharides.

Endo F_2 and Endo F_3

Endo F_2 and Endo F_3 appear homogeneous by SDS–PAGE following sulfopropyl chromatography at pH 4.5, and at this pH we have not observed any change in stability at 4° for over 1 year. The pH optimum for Endo F_2 and Endo F_3 is considerably lower than for Endo F_1; in acetate buffer the pH optimum for both enzymes is between pH 4.0 and 4.5, and at least 70% of the activity is present at pH 3.0. Above pH 6, the activity of Endo F_2 and Endo F_3 drops off sharply.

The substrate specificities of Endo F_2 and Endo F_3 are still incomplete and their profile is evolving as different substrates become available. Endo F_2 hydrolyzes both biantennary and high-mannose oligosaccharides (Fig. 1) but the enzyme prefers biantennary glycans by a factor of at least 20-fold. Endo F_2 does not cleave swainsonine- or ovalbumin (e.g., peak A or B)-type hybrid structures, or tri- and tetraantennary oligosaccharides. Endo F_3 also hydrolyzes biantennary oligosaccharides but the rate of hydrolysis is much slower than for Endo F_2, unless the core asparagine-proximal N-acetylglucosamine is substituted with an $\alpha(1 \rightarrow 6)$-fucose residue. As shown in Table II, the biantennary fibrinogen glycopeptide is hydrolyzed by Endo F_2 20-fold faster (40 vs 1.8) than by Endo F_3. However, the biantennary IgM glycopeptide is hydrolyzed by Endo F_3 threefold faster (68 vs 24) than Endo F_2. Thus the net effect of fucose is to increase the rate of hydrolysis of biantennary glycans by Endo F_3 at least 30-fold (68 vs 1.8). Endo F_3, unlike Endo F_2, is also capable of hydrolyzing asparagine-linked triantennary oligosaccharides (Fig. 1), and it is presumed that the presence of core fucose would have a similar enhancing effect on the rate of hydrolysis. Additional studies using defined glycopeptides will be needed to characterize these new enzymes completely.

Deglycosylation Protocol

Human transferrin (HTF) contains two asparagine-linked biantennary oligosaccharide chains (fucose free), one of which is relatively inaccessible and released slowly by endoglycosidases only after denaturation in SDS.

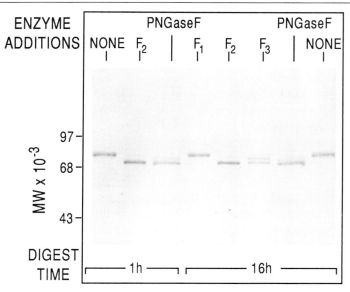

FIG. 4. Deglycosylation efficiency of PNGase F, Endo F_1, Endo F_2, and Endo F_3 on human transferrin. Enzyme digests (10 μl) were prepared according to the deglycosylation protocol (see text) and incubated at 37° for the indicated times. Reactions were stopped with a 2× concentrated sample buffer and boiled for 5 min. Aliquots of 8 μl were subjected to electrophoresis at 200 V for 1.5 hr through a 10% polyacrylamide gel [5% (w/v) stacking gel], and then stained with Coomassie blue.

We have used the HTF system to test the relative deglycosylation efficiency of the four oligosaccharide-cleaving enzymes, purified as described earlier, and monitored the results by classic band shifting[3,4] on SDS–PAGE (Fig. 4). A stock solution of heat-denatured HTF is prepared by dissolving 1 mg (98% purity; Sigma, St. Louis, MO) in 1 ml of 0.5% SDS and heating for 3 min at 90°. A typical reaction (37°) contained in a final volume of 10 μl: SDS heat-denatured HTF (1 μg/μl), 2 μl; 0.2 M buffer (for PNGase F, sodium phosphate, pH 8.6; for Endo F_1, sodium acetate, pH 5.8; for Endo F_2 or F_3, sodium acetate, pH 4.5), 4 μl; water, 2 μl; 10% (v/v) Nonidet P-40 (NP-40), 1 μl; and 1 μl of the appropriate endoglycosidase. The protein concentration of the enzyme stocks (A_{280}/ml) was PNGase F, 0.43; Endo F_1, 0.28; Endo F_2, 0.25; and Endo F_3, 0.25. Endo F_2 and PNGase F released both biantennary oligosaccharides from HTF within 1 hr, even when the enzyme stocks were diluted 10-fold (not shown). At 16 hr, there was no further change in the carbohydrate-depleted HTF band, demonstrating the protease-free quality of both Endo F_2 and PNGase F purified as described above. Deglycosylation of HTF by Endo F_3 was always incomplete regardless of the amount of enzyme used or the length of incubation; the enzyme appeared to release completely one of the biantennary oligosaccharides, and only partially removed the second oligosaccharide chain. Endo F_2 and Endo F_3 hydrolyze with equal efficiency the C-1 fucose-containing biantennary oligosaccharide from the M_r 60,000 Fabμ fragment of IgM (patient Ga) (data not shown), indicating that the core fucose-enhancing effect on Endo F_3 applies to glycoproteins as well as glycopeptides. Endo F_1, as indicated earlier, lacks biantennary-cleaving activity and, as expected, had no effect on the mobility of HTF.

New Applications: Oligosaccharide Remodeling

Principle

Hydrolysis of asparagine-linked glycans by PNGase F generates a 1-amino oligosaccharide that is relatively stable under alkaline conditions, and can be converted to a stable oligosaccharide thioamidine in good yield (50–80%) with the reagent 2-iminothiolane hydrochloride (2-IT) according to Scheme I.

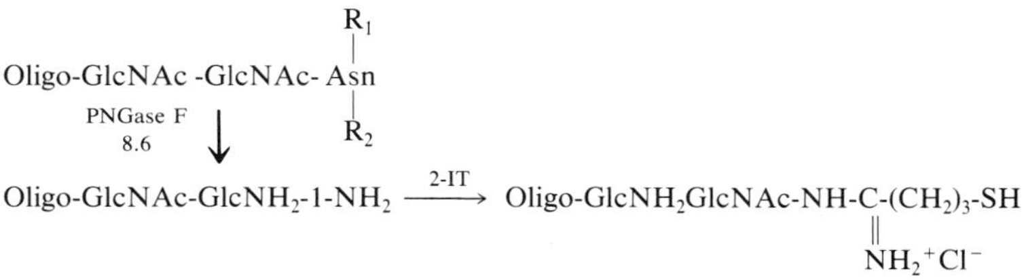

$$\text{Oligo-GlcNAc -GlcNAc- } \overset{\overset{\displaystyle R_1}{|}}{\underset{\underset{\displaystyle R_2}{|}}{\text{Asn}}}$$

$$\underset{8.6}{\overset{\text{PNGase F}}{\Big\downarrow}}$$

$$\text{Oligo-GlcNAc-GlcNH}_2\text{-1-NH}_2 \xrightarrow{\text{2-IT}} \text{Oligo-GlcNH}_2\text{GlcNAc-NH-C-(CH}_2)_3\text{-SH}$$

$$\underset{\text{NH}_2{}^+\text{Cl}^-}{\overset{\|}{}}$$

SCHEME I

The reactive sulfhydryl group can be alkylated with fluorescent reagents such as 1,5-IAEDANS to generate stable endoglycosidase substrates, resistant to PNGase F cleavage. Depending on the initial choice of asparagine-linked glycan, a specific probe can be developed to assay Endo F_1, Endo F_2, and/or Endo F_3 (see Fig. 5).

Methodology

A typical protocol using a didansylated fibrinogen glycopeptide, Dns-V-E-N(oligo)-K(Dns) is described. The glycopeptide (1.1 μmol) is lyophilized in a small conical glass tube, and 400 μl of 0.20 M sodium HEPES (pH 9.1) is added. The reaction is started by addition of 100 μl (83 μg) of a stock PNGase F preparation from the sulfopropyl column (A_{280}/ml = 1.5 = 830 μg/ml). Deglycosylation is monitored by paper chromatography[3] and the reaction is complete after 60–90 min. At this point the reaction is placed at 25° and 100 μl (50 μmol) of a freshly made 2-IT solution [0.5 M

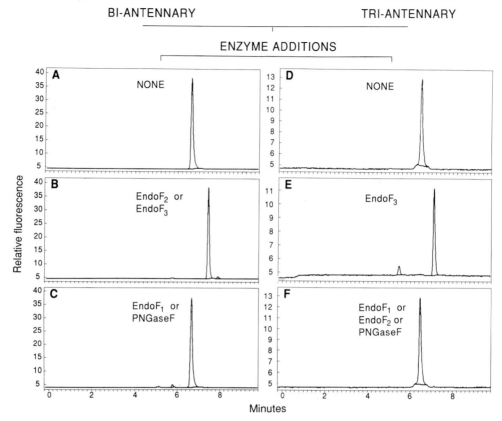

FIG. 5. HPLC assays of endoglycosidase digests, using flourescent bi- and triantennary [Oligo-NH]-S-AAEANS substrates. Enzyme digests (4 μl) were prepared as described under Assay Methods and incubated for 1 hr at 37° prior to HPLC.

in 0.2 M sodium HEPES (pH 9.1)] is added. Complete derivatization of the 1-aminooligosaccharide requires 1–2 hr. After this time, 20 μl of concentrated 2-mercaptoethanol is added and incubated 30 min to ensure complete reduction of any disulfides that may have formed under the alkaline conditions. The reaction is acidified with a few drops of 50% (v/v) acetic acid and chromatographed on a column (0.9 × 150 cm) of BioGel P-2 (200–400 mesh) in 0.1 M acetic acid–10% (v/v) methanol (flow rate, 12 ml/hr). The derivatized oligosaccharide is located by absorbance at 230 nm, by the phenol–sulfuric acid reagent for hexose, and by DTNB for free thiols. The appropriate fractions are combined, evaporated in a conical flask, and reacted in the same flask with 1,5-IAEDANS as follows:

The residue is dissolved in 0.25 ml of freshly made 0.2 M NaHCO$_3$ (pH 8.2) and 0.25 ml (2.5 μmol) of a 0.01 M 1,5-IAEDANS in 0.2 M NaHCO$_3$ (pH 8.2) is added. After 1 hr the reaction is applied to a column (0.9 $\times$ 150 cm) of Sephadex G-25 equilibrated in 50 mM NH$_4$HCO$_3$ (flow rate, 12 ml/hr). The highly fluorescent product in the void volume is collected and quantitated spectrophotometrically at 337 nm (6.3 $\times$ 10^{-3} cm^{-1} M^{-1}; Molecular Probes).

General Comments

It is not necessary that the glycopeptide substrate be dansylated, but free amino groups should be blocked (e.g., acetylation) because they will react with 2-IT directly. The amount of PNGase F used depends on the amount of glycopeptide and its quality as a substrate. Because the didansyl fibrogen glycopeptide has a short polypeptide chain, it is a relatively poor substrate, and the ratio of enzyme to substrate was increased to drive the reaction to completion in a short time. Efficient PNGase F substrates (1 μmol) can be deglycosylated with at least fivefold less enzyme.

Figure 5 demonstrates the ability of these new synthetic fluorescent substrates to discriminate the endoglycosidases, based on their known substrate specificity. Susceptibility to enzyme action by hydrolysis at the di-N-acetylchitobiose core is indicated by an increased retention time on HPLC. The fluorescent biantennary probe (Fig. 5A) was readily cleaved by Endo F$_2$ and Endo F$_3$ (Fig. 5B), but not by Endo F$_1$ or PNGase F (Fig. 5C). This probe does not conform to the substrate specificity requirements of the latter, because Endo F$_1$ is specific for high-mannose glycans and PNGase F requires a polypeptide backbone. For the same reasons, the fluorescent triantennary oligosaccharide can be cleaved only by Endo F$_3$ (Fig. 5E), and should be resistant to the other enzymes (Fig. 5F).

[5] Release of Oligosaccharides from Glycoproteins by Hydrazinolysis

By T. P. PATEL and R. B. PAREKH

Introduction

Hydrazinolysis predates most chemical and enzymatic methods as a technique for the nonselective and therefore uniform release of unreduced

N-linked oligosaccharides in high yield from glycoproteins.[1,2] It remains an important and powerful method for this purpose and has been reviewed previously in this series.[3] A reexamination and optimization of the method has been performed, and it was reported that hydrazinolysis can be an equally effective technique for the release of unreduced O-linked oligosaccharides from glycoproteins.[4]

The reaction mechanism by which hydrazinolysis leads to the release of N-linked oligosaccharides is still not fully elucidated.[3,5,6] The mechanism leading to the release of O-linked oligosaccharides has not yet been investigated at all, although a β elimination is suspected. Reaction conditions under which N- and O-linked oligosaccharides can be released from glycoproteins have therefore been established empirically after investigation of the release of such oligosaccharides from a range of glycoprotein standards,[4] whose glycosylation characteristics had been previously well defined. Despite the structural diversity of oligosaccharides and protein moieties associated with these standards, similar reaction conditions were found to allow essentially complete release of unreduced N- and O-linked oligosaccharides. Selective release of O-linked oligosaccharides with little release of N-linked oligosaccharides can be achieved with mild hydrazinolysis reaction conditions. Hydrazinolysis can therefore be used in three different ways, namely, to release all (N- and O-linked) oligosaccharides from a glycoprotein, to release selectively just O-linked oligosaccharides, and to release sequentially first O-linked, and then N-linked oligosaccharides.

As with any method of releasing oligosaccharides, the objective during the use of hydrazinolysis is to maximize the release of oligosaccharides while minimizing any undesired side reactions. This requires attention to be paid to the purity of the reagents used (and in particular the purity of the hydrazine), to preparation of the glycoprotein (or glycopeptide) sample, and to the reaction conditions used. In the hydrazinolysis procedure described here, reaction conditions optimized for glycoprotein samples in solution are described. Although equally effective as a method, different hydrazinolysis reaction conditions apply when samples are immobilized to solid supports.[6a]

[1] Y. Matsushima and N. Fujii, *Bull. Chem. Soc. Jpn.* **30**, 48 (1957).
[2] B. Bayard and J. Montreuil, *Colloq. Int. C.N.R.S.* **221**, 209 (1974).
[3] S. Takasaki, T. Mizuochi, and A. Kobata, this series, Vol. 83, p. 263.
[4] T. Patel, J. Bruce, A. H. Merry, J. C. Bigge, M. R. Wormald, and R. B. Parekh, *Biochemistry* **32**, 679 (1992).
[5] P. W. Tang and J. M. Williams, *Carbohydr. Res.* **121**, 89 (1983).
[6] B. Bendiak and D. A. Cumming, *Carbohydr. Res.* **144**, 1 (1985).
[6a] P. Goulding *et al.,* in preparation.

Sequence of Steps during Hydrazinolysis

Performing hydrazinolysis on glycoproteins involves the following general steps:

1. purification of reagents
2. preparation of sample
3. incubation of the sample with hydrazine to release oligosaccharides
4. removal of excess, unreacted hydrazine
5. re-N-acetylation of any de-N-acetylated amino groups
6. separation and recovery of released oligosaccharides from peptide (and glycopeptide) material
7. quantitative regeneration of intact reducing termini of released oligosaccharides

The performance of steps 1–7 will be described. Methods for subsequent analysis of the released oligosaccharides are discussed in [8]–[16] in this volume.

Procedure

Purification of Reagents/Resins

Hydrazine. The quality of hydrazine is critical if side reactions are to be minimized during hydrazinolysis, and in particular for the recovery of O-linked oligosaccharides. It is essential that the hydrazine to be used is metal free and essentially anhydrous (<1%, v/v). It is therefore preferable not to distill the hydrazine from calcium oxide or other inorganic desiccants, not to distill it as its azeotrope from toluene, and not to purify it by passage over anhydrous alumina.[7] The simplest purification process is a double partial-vacuum distillation under anhydrous argon. The freshly distilled hydrazine can be dispensed into (acid-rinsed and dried) glass ampoules, which are flame sealed for long-term storage (up to 12 months), and kept in the dark at ambient temperature. If commercial hydrazine is to be used, the process by which it was purified, date of preparation, and, if possible, water content should all be confirmed. The water content of a hydrazine preparation can be directly determined by gas chromatography.[7]

Acetic anhydride: Commercial high-performance liquid chromatography (HPLC)-grade reagent is suitable

n-Butanol, ethanol, methanol, and water: All at least HPLC grade

Saturated sodium bicarbonate: This should be prepared just before use

[7] E. W. Schmidt, "Hydrazine and Its Derivatives." Wiley, New York, 1984.

by adding water (at least HPLC grade, and ideally glass distilled) to analytical-grade sodium bicarbonate

Copper(II) acetate solution: A 1 mM solution of analytical-grade Cu(II) acetate in 1 mM acetic acid (analytical-grade acetic acid diluted with at least HPLC-grade water)

C_{18}-silica: Most commercial preparations are suitable. In particular, Sep-Pak cartridges (Millipore, Bedford, MA) are convenient. The C_{18}-silica should be generated by first washing with a $15\times$ column volume of methanol and then a $5\times$ column volume of water

Dowex AG 50-X12(H$^+$): This can be purchased from Bio-Rad (Richmond, CA) and should be generated according to manufacturer instructions. In particular, the prepared resin should be rinsed with water until the pH of the effluent equals that of the water eluant and no visible color continues to elute

The volumes of reagents, solutions, and resins required are discussed in subsequent sections.

Note. All glassware should be clean, preferably acid washed (sonicated for 10 min in 2 M nitric acid), thoroughly rinsed in water, and dried before use.

Preparation of Sample

Together with the quality of hydrazine reagent, the preparation of the glycoprotein or glycopeptide sample is crucial to the successful outcome of a hydrazinolysis. Many inorganic cations and anions can interfere with hydrazinolysis. Irrespective of their origin and nature, all samples should be rendered as free of salts as is practical, and thoroughly dried before addition of hydrazine.

The preferred approach for removing salts (and some detergents) from glycoproteins is microflow dialysis at 4° with an appropriate molecular weight cutoff membrane against a solution of water, dilute acid [such as 0.1% (v/v) trifluoroacetic acid or 1% (v/v) acetic acid], or volatile buffer [such as low (<0.1 M) molarity ammonium bicarbonate or ammonium acetate]. Flow dialysis is preferred over exchange dialysis wherever practical, due to its greater efficiency, speed, and practicality. The dialysis buffer should be chosen to maintain the sample in solution as much as possible, although this is not essential as shown by successful hydrazinolysis on crude or relatively insoluble samples. The dilute acids do not cause any detectable loss of acid-labile carbohydrate determinants, such as sialic acids, if dialysis is performed at 4°. The time of dialysis depends on the sample volume and the device being employed and should therefore be according to manufacturer guidelines. Microflow dialysis will also facilitate reduction in the content of many detergents, but will generally not remove them completely. The exclusion from the sample prior to dialysis of exogenous carbohydrate material is of particular importance. This is a particular problem when the final step in purification of the glycoprotein is its elution under harsh conditions (e.g., acid or base) from carbohydrate-based resins [e.g., IgG eluted from protein A–Sepharose or concanavalin A (ConA)–Sepharose/agarose under acidic conditions]. Some breakdown of the carbohydrate matrix occurs, leading to the inevitable recovery of a series of carbohydrate polymers that cannot be removed from the glycoprotein by dialysis. In such cases, a subsequent purification step should be introduced to separate matrix-derived polymers from glycoproteins, or an alternative matrix (e.g., silica or acrylamide) should be used. Alternatives to microflow dialysis include reversed-phase HPLC and gel filtration. Provided that volatile eluants are used, and the sample is filtered (0.2-μm pore size Teflon filter prewashed to remove any contaminating sucrose) prior to lyophilization, these alternatives are generally equally effective. Most detergents can be removed from glycoprotein samples by reversed-phase HPLC. Deoxycholate interferes with hydrazinolysis but, surprisingly, low levels ($\sim$ 1%) of sodium dodecyl sulfate (SDS) or Tween 80 can be tolerated during hydrazinolysis.[7a]

Salt-free samples should be transferred to clean (see above) glass reaction vessels prior to drying. These vessels can be ones that allow easy subsequent flame sealing, to allow hydrazinolysis to be performed under a controlled atmosphere, or that can be tightly sealed with Teflon-lined screw caps. The rate of uptake of water into a hydrazine solution through screw-cap vials is not significant and these are therefore acceptable and more practical alternatives to flame-sealing reaction vessels. Glycoprotein and multicomponent samples, such as membrane preparations, should be dried by lyophilization. Glycopeptide samples can be rotary evaporated prior to lyophilization. Lyophilization should be performed for at least 24 hr. Further lyophilization, such as incubation over P_2O_5 or activated charcoal, is not necessary.

Comment on Sample Preparation. Although sample preparation for hydrazinolysis is important, it can usually be done with relatively little effort. Microflow dialysis has, in our experience, always been suitable for all samples except those derived directly from SDS gels. In the latter case, reversed-phase HPLC is useful to separate glycoprotein from excess SDS. Hydrazinolysis has been successfully performed on a wide variety

[7a] P. Goulding, J. Bruce, T. Patel, and R. Paukh, in preparation.

of samples, including <0.1 µg of a highly purified glycopeptide,[8] plasma membrane extracts,[9] several grams of mouse brain crude homogenate,[10] and *Dictyostelium discoideum* membrane extracts.[11]

Release of Oligosaccharides[4]

Hydrazine (either freshly distilled or from a fresh vial) should be added to the thoroughly lyophilized salt-free glycoprotein or glycopeptide immediately on its removal from the freeze-drier. An oven-dried (rinsed first with water and then with methanol before drying) Hamilton syringe [glass barrel, gas-tight Teflon plunger, securely fitted stainless steel (not platinum) needle] is used for this purpose. The syringe is rinsed a few times with the hydrazine before adding the appropriate volume of hydrazine to the lyophilized sample. This addition can be performed in a fume hood with no special precautions to maintain an anhydrous/anoxic atmosphere, provided it is performed rapidly. The lyophilized sample and the hydrazine vial should not be open to the atmosphere for more than about 2 min. The volume of hydrazine added should be sufficient to generate a 5- to 25-mg/ml solution of glycoprotein or glycopeptide. As a minimum, 50–100 µl of hydrazine is a lower practical limit in most cases. Addition of a large excess of hydrazine does not usually create any problems. For hydrazinolysis of large amounts of material, concentrations above 25 mg/ml should be avoided. After adding the hydrazine, cap or seal the reaction vessel immediately, shake it gently, and transfer it to a preheated oven or reaction block. For the release of both N- and O-linked oligosaccharides, reaction conditions of 95° for 4 hr are suitable, and for the release of O-linked oligosaccharides only, reaction conditions of 60° for 5 hr are suitable. When sequential release of first O-linked and then N-linked oligosaccharides is to be performed, the reaction is first performed at 60° for 5 hr, released oligosaccharides are separated from glycopeptides, and the glycopeptide pool is separately reacted with hydrazine at 95° for 4 hr.

Removal of Excess Unreacted Hydrazine

At the end of the hydrazinolysis reaction, excess unreacted hydrazine should be removed under vacuum, using a high-vacuum pump fitted with

[8] R. B. Parekh, A. G. D. Tse, R. A. Dwek, A. F. Williams, and T. W. Rademacher, *EMBO J.* **6**, 1233 (1987).

[9] K. Yamashita, T. Ohkura, Y. Tachibana, S. Takasaki, and A. Kobata, *J. Biol. Chem.* **259**, 10834 (1984).

[10] D. R. Wing, T. W. Rademacher, M. C. Field, R. A. Dwek, B. Schmitz, G. Thor, and M. Schachner, *Glycoconj. J.* **9**, 293 (1992).

[11] S. Amatayakul-Chantler, M. A. J. Ferguson, R. A. Dwek, T. W. Rademacher, R. B. Parekh, I. E. Crandall, and P. C. Newell, *J. Cell Sci.* **99**, 485 (1991).

an activated charcoal/alumina trap. During this evaporation, the reaction vessel should be placed in a water bath so that its temperature does not exceed ~25°. Only if the sample does not thoroughly dry under vacuum (as can be decided by visual inspection) should an addition of toluene be made to assist removal of traces of residual hydrazine. This is usually necessary only when large amounts (more than ~50 mg) of glycoprotein are subjected to hydrazinolysis, and the toluene should be added in ~0.5- to 1-ml aliquots. If necessary for removal of final traces of residual hydrazine, the toluene should be anhydrous.

Re-N-acetylation

Place the dried reaction mixture on ice and add first the solution phase of ice-cold saturated sodium bicarbonate, immediately followed by acetic anhydride. Mix gently and incubate at room temperature for 10 min. Add a second equal aliquot of acetic anhydride and incubate for a further 20 min. It is not uncommon for some frothing to occur during re-N-acetylation. It should be watched, and controlled if necessary. The volumes of reagents added are calculated on the basis of the assumption that all possible amino groups are to be N-acetylated. In the case of glycoproteins, for practical purposes it can be assumed that all of these derive from amino acids. A volume of acetic anhydride equal to a $5\times$ molar excess over these amino groups should be added, and the volume of sodium bicarbonate is such as to render the final acetylation mixture 0.5 M in acetic anhydride. After the incubation, pass the acetylation mixture through a column of Dowex AG 50-X12 (H^+), of a volume sufficient to remove five times the content of sodium (assume the saturated sodium bicarbonate to be 1 M). The reaction vessel and column are rinsed with water ($5\times$ column volume), and the eluant and washings combined and rotary evaporated to dryness (<27°).

Separation and Recovery of Released Oligosaccharides

Released oligosaccharides can be separated from peptide-derived material in a variety of ways, including paper chromatography,[3] adsorption chromatography, and gel filtration.[12] For sample amounts of less than 10 mg, paper chromatography is recommended. For larger sample amounts (up to several grams), adsorption chromatography can be used. Gel filtration may also be used, but it must be remembered that smaller N- and O-linked oligosaccharides may not be well separated from peptide-derived

[12] M. Fukuda, T. Kondo, and T. Osawa, *J. Biochem. (Tokyo)* **80,** 1223 (1976).

material and some fractionation of the oligosaccharide pool may also occur.

Paper Chromatography. The N-acetylated mixture is dissolved in water, applied to Whatman (Clifton, NJ) 3MM paper, allowed to dry, and subjected to descending paper chromatography with *n*-butanol–ethanol–water, 4 : 1 : 1 (v/v/v) at 30° for 16–24 hr. Under these conditions, disaccharides and larger oligosaccharides are immobile and can be recovered from the origin by extraction with water. To minimize recovery of glucose- and xylose-containing contaminants derived from the paper, each paper strip can be prewashed in water.

Adsorption Chromatography. Extensively fined microcrystalline cellulose is used as resin. The cellulose is first ethanolized by incubation overnight in 50% (v/v) ethanol, then thoroughly fined (in water). This step is vital to achieve reasonable flow rates during subsequent chromatography. The column of cellulose (~2 ml settled bed for each 20 mg of hydrazinolysate) is extensively washed first with water and then with ethanol, and then equilibrated in butanol–ethanol–water, 4 : 1 : 0.5 (v/v/v). The sample is dissolved in a minimum amount of water, and this aqueous solution brought to 4 : 1 : 0.2 (v/v/v) by the addition, first, of 5 vol of ethanol and then of 20 vol of butanol. This solution is applied to the cellulose column, which is then washed with 10 column volumes of butanol–ethanol–water, 4 : 1 : 0.5 (to retain both N- and O-linked oligosaccharides) and then with two column volumes of ethanol. Oligosaccharides are eluted with two column volumes of water. A flow rate of about two column volumes per hour will provide good separation of peptide [which elutes in the load, butanol–ethanol–water (4 : 1 : 0.5) and ethanol] from oligosaccharides, which elute in water. As with paper chromatography, oligosaccharides of degree of polymerization ≥2 are recovered quantitatively.

When both N- and O-linked oligosaccharides have been released from a sample, either of the above procedures is suitable. When only O-linked oligosaccharides have been released, a prior reversed-phase step is performed to separate released O-linked oligosaccharides from *N*-glycopeptides. The N-acetylated hydrazinolysate is suspended in ~1 ml of water and applied to a column of C_{18}-silica (e.g., Sep-Pak cartridges). The column is washed with two column volumes of 5% (v/v) methanol in water and load, eluant, and washings are combined as the pool of released O-linked oligosaccharides. This pool is then subjected to paper or adsorption chromatography as described above. The C_{18}-silica column is eluted with methanol to recover bound N-glycopeptides, if desired, which can then be separately subjected to hydrazinolysis. After recovery from C_{18}-silica, paper, or cellulose chromatography, oligosaccharide pools should be filtered (0.2 μm, Teflon filter) prior to any subsequent steps.

Quantitative Regeneration of Oligosaccharide Reducing Termini

After N-acetylation, the oligosaccharide pools consist mainly of unreduced oligosaccharides with a small fraction present as the acetohydrazide derivative.[13] These latter are catalytically converted to the authentic unreduced oligosaccharides as follows. The oligosaccharide pool obtained after step 6 is rotary evaporated to dryness (at ≤27°), resuspended in ~0.5 ml of 1 mM Cu(II) acetate in 1 mM acetic acid, and incubated at room temperature for 1 hr (the precise volume of copper acetate solution is not important). The solution is then passaged through a column (~0.5 ml) of Dowex AG 50-X12 (H$^+$) and eluant and washings are combined, filtered, and, if necessary, the acetic acid removed by evaporation.

Comments on Hydrazinolysis Procedure

This hydrazinolysis procedure has to date been used as described with approximately 10 glycoproteins carrying both N- and O-linked oligosaccharides and over 50 glycoproteins carrying only N-linked oligosaccharides. It was found to release O-linked oligosaccharides with high yield (>90%) from all O-glycosylated glycoproteins under both reaction conditions. Of the N-glycosylated glycoproteins, high yield (>90%) of N-linked oligosaccharides was obtained with all but two. In the case of these two, harsher reaction conditions (95° for up to 8 hr) were required. Although each new glycoprotein sample may therefore require some optimization of reaction conditions, the empirically established reaction conditions described here should be suitable for most cases and constitute a good "first try." Should optimization of reaction conditions be required, this can be performed as described in Takasaki et al.[3] and Patel et al.[4]

Limitations to Use of Hydrazinolysis

This hydrazinolysis procedure has been tested only against N- and O-glycosylated glycoproteins containing oligosaccharides attached to protein through an N-acetylglucosamine to asparagine N-glycosidic linkage and through N-acetylgalactosamine or mannose to serine or threonine through an O-glycosidic linkage. It has not been used against "true" mucins and proteoglycans, and the behavior of the glyan moiety of glycosyl phosphatidyl inositol (GPI)-membrane anchors is not known. It should also be emphasized that although sialic acid linkages themselves are stable

---

[13] B. Bendiak and D. A. Cumming, *Carbohydr. Res.* **151**, 89 (1986).
[14] P. Hermentin, R. Witzel, R. Doenges, R. Bauer, H. Haupt, T. Patel, R. B. Parekh, and D. Brazel, *Anal. Biochem.* **206**, 419 (1992).

in hydrazinolysis,[4,14] substitutions at various positions in the sialic acid ring, and particular O-acyl substitutions, are not expected to be retained. Finally, substantial peptide bond cleavage occurs during hydrazinolysis and hydrazinolysis should therefore not be used when it is essential to recover both released oligosaccharides and intact peptide backbone.

Acknowledgments

The critical comments of David Ashford and Gordon Holt are gratefully acknowledged.

[6] Use of Lectins in Analysis of Glycoconjugates

By RICHARD D. CUMMINGS

Introduction

The use of lectins from plants and animals to study animal cell glyco-conjugates has a long and productive history.[1,2] Lectins, recognized by their ability to bind sugars and to agglutinate cells, have been used through much of this century to type blood. However, interest in both lectins and glycoconjugates was stimulated by findings about 30 years ago, showing that lectins from the red kidney bean *Phaseolus vulgaris* have mitogenic activity for lymphocytes[3,4] and that some tumor cells are more agglutinable with plant lectins than their nontransformed counterparts.[5] These and other such discoveries involving lectins prompted closer investigations into the nature and the functions of the glycoconjugates bound by the lectins. Thus, in many ways our current understanding of the structure and function of animal glycoconjugates is identified with studies using lectins.

Today, dozens of different lectins are commercially available and used to identify and separate cells, to assay for the activities of glycosyltransferases, to isolate glycoproteins and glycolipids, and to separate and purify oligosaccharides and glycopeptides by affinity chromatography. This latter technique has been expanded to include multiple lectins and is called

[1] H. Lis and N. Sharon, *Annu. Rev. Biochem.* **55**, 35 (1986).
[2] I. E. Liener, N. Sharon, and I. Goldstein, eds., "The Lectins: Properties, Functions and Applications in Biology and Medicine." Academic Press, Orlando, FL, 1986.
[3] D. A. Hungerford, A. J. Donally, P. O. C. Nowell, and S. Beck, *Am. J. Hum. Genet.* **11**, 215 (1959).
[4] P. C. Nowell, *Cancer Res.* **20**, 462 (1960).
[5] M. M. Burger and A. R. Goldberg, *Proc. Natl. Acad. Sci. U.S.A.* **57**, 359 (1967).

serial lectin affinity chromatography. This specific methodology is the subject of this chapter. The technique of serial lectin affinity chromatography allows for the fractionation of oligosaccharides based on their differential affinity for a series of immobilized lectins. The chromatography of glycoconjugates on immobilized lectins has been the subject of several reviews and articles.[6-11] The use of immobilized lectins to separate related oligosaccharides is derived from earlier observations by a number of laboratories.[12-14] Later, the combined use of many lectins was shown to be a powerful method for fractionating and purifying a number of complex oligosaccharides from cell-derived mixtures.[15]

The technique of serial lectin affinity chromatography for the purification of oligosaccharides has many advantages. The binding of an oligosaccharide to a particular lectin can provide clues about the structure of the oligosaccharide. The differential affinity of different oligosaccharides for different lectins can facilitate their rapid separation from each other. Finally, when used in conjunction with other techniques, such as gel filtration, high-performance liquid chromatography (HPLC), and paper chromatography, serial lectin affinity chromatography can provide purified oligosaccharides for structural analysis.

Factors Involved in Lectin–Ligand Binding

Most lectins by definition are multivalent proteins having multiple subunits. The interaction of a glycoconjugate with a lectin is governed by the binding affinity of each subunit for a glycoconjugate. However, if a single molecule of a glycoconjugate possesses several carbohydrate determinants bound by a lectin, then the affinity (or more correctly avidity) of the lectin for the glycoconjugate may be enhanced tremendously. This explains in some cases why it is difficult to dissociate an intact glycoprotein

[6] D. A. Blake and I. J. Goldstein, in "Methods in Enzymology" (V. Ginsburg, ed.), Vol. 83, p. 127. Academic Press, New York, 1982.

[7] J. Finne and T. Krusius, in "Methods in Enzymology" (V. Ginsburg, ed.), Vol. 83, p. 269. Academic Press, New York, 1982.

[8] T. Osawa and T. Tsuji, Annu. Rev. Biochem. 56, 21 (1987).

[9] R. K. Merkle and R. D. Cummings, in "Methods in Enzymology" (V. Ginsburg, ed.), Vol. 138, p. 232. Academic Press, Orlando, FL, 1987.

[10] R. D. Cummings, R. K. Merkle, and N. L. Stults, Methods Cell. Biol. 32, 142 (1989).

[11] R. D. Cummings, in "Glycobiology: A Practical Approach" (M. Fukuda, ed.). Oxford Univ. Press, Oxford (in press).

[12] S. Ogata, T. Muramatsu, and A. Kobata, J. Biochem. (Tokyo) 78, 687 (1975).

[13] T. Krusius, J. Finne, and H. Rauvala, FEBS Lett. 71, 117 (1976).

[14] K. Kornfeld, M. L. Reitman, and R. Kornfeld, J. Biol. Chem. 256, 6633 (1981).

[15] R. D. Cummings and S. Kornfeld, J. Biol. Chem. 257, 11235 (1982).

from an immobilized lectin in comparison to individual glycopeptides or oligosaccharides from that glycoprotein.

During affinity chromatography of glycoconjugates over columns of immobilized lectins it is important to consider several factors. The first and most important factor is the binding affinity of the lectin for oligosaccharides. In general oligosaccharides that bind to a lectin with a K_d of 10^4 or greater can be readily fractionated by lectin affinity chromatography. Of course, if the binding affinity is low, then one should use lectins conjugated at a higher density. Thus, the second factor to consider is the lectin density on the column matrix and the geometry of the column. At low lectin concentrations, it is advantageous to have a long and thin column of the immobilized lectin to maximize the interactions with glycoconjugates. With a high lectin concentration, however, column geometry is less of a concern. Some lectins that have a K_d of 10^5 or greater may be used at a concentration of 2–3 mg of conjugated lectin/ml, whereas those lectins with a K_d of 10^5 or lower for oligosaccharides should be conjugated at higher densities of 5–15 mg/ml.

A third factor to consider is the rate of binding of oligosaccharides to a lectin. Plant lectins usually bind comparatively slowly, and under solution conditions it can take minutes or hours for an equilibrium to be established. However, in lectin affinity chromatography the high concentration of immobilized lectin relative to the glycoconjugates passed through the columns promotes more rapid and quantitative binding. Still, affinity chromatography is best performed at relatively slow flow rates to maximize the interactions. A fourth factor to consider is the haptenic sugars used to elute oligosaccharides bound by immobilized lectins. For example, all of the lectins used in serial lectin affinity chromatography can interact with small molecular haptenic sugars and these are used to elute bound materials. However, the binding affinity of lectins to small haptens is usually much poorer than that toward complex glycoconjugates. In most cases it has been demonstrated that lectins recognize a complex determinant on a glycoconjugate rather than just a simple monosaccharide residue. Structurally different glycoconjugates may share a common determinant recognized by the lectin, but the presentation or accessibility of the determinant to the lectin may be dependent on other structural features of the glycoconjugate. From this explanation, it is easy to see how immobilized lectins can be used to separate closely related molecules. It also indicates that the ability of a lectin to be inhibited by a high concentration of small molecular weight hapten may not reveal anything substantial about the complex binding specificity of the lectin.

A fifth factor to consider is the capacity of the lectin column and the activity of the immobilized lectin. In many cases lectins are covalently coupled to CNBr-activated Sepharose or N-hydroxysuccinimide-activated supports. Coupling is usually done in the presence of high concentrations of low molecular weight haptens to protect the lectin-binding sites. However, these small haptens may not be able to fully protect the extended binding sites in some lectins, and the coupling may result in partial inactivation of the immobilized lectin. Unfortunately, it is difficult to quantify the amount of active lectin on the gel, because the oligosaccharides bound by the lectins with high affinity may be too expensive or precious to use in such a manner. In general, the activity of an immobilized lectin should be demonstrated by the investigator and column capacity may not be an issue until high amounts of samples are used. By using high amounts of samples it is possible to determine the functional capacity of the column directly. We have found that in general most lectin columns that have 5–10 mg of active lectin immobilized, for example, concanavalin A (ConA), lentil lectin, and the phytohemagglutinin L_4PHA, can bind from 25 to 100 nmol of glycoconjugate per milliliter of gel. This number varies somewhat depending on the activity and size of the lectin, but most plant lectins are on the order of 10^5 Da. A sixth factor to consider is the stability of the immobilized lectin and the longevity of the column. In our experience we always use lectin affinity columns for the separation only of isolated glycopeptides or oligosaccharides, thus avoiding the introduction of proteases or other enzymes or proteins that might precipitate or inactivate the column. Lectins are used with buffers that contain Ca^{2+} and Mg^{2+} because many of the lectins are metalloproteins and can slowly lose activity if constantly used in buffers lacking metals. It is also wise to include sodium azide in buffers used with lectins to prevent microbial contamination and to store lectin columns in the cold when not in use. Even with such precautions, however, columns eventually lose activity and may be useful for only a few months of heavy use. Investigators should routinely retest columns of immobilized lectins to assess their activity.

Principle of Method

The most popular method of lectin affinity chromatography is to prepare a small column containing 1–5 ml of immobilized lectin on a gel. The chromatography is conducted by gravity flow and the columns have a 15- to 20-ml buffer reservoir on top and are connected to a fraction collector at the outlet. Of course, if a larger amount of sample is to be used, then the columns should be proportionately larger. Glycopeptides or oligosaccharides are generally labeled in some way with either radioactive moieties on the sugars or the peptide or fluorescent tags on the peptides or reducing ends of the oligosaccharides. Unlabeled oligosaccharides can be used;

however, but their detection is often difficult because haptenic sugars are used for their elution from columns. In those cases the eluted oligosaccharides are separated from haptenic sugars by gel filtration and the amount of sugar in the oligosaccharide determined.

In many cases the expense of the lectins and haptenic sugars can be an important consideration. However, extremely small columns can be constructed with 0.5 to 1.0 ml of gel in a plastic disposable pipette or some other suitable material. The long, thin nature of the column is ideal for promoting lectin interaction with glycoconjugate and also allows the researcher to use much less total hapten for elution. These columns may have a low capacity but they are usually ideal for analytical quantities of material.

The binding of an oligosaccharide to an immobilized lectin may be of high, intermediate, or low affinity. We usually think of binding as being high affinity in nature, requiring haptenic sugars for elution; however, low-affinity binding can be just as useful. In that case the interactive oligosaccharides may be retarded in their elution from the column and well separated from the unbound material, and may not even require haptenic sugars for elution. Such methods have worked extremely well for several types of lectins.[9,14,16–18]

A number of plant and animal lectins are used for lectin affinity chromatography and these are used at different densities and in different column geometries, as indicated in Table I. Table I lists the names of the common lectins used routinely for serial lectin affinity chromatography, their abbreviations, common haptens, and information about the lectin specificities and the use of lectins. Figure 1 presents a more detailed view of the carbohydrate-binding affinity of lectins and the glycoconjugates to which they bind. Twenty-three different N-linked and O-linked oligosaccharides that bind to different lectins are indicated in Fig. 1. There are, of course, many more oligosaccharides but these are shown to illustrate the power of the technique and the differential affinities that even structurally related compounds have for different lectins. The important carbohydrate-binding determinants recognized by each of the lectins are discussed in more detail in literature cited in Table I and in several reviews on the subject.[8,11] It should be noted that most of the lectins used are especially useful for fractionating N-linked oligosaccharides. However, some lectins are also used for fractionating O-linked oligosaccharides. Many lectins can bind

[16] R. D. Cummings and S. Kornfeld, *J. Biol. Chem.* **257**, 11230 (1982).
[17] S. Narasimhan, J. C. Freed, and H. Schacter, *Biochemistry* **24**, 1694 (1985).
[18] W.-C. Wang and R. D. Cummings, *J. Biol. Chem.* **263**, 4576 (1988).

to determinants found in all classes of oligosaccharides in animal cell glycoconjugates. For example, lectins have been used to fractionate intact glycolipids[19] and oligosaccharides derived from glycolipids.[20]

The use of multiple lectins to achieve fractionation of a mixture of oligosaccharides is shown in Fig. 2. In this example, the mixture of oligosacchrides is applied to a column of ConA–Sepharose, which binds to certain high mannose-type and biantennary-type N-linked oligosaccharides. The material bound and unbound by the column is pooled and the salt and haptenic sugars are removed by gel filtration over a column of BioGel P-2 (Bio-Rad, Richmond, CA) or Sephadex G-10 or G-25 (Pharmacia, Piscataway, NJ) in 7% (v/v) n-propanol or 0.1 M pyridine–acetate (pH 5.6). The samples are dried, resuspended in buffer, and applied to a second column containing immobilized lentil lectin. The bound and unbound fractions are pooled, desalted, and further applied to a column of immobilized L₄PHA. This procedure is repeated again on a column of immobilized *Maackia amurensis* leukoagglutinin (MAL). With more complex mixtures it may be necessary to include even more lectins. In general, mixtures of oligosaccharides can be fractionated by serial lectin affinity chromatography over a group of five or six lectins within a few days. The oligosaccharides separated by this technique can then be structurally analyzed or further purified if necessary before initiating structural analyses. Because haptenic sugars are used, it is important to remove them completely before doing structural analysis on any samples. This is not a problem, however, for samples in which the sugars have been metabolically radiolabeled.

Procedure

Columns and lectins used for lectin affinity chromatography are generally available from a number of suppliers. However, the standard oligosaccharides used by researchers to optimize the method are expensive and available from only a few sources. In many cases, we have used the cited methods for isolating our own standards.[9,11] Nevertheless, even without a wide range of standards, it is possible to proceed with the technique, and for many lectins their stability and activity is not of great concern. Solutions of haptenic sugars should be made for each lectin on the basis of the information in Table I. In addition, each lectin column may have a different geometry, as indicated in Table I.

[19] V. Torres and D. F. Smith, *Anal. Biochem.* **170**, 209 (1988).
[20] V. Torres, D. K. McCrumb, and D. F. Smith, *Arch. Biochem. Biophys.* **262**, 1 (1988).

TABLE I

Immobilized Lectins Used for Affinity Chromatography of Oligosaccharides and Information about Their Use

Source of lectin	Common name or abbreviation	Hapten sugars used for elution	Column dimensions; conjugation density	Comments	Refs.
Canavalia ensiformis	Concanavalin A (ConA)	α-Methylglucoside (10 mM) and 100 mM α-methyl-mannoside prewarmed to 60°	2 ml (0.7 × 5 cm); 10–15 mg/ml	Biantennary-type glycopeptides bind and are eluted with low hapten concentration High mannose/hybrid-type glycopeptides bind and are eluted with high hapten concentration	12–14, 21
Pisum sativum	Pea lectin	α-Methylmannoside (100 mM)	2 ml (0.7 × 5 cm); 10–15 mg/ml	Only certain bi- and triantennary-type glycopeptides that contain a core fucose residue bind	14, 15
Lens culinaris	Lentil lectin	α-Methylmannoside (100 mM)	2 ml (0.7 × 5 cm); 10–15 mg/ml	Only certain bi- and triantennary-type glycopeptides that contain a core fucose residue bind	14, 15
Ricinus communis agglutinin	RCA-I	Lactose (100 mM)	20 ml (0.8 × 40 cm); 3–5 mg/ml	Glycopeptides with terminal galactosyl residues linked β1-4 bind with high affinity; using retardation chromatography, glycopeptides can be separated depending on the number or positions of these residues	14, 17, 22, 23
Phaseolus vulgaris leukoagglutinin	L₄PHA	GalNAc (400 mM) (see Comments)	0.5 × 40 cm; 3–5 mg/ml	Only certain tri- and tetraantennary-type glycopeptides interact; on most columns their elution is retarded and no hapten is needed; however, on denser columns the GalNAc hapten may be needed	16
Phaseolus vulgaris erythroagglutinin	E₄PHA	GalNAc (400 mM) (see Comments)	0.5 × 40 cm; 3–5 mg/ml	Only certain bisected bi- and triantennary-type glycopeptides interact; on most columns their elution is retarded and no hapten is needed; however, on denser columns the GalNAc hapten may be needed	16, 24
Sambucus nigra agglutinin	SNA	Lactose (100 mM) (see Comments)	0.6 × 17 cm; 2–3 mg/ml	Only certain sialylated glycopeptides with the terminal sequence NeuAcα2-6Gal (NAc) bind; their elution is retarded at low lectin density; 100 mM lactose is used to elute bound material at higher lectin densities	25, 26

Organism	Lectin	Eluent	Column; mg/ml	Comments	Ref.
Maackia amurensis leukoagglutinin	MAL	Lactose (100 mM) (see Comments)	0.6 × 17 cm; 2–3 mg/ml	Only certain sialylated glycopeptides with the terminal sequence NeuAcα2-3Gal β1-4GlcNAc bind; their elution is retarded at low lactose density; 100 mM lactose is used to elute bound material at higher lectin densities	18
Datura stramonium agglutinin	DSA	10 mg/ml of a mixture of chitobiose and chitotriose	0.3 × 14 cm; 3 mg/ml	Use a long, thin column with a small amount of gel to reduce the amount of lectin and hapten needed; the hapten can be recovered and reused	27, 28
Lycopersicon esculentum	Tomato lectin	10 mg/ml of a mixture of chitobiose and chitotriose	0.3 × 14 cm; 3–5 mg/ml	Use a long, thin column with a small amount of gel to reduce the amount of lectin and hapten needed; the hapten can be recovered and reused	29
Tetragonolobus purpureas agglutinin	Lotus lectin	Fucose (25 mM)	0.3 × 14 cm; 3–5 mg/ml	Outer fucose residues are required and the presence of outer sialic acid residues may interfere with binding	30
Griffonia simplicifolia I-B4	GS-I-B$_4$	Raffinose (0.1 mM), 10 mM raffinose	0.3 × 14 cm; 10 mg/ml	Oligosaccharides with one terminal α-galactosyl residue are retarded in elution; those with two or more can be step eluted from the column by increasing raffinose	31, 32
Wisteria floribunda agglutinin	WFA	GalNAc (5 mM)	0.3 × 14 cm; 3–5 mg/ml	Glycopeptides with terminal GalNAc linked β(1 → 4) bind tightly; whereas those with α(1 → 3) or β(1 → 3) linkages bind weakly	20, 33
Ulex europaeus I	UEA-I	Fucose (25 mM)	0.3 × 14 cm; 3–5 mg/ml	Outer fucose residues are required but the lectin has highest affinity for human blood group H-containing oligosaccharides	34
Helix pomatia agglutinin	HPA	GalNAc (50 mM)	0.3 × 14 cm; 10 mg/ml	Both N- and O-linked oligosaccharides with terminal α-linked GalNAc bind with high affinity	35
Artocarpus integrifolia lectin	Jacalin lectin	α-Methylgalactoside (100 mM)	0.3 × 14 cm; 4 mg/ml	Lectin binds to core Galβ1-3GalNAc structure of O-linked oligosaccharides and to terminal α-linked galactosyl residues on other oligosaccharides	36, 37
Phytolacca americana lectin	Pokeweed mitogen (PM)	Chitotriose (10 mM)	0.3 × 14 cm; 1–3 mg/ml	Lectin binds to a variety of oligosaccharides containing a branched chain where internal galactosyl residues are disubstituted at C-3 and C-6 with GlcNAc residues	38, 39

a

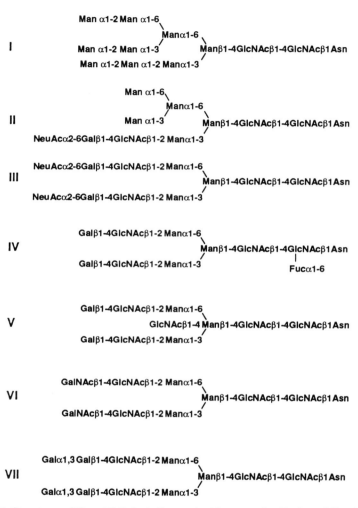

FIG. 1. Structures of N- and O-linked oligosaccharides recognized by immobilized lectins. (a) Oligosaccharides I–VII; (b) oligosaccharides VIII–XII; (c) oligosaccharides XIII–XVII; (d) oligosaccharides XVIII–XXIII. The structures of a number of representative oligosaccharides are shown and their behavior on columns of immobilized lectins is indicated on the right-hand side. +, strong binding; ±, weak binding. The lack of either sign indicates no significant binding.

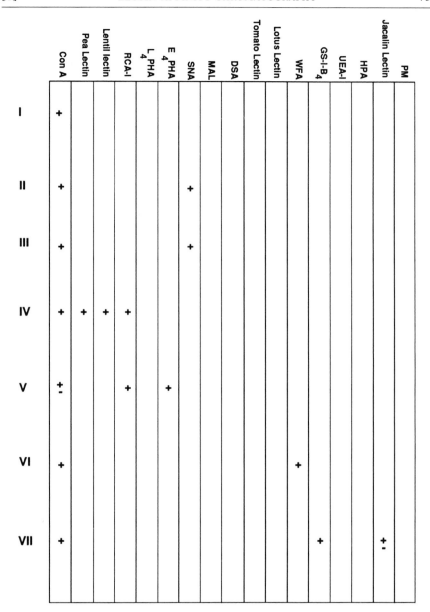

FIG. 1. (*continued*)

b

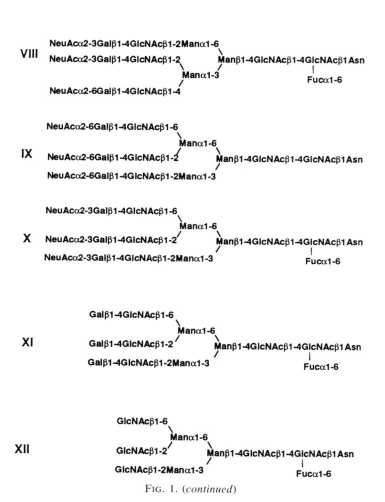

FIG. 1. (*continued*)

Stock Solutions

Tris-buffered saline (TBS) containing sodium azide: Dissolve 1.57 g of
Tris in 0.9 liter of deionized water along with 8.7 g of NaCl, 0.2 g of
$MgCl_2 \cdot 6H_2O$, 0.15 g of $CaCl_2$ dihydrate, and 0.2 g of NaN_3. Adjust
the pH to 8.0 and the volume to 1.0 liter. This gives a final TBS

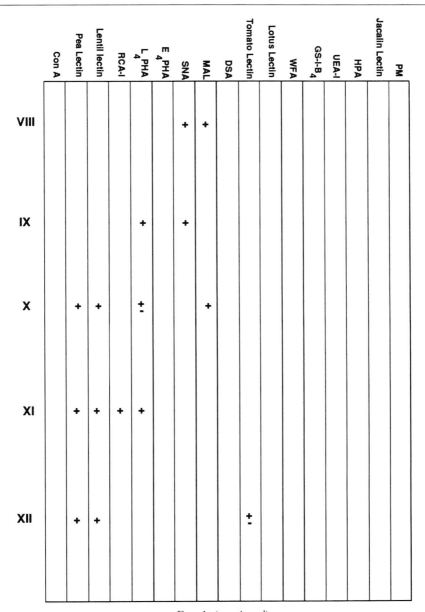

FIG. 1. (*continued*)

C

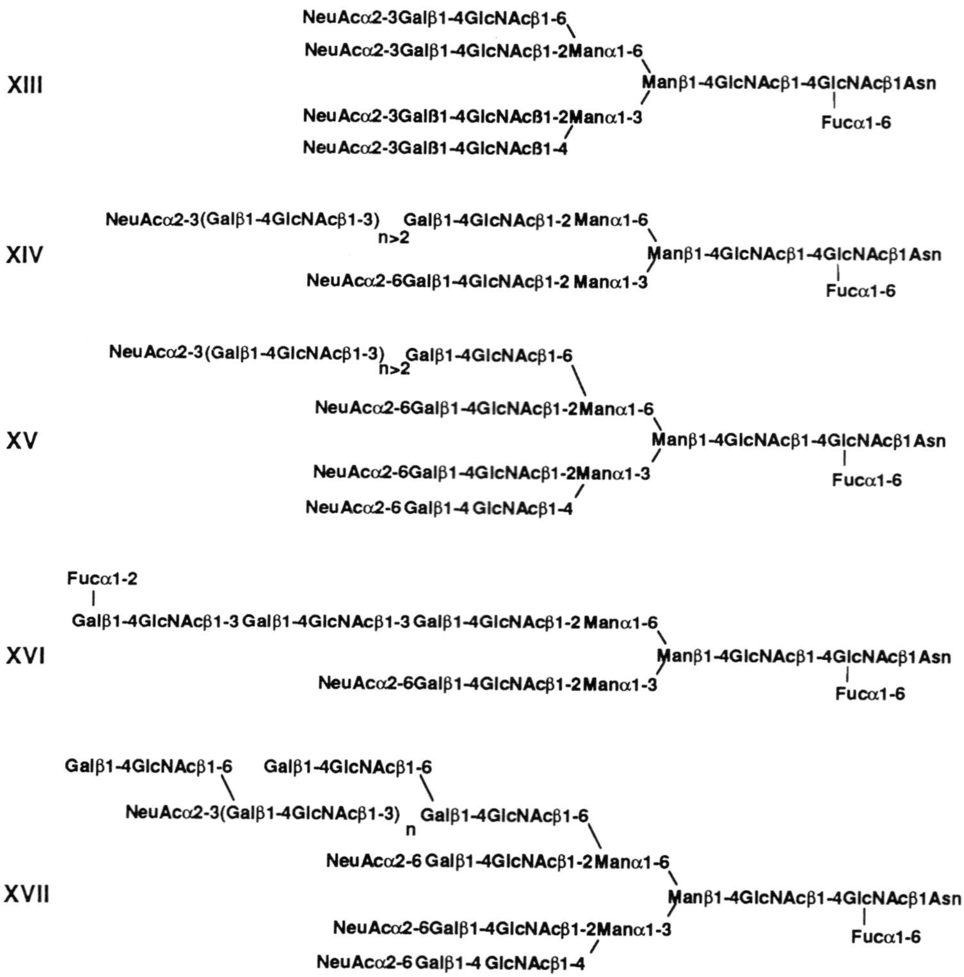

XIII

NeuAcα2-3Galβ1-4GlcNAcβ1-6

NeuAcα2-3Galβ1-4GlcNAcβ1-2Manα1-6

Manβ1-4GlcNAcβ1-4GlcNAcβ1Asn

NeuAcα2-3Galβ1-4GlcNAcβ1-2Manα1-3

Fucα1-6

NeuAcα2-3Galβ1-4GlcNAcβ1-4

XIV

NeuAcα2-3(Galβ1-4GlcNAcβ1-3)   Galβ1-4GlcNAcβ1-2Manα1-6
n>2

Manβ1-4GlcNAcβ1-4GlcNAcβ1Asn

NeuAcα2-6Galβ1-4GlcNAcβ1-2Manα1-3

Fucα1-6

XV

NeuAcα2-3(Galβ1-4GlcNAcβ1-3)   Galβ1-4GlcNAcβ1-6
n>2

NeuAcα2-6Galβ1-4GlcNAcβ1-2Manα1-6

Manβ1-4GlcNAcβ1-4GlcNAcβ1Asn

NeuAcα2-6Galβ1-4GlcNAcβ1-2Manα1-3

Fucα1-6

NeuAcα2-6 Galβ1-4 GlcNAcβ1-4

XVI

Fucα1-2

Galβ1-4GlcNAcβ1-3 Galβ1-4GlcNAcβ1-3 Galβ1-4GlcNAcβ1-2Manα1-6

Manβ1-4GlcNAcβ1-4GlcNAcβ1Asn

NeuAcα2-6Galβ1-4GlcNAcβ1-2Manα1-3

Fucα1-6

XVII

Galβ1-4GlcNAcβ1-6   Galβ1-4GlcNAcβ1-6

NeuAcα2-3(Galβ1-4GlcNAcβ1-3)   Galβ1-4GlcNAcβ1-6
n

NeuAcα2-6 Galβ1-4GlcNAcβ1-2Manα1-6

Manβ1-4GlcNAcβ1-4GlcNAcβ1Asn

NeuAcα2-6Galβ1-4GlcNAcβ1-2Manα1-3

Fucα1-6

NeuAcα2-6 Galβ1-4 GlcNAcβ1-4

FIG. 1. (*continued*)

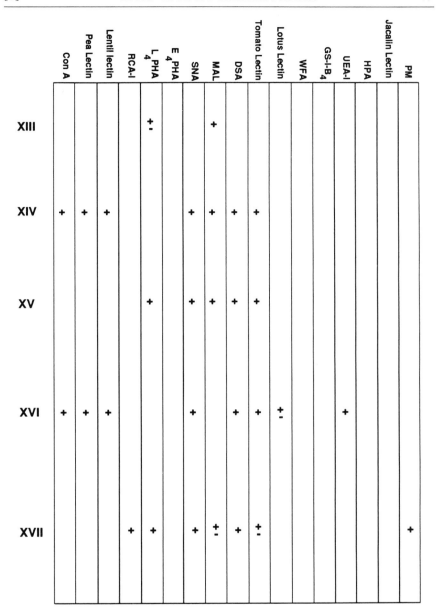

Fig. 1. (*continued*)

d

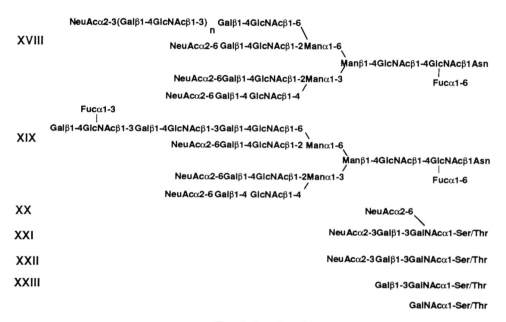

FIG. 1. (*continued*)

containing 10 mM Tris, 150 mM NaCl, 1 mM MgCl$_2$, 1 mM CaCl$_2$, and 0.02% (w/v) sodium azide. The solution is stable at room temperature for several weeks and in the refrigerator for several months
Haptenic solutions: Dissolve the haptenic sugars in TBS containing sodium azide to appropriate concentrations as indicated in Table I

Preparing Lectin Affinity Columns

If commercial columns are used they may already be fitted with a glass or plastic frit at the bottom and tube attachments at the top and bottom. Otherwise, columns can be constructed in the laboratory from glass tubes or plastic disposable pipettes of 1-, 5-, or 10-ml capacity. For example, a 1-ml column of lectin in a 1-ml disposable pipette has a column geometry

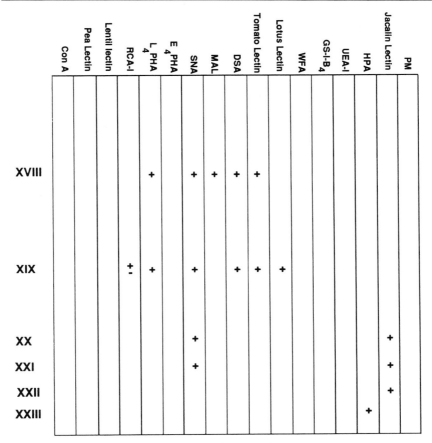

FIG. 1. (*continued*)

of 0.3 × 14 cm. Glass wool is used to plug the bottom of the column and adapters are added to tube fittings at the top and bottom. The column is filled with TBS and if desirable a small layer of washed sand (about 1/2 cm) may be added to the columns to separate the lectin gels from the glass wool. To the nearly filled column slowly pipette in the lectin gel solution until the proper bed height is obtained. This is done with the stopcock open on the column and buffer flowing. Care should be taken to avoid air bubbles by slowly adding gel to the column under flow conditions. If bubbles occur, a small wire or glass rod may be used to stir the gel and release the bubbles. Alternatively, air bubbles in the column may be dissipated by using degassed TBS buffer. Column packing is not a

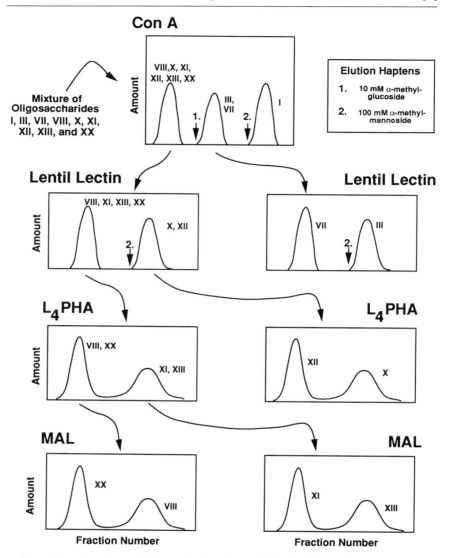

FIG. 2. Fractionation of oligosaccharides by serial lectin affinity chromatography. A mixture of oligosaccharides is applied to a column of ConA–Sepharose. Some oligosaccharides do not bind to the column whereas some bind and are eluted with α-methylglucoside and α-methylmannoside, respectively. The peaks of eluted oligosaccharides are pooled, concentrated, and the haptenic sugars removed. The samples are then applied to a column of lentil lectin–Sepharose. Again, some oligosaccharides do not bind whereas others bind and are eluted with α-methylmannoside. The bound and unbound materials from one of the lentil lectin–Sepharose columns are concentrated and the haptenic sugars removed. The

serious concern, but discontinuities caused by air bubbles and dry pockets should, of course, be avoided. After packing the column, it should be washed with at least 10 column volumes of TBS before use.

If commercial immobilized lectins are not available, then lectins may be covalently conjugated to Sepharose or agarose supports in the laboratory, following manufacturer instructions, and using haptenic sugars to protect the lectin activity. During the coupling, the haptenic sugars should be present at the concentrations indicated in Table I. The degree of conjugation of the lectin to a support should be defined and the immobilized lectin may then be added to columns as above. As an alternative method to covalent coupling we developed an affinity approach to attach lectins to supports. In this technique a lectin containing high mannose-type N-linked oligosaccharides can be immobilized to ConA–Sepharose[18] in a type of "sandwich." The conjugation is immediate and the amount of lectin conjugated to ConA–Sepharose can be precisely controlled. The immobilized lectins appear to be fully active and highly stable. Because ConA–Sepharose has such high avidity for glycoproteins, there is little leakage of the bound lectin from the ConA–Sepharose. However, when employing this conjugation method one should take care to analyze oligosaccharides that do not interact with ConA–Sepharose.

Sample Preparation

Samples of oligosaccharides for serial lectin affinity chromatography should be free of proteins if possible and devoid of any type of detectable enzyme activity, such as proteases or glycosidases. In general only clean samples are used to protect the lectins and prolong longevity of the columns. Samples are dissolved in 0.5 ml of TBS and applied directly to the columns. For lectin columns that retard the elution of oligosaccharides, for example, MAL, L_4PHA, and E_4PHA, sample volumes should be reduced to 0.1 ml. If large amounts of samples are used, then proportionately larger columns and application volumes should be used. Methods to prepare oligosaccharides from glycoproteins and glycolipids for analysis by serial lectin affinity chromatography are supplied in other chapters in this

---

samples are then applied to a column of L_4PHA–agarose. In this column under these conditions (see Table I) oligosaccharides interactive with the lectin are retarded in their elution and haptenic sugars are not required. The unbound and bound oligosaccharides are concentrated, desalted, and further applied to a column of MAL–agarose. The oligosaccharides interactive with this lectin are retarded in their elution under these conditions (see Table I) and no haptenic sugars are required. By this serial lectin affinity chromatography the mixture of starting oligosaccharides is purified.

volume and in previous volumes.[9,10] If samples are to be stored for any time before use they should either be frozen or stored in the presence of 50% (v/v) ethanol.

Chromatography of Samples

Columns of immobilized lectins are routinely stored at 4°. The chromatographic separations of oligosaccharides on lectin columns are performed at room temperature. Thus, both the buffers used for chromatography and the columns need to be prewarmed to room temperature about 30 min before use. After warming, wash the column with two to three column volumes of TBS. Collect 1.0-ml fractions from columns that contain a volume from 1 to 5 ml; the flow rate should be kept slow, at 0.5 to 1.0 ml/min. Usually, after application of the sample about 5–10 column volumes of the eluate is collected before a buffer containing haptenic sugars is added to elute bound material. However, as indicated in Table I, some immobilized lectins are used to isolate relevant oligosaccharides by retardation chromatography.[21-39] In those cases it may be necessary to collect up to 30 column volumes of eluate before adding haptenic sugars.

Portions of each fraction collected are removed for determination of radioactivity in the case of radiolabeled samples. For glycopeptides it

[21] J. U. Baenziger and D. Fiete, *J. Biol. Chem.* **254,** 2400 (1979).

[22] J. U. Baenziger and D. Fiete, *J. Biol. Chem.* **254,** 9795 (1979).

[23] C. A. Rivera-Marrero and R. D. Cummings, *Mol. Biochem. Parasitol.* **43,** 59 (1990).

[24] K. Yamashita, A. Hitoi, and A. Kobata, *J. Biol. Chem.* **258,** 14753 (1983).

[25] N. Shibuya, I. J. Goldstein, W. F. Broekaert, M. Nsimba-Lubaki, B. Peeters, and W. J. Peumans, *J. Biol. Chem.* **262,** 1596 (1987).

[26] N. Shibuya, I. J. Goldstein, W. F. Broekaert, M. Nsimba-Lubaki, B. Peeters, and W. J. Peumans, *Arch. Biochem. Biophys.* **254,** 1 (1987).

[27] R. D. Cummings and S. Kornfeld, *J. Biol. Chem.* **259,** 6253 (1984).

[28] K. Yamashita, K. Totani, T. Ohkura, S. Takasaki, I. J. Goldstein, and A. Kobata, *J. Biol. Chem.* **262,** 1602 (1987).

[29] R. K. Merkle and R. D. Cummings, *J. Biol. Chem.* **262,** 8179 (1987).

[30] S. Srivatsan, D. F. Smith, and R. D. Cummings, *J. Biol. Chem.* **267,** 20196 (1992).

[31] L. A. Murphy and I. J. Goldstein, *J. Biol. Chem.* **252,** 4739 (1977).

[32] I. J. Goldstein, D. A. Blake, S. Ebisu, T. S. Williams, and L. A. Murphy, *J. Biol. Chem.* **256,** 3890 (1981).

[33] J. Srivatsan, D. F. Smith, and R. D. Cummings, *Glycobiology* **2,** 445 (1992).

[34] P. A. Prieto, R. D. Cummings, and D. F. Smith, unpublished observations (1992).

[35] S.-I. Do and R. D. Cummings, *J. Biochem. Biophys. Methods* **24,** 153 (1992).

[36] S.-I. Do and R. D. Cummings, unpublished observations (1992).

[37] G. L. Hortin, *Anal. Biochem.* **191,** 262 (1990).

[38] T. Irimura and G. L. Nicolson, *Carbohydr. Res.* **120,** 187 (1983).

[39] H. Kawashima, S. Sueyoshi, H. Li, K. Yamamoto, and T. Osawa, *Glycoconjugate J.* **7,** 323 (1990).

may be possible to monitor absorbance at 280 nm. Fluorescently labeled glycopeptides or oligosaccharides can be monitored by a fluorescence spectrophotometer or visually observed under an ultraviolet light. Non-tagged samples can be detected by carbohydrate assays such as the phenol–sulfuric acid assay[40] or the ferricyanide reducing sugar assay.[41]

Peaks of eluted glycopeptides or oligosaccharides can be pooled, concentrated by rotary evaporation or lyophilization, and desalted by passage over a column (1 × 50 cm) of either Sephadex G-10 or G-25 or BioGel P-2 equilibrated either in water containing 7% n-propanol to inhibit bacterial growth or in 0.1 M pyridine–acetate (pH 5.6). After desalting and removal of haptenic sugars the samples in the fractions representing the void volume of the columns can be concentrated and reapplied to another column of immobilized lectin. The procedure is repeated until the serial lectin affinity chromatography is complete. It should be noted that valuable haptenic sugars, such as chitobiose or chitotriose, can be recovered following the chromatography on the desalting columns and reused.

Comments on Method

Lectin affinity chromatography can be used with any type of oligosaccharide sample, as long as the samples are free of protein and other macromolecular contaminants. In the case of intact glycoproteins, lectin affinity chromatography is useful for purification of glycoproteins. It does not reveal much about the structures of the oligosaccharides on the proteins because a glycoprotein may be highly glycosylated and a number of low-affinity interactions with lectins can promote binding of the glycoprotein to the lectin. For example, although lentil lectin binds with high affinity to N-linked oligosaccharides containing internal core fucosyl residues, it can bind weakly to an intact, heavily glycosylated protein lacking such structural features.

It is possible to use a variety of buffers other than TBS for lectin affinity chromatography, and phosphate-buffered saline is often used.[15,16] In general, any buffer should be fortified with metals and salt to prevent inactivation of the lectin. In some cases, researchers may want to use low ionic strength and low buffer concentrations to avoid excessive salts in the samples. Most lectins are remarkably stable under relatively harsh conditions and can withstand a wide variation in conditions, but this should be tested with standards to ensure that the lectin remains active under these modified conditions.

[40] M. Dubois, K. A. Gilles, J. K. Hamilton, P. A. Rebers, and F. Smith, *Anal. Chem.* **28,** 350 (1956).
[41] J. T. Park and M. J. Johnson, *J. Biol. Chem.* **181,** 149 (1949).

Variations in the Method

It is possible to couple lectins to other types of supports, such as HPLC supports, and to perform lectin affinity chromatography more rapidly and with greater precision.[42] The many advantages of such methods may well surpass those of currently used methods for many lectins. However, the current technique, employing gravity flow columns, presents certain advantages. The procedures outlined in this chapter are relatively inexpensive and multiple columns can be used simultaneously without refitting an instrument. Smaller amounts of lectin and gel can be used to fractionate complex mixtures efficiently. In most cases the techniques described in this chapter can afford considerable purification of complex oligosaccharides.

Acknowledgments

The author thanks Dr. Su-il Do and Jiunn-Chern Yeh for critical reading of the manuscript and Judy Gaar for assistance in preparing the manuscript. The author's work in this area has been supported by NIH Grant RO1 CA37626-08.

[42] E. D. Green and J. U. Baenziger, *J. Biol. Chem.* **263**, 25 (1988).

[7] Saccharide Linkage Analysis Using Methylation and Other Techniques

By Rudolf Geyer and Hildegard Geyer

Introduction

Studies on potential functions of glycoconjugates require a detailed knowledge of the structures of the carbohydrate moieties, that is, the sequence, linkage positions, anomeric configurations, and ring forms of the monosaccharide constitutents present, as well as their branching pattern and three-dimensional structure. Such information can be obtained by various methods, including ^{1}H nuclear magnetic resonance (NMR), fast atom bombardment or liquid secondary ion mass spectrometry (FAB-MS, LSIMS) of native or derivatized oligosaccharides, gas chromatography-mass spectrometry (GC-MS) of partially methylated alditol acetates (methylation analysis), selective chemical cleavage (e.g., by Smith degradation or acetolysis), sequential degradation with (linkage-specific) exoglycosidases, lectin affinity studies, or chromatographic procedures. Most of these techniques have been extensively reviewed in previous volumes of this series (see, e.g., Volumes 28, 50, 83, 138, 179, and 193). Because many complex carbohydrates of biological origin can be obtained in small quantities only, this chapter focuses primarily on a microscale version of methylation analysis that can be applied to (glycoprotein) glycans available in picomolar amounts. Furthermore, examples for the allocation of outer substituents by a combination of methylation analysis and preparative exoglycosidase digestion, as well as for the identification of high mannose-type oligosaccharide isomers by use of acetolysis, are presented.

Methylation Analysis

Methylation analysis is currently used in many laboratories for linkage analysis of complex carbohydrates. The method dates back to the work of Hakomori,[1] who introduced the use of the Corey–Chaykovsky base[2] for permethylation of glycolipids. Usually, the partially methylated alditol acetates (PMAAs) obtained after hydrolysis, reduction, and peracetylation are then separated by gas chromatography (GC) and identified by their retention times as well as by electron ionization (EI)–MS on the basis of their characteristic fragmentation pattern,[3–8] thus revealing unambiguous information on the linkage positions of the monosaccharide constituents present. Except for several improvements in the method, primarily by use of other bases for deprotonation of the oligosaccharide,[9–17] capillary columns for separation of PMAAs and chemical ionization (CI)–MS,[18–22]

[1] S.-I. Hakomori, *J. Biochem. (Tokyo)* **55,** 205 (1964).
[2] E. G. Corey and M. Chaykovsky, *J. Am. Chem. Soc.* **84,** 866 (1962).
[3] H. Björndal, C. G. Hellerqvist, B. Lindberg, and S. Svensson, *Angew. Chem., Int. Ed. Engl.* **9,** 610 (1970).
[4] B. Lindberg, this series, Vol. 28, p. 178.
[5] K. Stellner, H. Saito, and S.-I. Hakomori, *Arch. Biochem. Biophys.* **155,** 464 (1973).
[6] G. O. H. Schwarzmann and R. W. Jeanloz, *Carbohydr. Res.* **34,** 161 (1974).
[7] P.-E. Jansson, L. Kenne, H. Liedgren, B. Lindberg, and J. Lönngren, *Chem. Commun., Univ. Stockholm* **8,** 1 (1976).
[8] C. G. Hellerqvist, this series, Vol. 193, p. 554.
[9] J. Finne, T. Krusius, and H. Rauvala, *Carbohydr. Res.* **80,** 336 (1980).
[10] L. R. Phillips and B. A. Fraser, *Carbohydr. Res.* **90,** 149 (1981).
[11] I. Ciucanu and F. Kerek, *Carbohydr. Res.* **131,** 209 (1984).
[12] A. B. Blakeney and B. A. Stone, *Carbohydr. Res.* **140,** 319 (1985).
[13] A. Isogai, A. Ishizu, J. Nakano, S. Eda, and K. Kato, *Carbohydr. Res.* **138,** 99 (1985).
[14] J. Paz Parente, P. Cardon, Y. Leroy, J. Montreuil, B. Fournet, and G. Ricart, *Carbohydr. Res.* **141,** 41 (1985).
[15] A. Gunnarsson, *Glycoconjugate J.* **4,** 239 (1987).
[16] A. L. Kvernheim, *Acta Chem. Scand., Ser. B* **41,** 150 (1987).
[17] G. Larson, H. Karlsson, G. C. Hansson, and W. Pimlott, *Carbohydr. Res.* **161,** 281 (1987).

this methodology has remained basically the same, because of its general applicability.

A number of factors, however, have been reported to influence the recovery of PMAAs, including undermethylation,[23] side reactions occurring during methylation,[23,24] liberation of substituents because of the strong alkaline conditions used,[3,4,25–28] degradation or de-O-methylation of the methylated sugar derivatives during acid hydrolysis,[29–32] selective losses of volatile components during evaporation of solvents,[33] loss of material by adsorption to glassware, and decomposition of the PMAAs (especially of hexosamine derivatives) during GC analysis.[18,34] Furthermore, impurities derived from solvents or reagents may impair the identification and quantitation of PMAAs during GC-MS analysis.

To apply methylation analysis to glycoprotein glycans available in small amounts only, a microscale version of this technique has been established by minimizing the amounts of solvents and reagents used and by increasing the sensitivity of detection by use of capillary columns, ammonia CI-MS, and selected ion monitoring (SIM) of the ions obtained.[21,35] The method described in this article has been successfully applied to N- and O-glycans released from glycopeptides either enzymatically or chemically,[36–51] as well as to glycopeptides[52–54] and glycolipids.[55–58] In the

[18] S. B. Leverly and S.-I. Hakomori, this series, Vol. 138, p. 13.

[19] M. McNeil and P. Albersheim, *Carbohydr. Res.* **56**, 239 (1977).

[20] R. A. Laine, *Anal. Biochem.* **116**, 383 (1981).

[21] R. Geyer, H. Geyer, S. Kühnhardt, W. Mink, and S. Stirm, *Anal. Biochem.* **133**, 197 (1983).

[22] V. N. Reinhold, this series, Vol. 138, p. 59.

[23] U. Zähringer and E. T. Rietschel, *Carbohydr. Res.* **152**, 81 (1986).

[24] W. S. York, L. L. Kiefer, P. Albersheim, and A. G. Darvill, *Carbohydr. Res.* **208**, 175 (1990).

[25] O. Holst, H. Moll, and H. Brade, *Carbohydr. Res.* **188**, 219 (1989).

[26] B. Weske, R. D. Dennis, F. Helling, M. Keller, G. A. Nores, J. Peter-Katalinic, H. Egge, U. Dabrowski, and H. Wiegandt, *Eur. J. Biochem.* **191**, 379 (1990).

[27] U. Zähringer, H. Moll, and E. T. Rietschel, *Carbohydr. Res.* **196**, 147 (1990).

[28] F. Helling, R. D. Dennis, B. Weske, G. Nores, J. Peter-Katalinic, U. Dabrowski, H. Egge, and H. Wiegandt, *Eur. J. Biochem.* **200**, 409 (1991).

[29] J. Conchie, A. J. Hay, and J. A. Lomax, *Carbohydr. Res.* **103**, 129 (1982).

[30] M. Caroff and L. Szabó, *Biochem. Biophys. Res. Commun.* **89**, 410 (1979).

[31] M. Caroff and L. Szabó, *Carbohydr. Res.* **84**, 43 (1980).

[32] B. Nilsson, *Glycoconjugate J.* **2**, 335 (1985).

[33] H. O. Bouveng and B. Lindberg, *Methods Carbohydr. Chem.* **5**, 296 (1965).

[34] I. Funakoshi and I. Yamashina, *Anal. Biochem.* **107**, 265 (1980).

[35] R. Geyer, H. Geyer, S. Kühnhardt, W. Mink, and S. Stirm, *Anal. Biochem.* **121**, 263 (1982).

[36] S. Diabaté, R. Geyer, and S. Stirm, *Eur. J. Biochem.* **139**, 329 (1984).

[37] R. Geyer, H. Geyer, H. Egge, H.-H. Schott, and S. Stirm, *Eur. J. Biochem.* **143**, 531 (1984).

[38] R. Geyer, H. Geyer, S. Stirm, G. Hunsmann, J. Schneider, U. Dabrowski, and J. Dabrowski, *Biochemistry* **23**, 5628 (1984).

case of glycoproteins (or glycopeptides containing large peptide moieties) and (lipo)polysaccharides, however, modifications of this methodology may be required.

Reagents

All reagents should be of high quality: butyllithium, 1.6 M in hexane (Fluka, Ronkonkoma, NY), dimethyl sulfoxide (DMSO) (Fluka; >99.5%, <0.01% water, stored above molecular sieve), methyliodide (Roth, Karlsruhe, Germany; 99.5%, freshly distilled before use), acetonitrile (Roth), chloroform, dichloromethane, and methanol (Promochem, Wesel, Germany; nanograde), sulfuric acid, acetic acid, and sodium hydroxide (Merck, Rahway, NJ; suprapurum), pyridine (Fluka; >99.8%, <0.005% water, stored above molecular sieve), acetic anhydride (Supelco, Belle-

[39] H. Niemann, R. Geyer, H.-D. Klenk, D. Linder, S. Stirm, and M. Wirth, *EMBO J.* **3,** 665 (1984).

[40] H. S. Conradt, R. Geyer, J. Hoppe, L. Grotjahn, A. Plessing, and H. Mohr, *Eur. J. Biochem.* **153,** 255 (1985).

[41] H. Niemann, J. Dabrowski, U. Dabrowski, R. Geyer, W. Keil, H.-D. Klenk, and S. Stirm, *Eur. J. Biochem.* **146,** 523 (1985).

[42] R. Geyer, S. Diabaté, H. Geyer, H.-D. Klenk, H. Niemann, and S. Stirm, *Glycoconjugate J.* **4,** 17 (1987).

[43] K.-H. Strube, H.-H. Schott, and R. Geyer, *J. Biol. Chem.* **263,** 3762 (1988).

[44] K.-H. Strube and R. Geyer, *Eur. J. Biochem.* **179,** 441 (1989).

[45] G. Pfeiffer, M. Schmidt, K.-H. Strube, and R. Geyer, *Eur. J. Biochem.* **186,** 273 (1989).

[46] H. Geyer, R. Kempf, H.-H. Schott, and R. Geyer, *Eur. J. Biochem.* **193,** 855 (1990).

[47] R. Geyer, J. Dabrowski, U. Dabrowski, D. Linder, M. Schlüter, H.-H. Schott, and S. Stirm, *Eur. J. Biochem.* **187,** 95 (1990).

[48] P. Wendorf, D. Linder, A. Sziegoleit, and R. Geyer, *Biochem. J.* **278,** 505 (1991).

[49] G. Orberger, R. Geyer, S. Stirm, and R. Tauber, *Eur. J. Biochem.* **205,** 257 (1992).

[50] G. Pfeiffer, U. Dabrowski, J. Dabrowski, S. Stirm, K.-H. Strube, and R. Geyer, *Eur. J. Biochem.* **205,** 961 (1992).

[51] H. Geyer, C. Will, H. Feldmann, H.-D. Klenk, and R. Geyer, *Glycobiology* **2,** 299 (1992).

[52] M. Schlüter, D. Linder, R. Geyer, G. Hunsmann, J. Schneider, and S. Stirm, *FEBS Lett.* **169,** 194 (1984).

[53] M. Schlüter, D. Linder, and R. Geyer, *Carbohydr. Res.* **138,** 305 (1985).

[54] W. Keil, R. Geyer, J. Dabrowski, U. Dabrowski, H. Niemann, S. Stirm, and H.-D. Klenk, *EMBO J.* **4,** 2711 (1985).

[55] R. D. Dennis, R. Geyer, H. Egge, J. Peter-Katalinic, S.-C. Li, S. Stirm, and H. Wiegandt, *J. Biol. Chem.* **260,** 5370 (1985).

[56] R. D. Dennis, R. Geyer, H. Egge, H. Menges, S. Stirm, and H. Wiegandt, *Eur. J. Biochem.* **146,** 51 (1985).

[57] R. Jennemann, B. Felding-Habermann, R. Geyer, S. Stirm, and H. Wiegandt, *Arch. Biochem. Biophys.* **258,** 240 (1987).

[58] G. Ritter, W. Krause, R. Geyer, S. Stirm, and H. Wiegandt, *Arch. Biochem. Biophys.* **257,** 370 (1987).

fonte, PA), and sodium borohydride (Serva, Heidelberg, Germany). Water should be deionized and purified by a Milli-Q system (Millipore, Bedford, MA) or doubly distilled with a quartz still. All glassware should be either acid washed or heated to 500° for about 12 hr in an annealing furnace after cleaning in a dishwasher.

Preparation of Lithium Methylsulfinyl Carbanion

Several bases promoting formation of polyalkoxide ions have been reported, including sodium,[1] potassium,[9,10] or lithium methylsulfinyl carbanion[12,14,16] and powdered NaOH.[11,13,15,17] Furthermore, methylation of carbohydrates by methyltrifluoromethane sulfonate in trimethyl phosphate has been described.[59] In the procedure detailed below, lithium dimethylsulfinyl carbanion is used because this reagent gives less background in the GC chromatogram than does the sodium or potassium base. Sodium or potassium hydride is dispersed in oil whereas the lithium base can be prepared by using butyllithium, that is, a starting reagent available in high purity.[14] All operations must be conducted under a dry inert atmosphere.

Procedure

1. Flush a 500-ml, three-necked, round-bottom flask equipped with a magnetic stirring bar, a 250-ml dropping funnel (closed with a ground glass stopper) and two valves, which are connected either with an oil pump or an argon bottle, three times with argon.

2. Introduce 40 ml of DMSO into the flask while continuously flushing the vessel with argon.

3. Flush the flask twice with argon under reduced pressure.

4. Introduce 50 ml of butyllithium (in hexane, cooled to 4°) into the dropping funnel under constant flushing with argon.

5. Add butyllithium gradually with stirring within approximately 20 min. The color of the solution should be pale yellow-green.

6. Carefully remove residual hexane and butane formed during the reaction, using a vacuum pump at about 40°.

7. Leave the flask overnight at room temperature in a desiccator flushed with argon.

8. Take 0.3-ml aliquots of the clear supernatant (under argon) and introduce into 1-ml Teflon-lined, screw-capped test tubes.

9. Store the reagent at −20° under argon atmosphere and thaw just prior to use. Under these conditions, the base is stable for at least 6 months.

---

[59] P. Prehm, *Carbohydr. Res.* **78**, 372 (1980).

Preparation of Samples

Reducing oligosaccharides should be converted to the corresponding oligosaccharide alditols prior to methylation in order to prevent "alkaline peeling" reactions[8] (for details of the procedure, see Reduction, below). Reduction can also be used for radiolabeling of the oligosaccharides,[60] which facilitates their detection during subsequent purification steps. To remove salts and residual matrix particles [e.g., from the preceding high-performance liquid chromatography (HPLC) separations] that may interfere with the methylation reaction, samples are desalted and purified by chromatography through a BioGel (Bio-Rad, Richmond, CA) P-2 column. In our experience, desalting with ion-exchange resins is less advantageous, because this procedure may lead to more contaminant peaks in the GC chromatogram than BioGel P-2 chromatography. Small, nonsialylated O-glycans comprising fewer than four monosaccharide units, however, must be desalted by ion-exchange resins.

Procedure

1. Pass oligosaccharide alditol samples through a column (1 × 25 cm) of BioGel P-2 (-400 mesh; Bio-Rad) eluted with water.
2. Collect fractions of about 0.8 to 1.0 ml.
3. If the oligosaccharides are radiolabeled, trace their elution volume by liquid scintillation counting of aliquots of each fraction, pool corresponding fractions, and lyophilize. In the case of unlabeled glycans, calibrate the column with oligosaccharide standards prior to use and pool corresponding fractions.
4. Dissolve residues in 0.1–0.2 ml of water, transfer samples into conical 0.3-ml, Teflon-lined, screw-capped microvials (Zinsser, Frankfurt, Germany), lyophilize (with the cap loosely applied), and store at 35° for approximately 20 hr in vacuo over fresh P_2O_5 in a desiccator.
5. Flush the desiccator with dry argon and close the vials.

Micromethylation

To reduce contamination of the sample, the amount of reagents used must be minimized when microgram quantities of complex carbohydrates are subjected to methylation. Furthermore, temperature should be carefully controlled during the entire procedure and should not exceed 25–30°. For that purpose, the reaction mixture is frozen prior to addition of methyl iodide and the temperature of the sonication bath is maintained at about 25° during operation by aid of a cooling coil. Under these conditions,

---

[60] S. Takasaki and A. Kobata, this series, Vol. 50, p. 50.

formation of degradation products during sonication can be kept to a minimum. Permethylated oligosaccharides can be separated from reagents and methylation by-products by dialysis, partitioning between chloroform and water, chromatography through a Sep-Pak C_{18} cartridge[61,62] or gel filtration using a Sephadex LH-20 column.[5] We prefer the use of cartridges because of their easy handling. Furthermore, smaller permethylated oligosaccharides (e.g., disaccharides) can be most efficiently purified by this procedure.

Procedure

1. Add 50 μl of DMSO with a dry microsyringe through the Teflon-lined septum of the microvial to the dried sample, and dissolve the sample by sonication for 60 min at 25°.

2. Add 50 μl of lithium methylsulfinyl carbanion solution in the same manner and sonicate again for 60 min.

3. Store the sample for at least 15 min at $-20°$.

4. Introduce methyl iodide (50 μl, freshly distilled) through the septum and sonicate for another 60 min.

5. In between, precondition a cartridge filled with Chromabond C_{18}ec (100 mg; Macherey & Nagel, Düren, Germany) by sequential treatment with 5 ml of water, 5 ml of acetonitrile, 5 ml of methanol, and 10 ml of water.

6. Dilute the permethylation reaction mixture (150 μl) with the same volume of water and apply directly onto the cartridge. Elute with 5 ml of water, 2 ml of 15% aqueous acetonitrile, 2 ml of 30% aqueous acetonitrile, and 2 ml of acetonitrile. In the case of radiolabeled glycans, fractions may be pooled according to the distribution of radioactivity. Permethylated oligosaccharides and glycopeptides are usually recovered in the acetonitrile fraction. Small glycans (e.g., disaccharides) or sulfated species, however, elute in the 15–30% fractions. In the case of glycolipids, additionally wash the cartridge with 2 ml of methanol and elute permethylated material with 2 ml of chloroform.

7. Combine oligosaccharide-containing fractions into a Teflon-lined, screw-capped Pyrex glass tube (1.4 × 10 cm) with a round bottom and dry with a vacuum concentrator (Model RC 10.10; Jouan). All further reactions are carried out in the same tube.

[61] T. J. Waeghe, A. G. Darvill, M. McNeil, and P. Albersheim, *Carbohydr. Res.* **123,** 281 (1983).

[62] A. J. Mort, S. Parker, and M.-S. Kuo, *Anal. Biochem.* **133,** 380 (1983).

Hydrolysis

Acid hydrolysis has been reported to provoke partial degradation[29] and de-O-methylation especially of permethylated N-acetylhexosaminitols (HexNAcOH),[30-32] which may result in the formation of different PMAA derivatives of HexNAcOH lacking, in part, a methyl group at C-1 or C-3. Consequently, HexNAcOH-derived PMAAs may be detected and quantified with lower sensitivity and accuracy. Methanolysis is known to overcome this problem, but usually more than one partially methylated methylglycoside is formed from each monosaccharide constituent, which may complicate the separation, detection, and quantitation of sugar derivatives obtained from small amounts of complex carbohydrates. To decrease the complexity of the resulting gas chromatogram and to lower the detection limit, permethylated oligosaccharides are, therefore, subjected to acid hydrolysis.[18]

Procedure

1. Add 500 μl of 0.5 N sulfuric acid in 85% (v/v) acetic acid to the dry methylated sample and flush the tube with argon.

2. Heat the tightly closed tube for 16 hr at 80° in a GC oven.

3. After cooling in an ice bath, add 550 μl of 0.5 N sodium hydroxide and evaporate in a speed vacuum concentrator at room temperature. Dissolve the residue in 500 μl of water, lyophilize, and repeat this procedure once.

Reduction

1. Dissolve the methylated sugars in 200 μl of water and add aqueous NaBH$_4$ (10 mg/ml) in three portions of 150 μl. [*Note:* In contrast to EI-MS, resulting PMAAs are registered only by their quasimolecular ions in subsequent CI-MS analysis (see below). Therefore, reduction with sodium borodeuteride gives no further structural information.]

2. Verify alkaline conditions (pH 10–12) and leave the sample at room temperature overnight.

3. Destroy excess reducing agent by dropwise addition of 2 N acetic acid (resulting pH, ~5) and evaporate under reduced pressure.

4. Dissolve the residue in 100 μl of water, add 1 ml of methanol containing 1% acetic acid, and evaporate under a gentle stream of nitrogen at 30° to remove boric acid. Repeat methanol addition and evaporation four times.

5. Dry the sample *in vacuo* over P$_2$O$_5$ for at least 1 hr.

Acetylation

Acetylation of partially methylated alditols is achieved by addition of acetic anhydride. In the case of submicrogram quantities, the presence of pyridine in the reaction mixture turns out to be crucial for quantitative yields. Recoveries of PMAAs appear to be higher when pyridine and acetic anhydride are added in equal proportions, because salts are almost completely dissolved under these conditions. In this case, however, heating should be avoided, because elevated temperature appears to promote formation of by-products. Instead, a prolonged reaction time at room temperature should be allotted. After peracetylation, evaporation of reagents must be carried out with great care in order to prevent partial loss of volatile PMAAs.

Procedure

1. Add 400 μl of pyridine and 400 μl of acetic anhydride to the dried sample, flush with argon, and seal the test tube.

2. Ensure maximum solution of the sample by sonication at 25° and leave at room temperature for 16 hr.

3. Remove excess reagent using a stream of nitrogen (~25°).

4. Add 4 ml of dichloromethane, wash the organic phase three times with 2 ml of water, and discard the aqueous phases.

5. Transfer the organic phase to a conical-bottom, screw-capped Pyrex tube, evaporate the solvent with nitrogen, concentrate the methylated products at the bottom of the vial cone by rinsing the glass walls carefully with 1 ml of methanol, and evaporate to dryness under nitrogen stream (*Note:* Use of methanol instead of dichloromethane at this step ensures the complete removal of residual water by azeotropic effect.)

6. For GC-MS analysis, dissolve the sample in 50–100 μl of dichloromethane.

Instrumentation for Gas Chromatography-Mass Spectrometry

Equipment for GC-MS analyses can be obtained from a variety of suppliers. We use a Finnigan (San Jose, CA) 4021 GC-MS system combined with a Teknivent (Maryland Heights, MO) Vector II data system operating with a personal computer. To avoid contact of the PMAAs with hot metal surfaces, the system is equipped with an all-glass, moving needle injector.[63] The fused silica capillary column is introduced up to the narrowing of the injector so that the dead volume is negligible. With this type of injector, even larger amounts of sample (i.e., 50 μl) can be successively applied (in 5-μl portions) to the tip of the needle without leading to a huge

[63] P. M. J. Van den Berg and T. P. H. Cox, *Chromatographia* **5**, 301 (1972).

solvent peak because the solvent is removed each time under a stream of helium. This is of particular importance, when extremely small amounts of material must be analyzed. Furthermore, PMAA standards can be loaded together with the sample for identification of certain PMAAs by cochromatography with authentic standards.

Separation of Partially Methylated Alditol Acetates by Capillary Gas Chromatography

Separation of PMAAs and partially methylated and acetylated methylglycosides by GC has been achieved by a variety of glass or fused silica capillary columns.[8,18,21,35,61,64–66] In comparison to wall-coated glass columns, however, fused silica-bonded phase columns can be handled more easily; they give a better resolution, lead to more reproducible retention times, and have a longer life time. We usually use two fused silica bonded phase capillary columns of different polarity and selectivity (60-m DB-1 and 30-m DB-210; ICT, Frankfurt, Germany), with a film thickness of 0.25 μm, which are introduced directly into the ion source. Temperatures of the injector block and the transfer line are 250 and 270°, respectively. Helium is used as carrier gas, applying a head pressure of 30 psi (DB-1) or 12 psi (DB-210).

Procedure

1. Apply the sample (usually 3–5 μl) at room temperature with a microsyringe through the septum of the injector to the tip of the injection needle, evaporate the solvent with a stream of helium (3 min), and close the vent valve.

2. Wait 30 sec, then inject the sample by lowering the tip of the injection needle into the injector block and start the temperature program and data system. For separation on the DB-1 column, use a temperature gradient of 130–200° at 1.2°/min and 200–250° at 5°/min; for chromatography on the DB-210 column, use a gradient of 130–200° at 1.5°/min and 200–250° at 3°/min. In both cases, maintain the final temperature (250°) for 10 min.

Interpretation. Relative retention times (RRTs) of individual PMAAs are calculated by comparison with the retention times of 2,3,4,6-Glc (early standard) and 2,3-Glc (late standard) coinjected with the PMAAs to be studied.[8] Representative RRTs for PMAAs from the neutral sugar constituents of glycoprotein glycans on DB-1 and DB-210 columns are given in

[64] N. Shibuya, *J. Chromatogr.* **208**, 96 (1981).
[65] B. Fournet, G. Strecker, Y. Leroy, and J. Montreuil, *Anal. Biochem.* **116**, 489 (1981).
[66] J. A. Lomax, A. H. Gordon, and A. Chesson, *Carbohydr. Res.* **138**, 177 (1985).

Table I. Furthermore, RRTs of two Glc derivatives (2,3,4-GlcOH, 2,3,6-GlcOH) commonly observed as contaminants in methylation analysis are included. Although the exact RRT values may vary when different experimental conditions (e.g., different temperature programs) are used, the order of elution will be identical provided that columns with similar liquid phases are used. Nonetheless, the researcher is advised to prepare a collection of PMAA standards[67,68] in order to establish one's own list of RRTs.

Table I reveals that most of the PMAAs used can be separated on the DB-1 column. The only exceptions are 2,4,6- and 3,4,6-GalOH, 2,4,6-ManOH, and 2,3,6-GlcOH; the latter is often found due to cellulose contaminants in the methylation sample. To overcome this problem, a DB-210 column is also used, which enables the separation of these derivatives, their unambiguous identification, and quantitation. By using the two columns mentioned, all PMAAs listed in Table I can be easily identified by their RRTs. It should be pointed out, however, that the recovery of hexosamine derivatives is significantly lower when a DB-210 column is used. Therefore, quantitation of these derivatives should be based only on data obtained by chromatography on DB-1.

Monitoring of Partially Methylated Alditol Acetates by Chemical Ionization-Mass Spectrometry

In contrast to methane or isobutane CI mass spectra, in which fragment ions are more abundant,[8,19,20] ammonia CI spectra of PMAAs are dominated by molecular weight-related ions.[21,22] As illustrated in Fig. 1, hexose derivatives lead predominantly to an $[M + NH_4]^+$ ion accompanied, in part, by a small signal (usually less than 10% of that of $[M + NH_4]^+$) corresponding to $[(M + H) - 60]^+$ (Fig. 1A–D), which is not detectable when small amounts of sample are applied. N-Acetylhexosamine derivatives produce almost exclusively $[M + H]^+$ and $[M + NH_4]^+$ ions (Fig. 1E–H) under the conditions used (for ions obtained from other PMAAs, see Table I).

Similarly, peracetylated deoxyhexitol and hexitol derivatives obtained during carbohydrate constituent analysis lead to $[M + NH_4]^+$ ions, whereas corresponding aminohexitols form both $[M + H]^+$ and $[M + NH_4]^+$ ions (data not shown). No differences are observed when mass spectra are taken from the leading, middle, and trailing portion of the peaks. Thus, single-ion monitoring of the major molecular weight-related

[67] E. I. Oakley, D. F. Magin, G. H. Bokelman, and W. S. Ryan, Jr., *J. Carbohydr. Chem.* **4,** 53 (1985).
[68] S. H. Doares, P. Albersheim, and A. G. Darvill, *Carbohydr. Res.* **210,** 311 (1991).

TABLE I
RELATIVE RETENTION TIMES OF PARTIALLY METHYLATED ALDITOL ACETATES AND
QUASIMOLECULAR IONS OBTAINED IN AMMONIA CHEMICAL IONIZATION MASS SPECTROMETRY

| No. | Peracetylated derivative of[b] | Relative retention time on[a] | | Ions monitored |
		DB-1 column	DB-210 column	
1	2,3,4-FucOH	0.785	0.831	$[M + NH_4]^+$, $[M + H]^{+c}$, $[(M + H) - 60]^{+c}$
2	2,4-GalOH	1.511	1.862	$[M + NH_4]^+$
3	3,6-GalOH	1.381	1.723	$[M + NH_4]^+$, $[(M + H) - 60]^{+c}$
4	4,6-GalOH	1.372	1.639	$[M + NH_4]^+$, $[(M + H) - 60]^{+c}$
5	2,3,4-GalOH	1.323	1.600	$[M + NH_4]^+$, $[(M + H) - 60]^{+c}$
6	2,4,6-GalOH	1.232	1.316	$[M + NH_4]^+$, $[(M + H) - 60]^{+c}$
7	3,4,6-GalOH	1.232	1.437	$[M + NH_4]^+$, $[(M + H) - 60]^{+c}$
8	2,3,4,6-GalOH	1.046	1.077	$[M + NH_4]^+$
9	2,3-GlcOH	1.442	1.845	$[M + NH_4]^+$
10	2,3,4-GlcOH	1.256	1.476	$[M + NH_4]^+$, $[(M + H) - 60]^{+c}$
11	2,3,6-GlcOH	1.217	1.469	$[M + NH_4]^+$, $[(M + H) - 60]^{+c}$
12	2,3,4,6-GlcOH	1.000	1.000	$[M + NH_4]^+$, $[(M + H) - 60]^+$
13	2-ManOH	1.556	1.977	$[M + NH_4]^+$, $[(M + H) - 60]^{+c}$
14	2,3-ManOH	1.429	1.779	$[M + NH_4]^+$
15	2,4-ManOH	1.480	1.810	$[M + NH_4]^+$, $[(M + H) - 60]^{+c}$
16	2,6-ManOH	1.331	1.609	$[M + NH_4]^+$, $[(M + H) - 60]^{+c}$
17	3,4-ManOH	1.460	1.901	$[M + NH_4]^+$, $[(M + H) - 60]^{+c}$
18	3,6-ManOH	1.385	1.737	$[M + NH_4]^+$, $[(M + H) - 60]^{+c}$
19	4,6-ManOH	1.351	1.591	$[M + NH_4]^+$
20	2,3,4-ManOH	1.261	1.473	$[M + NH_4]^+$, $[(M + H) - 60]^{+c}$
21	2,3,6-ManOH	1.201	1.397	$[M + NH_4]^+$, $[(M + H) - 60]^+$
22	2,4,6-ManOH	1.216	1.286	$[M + NH_4]^+$, $[(M + H) - 60]^+$
23	3,4,6-ManOH	1.190	1.377	$[M + NH_4]^+$, $[(M + H) - 60]^{+c}$
24	2,3,4,6-ManOH	1.006	0.993	$[M + NH_4]^+$, $[(M + H) - 60]^{+c}$
25	4,6-GalN(Me)AcOH	1.794	2.286	$[M + H]^+$, $[M + NH_4]^+$
26	1,4,5-GalN(Me)AcOH	1.636	nd[d]	$[M + H]^+$, $[M + NH_4]^+$
27	3,4,6-GalN(Me)AcOH	1.686	2.187	$[M + H]^+$, $[M + NH_4]^+$
28	1,4,5,6-GalN(Me)AcOH	1.384	nd	$[M + H]^+$, $[M + NH_4]^{+e}$
29	3-GlcN(Me)AcOH	1.844	nd	$[M + H]^+$, $[M + NH_4]^+$
30	6-GlcN(Me)AcOH	1.819	nd	$[M + H]^+$, $[M + NH_4]^+$
31	3,6-GlcN(Me)AcOH	1.732	2.281	$[M + H]^+$, $[M + NH_4]^+$
32	4,6-GlcN(Me)AcOH	1.784	2.385	$[M + H]^+$, $[M + NH_4]^+$
33	1,3,5-GlcN(Me)AcOH	1.638	2.017	$[M + H]^+$, $[M + NH_4]^+$
34	3,4,6-GlcN(Me)AcOH	1.618	2.083	$[M + H]^+$, $[M + NH_4]^+$
35	3,5,6-GlcN(Me)AcOH	1.590	nd	$[M + H]^+$, $[M + NH_4]^+$
36	1,3,5,6-GlcN(Me)AcOH	1.340	1.533	$[M + H]^+$, $[M + NH_4]^{+e}$

[a] Retention times relative to 2,3,4,6-GlcOH (12; early standard) set to 1.000 and to 2,3-GlcOH (9; late standard) set to 1.442 (DB-1) or 1.845 (DB-210).

[b] 2,3,4-FucOH, 2,3,4-tri-O-methyl-L-fucitol; 3-GlcN(Me)AcOH, 2-deoxy-2-(N-methyl)acetamido-3-mono-O-methyl-D-glucitol; and so on.

[c] Less than 10% of $[M + NH_4]^+$.

[d] nd, Not determined.

[e] Less than 10% of $[M + H]^+$.

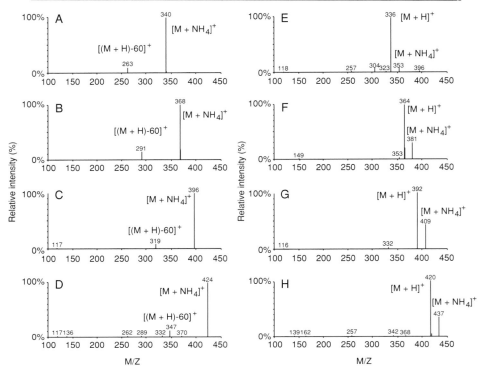

FIG. 1. Ammonia chemical ionization mass spectra of 2,3,4,6-ManOH (A), 2,4,6-ManOH (B), 2,4-ManOH (C), 2-ManOH (D), 1,3,5,6-GlcN(Me)AcOH (E), 3,4,6-GlcN(Me)AcOH (F), 3,6-GlcN(Me)AcOH (G), and 3-GlcN(Me)AcOH (H). For experimental details, see text.

ions allows a sensitive and highly selective detection of PMAAs. To ensure the selectivity of detection, the number of ions registered should be kept to a minimum. Therefore, at given time intervals only selected sets of ions are monitored depending on the RRTs of the respective PMAAs. For similar reasons, oligosaccharides under study are not reduced with NaB^2H_4 for tagging of the reducing sugar and isotope peaks of PMAA quasimolecular ions are also not registered.

Sensitivity of detection is also influenced by the pressure of the reagent gas in the ion source. Gradual increase in the pressure of ammonia in the range from 0.15 to 1.0 torr revealed maximum signal intensity at 0.6 torr. In comparison to the results obtained at 0.15 torr, a 4- to 6-fold (hexose derivatives) or about 10-fold increase (*N*-acetylhexosamine derivatives) was observed. Thus, the peak ratios of PMAAs may be influenced by the pressure of the reagent gas. Chemical noise, however, is also increased under these conditions, sometimes complicating the integration of small peaks. Therefore, we prefer a CI gas pressure of 0.45 torr.

Experimental conditions used for ammonia CI-MS are as follows: temperature of the ion source, 100°; pressure of the reagent gas in the ion source, 0.45 torr; electron energy, 150 eV; emission current, 0.3 mA; multiplier voltage, 2.0 kV; scan rate, 1 sec/scan. At time interval I (0–41 min, DB-1; 0–28 min, DB-210) the ions monitored are $[M + NH_4]^+$ (tri-O-methyldeoxyhexose derivatives, m/z 310) and tetra-O-methylhexoses, m/z 340), at time interval II (41–57.4 min, DB-1; 28–49.8 min, DB-210) the ions monitored are $[M + NH_4]^+$ (tri-, di-, and mono-O-methylhexoses, m/z 368, 396, and 424) and $[M + H]^+$ (tetra-O-methylaminohexitol derivatives, m/z 336), and at time interval III (57.4–70 min, DB-1; 49.8–70 min, DB-210) the ions monitored are $[M + H]^+$ and $[M + NH_4]^+$ (tri-, di-, and mono-O-methylaminohexoses as well as tri-O-methylaminohexitol derivatives, m/z 364, 381, 392, 409, 420, and 437).

Applications

Demonstration of Sensitivity. To illustrate the sensitivity of the method described, a series of methylation analyses has been carried out in parallel, using 10 ng (about 6 pmol) up to 2.5 μg (about 1.5 nmol) of a reduced, biantennary complex-type glycoprotein N-glycan [Galβ4GlcNAcβ2Man-α3(Galβ4GlcNAcβ2Manα6)Manβ4GlcNAcβ4GlcNAcOH] as model component.

As shown in Fig. 2A, methylation analysis of 2.5 μg of sample and CI-MS monitoring of resulting PMAAs demonstrated the presence of terminal Gal (2,3,4,6-GalOH), 2-substituted Man (3,4,6-ManOH), 4-substituted GlcNAcOH [1,3,5,6-GlcN(Me)AcOH], 3,6-disubstituted Man (2,4-ManOH), and 4-substituted GlcNAc [3,6-GlcN(Me)AcOH] in peak ratios approaching 1.6 : 2.3 : 0.8 : 1.0 : 3.3, consistent with the structure indicated above. De-O-methylation products of 1,3,5,6-GlcN(Me) AcOH such as 3,5,6-GlcN(Me)AcOH were observed in trace amounts only. EI-MS resulted in considerably lower peak intensities even if twice the amount of sample was applied. In particular, 1,3,5,6-GlcN(Me)AcOH was hardly detectable under these conditions (Fig. 2B). CI-MS analyses of smaller quantities of sample (e.g., 50–100 ng, about 30–60 pmol) revealed in principle similar results as the 2.5-μg sample (Fig. 2A), although sometimes variations of peak ratios were observed. Each PMAA, however, could be unambiguously identified (see, e.g., Fig. 2C), demonstrating that even such small amounts of glycoprotein glycans can be at least qualitatively characterized by this method. In conjunction with chromatographic size determination and exoglycosidase studies, these qualitative

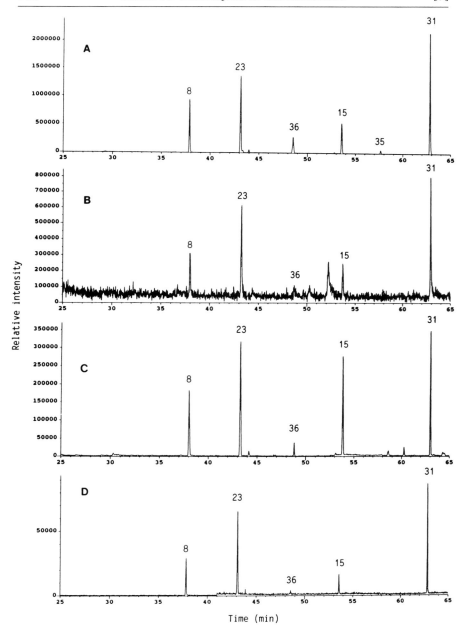

data provide useful information on the structure of the oligosaccharide under study. When lesser amounts of sample (e.g., 10 ng) were subjected to methylation, increasing numbers of contaminant peaks appeared, which could be distinguished from PMAAs or identified as glucose derivatives on the basis of their quasimolecular ions and/or RRTs. Although the PMAAs expected were still clearly detectable, such small amounts of sample seem to be the utmost limit for application of this method.

To demonstrate the decomposition of PMAAs during GC-MS analysis, the sugar derivatives obtained from the 2.5-μg sample (cf. Fig. 2A) were diluted and again analyzed by capillary GC/ammonia CI-MS. Applying 1/20 the amount used in Fig. 2A, the chromatogram revealed a significant reduction in signal intensity of 1,3,5,6-GlcN(Me)AcOH, demonstrating that peak ratios are influenced by partial degradation of certain PMAAs (especially N-acetylhexosamine derivatives) during GC separation. Because this effect is more pronounced when small amounts of sample are analyzed, such micromethylation results must be carefully interpreted and confirmed by other micromethods such as chromatographic analyses, lectin-binding studies, or enzymatic sequencing (see elsewhere in this volume).

Allocation of Outer Substituents. In combination with partial degradation by linkage- and position-specific exoglycosidases, micromethylation analysis can also be used for allocation of outer substituents such as α-galactosyl residues,[38,45] type 1 (Galβ3GlcNAcβ) lactosamine chains,[50] GalNAcβ4GlcNAcβ antennae,[50] blood group determinants[48] as well as incomplete branches in glycoprotein glycans.[42,50] To illustrate this approach, a radiolabeled, fucosylated, biantennary complex-type glycan comprising one type 2 (Galβ4GlcNAcβ-) lactosamine antenna and one GalNAcβ4GlcNAcβ unit was degraded with β-N-acetylhexosaminidase

---

FIG. 2. Micromethylation analysis of a biantennary complex-type oligosaccharide alditol. Different amounts of oligosaccharide alditol were permethylated, hydrolyzed, reduced, and peracetylated. PMAAs obtained were separated on a fused silica bonded phase DB-1 column (0.25-mm i.d., 60-m length). (A) Oligosaccharide alditol methylated (2.5 μg; 1.5 nmol), 2 μl/100 μl of sample injected. PMAAs were detected by selective ion monitoring after CI with ammonia. (B) Analysis of twice the amount of sample as in (A) by EI-MS (temperature of ion source, 260°; ionization energy, 70 eV). (C) Fifty nanograms (30 pmol) of oligosaccharide alditol methylated; 10 μl/20 μl of sample injected, detection of PMAAs as in (A). (D) Same sample as in (A), 1 μl/1000 μl injected, detection of PMAAs as in (A). Peaks identified are (**8**) 2,3,4,6-GalOH, (**23**) 3,4,6-ManOH, (**36**) 1,3,5,6-GlcN(Me)AcOH, (**15**) 2,4-ManOH, (**35**) 3,5,6-GlcN(Me)AcOH, and (**31**) 3,6-GlcN(Me)AcOH (see Table I). For experimental details, see text.

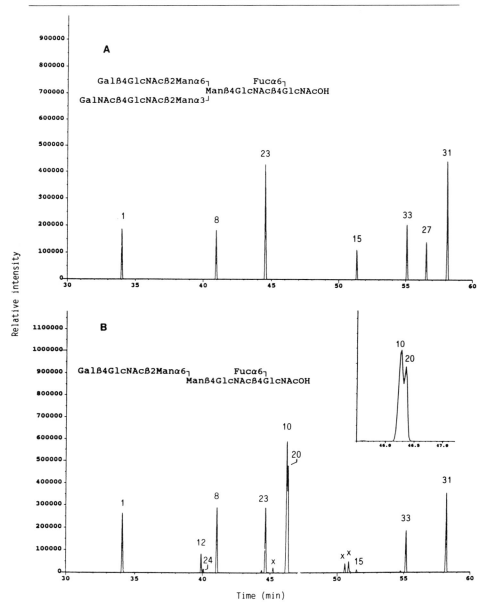

Fig. 3. Allocation of outer substituents in glycoprotein glycans by combined exoglycosidase digestion and methylation analysis. Aliquots (about 2 μg) of a radiolabeled biantennary complex-type glycan comprising one type-2 N-acetyllactosamine antenna and one GalNAcβ-4GlcNAc unit were subjected to micromethylation analysis before (A) and after (B) sequential

and α-mannosidase from jack beans.[69] Size determination of the product on a calibrated BioGel P-4 column[70] confirmed that two N-acetylhexosamines and one hexose residue were released by enzyme treatment. Starting material and reaction product were then subjected to methylation analysis. As shown in Fig. 3A, the analysis revealed the presence of about one terminal Fuc (2,3,4-FucOH), terminal Gal (2,3,4,6-GalOH), 3,6-disubstituted Man (2,4-ManOH), 4,6-disubstituted GlcNAcOH [1,3,5-GlcN(Me)AcOH] and terminal GalNAc [3,4,6-GalN(Me)AcOH], two 2-substituted Man (3,4,6-ManOH), and two to three 4-substituted GlcNAc residues [3,6-GlcN(Me)AcOH] in the starting material, in agreement with the structure proposed. In contrast, the truncated oligosaccharide was characterized by the presence of only one 2-substituted Man, one 6-substituted Man (2,3,4-ManOH), and about two 4-substituted GlcNAc as well as by the lack of 3,6-disubstituted Man and terminal GalNAc residues (Fig. 3B; the trace amounts of 2,3,4,6-ManOH and 2,4-ManOH are probably due to incomplete removal of Man during treatment with α-mannosidase). From the conversion of 3,6-disubstituted Man to 6-substituted Man, it can therefore be concluded that the GalNAcβ-4GlcNAcβ2Manα branch is linked to C-3 of the branching Man residue in the original oligosaccharide. Terminal (2,3,4,6-GlcOH) and 6-substituted Glc residues (2,3,4-GlcOH) also found result from isomaltosyl oligosaccharide standards introduced into the sample by calibration of the BioGel P-4 column with glucose oligomers.

The example demonstrates that the linkage position of outer substituents of glycoprotein glycans available in small amounts can be successfully assigned only by a combination of exoglycosidase digestion and micromethylation analysis.

[69] Y.-T. Li and S.-C. Li, this series, Vol. 28, p. 702.
[70] K. Yamashita, T. Mizuochi, and A. Kobata, this series, Vol. 83, p. 105.

degradation with β-N-acetylhexosaminidase and α-mannosidase and size determination on a calibrated BioGel P-4 column. PMAAs obtained after methylation, hydrolysis, reduction, and peracetylation were analyzed by capillary GC-MS as in Fig. 2A. Peaks identified are (1) 2,3,4-FucOH, (8) 2,3,4,6-GalOH, (23) 3,4,6-ManOH, (15) 2,4-ManOH, (33) 1,3,5-GlcN(Me)AcOH, (27) 3,4,6-GalN(Me)AcOH, (31) 3,6-GlcN(Me)AcOH, (24) 2,3,4,6-ManOH, and (20) 2,3,4-ManOH (see Table I). Peaks designated (12) and (10) in (B) represent 2,3,4,6-GlcOH and 2,3,4-GlcOH, respectively, which are derived from isomaltosyl oligosaccharides introduced into the sample during size determination of the truncated glycan. Peaks designated (×) represent unknown impurities. The inset in (B) demonstrates the partial separation of 2,3,4-GlcOH (10) and 2,3,4-ManOH (20) under the experimental conditions used (see text).

Limitations

The method outlined previously has several limitations.

1. Because sugar residues are identified solely by their quasimolecular ions and RRTs, PMAA standards for all monosaccharide constituents of a given oligosaccharide must be available for calibration of the GC column. Therefore this line of approach is not applicable in the case of oligo- or polysaccharides containing rare or novel monosaccharides. Furthermore, monosaccharide constituents may occur as furanoses, which are not considered in this article.

2. Determination of sialic acid residues often occurring in glycoprotein glycans is not possible by this method. The presence and linkage positions of these sugar constituents must be indirectly established by methylation analysis before and after desialylation. To obtain direct proof for the presence of sialic acid, methylated oligosaccharides must be degraded by methanolysis.[65] After peracetylation, resulting methylglycosides can be similarly analyzed by GC-MS, thus allowing the determination of neutral sugar methylglycosides and sialic acid methyl ester methylglycosides in one run. With this methodology, however, many sugar derivatives lead to more than one peak in the chromatogram, which may impair their separation and quantitation as well as the sensitivity of detection when small amounts of complex carbohydrates are used.

3. To facilitate their detection during chromatographic steps (e.g., sample preparation, solid-phase extraction), oligosaccharides should be radiolabeled either by metabolic labeling or by reduction with NaB^3H_4.[60] Alternatively, oligosaccharides can be fluorescently tagged.[71] In this case, however, the innermost GlcNAc residue is not registered in GC-MS analysis.

4. Under the experimental conditions described, alkali-labile substituents are lost during methylation. Therefore linkage positions of O-acyl groups,[3,4] phosphate residues,[27] as well as substituents linked to the carbohydrate chain via phosphate diester bonds (e.g., 2-aminoethyl phosphate residues),[26,28] cannot be determined by this method. In contrast, ester-linked sulfate groups were found to resist the strong alkaline conditions used.[45,72]

Acetolysis

Acetolysis is a useful supplementary method for structural studies on high mannose-type glycans.[73–75] The method is based on the observation

[71] S. Hase, this volume [13].

[72] G. Pfeiffer, S. Stirm, R. Geyer, K.-H. Strube, A. A. Bergwerff, J. P. Kamerling, and J. F. G. Vliegenthart, *Glycobiology* **2**, 411 (1992).

[73] J. Kocourek and C. E. Ballou, *J. Bacteriol.* **100**, 1175 (1969).

that, under controlled hydrolysis conditions, $\alpha(1 \rightarrow 6)$ linkages between mannose units can be preferentially cleaved in peracetylated oligosaccharides. Following de-O-acetylation, resulting oligosaccharide fragments can be analyzed by paper chromatography, gel filtration, and/or HPLC, thus allowing an unambiguous identification of the isomeric structure(s) of the glycan fraction under study.

To minimize cleavage of linkages other than Man(α1-6)Man, several improvements of the original procedure have been described, using a different ratio of acetic anhydride, acetic acid, and concentrated sulfuric acid,[76] as well as a lower reaction temperature and shorter incubation time.[77] By using conditions referred to as incipient acetolysis, it could be further demonstrated that the outer Man(α1-6)Man linkages of high mannose-type N-glycans are more rapidly cleaved than inner $\alpha(1 \rightarrow 6)$ linkages, thus generating intermediates that allow "a more definite interpretation of the fragmentation pattern."[77]

Starting from high mannose-type oligosaccharide alditols metabolically labeled with [³H]- or [¹⁴C]mannose, the procedure outlined below can be adopted to isomeric structure determination of minute amounts of this type of glycan.[36,42,43,78] Otherwise, oligosaccharides as well as fragments obtained thereof must, on acetolysis, be radiolabeled by reduction with NaB³H₄ or fluorescently tagged in order to ensure their sensitive detection in subsequent analysis.[60,71]

Reagents

All reagents should be of high quality (see above).

Procedure

1. Transfer the desalted oligosaccharide alditol (at least 5000 cpm) into a conical, Teflon-lined, screw-capped Pyrex test tube and dry the sample *in vacuo* over P_2O_5 overnight.

2. Flush the desiccator with argon and seal the vial.

3. Peracetylate the sample, using either the method described above (see Acetylation, steps 1–3) or the procedure outlined by Dell.[79]

---

[74] C. E. Ballou, *Adv. Enzymol. Relat. Areas Mol. Biol.* **40,** 239 (1974).

[75] T. Tai, K. Yamashita, M. Ogata-Arakawa, N. Koide, T. Muramatsu, S. Iwashita, Y. Inoue, and A. Kobata, *J. Biol. Chem.* **250,** 8569 (1975).

[76] E. Li and S. Kornfeld, *J. Biol. Chem.* **254,** 2754 (1979).

[77] I. K. Vijay and G. H. Perdew, *FEBS Lett.* **139,** 321 (1982).

[78] E. Bause, W. Breuer, J. Schweden, R. Roser, and R. Geyer, *Eur. J. Biochem.* **208,** 451 (1992).

[79] A. Dell, this series, Vol. 193, p. 157.

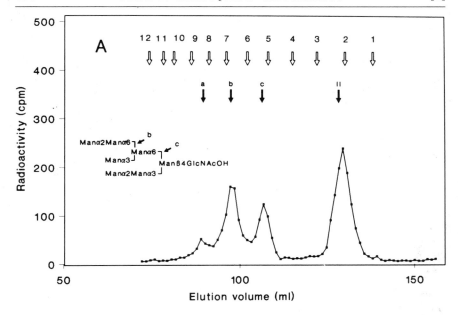

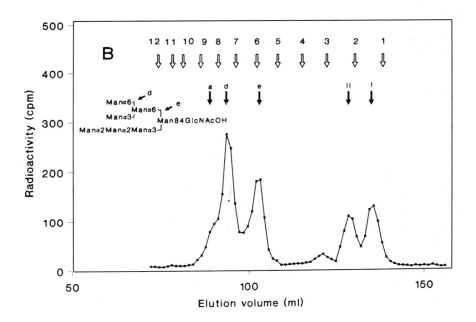

4. Dissolve the dry peracetylated sample in 100 μl of an 11 : 10 : 1 (v/v/v) mixture of acetic acid–acetic anhydride–concentrated sulfuric acid and incubate at 35° for 5 hr.

5. Add 1 ml of water, extract the aqueous phase with 1 ml of chloroform, transfer the organic layer into a conical screw-capped glass tube, wash three times with water (2 ml each), and evaporate under a stream of nitrogen.

6. For de-O-acetylation, add 200 μl of 0.1 M aqueous sodium hydroxide and leave at room temperature for 30 min.

7. Adjust the pH to 8–9 by dropwise addition of 1 M acetic acid (approximately 20 μl) and reduce acetolysis fragments with $NaBH_4$ as described above (see Reduction, steps 1–4).

8. Deionize the sample by passage through a column (0.5 × 6 cm) of mixed bed resin (Amberlite AG MB-3; Serva), using water for elution, and lyophilize.

9. Analyze acetolysis fragments by gel filtration[70] and/or HPLC, using a LiChrosorb-Diol (Merck) column.[36,42,43] (If acetolysis fragments are to be analyzed only by gel filtration, the reduction described in step 7 can be omitted.)

Illustrative Example

An application of the procedure outlined above is illustrated in Fig. 4. Two isomers of high mannose-type glycans with seven mannosyl residues (Man$_7$GlcNAcOH), separated by high-pH anion-exchange chromatography,[80] were subjected to acetolysis. Following de-O-acetylation and reduction (see above), fragments obtained were characterized by gel filtration, using a calibrated BioGel P-4 column.[70] The results revealed that, apart

[80] G. Pfeiffer, H. Geyer, R. Geyer, I. Kalsner, and P. Wendorf, *Biomed. Chromatogr.* **4,** 193 (1990).

FIG. 4. Analysis of two structural isomers of high mannose-type glycans with seven mannosyl residues. Oligosaccharide alditols, metabolically labeled with [2-^{3}H]mannose, were peracetylated and treated with acetic acid–acetic anhydride–concentrated sulfuric acid, 11 : 10 : 1 (v/v/v) at 35° for 5 hr. The reaction products were de-O-acetylated, reduced with $NaBH_4$, and chromatographed on a BioGel P-4 column (−400 mesh, 1.6 × 60 cm) at 55° with aqueous 0.02% (w/v) sodium azide as eluant. Fractions (0.4 ml) were collected at 0.2 ml/min and monitored for radioactivity. (A) and (B) Elution profiles of acetolysis fragments obtained from the isomers indicated. Elution volumes of starting material (a), Man$_5$GlcNAcOH (b), Man$_3$GlcNAcOH (c), Man$_6$GlcNAcOH (d), Man$_4$GlcNAcOH (e), ManOH (I), and ManManOH (II) are indicated by bold arrows. Numbers (1–12) over arrows: elution volumes of glucose oligomers with 1–12 glucose units.

from small amounts of starting material, the isomer in Fig. 4A led to the fragments ManManOH, Man$_3$GlcNAcOH, and Man$_5$GlcNAcOH, whereas acetolysis of the structural isomer shown in Fig. 4B yielded the fragments ManOH, ManManOH, Man$_4$GlcNAcOH, and Man$_6$GlcNAcOH, demonstrating that the two glycans have the isomeric structures proposed.

Acknowledgments

The authors are indebted to their mentor Professor Stephan Stirm and express their gratitude to Peter Kaese, Siegfried Kühnhardt, and Werner Mink for methylation and GC-MS analyses and to Ines Jacobi and Günter Pfeiffer for practical help. This work was supported by the Deutsche Forschungsgemeinschaft (Sonderforschungsbereich 272, Teilprojekt Z2).

[8] Mass Spectrometry of Carbohydrate-Containing Biopolymers

By ANNE DELL, ANDREW J. REASON, KAY-HOOI KHOO,
MARIA PANICO, ROY A. MCDOWELL, and HOWARD R. MORRIS

Introduction

Of the many mass spectrometric techniques that have been applied to carbohydrate-containing biopolymers, gas chromatography-mass spectrometry (GC-MS) and fast atom bombardment-mass spectrometry (FAB-MS)[1] are by far the most important with respect to the amount of structural information they can provide, and the range of biological problems to which they have been applied. Together, these two technologies are capable of defining the complete primary structure of an oligosaccharide or glycoconjugate.[2] In selected areas of carbohydrate structure analysis, most notably the glycoprotein field, the more recently introduced technique of electrospray-mass spectrometry (ES-MS)[3] has become a powerful adjunct to FAB-MS.

This chapter covers practical aspects of FAB-MS and ES-MS, including data interpretation, and describes examples of experimental strategies,

[1] M. Barber, R. S. Bordoli, R. D. Sedgwick, and A. N. Tyler, *Nature (London)* **293**, 270 (1981).

[2] K. H. Khoo, R. M. Maizels, A. P. Page, G. W. Taylor, N. B. Rendell, and A. Dell, *Glycobiology* **1**, 163 (1991).

[3] J. B. Fenn, M. Mann, C. K. Meng, S. F. Wong, and C. M. Whitehouse, *Science* **246**, 64 (1989).

incorporating all three mass spectrometric methods, which are best suited to the analysis of glycoproteins, complex carbohydrates (including glycolipids), and glycosaminoglycans, respectively. Gas chromatography-mass spectrometry methodology is described in [7] this volume. The reader should refer to [25] [glycosylphosphatidylinositol (GPI) anchors] and [28] (neoglycolipids) for additional applications of these powerful mass spectrometric techniques.

The principles of FAB-MS and ES-MS have been well documented elsewhere[4] and, except where they are relevant to practical considerations, are not discussed herein.

Choice of Instrumentation

Fast Atom Bombardment-Mass Spectrometry

A high-field[5] sector mass spectrometer is essential for FAB-MS work if the capabilities of the technique are to be fully exploited with respect to both sensitivity and molecular size. A mass range of about 5000 Da at full sensitivity is sufficient for the vast majority of carbohydrate problems. However, some laboratories may need to invest in large-geometry high-field instruments (e.g., the VG Analytical ZAB-SE, which has a range of 15,000 Da), if high molecular weight molecules such as multiantennary lactosaminoglycan-containing N-glycans comprise a significant proportion of the samples requiring FAB analysis. Although often used for biopolymer work, expensive tandem instrumentation (MS/MS instrumentation)[6] is not a requirement for successful carbohydrate analysis. This is because (1) derivatization/FAB-MS and/or hydrolysis/FAB-MS strategies (see later) are often more effective in providing structurally informative ions than are collisional activation MS/MS techniques, and (2) B/E linked scanning on a two-sector instrument is a viable alternative to tandem MS/MS for defining the daughters belonging to the various parents in a mixture[7]; it should be noted that the low parent ion resolution inherent in the B/E experiment rarely poses problems because the probability of two parent ions of interest being close together is actually very low indeed in a real mixture experiment.

[4] J. A. McCloskey, ed., "Methods in Enzymology," Vol. 193. Academic Press, San Diego, 1990.
[5] H. R. Morris, A. Dell, and R. McDowell, *Biomed. Mass Spectrom.* **8,** 463 (1981).
[6] F. W. McLafferty, ed., "Tandem Mass Spectrometry." Wiley, New York, 1983.
[7] H. R. Morris, A. Chatterjee, M. Panico, B. S. Green, and B. S. Hartley, *Rapid Commun. Mass Spectrom.* **3,** 110 (1989).

Atom (normally xenon gas) and ion (either xenon or cesium) guns are both suitable for carbohydrate FAB-MS and give comparable data from most sample types. However, the cesium ion gun is operated at significantly higher voltages (typically 30 kV) than those used for atom guns (typically 8–10 kV) and this can be an important factor in achieving the highest sensitivity, especially for high molecular weight samples. When ion guns are used the technique is sometimes called liquid secondary ion mass spectrometry (SIMS) instead of FAB-MS, but the result is the same.

Electrospray-Mass Spectrometry

In this technique a stream of liquid containing the sample of interest is injected directly into the ES source where the sample molecules are stripped of solvent, leaving them as multiply charged species whose charges reflect the number of functional groups that can be protonated (positive ion mode) or deprotonated (negative ion mode) at the pH of the carrier liquid. In most ES-MS experiments the m/z values encountered are compatible with analysis by quadrupole analyzers, which are considerably less expensive than the magnetic sector analyzers required for FAB-MS. Electrospray-mass spectrometry is being used extensively in the protein field, in which the ubiquitous presence of basic side chains results in low m/z values even for proteins of 100 kDa. In the carbohydrate area, ES-MS has been successfully applied to the characterization of glycopeptides together with a limited number of glycoproteins,[8,9] and shows considerable promise for these classes of polymer. Electrospray-mass spectrometry on quadrupole instruments is less useful, however, for the wide range of carbohydrate polymers that give predominantly singly charged molecular ions. When large (more than 3000 Da), these samples are outside the mass range of quadrupole instruments, and when small (less than 3000 Da), spectral quality is rarely superior to that achieved in an FAB-MS experiment. For these classes of carbohydrates, the availability of ES sources on magnetic sector instruments is a promising development, in principle allowing ES-MS to be performed on instruments with the same mass range and resolution capabilities as those used for FAB-MS. Whether ES-MS will prove more powerful than FAB-MS remains to be seen.

Fast Atom Bombardment-Mass Spectrometry Procedure

1. Dissolve the sample in 5% (v/v) aqueous acetic acid, methanol, or a mixture of the two, depending on solubility, to such a concentration

[8] M. Rassouli, H. Sambasivam, P. Azadi, A. Dell, H. R. Morris, A. Nagpurkar, S. Mookerjea, and R. K. Murray, *J. Biol. Chem.* **267,** 2947 (1992).
[9] K. L. Duffin, J. K. Welply, E. Huang, and J. D. Henion, *Anal. Chem.* **64,** 1440 (1992).

that a 1-μl aliquot of the sample contains the desired amount of sample for FAB-MS analysis. The quantity of sample required will depend on a number of factors, including the carbohydrate class, the mass range required, and the type of derivative, but is typically in the range of 0.1–5 μg.

2. Using a micropipette, smear about 1 μl of the matrix onto the metal target on the end of the FAB probe. Monothioglycerol (TG) is the most versatile matrix for carbohydrate FAB-MS and is suitable for all compound classes, both underivatized and derivatized, and for both positive and negative ion modes. The majority of FAB experiments will be successful with this matrix. Failed experiments are usually the result of impure samples and are only rarely caused by TG being an unsuitable matrix. Three other matrices are, however, useful for specific applications: (1) glycerol, which sometimes results in better quality data from polar substances; (2) m-nitrobenzyl alcohol (mNBA), which is a suitable alternative to TG for hydrophobic samples, especially if they are slightly salty, because this matrix promotes cationization of carbohydrates rather than protonation; (3) triethylene glycol monobutyl ether, which is compatible with nonpolar solvents such as chloroform, and is therefore valuable for hydrophobic samples, for example, carbohydrates heavily acylated with long-chain fatty acids that are insoluble in methanol.

3. Carefully load 1 μl of the sample solution onto the surface of the matrix, using a micropipette or syringe.

4. Introduce the probe into the FAB source but do not fully insert into the atom/ion beam at this stage.

5. Ensure that the computer is ready to acquire data. Initiate scanning. Fully insert the probe just before the magnet resets for a new scan and acquire data immediately this scan begins. The best quality data are usually obtained within the first 30 sec of atom/ion impact. If a wide mass range needs to be covered, for example, m/z 5000 to 200, it may occasionally be necessary to scan the high and low ends of the mass range in two separate experiments in order to ensure that spectral quality is maintained over the complete mass range.

Electrospray-Mass Spectrometry Procedure

The following protocol routinely produces good quality data from a VG Bio Q instrument in our laboratory. Optimal running conditions, for example, flow rates, for other instrument types may vary slightly from those given.

1. Dissolve the sample in its own best solvent, preferably 5% aqueous acetic acid, for work in the positive ion mode, or a mild base such as

triethylamine for negative ionization studies. The volume of solvent used should be the minimum required to dissolve the sample completely.

2. Dilute the sample with an organic solvent, usually methanol or acetonitrile, and with the carrier solution (see step 3) when necessary (this depends on the volume used in step 1), to provide a 50% organic solution and a final sample concentration of 1–50 pmol/μl.

3. Inject 10 μl of sample solution from step 2 into a flow of approximately 3 μl/min of carrier solution made up from 0.1% aqueous acetic acid–methanol (1 : 1, by volume) for positive ion work, or 0.1% aqueous triethylamine (or 10 mM aqueous ammonium acetate)–acetonitrile (1 : 1, by volume) for negative ion work. Other solvent systems are possible provided they achieve protonation (or deprotonation) assistance, and allow microdroplet formation at the end of the capillary injector (a function of the surface tension of the liquid). A major consideration is the solubility of the sample in the aqueous–organic mixture. Poorly soluble samples may block the capillary inlet or prevent adequate spray formation.

4. Collect data over the lifetime of the sample, ~ 3 min under the above conditions. If poor-quality data are obtained check the pump pressure readout to identify possible blockage. If this is not the case, redissolve the sample in stronger solvents, for example, formic acid, prior to repeating the above steps with and without the assistance of nebulizing gas. If the experiment still fails it may be necessary to purify the sample further prior to repeating the ES-MS analysis. In most cases, however, it is more productive to pursue an alternative strategy such as derivatization/FAB-MS when ES-MS experiments on underivatized samples are unsuccessful.

Choosing Suitable Derivatives for Fast Atom Bombardment-Mass Spectrometry

Derivatization underpins the majority of FAB-MS strategies in the carbohydrate area. This is not, however, because of the intractability of underivatized carbohydrates to FAB-MS, as frequently assumed by mass spectrometry specialists who use commercial carbohydrates as test compounds for new ionization techniques. Instead, derivatives are chosen because (1) they are easier to obtain free from salt impurities, which may prejudice the MS experiment, (2) sensitivity is significantly improved when hydrophobic moieties are present, and (3) the fragmentation pathways of derivatives are predictable and lead to abundant fragment ions that can be assigned unambiguously; underivatized samples, on the other hand, are less predictable in their fragmentation behavior and frequently give fragment ions that are derived from multiple cleavages and cannot, therefore, be attributed to a unique structure.

The precise choice of derivative depends on the particular problem under investigation and the reader is referred to later sections that describe suitable strategies for three different classes of carbohydrate for specific guidance on this matter. However, some general remarks are included here, together with protocols for the most important derivatives.

Derivatization methods can be broadly divided into two categories: (1) "tagging" of reducing ends, and (2) protection of most or all functional groups.

Tags were originally introduced for the purpose of improving sensitivity,[10] for example, the pentafluorobenzyloxime derivative that was successfully used to characterize low-microgram quantities of elicitor-active oligogalacturonides isolated from plant cell walls.[11] Subsequently, a variety of tags have been exploited both for their utility as chromophores during chromatographic purifications and for their FAB-MS properties.[12] Experimental conditions for the most commonly used reducing end derivative, namely p-aminobenzoic acid ethyl ester (ABEE), are given below. The fluorescently tagged saccharides described in [13] are also excellent candidates for FAB-MS analyses.

Procedure for p-Aminobenzoic Acid Ethyl Ester Tagging

1. Dry the sample in a glass tube with a Teflon-lined screw cap.
2. Take up the sample in 10 μl of water.
3. Add 40 μl of the derivatizing reagent, which contains 35 mg of ABEE, 3.5 mg of sodium cyanoborohydride, 41 μl of glacial acetic acid, and 350 μl of methanol (prepared by adding the solids to the two liquids).
4. Incubate the reaction mixture at 80° for 1 hr.
5. Add 1 ml of water to terminate the reaction.
6. Extract the excess ABEE five times, using 1-ml washes of ether each time.
7. Dry the bottom aqueous layer containing the tagged sample.

Protection of functional groups by permethylation or peracetylation is by far the most important type of derivatization used in carbohydrate FAB-MS. Indeed, even tagged molecules should be further derivatized, if possible, in order to exploit the characteristic A-type fragmentation (see Scheme I) that occurs once the hydroxyl groups are protected. If methyl or acetyl substituents are likely to be present in the natural product, deuterated counterparts of the derivatizing reagents are used.

[10] A. Dell, H. R. Morris, and H. Egge, *Int. J. Mass Spectrom. Ion. Phys.* **46,** 414 (1983).
[11] E. A. Nothnagel, M. McNeil, P. Albersheim, and A. Dell, *Plant Physiol.* **71,** 916 (1983).
[12] W. T. Wang, N. C. Ledonne, B. J. Ackerman, and C. C. Sweeley, *Anal. Biochem.* **141,** 366 (1984).

Unless sulfate groups are known or likely to be present, the sodium hydroxide procedure is the method of choice for permethylating saccharides and glycoconjugates. It is experimentally simple and gives high yields. This procedure is not, however, suitable for sulfated samples because the charged products are not readily partitioned into the organic phase at the end of the reaction. These samples require Hakomori permethylation, using the short reaction conditions originally introduced by our laboratory,[13] and which preserve the sulfates intact.[14]

Stock Reagents for Permethylation

Anhydrous Dimethyl Sulfoxide

Prepare anhydrous dimethyl sulfoxide (DMSO) as follows: Stand commercial DMSO over calcium hydride overnight or longer in a large, round-bottom flask. Decant the treated DMSO into a clean, dry flask and add fresh calcium hydride. Distill the treated DMSO under reduced pressure, using an oil pump. The pressure should be sufficiently low to allow distillation under about 50° in order to avoid thermal decomposition. Store the redistilled DMSO in a glass stoppered flask over fresh calcium hydride at room temperature. Keep the stopper tight and replace immediately each time the DMSO is dispensed.

Oil-Free Sodium Hydride

Oil-free sodium hydride (NaH, a dark gray powder when it is active) is prepared from the commercial oil suspension by washing with sodium-dried ether. The latter is obtained by carefully adding sodium wire from a sodium press to a previously unused bottle of analytical-grade ether. (*Caution:* Sodium will react explosively on contact with water.) The treated ether should be left to stand overnight before use. Alternatively, a commercial supply of good-quality sodium-dried ether may be used directly provided it has not been exposed to air previously. The oil suspension of NaH is washed as follows.

1. Place about 500 g of the suspension in a 500-ml round-bottom flask in a fume hood.
2. Swirl the flask and allow the NaH to settle. Decant the ether into a waste container and add fresh ether. Repeat at least 10 times. After the

[13] H. R. Morris, M. R. Thompson, D. T. Osuga, A. I. Ahmed, S. M. Chan, J. R. Vandenheede, and R. E. Feeney, *J. Biol. Chem.* **253**, 5155 (1978).
[14] A. Dell, H. R. Morris, F. Greer, J. M. Redfern, M. E. Rogers, G. Weisshaar, J. Hiyama, and A. G. C. Renwick, *Carbohydr. Res.* **209**, 33 (1991).

final wash, place the flask in a large desiccator and connect it to a vacuum pump to remove the residual ether.

3. Transfer the dry, clean NaH to a dry, screw-capped polypropylene jar and store at room temperature in a dry atmosphere.

4. Dispose of waste ether by gentle evaporation under a stream of nitrogen and destroy residual NaH in the waste ether by slow dropwise addition of ethanol. (*Caution:* Sodium hydride reacts explosively with water.)

5. Discard the NaH if it turns into white powder on prolonged storage or contact with moisture.

Procedure for NaOH Permethylation

1. Dry the sample to be permethylated in a glass tube with a Teflon-lined screw cap.

2. Place five pellets of sodium hydroxide in a dry mortar and add approximately 3 ml of dry DMSO, using a Pasteur pipette. Use a pestle to grind the pellets until a slurry is formed with DMSO. This should be done as quickly as possible to minimize absorption of moisture from the atmosphere.

3. Take up 0.5–1.0 ml of the DMSO–NaOH slurry with a Pasteur pipette and add to the sample.

4. Add about 0.5 ml of methyl iodide and mix vigorously. Place the reaction mixture on an automatic shaker for 10 min at room temperature.

5. Quench the reaction by careful, dropwise addition of about 1 ml of water with constant shaking between additions to lessen the effects of the highly exothermic reaction.

6. Add 2 ml of chloroform, mix thoroughly, and allow the mixture to settle into two layers. Centrifuge the mixture to assist better separation, if necessary.

7. Remove and discard the upper aqueous layer and wash the lower chloroform layer several times with water, until the water being removed is completely clear.

8. Dry the chloroform layer under a stream of nitrogen and dissolve the residue in methanol for FAB-MS analysis.

Procedure for Short Hakomori Permethylation

Prepare the methylsulfinyl methanide or "dimsyl" anion base as follows:

1. Place two small microspatula scoops of NaH in the bottom of a Quickfit tube (10 × 1 cm) and add approximately 3 ml of dry DMSO with

a Pasteur pipette. Mix gently to ensure an even suspension. Remove the stopper and place the tube in an oven at 90°.

2. Remove the tube when the color of the reaction mixture just turns from gray to light honey brown, usually after about 10 min. If it is then still gray with little NaH remaining on the bottom, an additional microspatula of NaH should be added and the tube returned to the oven for a further 5- to 10-min incubation.

3. Lightly stopper the tube, allowing any remaining effervescence to escape, and leave the reaction mixture to cool for a few minutes.

4. Centrifuge the mixture with a bench-top centrifuge at 3000 rpm for 10 min. The supernatant DMSO solution containing the dimsyl anion should be crystal clear and honey colored; excess NaH forms a pellet at the bottom of the tube and should not be disturbed when handling the tube.

5. Add a drop of the base to a few crystals of triphenylmethane, which will turn scarlet red if the base is active.

6. Use the freshly prepared base without delay. The amounts given above are usually sufficient for permethylating two samples and it is preferable to prepare additional tubes of base than to scale up the preparation for more samples.

Permethylation of the sample is achieved as follows.

1. Dry the sample to be permethylated in a glass tube with a Teflon-lined screw cap.

2. Redissolve the sample in 2 drops of dry DMSO solution (clean Pasteur pipettes are used for dispensing all liquid reagents in the permethylation procedures).

3. Add 5 drops of the freshly prepared dimsyl base, followed by 3 drops of methyl iodide, and mix gently.

4. Leave standing at room temperature for 5 min and then add a further 20 drops of the dimsyl base and mix thoroughly.

5. Withdraw 1–2 μl of the reaction mixture to check for excess base by adding to a few crystals of triphenylmethane. If the crystals turn scarlet, add about 10 drops or more of methyl iodide to the reaction mixture. If excess base is not observed, that is, the triphenylmethane does not turn scarlet, add more base until the test shows it to be in excess before adding the methyl iodide.

6. Leave the reaction mixture standing at room temperature for 10–20 min.

7. Stop the reaction by slowly adding 1 ml of water with constant shaking. The solution becomes cloudy and excess droplets of methyl iodide (yellow) settle to the bottom of the tube.

8. Bubble nitrogen gas into the mixture until all droplets of excess

methyl iodide are evaporated off and the sample solution becomes clear again.

    9. Purify the sample with a C_{18}Sep-Pak cartridge as follows.

        a. Condition the Sep-Pak cartridge by sequential washing with water (5 ml), acetonitrile (5 ml), ethanol (5 ml), and water (10 ml).

        b. Load the permethylation solution directly onto the Sep-Pak.

        c. Elute sequentially with 5 ml of water, 2 ml each of 15, 35, 50, and 75% aqueous acetonitrile, and 2 ml of acetonitrile.

        d. Dry each fraction with a Speedivap (Savant, Hicksville, NY) or equivalent instrument.

    10. Analyze the 50 and 75% acetonitrile fractions by positive FAB-MS and the 15 and 35% fractions by negative FAB-MS. Sulfated samples elute in the 15 and 35% acetonitrile fractions. All other saccharides, except those that are very small, are retained until the 50 or 75% acetonitrile washes.

Procedures for Peracetylation

Peracetylation can be carried out by either acidic or basic catalysis, although the former must be chosen if base-labile substituents such as acetyl need to be retained,[15] and the latter must be used to preserve acid-labile groups such as sulfate.[16] Deuterated reagents are used when natural O-acetyl groups are anticipated in the saccharide.

Acid-Catalyzed Peracetylation

    1. Prepare the acetylating reagent by adding 1 vol of glacial acetic acid or deuteroacetic acid to 2 vol of trifluoroacetic anhydride in a glass tube. Mix and leave standing at room temperature for a few minutes, until the heat from the exothermic reaction has dissipated.

    2. Add 100–200 μl of the reagent to the dried sample in a glass tube with a Teflon-lined screw cap and leave standing at room temperature for 10 min.

    3. Remove the reagent under a stream of nitrogen in a fumehood.

    4. Redissolve the peracetylated sample in 1 ml of chloroform, wash several times with water, and then dry under a stream of nitrogen.

    5. Redissolve the sample in methanol for FAB-MS analysis.

Base-Catalyzed Peracetylation

This can be carried out with either pyridine or 1-methylimidazole as the base catalyst. Both procedures are given below.

[15] A. Dell and P. R. Tiller, *Biochem. Biophys. Res. Commun.* **135**, 1126 (1986).

[16] K.-H. Khoo, H. R. Morris, R. A. McDowell, A. Dell, M. Maccarana, and U. Lindahl, *Carbohydr. Res.* **244**, 205 (1993).

Pyridine base:

1. Dissolve the sample to be peracetylated in 50 μl of pyridine in a glass tube fitted with a Teflon-lined screw cap.

2. Add 50 μl of acetic anhydride or d_6-(acetic anhydride) to the sample solution and mix thoroughly.

3. Incubate at 80° in a heating block for 2 hr.

4. Remove the reagent under a stream of nitrogen in a fumehood. If the sample was contaminated with salt prior to derivatization, the products are dissolved in chloroform, salts are extracted with water, and the chloroform is removed under a stream of nitrogen.

5. Redissolve the sample in methanol for FAB-MS analysis.

1-Methylimidazole base:

1. Dissolve the sample to be peracetylated in 5 μl of 1-methylimidazole in a glass tube fitted with a Teflon-lined screw cap or a small (1-ml) Reacti-vial (Pierce, Rockford, IL).

2. Add 25 μl of acetic anhydride or d_6-(acetic anhydride) to the sample solution and mix thoroughly.

3. Leave standing at room temperature for 30–60 min.

4. Remove the reagent under a stream of nitrogen in a fume hood. A few drops of methanol may be added to help remove the residual trace of 1-methylimidazole. If salt is present, the products are disssolved in chloroform, salts are extracted with water, and the chloroform is removed under a stream of nitrogen.

5. Redissolve the sample in methanol for FAB-MS analysis.

Fast Atom Bombardment-Mass Spectrometry Fragmentation

Two fragmentation pathways dominate carbohydrate FAB-MS, namely, A-type cleavage and β cleavage.[17] These are illustrated in Scheme 1. A-Type cleavage occurs on the nonreducing side of glycosidic bonds to give oxonium-type fragment ions and is the dominant pathway for permethyl and peracetyl derivatives. Cleavage is favored at amino sugar residues. A-Type fragment ions are extremely helpful for structure assignments. β Cleavage also involves glycosidic fission but, because a hydrogen transfer accompanies bond breakage, no charge is produced at the actual cleavage site as occurs during A-type cleavage. The charge on a β-cleavage fragment ion results from protonation or cationization. Fragment ions can be reducing or nonreducing, depending on which of the two bonds to the glycosidic oxygen is broken. β-Cleavage ions are commonly produced by native saccharides and glycoconjugates and by samples in which the

[17] A. Dell, *Adv. Carbohydr. Chem. Biochem.* **45**, 19 (1987).

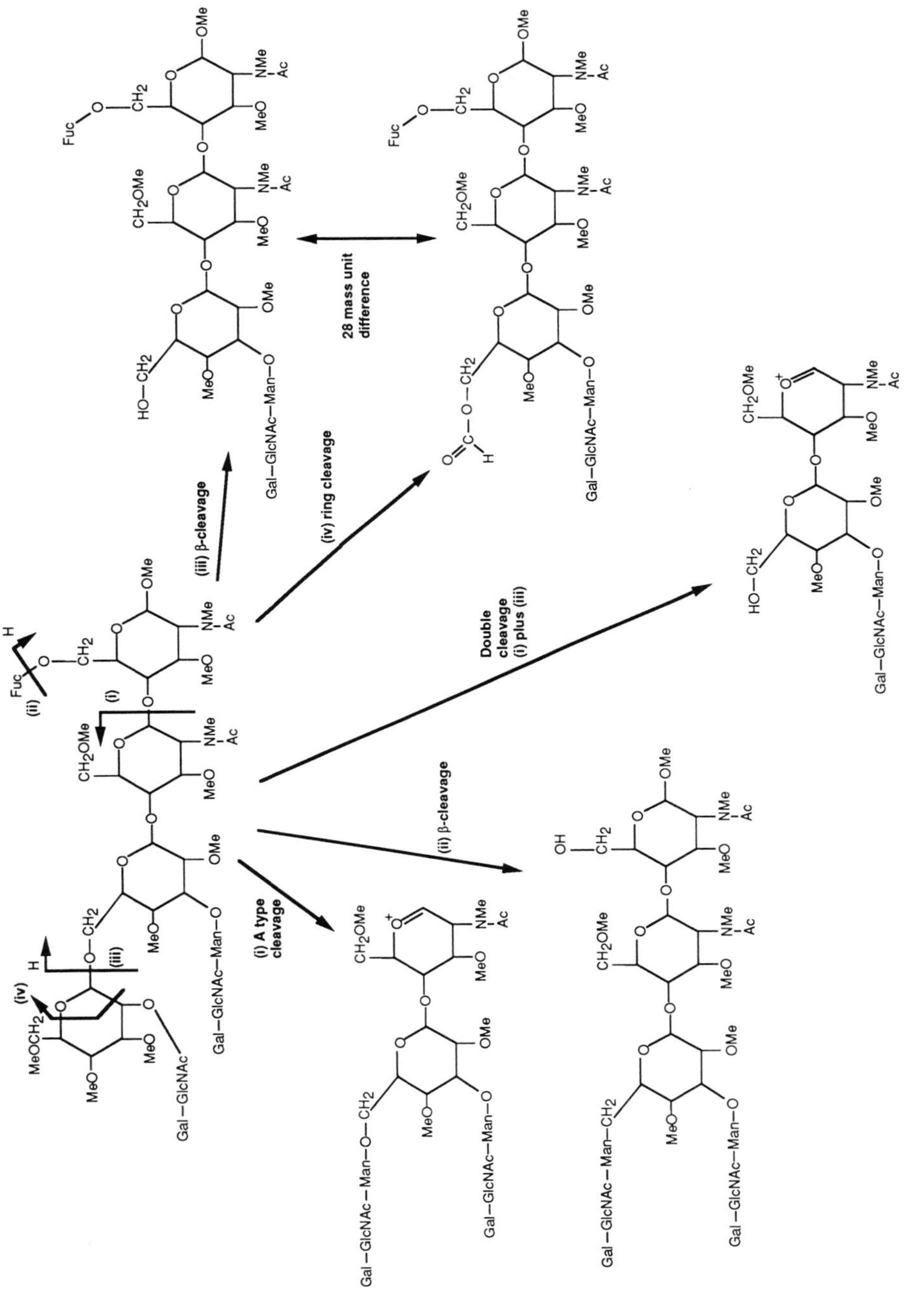
(iii) β-cleavage
(iv) ring cleavage
28 mass unit difference
Double cleavage (i) plus (iii)
(ii) β-cleavage
(i) A type cleavage

SCHEME I

charge is localized to a particular site in the molecular ion, for example, a negative charge on a sulfated antenna or a positive charge on the nitrogen of an ABEE tag.

"Double-cleavage" ions are formed when two or more cleavages occur simultaneously in different parts of the molecular ion, for example, A-type cleavage plus β cleavage (see Scheme 1), or two β cleavages. They are usually less abundant than ions resulting from single-cleavage events. Less common fragmentation pathways include ring cleavages (see pathway iv, Scheme 1) and losses of substituents such as methanol and water.

Calculating Masses of Molecular and Fragment Ions

All necessary data for calculating the masses of molecular and fragment ions of saccharides (excluding those with less common sugar residues) are given in Tables I and II. The following points should be taken into account when using Tables I and II.

1. Both accurate and average masses are given. The former are used to assign the ^{12}C peak in resolved spectra, the latter when assigning the centroid of an unresolved cluster.

2. $[M + H]^+$ values are calculated by adding the sum of the protonated reducing and nonreducing end increments (Table II) to the sum of the residue masses.

3. The mass of an A-type fragment ion is calculated by adding the nonreducing end increment (Table II) to the sum of the residue masses.

4. The mass of an ion derived from one or more β cleavages is the same as that of a quasimolecular ion of the same sugar composition except that, in the case of derivatives, the incremental mass of one functional group is subtracted for each β-cleavage event that has occurred to give the fragment ion.

5. The mass of a double-cleavage ion resulting from a combination of one A-type and one β cleavage is calculated by adding one hydrogen to the sum of the residue masses.

Structural Information Deduced from Fast Atom Bombardment- and Electrospray-Mass Spectrometry Data

A wealth of structural data can be derived from FAB-MS and ES-MS experiments, provided appropriate analytical strategies, for example, derivatization, are pursued. A summary of the type of structural information afforded by all saccharide and glycoconjugate classes, together with guidance to the methodology that produces it, is given next. Specific examples are covered in the strategy sections that follow.

TABLE I
RESIDUE MASSES OF SUGARS COMMONLY FOUND IN GLYCOCONJUGATES

Monosaccharides	Native		Permethylated		Deuteromethylated		Peracetylated		Deuteroacetylated	
	Accurate mass	Average mass	Accurate mass	Average mass	Accurate mass	Average mass	Accurate mass	Average mass	Accurate mass	Average mass
Pentose (Pent)	132.0432	132.1161	160.0736	160.1699	166.1112	166.2069	216.0634	216.1907	222.1010	222.2277
Deoxyhexose (deoxyHex)	146.0579	146.1430	174.0892	174.1968	180.1269	180.2237	230.0790	230.2176	236.1167	236.2545
Hexose (Hex)	162.0528	162.1424	204.0998	204.2230	213.1563	213.2785	288.0845	288.2542	297.1410	297.3097
Hexuronic acid (HexA)	176.0321	176.1259	218.0790	218.2066	227.1355	227.2620	260.0532	260.2005	266.0909	266.2375
Heptose (Hept)	192.0634	192.1687	248.1260	248.2762	260.2013	260.3501	360.1056	360.3178	372.1810	372.3917
N-Acetylhexosamine (HexNAc)	203.0794	203.1950	245.1263	245.2756	254.1828	254.3311	287.1005	287.2695	293.1382	293.3065
2-Keto-3-deoxyoctonate (KDO)	220.0583	220.1791	276.1209	276.2866	288.1962	288.3605	346.0900	346.2909	355.1465	355.3464
Muramic acid (Mur)	275.1005	275.2585	317.1475	317.3392	326.2039	326.3946	317.1111	317.2958	320.1299	320.3143
N-Acetylneuraminic acid (NeuAc)	291.0954	291.2579	361.1737	361.3923	376.2678	376.4847	417.1271	417.3698	426.1836	426.4252
N-Glycolylneuraminic acid (NeuGc)	307.0903	307.2573	391.1842	391.4186	409.2972	409.5295	475.1326	475.4064	487.2079	487.4804

TABLE II
MASSES OF NONREDUCING AND REDUCING END MOIETIES

Terminal group	Native		Permethylated		Deuteromethylated		Peracetylated		Deuteroacetylated	
	Accurate mass	Average mass	Accurate mass	Average mass	Accurate mass	Average mass	Accurate mass	Average mass	Accurate mass	Average mass
Nonreducing end	1.0078	1.0079	15.0235	15.0348	18.0423	18.0533	43.0184	43.0452	46.0372	46.0372
Free reducing end	17.0027	17.0073	31.0184	31.0342	34.0372	34.0527	59.0133	59.0446	62.0321	62.0631
Reduced reducing end	19.0184	19.0232	47.0497	47.0770	53.0874	53.1139	103.0395	103.0978	109.0772	109.1347
Sum of terminal masses (including the proton) for molecules that are not reduced	19.0184	19.0232	47.0497	47.0769	53.0873	53.1139	103.0395	103.0977	109.0772	109.1347
Sum of terminal masses (including the proton) for reduced molecules	21.0340	21.0391	63.0810	63.1197	72.1375	72.1751	147.0657	147.1509	156.1222	156.2063

Composition

Quasimolecular ions derived from both FAB-MS and ES-MS experiments permit compositional assignments, including functional groups and aglycone type, but excluding the differentiation of isomeric sugars.

Assigning Sites of Glycosylation and Mapping Glycan Populations at Individual Glycosylation Sites in Glycoproteins

Fast atom bombardment-mass spectrometry and ES-MS are powerful tools for glycoprotein characterization. More details are given in the section entitled Strategies for Glycoprotein Analysis.

Sequencing

Fast atom bombardment-mass spectrometry is widely used to sequence all classes of carbohydrate-containing polymers, including glycolipids, glycoproteins, glycopeptidolipids, glycosaminoglycans, lipopolysaccharides, and polysaccharides. Similar sequencing strategies are used, irrespective of compound class. Sequence information is derived from fragment ions produced by the intact material and from quasimolecular and fragment ions produced by enzymatic or chemical manipulation of the sample. Exoglycosidase digests (see [17]) can, for example, be conveniently monitored by removing aliquots, drying them, peracetylating or permethylating and analyzing the products by FAB-MS. A variety of chemical hydrolyses are amenable to monitoring by FAB-MS, including aqueous acid hydrolysis, methanolysis, and acetolysis. In most cases, aliquots of the hydrolysis reaction can be loaded directly into the matrix for FAB-MS analysis. A particularly valuable procedure is time-course methanolysis of permethylated samples, which is useful for defining sequences, branching patterns, and locations of substituents.[16,18] During methanolysis the permethylated sample is gradually degraded, giving methyl glycosides at the released reducing termini. Each released glycosidic oxygen and each hydrolyzed substituent becomes a free hydroxyl group. The number of free hydroxyl groups present in each methanolysis fragment is used to define sequence and branching. For example, hydrolytic removal of one or more residues without the generation of a free hydroxyl group indicates that those residues are at the reducing end in the intact oligosaccharide. Similarly, the production of two free hydroxyl groups indicates the simultaneous hydrolysis of two different branches. It is sometimes useful to use deuterated methanol for the methanolysis

---

[18] J. R. Woodward, D. Craik, A. Dell, K.-H. Khoo, S. L. A. Munro, A. E. Clarke, and A. Bacic, *Glycobiology* **2**, 241 (1992).

reaction in order to differentiate newly formed methylglycosides from the methylglycoside of the parent molecule.

Methanolysis Procedure. Dry HCl gas is bubbled into 0.5–1.0 ml of methanol or deuteromethanol until the solution becomes hot to the touch (approximately 1 M in HCl). After cooling, a 10- to 20-μl aliquot of this reagent is added to the permethylated sample, which has been dried down in a small Reacti-vial (Pierce). The sample is incubated at 40° in a heating block and a 1-μl aliquot is removed after a few minutes and analyzed immediately by FAB-MS. Further aliquots are taken at suitable time intervals and the temperature is increased to 60° if hydrolysis is proceeding slowly. The most appropriate time course will vary according to the type of sample and is selected as the reaction proceeds. The time that is allowed to elapse between aliquots is based on the amount of degradation observed at the early time points. Typically, aliquots are removed after 2, 15, and 40 min. Each time an aliquot is removed, especially at the higher temperatures, 1–5 μl of fresh reagent should be added to ensure that the sample does not become dry.

Linkage Assignment

In the majority of studies, linkage assignment is most conveniently carried out by methylation analysis procedures (see [7]). However, FAB-MS provides useful linkage information in the following cases: (1) The A-type ions formed by cleavage at GlcNAc in permethylated oligosaccharides undergo facile loss of the substituent at position 3, allowing type 1 and type 2 chains to be differentiated, and defining fucosylation at this site[17,19]; (2) FAB-MS can be used[20] to analyze the products of periodate degradation after conversion to a suitable derivative, normally permethyl or peracetyl. The FAB data show which residues are susceptible to periodate cleavage, allowing linkage information to be deduced; (3) under certain circumstances fragment ions formed by ring cleavages are helpful in defining linkages. The principle is illustrated in Scheme 2, which shows the two modes of ring cleavage that can occur at the β-mannose of a trimannosyl core. Note that the 3-arm gives a single fragment ion from ring cleavage, whereas the 6-arm gives two fragment ions. These ring cleavages are most commonly observed in negative spectra of sulfated N-glycans,[14] and in positive spectra of acetylated high mannose structures run in an mNBA matrix, where they help to define which antenna(e) carry

---

[19] J. Peter-Katalinic and H. Egge, *Mass Spectrom. Rev.* **6,** 331 (1987).
[20] A.-S. Angel and B. Nilsson, *in* "Methods in Enzymology" (J. McCloskey, ed.), Vol. 193, p. 587. Academic Press, San Diego, 1990.

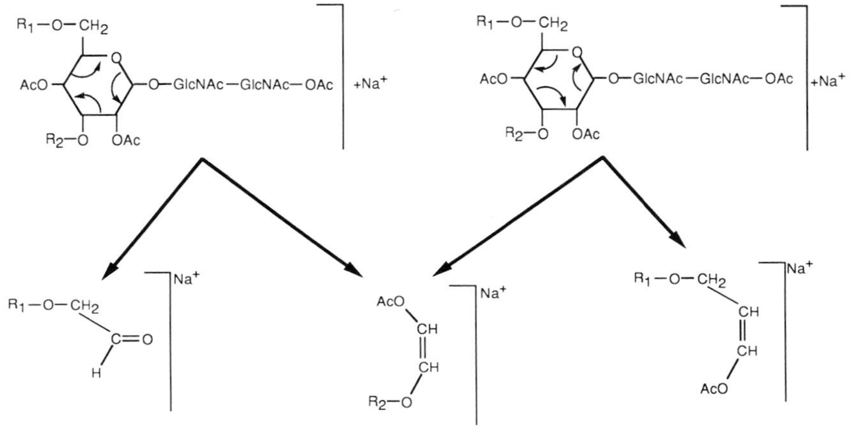

R_1-O-CH_2

AcO —O—GlcNAc—GlcNAc—OAc $+Na^+$

R_2-O OAc

R_1-O-CH_2

AcO —O—GlcNAc—GlcNAc—OAc $+Na^+$

R_2-O OAc

R_1-O-CH_2

$C=O$

H

Na^+

AcO

CH

CH

R_2-O

Na^+

R_1-O-CH_2

CH

CH

AcO

Na^+

SCHEME II

the sulfate(s) and branching patterns of the mannosyl residues,[21] respectively.

Assigning Anomeric Configurations

In many structural studies anomeric configurations are conveniently assigned by digesting with specific glycosidases or by nuclear magnetic resonance (NMR) methods. However, if the appropriate enzymes are unavailable and there is insufficient material for NMR, alternative approaches must be used. The chromium trioxide oxidation strategy of Angyal and James[22] offers one such alternative and FAB-MS allows this procedure to be used at high sensitivity.[23] Linkage assignments using this procedure exploit the difference in rates of oxidation of α- and β-linked residues in acetylated oligosaccharides. β linkages are rapidly oxidized and the progress of oxidation can be determined by monitoring the molecular weight of the products. An increment of 14 mass units occurs each time a glycosidic bond is oxidized. The site(s) of oxidation is established from the fragmentation pattern. Other reactions can also occur, for example, oxidation of methyl groups and of C-6 of pyranoses, therefore data need careful interpretation. This methodology has been used successfully to characterize naturally methylated O-glycans in nematode glycoproteins.[2]

[21] A. J. Reason, A. Dell, P. A. Romero, and A. Herscovics, *Glycobiology* **1**, 387 (1991).
[22] S. J. Angyal and K. James, *Aust. J. Chem.* **23**, 1209 (1970).
[23] K.-H. Khoo and A. Dell, *Glycobiology* **1**, 83 (1990).

Procedure for Chromium Trioxide Oxidation

1. Dissolve the perdeuteroacetylated sample (5–10 μg) in 100 μl of glacial acetic acid in a Teflon-lined, screw-capped glass tube. Peracetylated samples can also be used but ambiguities may arise with the interpretation of the FAB-mass spectra of the oxidized sample.

2. Add approximately 10 mg of chromium trioxide with a microspatula.

3. Shake the reaction mixture vigorously at 50° for 1–2 hr.

4. Quench the reaction with 2 ml of water, which should then yield a clear, orange-yellow solution.

5. Add 2 ml of chloroform, mix, and allow to settle into two layers.

6. Discard the upper aqueous layer.

7. Wash the chloroform layer several times with water, until the upper aqueous layer is colorless.

8. Dry the chloroform layer under a stream of nitrogen.

9. Redissolve the sample in methanol for FAB-MS analysis.

Defining O-GlcNAc Attachment Sites in Nuclear and Cytoplasmic Glycoproteins

Despite the difficulties associated with the analysis of O-GlcNAc in nuclear and cytoplasmic glycoproteins, which result from limited sample availability, substoichiometric glycosylation, and intractable serine/threonine-rich glycopeptides, MS is beginning to have a significant impact in this field.[24–26] The intact glycoprotein is first examined by ES-MS. If successful, this experiment will give an indication of the degree of glycosylation, but not of course the substitution position(s). The glycoprotein is then subdigested, usually with two or more enzymes to ensure that putative glycopeptides are relatively small. When the protein sequence is known, the choice of enzyme is not difficult because potential regions of glycosylation, which are known to contain proline and to be serine/threonine rich, are easy to predict. If the protein sequence is not known, sequential digestion with cyanogen bromide, trypsin, and proline-specific enzyme will give suitably small glycopeptides in most cases. The mixture of peptides and glycopeptides is then separated by high-performance liquid chromatography (HPLC) and each fraction is screened by FAB-MS. Putative glycopeptides and their nonglycosylated counterparts, which often

[24] A. J. Reason, I. P. Blench, R. S. Haltiwanger, G. W. Hart, H. R. Morris, M. Panico, and A. Dell, *Glycobiology* **1**, 585 (1991).

[25] E. P. Roquemore, A. Dell, H. R. Morris, M. Panico, A. J. Reason, L.-A. Savoy, G. J. Wistow, J. S. Zigler, B. J. Earles, and G. W. Hart, *J. Biol. Chem.* **267**, 555 (1992).

[26] A. J. Reason, H. R. Morris, M. Panico, R. H. Treisman, R. Marais, R. S. Haltiwanger, G. W. Hart, and A. Dell, *J. Biol. Chem.* **267**, 16911 (1992).

coelute, are identified by the presence of pairs of quasimolecular ions differing in mass by 203 mass units, the residue mass of GlcNAc. Assignments are confirmed by mass shifts after propionylation,[24] a derivatization step that results in enhanced abundance of glycopeptide-derived ions. Propionylation is effected by adding 10 μl of reagent, prepared by mixing 2 vol of trifluoroacetic anhydride with 1 vol of propionic acid, to the dried sample. Aliquots of the reaction mixture are loaded directly into the matrix for FAB analysis. Reaction times of about 1 min give the best data.

Examples of Structural Protocols

Strategies for Glycoprotein Analysis

The following general strategies are used to maximize the amount of structural detail that can be defined by MS procedures.[27-29] If partial structural information is all that is required, many of the steps can be omitted, for example, (1) if the glycoprotein is being screened for the existence of specific nonreducing epitopes it may suffice to permethylate an unpurified pronase digest of the glycoprotein,[30] or (2) if no N-glycans are present the intact glycoprotein can be reductively eliminated directly.

Studying Glycans without Site Analysis. The glycoprotein is first screened by ES-MS. This experiment may give useful information that will help to establish the extent of glycosylation. However, many heterogeneous glycoproteins are intractable to ES-MS or give unresolved spectra that are of little use. The glycoprotein is then reduced and carboxymethylated to protect S–S bridges and digested sequentially with one or more proteolytic enzymes. Fast atom bombardment-mass spectrometry and ES-MS on a portion of the product mixture will give an indication of the efficiency of the digestion and may afford valuable molecular weight data on peptides/glycopeptides of interest. Digestion with peptide N-glycosidase yields a mixture of peptides, O-linked glycopeptides (if O-glycans

[27] H. R. Morris, A. Dell, M. Panico, R. McDowell, and A. Chatterjee, *in* "Methods in Protein Sequencing" (Wittmann-Liebald, ed.), p. 206. Springer-Verlag, Heidelberg, 1989.

[28] H. R. Morris, A. Dell, M. Panico, J. E. Thomas-Oates, M. E. Rogers, R. McDowell, and A. Chatterjee, *in* "Mass Spectrometry of Biological Materials" (C. N. McEwen and B. S. Larsen, eds.), p. 137. Dekker, New York, 1990.

[29] A. L. Chan, H. R. Morris, M. Panico, A. T. Etienne, M. E. Rogers, P. Gaffney, L. Creighton-Kempsford, and A. Dell, *Glycobiology* **1**, 173 (1991).

[30] A. Dell, R. A. McDowell, and M. E. Rogers, *in* "Post-Translational Modifications of Proteins" (J. J. Harding and M. C. C. Crabbé, eds.), p. 185. CRC Press, Boca Raton, FL, 1991.

TABLE III
A-TYPE IONS CORRESPONDING TO
COMMON NONREDUCING STRUCTURES
IN PERMETHYLATED N-Glycans

Accurate mass	A-type ion
260.1498	HexNAc$^+$
376.1972	NeuAc$^+$
464.2496	HexHexNAc$^+$
505.2761	HexNAc$_2^+$
638.3388	FucHexHexNAc$^+$
668.3494	Hex$_2$HexNAc$^+$
679.3653	FucHexNAc$_2^+$
709.3759	HexNAcHexHexNAc$^+$
812.4280	Fuc$_2$HexHexNAc$^+$
825.4233	NeuAcHexHexNAc$^+$
866.4498	NeuAcHexNAc$_2^+$
913.4757	Hex$_2$HexNAc$_2^+$
999.5125	NeuAcFucHexHexNAc$^+$

are present), and N-glycans. The mixture is screened again by ES-MS and FAB-MS and then fractionated, using a Sep-Pak cartridge, into a void volume of unbound material (the N-glycans) and an included volume containing the peptides and glycopeptides. A suitable amount (this will depend on the quantity of starting material; 5–10 μg is desirable) of the void volume fraction is permethylated and a portion is analyzed by FAB-MS. A further portion is subjected to linkage analysis by GC-MS. Methanolysis/ FAB-MS can be usefully performed on any remaining permethylated material. The rest of the void volume fraction can be purified, if desired, by chromatographic procedures described elsewhere in this volume [11–16]. The purified oligosaccharides can then be analyzed as their permethyl or peracetyl derivatives before and after chemical or enzymatic digestions.

The included fraction from the Sep-Pak purification is subjected to β elimination to release the O-glycans, which are purified on a short Dowex column and peracetylated or permethylated prior to analysis by the procedures given above for the N-glycans.

The positive FAB spectra of the permethyl derivatives are characterized by abundant quasimolecular ions and A-type fragment ions, the latter resulting from cleavage at the amino sugar residues. The most commonly encountered A-type ions of masses below m/z 1000 are given in Table III. These define the nonreducing epitopes in the glycans. The compositions of higher mass A-type ions are readily calculated by adding residue incre-

TABLE IV

MASSES OF QUASIMOLECULAR IONS OF PERMETHYLATED N–GLYCANS COMMONLY FOUND IN RECOMBINANT GLYCOPROEINS

Mass		Molecular ion	
Accurate	Average	[M + H]$^+$	Description
2048.0539	2049.2945	Hex$_5$HexNAc$_4$	Unmodified biantennary
2409.2276	2410.6868	NeuAcHex$_5$HexNAc$_4$	Monosialylated biantennary
2583.3168	2584.8836	NeuAcFucHex$_5$HexNAc$_4$	Monosialylated fucosylated biantennary
2770.4013	2772.0791	NeuAc$_2$Hex$_5$HexNAc$_4$	Disialylated biantennary
2944.4905	2946.2759	NeuAc$_2$FucHex$_5$HexNAc$_4$	Disialylated fucosylated biantennary
2497.2800	2498.7931	Hex$_6$HexNAc$_5$	Ummodified triantennary
2858.4537	2860.1854	NeuAcHex$_6$HexNAc$_5$	Monosialylated triantennary
3032.5429	3034.3822	NeuAcFucHex$_6$HexNAc$_5$	Monosialylated fucosylated triantennary
3219.6274	3221.5777	NeuAc$_2$Hex$_6$HexNAc$_5$	Disialylated triantennary
3393.7166	3395.7745	NeuAc$_2$FucHex$_6$HexNAc$_5$	Disialylated fucosylated triantennary
3580.8011	3582.9700	NeuAc$_3$Hex$_6$HexNAc$_5$	Trisialylated triantennary
3754.8903	3757.1668	NeuAc$_3$FucHex$_6$HexNAc$_5$	Trisialylated fucosylated triantennary
2946.5061	2948.2917	Hex$_7$HexNAc$_6$	Ummodified tetraantennary
3307.6798	3309.6840	NeuAcHex$_7$HexNAc$_6$	Monosialylated tetraantennary
3481.7690	3483.8808	NeuAcFucHex$_7$HexNAc$_6$	Monosialylated fucosylated tetraantennary
3668.8535	3671.0763	NeuAc$_2$Hex$_7$HexNAc$_6$	Disialylated tetraantennary
3842.9427	3845.2731	NeuAc$_2$FucHex$_7$HexNAc$_6$	Disialylated fucosylated tetraantennary
4030.0265	4032.4686	NeuAc$_3$Hex$_7$HexNAc$_6$	Trisialylated tetraantennary
4204.1157	4206.6654	NeuAc$_3$FucHex$_7$HexNAc$_6$	Trisialylated fucosylated tetraantennary
4391.2002	4393.8609	NeuAc$_4$Hex$_7$HexNAc$_6$	Tetrasialylated tetraantennary
4565.2894	4568.0577	NeuAc$_4$FucHex$_7$HexNAc$_6$	Tetrasialylated fucosylated tetraantennary

ments (Table I) to the Table III values. Many recombinant glycoproteins are characterized by the presence of sialylated bi-, tri-, and tetraantennary complex structures, and, for convenience, the masses of structures frequently observed are given in Table IV. Other compositions can be calculated by using Tables I and II.

If sulfated components are present, spectra are also acquired in the negative ion mode. Sulfated components give abundant quasimolecular ions and characteristic clusters of sulfated fragment ions. These clusters include ring cleavage ions, which are sometimes helpful in defining the linkage sites of sulfated antennae (see section on Linkage Assignment).

The FAB-MS and GC-MS experiments give information on (1) the degree of heterogeneity and kinds of glycosylation, for example, high mannose, hybrid, and complex, and the composition in terms of hexoses, N-acetylaminohexoses, deoxyhexoses, sialic acids, and so on of each glycan chain, (2) sequences and branching patterns of the glycans, (3) the number and lengths of antennae in complex N-glycans and the patterns of substitution with fucose, sialic acids, or other capping groups such as sulfate, (4) the sequences of all nonreducing structures in complex-type N-glycans, irrespective of their overall size or the amount of heterogeneity, (5) the presence of lactosaminoglycan chains in N- or O-linked glycans and the number of repeats, (6) the fidelity of the carbohydrate structures of recombinant glycoproteins as compared with naturally occurring substances, and (7) abnormalities in naturally occurring structures resulting from defects in biosynthesis.

Additional points: Fast atom bombardment-mass spectrometry of the permethyl derivative of high-mannose structures gives compositional information. If branching patterns are being investigated these glycans should be analyzed in an *m*NBA matrix after peracetylation.[21] Alternatively, after tagging with ABEE the pattern of β-cleavage ions can be used to deduce the arrangement of the mannose residues.[31]

When permethylated glycans that carry sulfate substituents are analyzed in the positive ion mode, free hydroxyl groups are observed in place of the sulfates. These are the result of facile β cleavage of the sulfate ester linkage.

Analyzing Glycans at Individual Sites in Glycoproteins. The previous strategy is readily modified if the glycans at individual sites are being studied. Prior to the peptide N-glycosidase digestion the mixture of peptides and glycopeptides is fractionated by the chromatographic procedures (normally reversed-phase HPLC) described elsewhere in this volume and the individual glycopeptides are analyzed as described for the glycopeptide mixtures. Electrospray-mass spectroscopy can be particularly useful here in locating glycopeptides that would normally carry between, say, 2 and 10 charges, by spotting mass differences in satellite signals of different charge states corresponding to related glycoforms.[32] For example, a glycopeptide containing a complex biantennary structure, a proportion of which is partially lacking one sialic acid, will be spotted by the presence of satellite signals of decreasing mass difference, that is, 97, 72.765, 58.2 for charges of 3, 4, and 5 respectively. Other related glycoforms may be spotted in a similar manner by dividing the presumed charge state into the masses of saccharide units.

---

[31] L. M. Hernandez, L. Ballou, E. Alvarado, B. L. Gillece-Castro, A. L. Burlingame, and C. E. Ballou, *J. Biol. Chem.* **264,** 11849 (1989).

[32] H. R. Morris, L.-A. Savoy, and M. Panico, unpublished data.

Strategies for Characterizing Complex Carbohydrates

The following general strategy is suited to the analysis of oligosaccharides, glycolipids, peptidolipids, and related substances. Because a variety of methods are available for degrading carbohydrate biopolymers, this FAB-MS/GC-MS methodology is also applicable to polysaccharides, lipopolysaccharides, and so on. If sample quantities are limiting it may not be practicable to include all steps. In such cases steps 4 and 5 are given priority.

1. A portion of the sample is analyzed by both positive and negative FAB-MS. Probable data: Compositional information and, depending on the sample, some sequence.

2. A portion of the sample is subjected to sugar analysis, using GC and/or GC-MS.

3. A portion of the sample is subjected to a time-course methanolysis monitored by FAB-MS. Probable data: Types of functional groups and the residues to which they are attached; some sequence data.

4. A portion of the sample is perdeuteroacetylated and analyzed by both positive and negative FAB-MS. Probable data: Compositional information, sequence data, locations of functional groups such as acetyl.

5. A portion of the sample is permethylated and analyzed by positive FAB-MS (unless sulfate is expected when the negative mode is also used). Probable data: Compositional information, sequence data, linkage information in certain cases, for example, the attachment site of a glycosyl chain to a HexNAc residue (see above).

6. A portion of the permethylated sample is hydrolyzed, reduced, acetylated, and analyzed by GC-MS to determine linkages.

7. A portion of the permethylated sample is subjected to time-course methanolysis monitored by FAB-MS. Probable data: Sequence data, branching patterns.

8. Additional experiments such as periodate oxidation and chromium trioxide oxidation are helpful in specific cases (see above).

Strategy for Analyzing Glycosaminoglycans

The following FAB-MS experiments define the number of repeats in each oligosaccharide, the number of sulfate groups present, and the residues to which they are attached.[16,33]

---

[33] A. Dell, M. E. Rogers, J. E. Thomas-Oates, T. N. Huckerby, P. N. Sanderson, and I. A. Nieduszynski, *Carbohydr. Res.* **179,** 7 (1988).

1. The glycosaminoglycan is converted to oligosaccharides by enzymes or by deaminative cleavage with nitrous acid.

2. A portion of each purified oligosaccharide is analyzed by FAB-MS in the negative ion mode. If successful (impurities may prevent the formation of sample-derived ions), this experiment gives quasimolecular ions together with fragment ions resulting from loss of sulfite. When relatively large quantities of highly purified material are analyzed, sequence ions may also be present. In this experiment it is not possible to distinguish between fragment ions formed by loss of sulfite and quasimolecular ions of components with fewer sulfate moieties.

3. A portion of each sample is subjected to short Hakomori permethylation and a fraction is analyzed by FAB-MS in the negative ion mode. In addition to quasimolecular ions for the intact oligosaccharide, and fragment ions formed from loss of sulfite, the FAB spectra contain quasimolecular ions of truncated molecules resulting from base-catalyzed lyase-like cleavage at the uronic acids and some sequence ions. Both of the latter help to define the residues that are sulfated.

4. The remainder of the permethylated sample is subjected to methanolysis, monitored by FAB-MS in the positive ion mode. The number of free hydroxyl groups in each methanolyzed fragment is indicative of the pattern of sulfate substitution.

5. A portion of each sample is deuteroacetylated by basic catalysis and analyzed by negative FAB-MS. This gives compositional and sequence information. The deuterated derivative is chosen in order to minimize ambiguities in mass intervals that may arise from the peracetyl derivative.

[9] ¹H Nuclear Magnetic Resonance Spectroscopy of Carbohydrate Chains of Glycoproteins

By HERMAN VAN HALBEEK

Introduction

Nuclear magnetic resonance (NMR) spectroscopy provides a powerful, nondestructive means to characterize glycoprotein carbohydrates structurally, and has become an integral part of the current methodology of protein glycosylation site mapping.[1,2] This chapter discusses ¹H NMR

[1] M. W. Spellman, *Anal. Chem.* **62**, 1714 (1990).
[2] H. van Halbeek, *Methods Mol. Biol.* **17**, 115 (1993).

spectroscopy as a method for the characterization of the primary structure of N-type and O-type oligosaccharide chains of glycoproteins. We emphasize the practical aspects of the method, including sample preparation, recording a spectrum (data acquisition), and data processing. The usefulness of the structural reporter group concept for interpretation of the ^{1}H NMR spectrum of a carbohydrate in terms of its primary structure is outlined. Ongoing efforts aimed at automation of NMR data processing and spectral interpretation are mentioned. Throughout this chapter, we illustrate the application of ^{1}H NMR spectroscopy for the structural elucidation of the carbohydrate chains of human neutrophil elastase[3,4] (HNE)[N] and of recombinant human erythropoietin[5,6] (rEPO)[N,O] expressed in Chinese hamster ovary (CHO) cells.

The fundamental message of this chapter is that the ^{1}H NMR spectrum of an oligosaccharide or glycopeptide represents an "identity card" of the carbohydrate. An ^{1}H NMR study may suffice for primary structure determination if the oligosaccharide itself, or a compound having a closely related structure, has been characterized previously. Fingerprinting a carbohydrate through ^{1}H NMR spectroscopy, which operates at radio frequencies (rf) of 500 or 600 MHz, requires at least 10 nmol of pure oligosaccharide or glycopeptide. When applying artificial intelligence pattern recognition techniques for spectral interpretation, the sensitivity of the method may (at least for carbohydrates previously characterized) be increased by two or three orders of magnitude.

Principle of Method

This section provides a brief survey of the principal parameters that determine the appearance of an NMR spectrum and contain information on the structure of the carbohydrate under investigation. We start with the description of some elementary features of an NMR experiment (see Fig. 1).

Basic Features

Nuclear magnetic resonance spectroscopy studies the interaction of magnetically active, atomic nuclei with radio-frequency electromagnetic radiation in a strong, external magnetic field. The magnetically active nuclei ("spins") in carbohydrates that are of practical interest to NMR

[3] W. Bode, R. Huber, E. Meyer, J. Travis, and S. Neumann, *EMBO J.* **51,** 2453 (1986).
[4] W. Watorek, H. van Halbeek, and J. Travis, *Biol. Chem. Hoppe Seyler* **374,** 385 (1993).
[5] E. Watson, A. Bhide, and H. van Halbeek, submitted for publication (1993).
[6] M. Takeuchi and A. Kobata, *Glycobiology* **1,** 337 (1991).

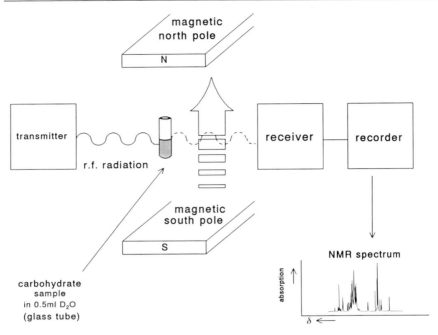

FIG. 1. Diagram of an NMR experiment. The magnetic field $\mathbf{B}_0$ is in the vertical direction (along the z axis); it is actually generated by the electric current in a superconducting solenoid.

spectroscopy include the ^{1}H and ^{13}C nuclei (others, such as ^{17}O, ^{14}N, and ^{15}N nuclei, are of little practical importance; in phosphorylated carbohydrates, ^{31}P nuclei are valuable additional probes to NMR studies). Many nuclei abundant in carbohydrates (such as ^{12}C and ^{16}O) are nonmagnetic and cannot be studied by NMR spectroscopy. Both ^{1}H and ^{13}C are spin-$\frac{1}{2}$ nuclei, that is, the orientation of the net magnetic dipole vector of the nucleus in an external magnetic field is either parallel or antiparallel to the direction of the external ($\mathbf{B}_0$) field. The net macroscopic magnetization of a sample in equilibrium is parallel to the $\mathbf{B}_0$ field axis, because parallel orientation of an individual spin is energetically favored over antiparallel orientation; therefore, when a large number of ^{1}H or ^{13}C nuclei are placed in a strong external magnetic field, most spins will orient parallel to the external field in the thermodynamic equilibrium situation (see Fig. 2a). Further discussion is limited to ^{1}H nuclei ("protons"), mainly because the combination of their natural abundance (99.98 vs 1.11% for ^{13}C) and their inherently strong magnetic dipole moment (four times stronger than that of ^{13}C) makes the NMR method far more sensitive for the observation

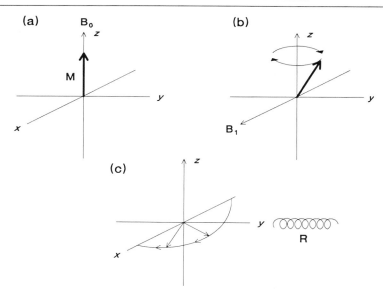

FIG. 2. (a) Thermodynamic equilibrium of the macroscopic magnetization (**M**) of a sample in a magnetic field (**B**$_0$) in the z direction. (b) The effect of a pulse of radio-frequency radiation on **M**. The pulse **B**$_1$ tips the magnetization in the yz plane. (c) After the pulse, the magnetization returns to its initial value. The components of **M** in the xy plane "fan out," each with their typical precession frequency. The resulting induced magnetization is measured in a receiver coil (R).

of ^{1}H nuclei than for any other type of nuclei present in biomolecules in natural abundance.

It is possible to cause nuclear spins to resonate, that is, to change their orientation from parallel to antiparallel (and vice versa) with respect to the external magnetic field. In the resulting nonequilibrium situation, the macroscopic magnetization of the sample no longer coincides with the **B**$_0$ axis (Fig. 2b). To accomplish "resonance" (spin-flip), energy is needed. Energy is put into the assembly of spins through a short pulse of radio-frequency electromagnetic radiation, with frequencies[7] typically on the order of 200 to 600 MHz. The precise frequency required for resonance is directly proportional to the strength of the **B**$_0$ field; magnetic fields used for NMR spectroscopy are typically between 5 and 15 T. When

[7] According to Planck's equation, $E = h\nu$ (where E stands for energy, ν for frequency of the radiation). Radio frequencies (rf) are on the order of 200 to 600 MHz [1 MHz (megahertz) $= 10^6$ Hz (hertz); 1 Hz $= 1$ cps (cycle per second)], corresponding to wavelengths between 1.5 m and 50 cm.

the sample is irradiated with a pulse of radio-frequency radiation, all possible NMR resonances are simultaneously excited. After the pulse, the spins try to restore ("relax" to) their equilibrium orientation, such that the macroscopic magnetization of the sample is parallel to B_0 again. Each proton spin does so by a spiral movement, called precession, around the B_0 field axis (Fig. 2c). Each proton has its own characteristic precession frequency (frequencies). The frequency of precession is determined by (1) the (relative) electron density of the proton in the molecule, and (2) interaction ("coupling") with other protons in the same molecule.

An audio receiver (Figs. 1 and 2c) registers the superposition of all the precession frequencies of all the protons in the complex carbohydrate molecule. The "interference pattern" of induced magnetization obtained directly in this way is referred to as free induction decay (FID). This NMR signal in the time domain is analyzed by a computer, typically by Fourier transformation (FT), to obtain the spectrum in the frequency domain, presenting to us a pattern of absorption of radiation as a function of frequency, and thus an "electron density map" of the protons in the carbohydrate molecule. Higher resolution and better sensitivity [signal-to-noise (S/N) ratio] are obtained in an NMR spectrum when recorded in a stronger external magnetic field, corresponding to a larger energy gap between parallel and antiparallel orientation of a spin in the B_0 field, and thus to a higher frequency necessary for spin-flip. Further information on the basic principles of NMR may be found in general textbooks on NMR spectroscopy,[8-10] as well as in most general physical chemistry textbooks.[11]

^1H Nuclear Magnetic Resonance of Carbohydrates

Carbohydrate molecules contain quite a few protons (^1H nuclei). Each constituting hexosyl residue (such as the galactosyl residue in Fig. 3) has 10 or 11 protons, 7 of which are involved in CH linkages, the others in OH groups. Each proton in a glycosyl residue has its own characteristic environment, and thus, electron density. For primary structural characterization of a carbohydrate by ^1H NMR spectroscopy, we usually are interested only in those protons that are directly linked to carbon atoms. Those

[8] A. E. Derome, "Modern NMR Techniques for Chemistry Reserch." Pergamon, Oxford, 1987.

[9] J. K. M. Sanders and B. K. Hunter, "Modern NMR Spectroscopy: A Guide for Chemists." Oxford Univ. Press, Oxford, 1987.

[10] H. Friebolin, "Basic One- and Two-Dimensional NMR Spectroscopy." VCH, New York, 1991.

[11] R. A. Alberty, "Physical Chemistry." Wiley, New York, 1987.

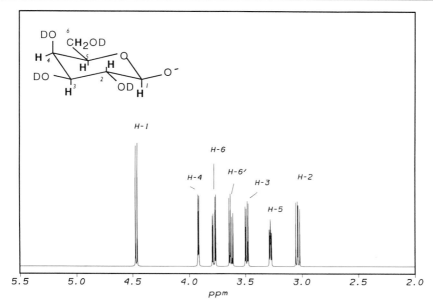

FIG. 3. Computer-simulated ¹H NMR spectrum of a β-linked galactosyl residue in D₂O.

in OH groups are "exchangeable" protons; they do not have a well-defined environment because they are in fast exchange with solvent protons, resulting in broad lines in the NMR spectrum.[12,13] Therefore, the carbohydrate sample to be examined by NMR is dissolved in D₂O (see Fig. 1) (no chemical derivatization is necessary). The OH groups are converted into OD (Fig. 3), thus effectively removing any OH signals from the spectrum. Working in D₂O also eliminates the problems associated with observing signals of carbohydrate protons, present in submillimolar amounts, in the presence of a vast excess (110 M) of H₂O protons. In practice, a residual solvent (HOD) signal is detected in the ¹H NMR spectrum of an oligosaccharide in D₂O because the solvent is not 100% deuterated.

The result of an ¹H NMR study most familiar to the glycobiologist is a spectrum of the carbohydrate, an assembly of peaks (signals) that can be thought of as a distribution curve of protons as a function of their structural environment (electron density) in the molecule. A typical example of an ¹H NMR spectrum of a glycoprotein-derived N-type oligosaccha-

[12] L. Poppe and H. van Halbeek, *J. Am. Chem. Soc.* **113**, 363 (1991).
[13] L. Poppe, R. Stuike-Prill, B. Meyer, and H. van Halbeek, *J. Biomol. NMR* **2**, 109 (1992).

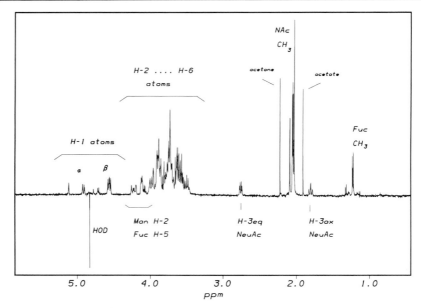

FIG. 4. A 500-MHz ^{1}H NMR spectrum of a diantennary N-type oligosaccharide alditol obtained from recombinant human erythropoietin expressed in CHO cells. The structural reporter group regions are expanded in Fig. 7a.

ride alditol recorded in D_2O solution at a frequency of 500 MHz is shown in Fig. 4.

Nuclear Magnetic Resonance Parameters

Each signal in the spectrum is characterized by a set of parameters, and each parameter holds specific information about the structural environment of the corresponding protons. Thus, the total spectrum is characteristic for the structure of the carbohydrate. The NMR parameters most relevant to the primary structural analysis of carbohydrates are listed in Table I.

The chemical shift (δ) defines the location of an NMR line (signal) along the rf axis (horizontal axis). It is measured relative to a reference compound [for underivatized carbohydrates in aqueous solution, usually a water-soluble derivative of tetramethylsilane, e.g., 4,4-dimethyl-4-sila-pentane 1-sulfonate (DSS)] (Fig. 5). In frequency units (hertz) the chemical shift is proportional to the strength of the applied static magnetic field ($\mathbf{B}_0$), and chemical shifts are quoted in parts per million (ppm). The chemi-

TABLE I
NUCLEAR MAGNETIC RESONANCE PARAMETERS

NMR parameter	Symbol	Unit	Type of structural information
Chemical shift	δ	ppm[a]	Relative electron density
Coupling constant	J	Hz	Number of neighboring protons; dihedral angle (HCC′H′)
Intensity	—	—	Relative number of protons with given δ
Linewidth	$\Delta\nu$	Hz	Mobility of proton; flexibility of the structure

[a] Parts per million (of the spectrometer frequency ν).

cal shift δ is related primarily to the chemical structure of the molecule studied.

Spin–spin coupling constants (J) characterize scalar (through bond) interactions between nuclei linked via a small number of covalent bonds in a chemical structure. The magnitude of J is field independent and customarily is quoted in hertz. Spin couplings are manifested in the spectrum by (partially) resolved fine structure (''splitting''; see Fig. 5) of the individual resonance lines. For example, the signal of an anomeric proton of a glycosyl residue has a doublet fine structure (see also Fig. 3).

The NMR line intensity (actually, the area under the signal; see Fig. 5) reflects, in a normal NMR spectrum, the relative number of nuclei manifested by this resonance line.

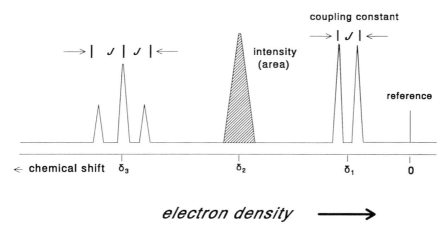

FIG. 5. NMR parameters relevant to the primary structural analysis of carbohydrates.

Other important NMR parameters are as follows.

1. The nuclear Overhauser effect (NOE) is due to dipolar (through space) interactions between different nuclei and correlated with the inverse sixth power of the internuclear distance. The NOE does not affect the appearance of a normal spectrum. Special NMR experiments need to be performed to measure this effect. The NOE is the most valuable parameter in conformational analysis of carbohydrates in solution by NMR (see section on Solution Conformation Analysis by Nuclear Magnetic Resonance).

2. The longitudinal relaxation time (T_1, usually specified in seconds) describes the rate at which the magnetization $\mathbf{M}$ returns to the thermodynamic equilibrium orientation along $\mathbf{B_0}$ after an rf pulse (Fig. 2c). T_1 is correlated with the overall rotational tumbling of the carbohydrate molecule in solution and may be further affected by intramolecular flexibility in flexible structures.

3. The transverse relaxation time (T_2, also in seconds) describes the decay rate of the effective magnetization observed in the receiver after an rf pulse (see Fig. 2). T_2 is correlated with dynamic properties in the carbohydrate molecule under study. For example, T_2 decreases monotonically with increasing molecular size. This presents a limiting factor for high-resolution NMR on carbohydrate polymers. The linewidth is inversely proportional to T_2.

Interpretation of an ^{1}H NMR spectrum such as depicted in Fig. 4 requires the cataloging of signals by position and pattern; the positions (δ values) are usually compiled in tables. These values are then compared with literature data on similar oligosaccharides and/or glycopeptides. The positions (δ) and patterns (J, linewidth, intensity) of the signals in the ^{1}H NMR spectrum of an oligosaccharide are characteristic for the primary structure of the compound under investigation. The crux of the structural reporter group concept is that it is sufficient to inspect only certain regions in the spectrum (see Fig. 4) to ascertain the primary structure of the carbohydrate from its NMR spectrum. Bearing in mind that each proton in the carbohydrate molecule has its own signal in the spectrum, the ^{1}H NMR spectrum of an oligosaccharide is essentially a superposition of the spectra of the individual glycosyl residues (such as that of the galactosyl residue in Fig. 3), slightly modified because of their linkages to each other. A few empirical and semiempirical "rules" for interpretation of ^{1}H NMR spectra of carbohydrates are as follows.

1. The nonanomeric ring-skeleton protons in a glycosyl residue have similar electron densities, and thus chemical shifts; their signals all appear

between 3 and 4 ppm (Figs. 3 and 4). In the spectrum of an oligosaccharide (already at the size of of a trisaccharide) the signals of the skeleton protons become partly overlapping; they add up to a broad envelope of signals in the middle of the spectrum, with little fine structure and therefore little readily accessible structural information. However, anomeric protons (H-1) show signals more to the left (at higher chemical shift), because their electron density is typically lower than that of the other skeleton protons. Anomeric proton signals (doublets) typically are found between δ 4.3 and 5.6 ppm.

2. A proton in equatorial position at a certain carbon atom in a pyranose ring has a chemical shift ~0.5 ppm more to the left than the corresponding axial proton attached to the same carbon. Therefore, for residues involved in α-glycosidic linkages (H-1 equatorial) δ H-1 is ~4.9–5.6 ppm, whereas for β-anomeric linkages (H-1 axial) δ H-1 is ~4.3–4.7 ppm. For underivatized carbohydrates in D_2O at ambient temperature, these positions correspond to the near left and the near right of the HOD signal at δ ~4.80, respectively. Other illustrations of this rule are: Man H-2 (equatorial) signals are typically found ~0.5 ppm more to the left than Glc or Gal H-2 (axial) signals (see Fig. 4). Also, Gal H-4 protons resonate ~0.5 ppm more to the left than Glc(NAc) H-4 protons (cf. Fig. 3).

3. Signals in NMR spectra show "splitting" due to the interaction ("spin–spin coupling") of protons with other protons in their environment. The number of lines in a signal is 2^n, where n is the number of protons attached to adjacent carbons. Each anomeric proton has only one neighboring proton (namely, H-2), and thus resonates as a doublet. The splitting between the various lines in a multiplet (the "coupling constant"; see Fig. 5) reflects (among other things) the relative orientation of neighboring protons. The coupling constant $^3J_{12}$ for two neighboring protons (H-1 and H-2, separated from each other by three covalent bonds) that are both in axial orientation is 7–8 Hz (i.e., relatively large), whereas coupling constants $^3J_{12}$ between two neighboring protons, one of which is axial and the other equatorial, or between two equatorial protons, are 3–4 Hz (relatively small). Thus, the anomeric configuration of a glycosyl residue can be inferred both from the chemical shift and the coupling constant of its anomeric proton signal in the ^{1}H NMR spectrum.

4. When a glycosyl residue is linked at a certain carbon position (i.e., has either another glycosyl residue or a noncarbohydrate group attached at that position), the signal of the proton attached to this carbon tends to shift to the left in the spectrum by ~0.5 ppm. Therefore, signals of nonanomeric protons that are attached to linkage sites usually appear outside the envelope resonance of skeleton protons (Fig. 4). They become individually observable "structural reporter" signals. Thus, linkage posi-

tions may be deduced from an ^{1}H NMR spectrum, if the ring skeleton proton signals that emerge in the region between δ 4.0 and 4.35 can be assigned to certain protons.

The detailed interpretation of the ^{1}H NMR spectrum depicted in Fig. 4 in terms of the oligosaccharide structure is discussed below (see Recording an ^{1}H Nuclear Magnetic Resonance Spectrum).

Procedure

Materials

Glycoproteins generally are too large to be studied as intact macromolecules by ^{1}H NMR spectroscopy for the structure of their carbohydrates. The spectra of intact glycoproteins show mostly fairly wide lines and lack the resolution needed for detailed structural analysis. The microheterogeneity of the sample and the severe overlap of resonances, including the overlap of carbohydrate and protein signals, render spectral analysis of an intact glycoprotein as yet impossible. Degradation to partial structures is mandatory when detailed primary structural information on oligosaccharides is desired. Partial structures suitable for NMR spectroscopy are (1) glycopeptides, (2) oligosaccharides, and (3) reduced oligosaccharides.

Glycopeptides are the partial structures of choice to preserve information on the glycosylation site in the protein. Glycopeptides can be generated by specific (e.g., trypsin, chymotrypsin) or nonspecific (pronase, pepsin) proteolytic digestion of the protein portion of the glycoprotein. Although larger peptides can be handled, NMR spectroscopic analysis is facilitated when the glycopeptides (once purified from the peptides) have a peptide chain no longer than ~10 amino acids. Peptide heterogeneity usually is not an insurmountable problem because only the positions of the NMR signals of the glycosyl residue directly attached to the peptide are affected. However, for glycopeptides to be analyzed successfully, they should contain only one glycosylation site; the heterogeneity in the carbohydrate structure at that site then can be characterized adequately by ^{1}H NMR spectroscopic analysis (see below, Fig. 8).[4]

Oligosaccharides of the N type are generated[14] by enzymatic release from the glycoprotein (sometimes after denaturation) or from (tryptic) glycopeptides, by N-glycanase (also known as PNGase F) or by an endoglycosidase such as Endo H or Endo F (see [4] in this volume).

[14] A. L. Tarentino and T. H. Plummer, Jr., this volume [4].

N-Glycanase cleaves fairly aspecifically the N-glycosidic linkage between core residue GlcNAc-**1** and asparagine, resulting in a reducing oligosaccharide with an intact *N,N'*-diacetylchitobiose moiety. The Endo enzymes cleave the linkage between the two GlcNAc residues in the core, producing an oligosaccharide that ends in just one GlcNAc residue at the reducing end. The Endo enzymes vary in specificity.[14] The advantages of analyzing oligosaccharides over glycopeptides are (1) oligosaccharides show no signal overlap in the NMR spectrum with amino acid protons, and (2) they are easier than glycopeptides to purify[15,16] to homogeneity of carbohydrate chain, which is reflected in the NMR spectrum. The drawback of analyzing oligosaccharides is the loss of information about the glycosylation site and, sometimes, the anomerization of the reducing oligosaccharide. The oligosaccharide in solution is present as a mixture of the α and β anomers, and this may affect the NMR spectrum to such an extent that the interpretation becomes cumbersome.

To avoid the anomerization effect on the ^{1}H NMR spectrum, to simplify subsequent purification procedures, or to render the oligosaccharide amenable to techniques of structural analysis other than NMR, the oligosaccharide may be reduced by $NaBH_4$ (or $NaBD_4$ for mass spectrometry) to its corresponding oligosaccharide alditol. Alternatively, N-type oligosaccharide alditols may be obtained by hydrazinolysis of the glycoprotein followed by reduction (see [5] in this volume).[17] To enhance sensitivity of detection during purification of the oligosaccharides, the reduction may be carried out with $NaBT_4$, thereby incorporating a radioactive label in the compound. Neither the deuterium nor the tritium label in the oligosaccharide alditols has an adverse effect on ^{1}H NMR spectroscopic analysis of the compounds.

Serine- or threonine-linked carbohydrates having α-linked GalNAc in the carbohydrate–peptide linkage region ("mucin-type" O-linked oligosaccharides) are most conveniently studied by ^{1}H NMR spectroscopy as oligosaccharide alditols. Mucin-type glycopeptides, ordinarily containing clusters of carbohydrates (i.e., having more than one occupied glycosylation site), are not suitable for NMR analysis. O-Type oligosaccharide alditols are typically generated by reductive alkalinolysis (a β-elimination reaction in the presence of a reducing agent).[18,19]

[15] A. Kobata, this volume [11].

[16] M. R. Hardy and R. R. Townsend, this volume [12].

[17] T. P. Patel and R. B. Parekh, this volume [5].

[18] R. D. Cummings, R. K. Merkle, and N. L. Stults, *Methods Cell Biol.* **32**, 141 (1989).

[19] M. Lhermitte, H. Rahmoune, G. Lamblin, P. Roussel, A.-M. Strang, and H. van Halbeek, *Glycobiology* **1**, 277 (1991).

Evidently in all cases the partial degradation technique and subsequent chemical modifications applied should not affect the structure of the carbohydrate. Desialylation, defucosylation, deacetylation, desulfation, and nonspecific cleavage at the reducing end are known to occur when chemical methods are used for liberation of the oligosaccharide or when enzymatic methods are followed by workup procedures under conditions not mild enough (pH, temperature, chromatographic conditions for purifications) to keep the carbohydrate structures intact.

Finally, the (reduced) oligosaccharides must be purified before undergoing NMR analysis. It is important to remove any (proton-containing and non-proton-containing) contaminants and to purify the carbohydrates to virtual homogeneity (in terms of charge and size). The most frequently observed noncarbohydrate contaminants that can disturb the NMR spectrum are salts and/or buffers [acetate, lactate (from the fingers of the experimenter), sodium dodecyl sulfate (SDS) and ethylenediaminetetraacetic acid (EDTA)]. Even salts that do not contain protons (such as NaCl and phosphates) will impair the NMR spectrum (line broadening) and shift the HOD peak (effect on dielectric constant and pH of the solution). One should be aware that carbohydrates monitored during purification only by their radioactive label are usually the ones most seriously contaminated with nonradioactive but NMR-disturbing substances.

The amount of pure oligosaccharide needed to obtain an ^{1}H NMR spectrum in less than 6 hr of data acquisition time depends on the magnetic field strength of the spectrometer available and the sensitivity of the receiver coil in the probe. As a rule of thumb, 15 nmol of oligosaccharide (corresponding to 25 to 30 μg of a decasaccharide) is the minimum requirement for primary structural analysis on a 500-MHz spectrometer. For analysis at 600 MHz, 7 to 10 nmol (15 to 20 μg) is sufficient, whereas 35 to 40 nmol is needed at 400 MHz, and about 80 nmol at 300 MHz.

Characteristically, a glycoprotein does not have a single carbohydrate structure attached to a specific glycosylation site. This microheterogeneity, illustrating the phenomenon that a glycoprotein can take on various glycoforms, greatly complicates the structural characterization of glycoprotein carbohydrates. However, because NMR spectroscopy can handle mixtures of closely related structures, we often can cope with heterogeneity without resorting to rigorous purification of a glycoprotein into its individual glycoforms. Evidently, the glycoprotein must be purified to homogeneity, because the presence of contaminating (glyco)proteins must be eliminated before tackling the structural analysis of the carbohydrates. The molecular weight of the glycoprotein and the number of glycosylation sites occupied (i.e., the carbohydrate content) determine how much sample is needed to begin structural analysis. Ordinarily, about 20 mg of pure glycoprotein starting material is needed.[20,21]

Other materials needed for NMR analysis of oligosaccharides are a deuterated solvent (D_2O for underivatized glycoprotein oligosaccharides) and an NMR tube. In using D_2O for exchange of OH and NH protons against deuterium atoms, it is important to remember that the actual spectrum is recorded with a solution of the compound in 0.4 or 0.5 ml of D_2O of the highest available deuteration grade (in practice, >99.99%), but the initial exchange steps can be performed with 99.8% D_2O. Also, the deuteration percentage on the label of the bottle or ampoule is realistic only if the D_2O is handled correctly (see Sample preparation, below).

Typically, ^{1}H NMR analyses of oligosaccharide samples are carried out in 5-mm NMR tubes (5 mm in diameter, typically 8 in. in length). It is vital to purchase quality glass tubes. Wilmad 535-PP tubes (Wilmad, Buena, NJ) are recommended for analysis at 500 or 600 MHz; Wilmad 528-PP tubes are acceptable. Poor-quality spectra may be obtained on an otherwise excellent NMR spectrometer and valuable instrument time may be wasted by using tubes of inferior quality.

Sample Preparation

A glycopeptide or oligosaccharide sample submitted for NMR analysis is stored in the dry state at $-20°$ until use. The first steps in the actual preparation of the sample for NMR analysis are proton exchanges in D_2O. The sample is dissolved twice in D_2O (99.8 and 99.96% D, respectively) at room temperature and pD 6–7 with intermediate lyophilization. The purpose of the exchange treatments is the complete conversion of OH and NH groups in the constituent glycosyl residues into OD and ND (chemical exchange), and the preparation of an eventual solution of the carbohydrate with as low as possible a residual amount of HOD. The exchanges usually are performed in small glass vials; some researchers prefer to perform the exchange directly in the NMR tube. Each step in the exchange procedure (dissolving the sample in ~0.5 ml of D_2O, allowing the exchange process to take place, and subsequent lyophilization) may take 6 to 8 hr. Lyophilization can be replaced by flash evaporation. Check the pH (pD) of the solution immediately after dissolving the sample for the first exchange step. If the pH does not fall between 6 and 8, adjust it

[20] M. W. Spellman, C. K. Leonard, L. J. Basa, I. Gelineo, and H. van Halbeek, *Biochemistry* **30**, 2395 (1991).

[21] M. W. Spellman, L. J. Basa, C. K. Leonard, J. A. Chakel, J. V. O'Connor, S. Wilson, and H. van Halbeek, *J. Biol. Chem.* **264**, 14100 (1989).

with dilute DCl or NaOD. Remember that the glycosidic linkages of sialic acid residues tend to hydrolyze at pH 5 or lower, whereas esters such as *O*-acetates may be cleaved under both acidic (pH < 5) and basic (pH > 8) conditions.

One last purification step critical to the quality of the NMR spectrum is removal of paramagnetic impurities (metal ions). This step typically is performed in the NMR laboratory prior to the recording of the NMR spectrum. Chelex is the best material to use for the routine removal of paramagnetic ions; a few particles are sufficient to sequester the metals. Incubate Chelex with sample in the small vial during the first and/or second exchange step for 30 min, under slow, continuous swirling to remove the paramagnetics, and then pipette off the solution. In this way, a 0.5-ml sample can be processed with minimal loss from adsorption and with no significant dilution.[22]

Next, the sample is dissolved three more times in highest quality D_2O (>99.99% D) at room temperature and pD 6, with intermediate lyophilization. Altogether, a total of at least 48 hr is necessary to prepare a sample for NMR analysis, timed from its arrival in the NMR laboratory.

Additional Remarks. Several companies market D_2O of the quality (deuteration grade, free of paramagnetic impurities, etc.) required for this type of NMR analysis. It is best to purchase D_2O (especially the 99.996% D-grade D_2O) in small ampoules (0.5 to 1.0 ml) rather than in large bottles (over 10 ml). In any case, the D_2O should be handled in a dry atmosphere so as to preserve its quality after the container is opened. Perform the exchanges in a glove box, maintaining humidity at less than 7 or 8%. Allow samples that have been stored in a freezer to warm to room temperature before dissolving them in D_2O. Prerinse syringes and pipette tips in D_2O just prior to use. Moisture in the air is the NMR spectroscopist's worst enemy, therefore, it is imperative to prevent contact of the sample with the air.

Prior to ^1H NMR spectroscopic analysis, the sample is redissolved in 0.4 to 0.5 ml of D_2O (99.996% D) and transferred into a 5-mm NMR tube. The actual volume depends on the length of the rf receiver coil in the probe that the NMR spectrometer uses. The sample should be dissolved at least a few hours before the actual spectrum is recorded, to ensure complete solvation of the molecules. This time lapse significantly improves the quality of the NMR spectrum. When transferring the sample into the NMR tube, filter it (over cotton wool, prerinsed with highest quality D_2O) to remove any insoluble particulates. Check the pH of the resulting solution, either before or after the NMR spectrum is run, by putting a

[22] N. J. Oppenheimer, this series, Vol. 176, p. 78.

droplet on pH paper. The pH (pD) of the solution should be between 6 and 8. It is not necessary to degas the solution and/or seal the NMR tube for the type of NMR experiments described here.

Acetone is the most widely used internal standard for ^{1}H NMR spectroscopy in D_2O; its chemical shift is δ 2.225 ppm (referenced to DSS in D_2O at 0 ppm; see Fig. 4). Typically, the carefully cleaned NMR tube used for the analysis contains a trace amount of acetone. Of course, 1 to 10 μl of acetone may be added as a standard, before the NMR experiment, from a syringe to the oligosaccharide solution. Sometimes the sample contains a small amount of free acetate. The acetate methyl protons will show up in the ^{1}H NMR spectrum as a singlet at δ 1.908 ppm (see Fig. 4), and may be used as an alternative internal standard for chemical shift calibration unless the pD of the solution deviates significantly from the pD 6 to 8 range. Occasionally, deuterated acetone (acetone-d_6, CD_3COCD_3) (~50 μl) is added to the solution for lock purposes (see the next section); the residual acetone-d_5 (CD_3COCD_2H) gives rise to a multiplet, centered around 2.167 ppm, and provides another means for the calibration of chemical shifts.

Nuclear Magnetic Resonance Spectrometer

^{1}H NMR spectroscopy is performed on a pulse-FT NMR spectrometer, operating at a radio frequency in the range of 300 to 600 MHz for ^{1}H. For this type of analysis, there are no major differences in performance of NMR spectrometers of different manufacturers (Bruker, General Electric, JEOL, Varian) of the same field strength. The spectra shown in this chapter were recorded on a Bruker AM-500 spectrometer (operating at 500 MHz for ^{1}H) interfaced with an Aspect-3000 computer, or on a Bruker AMX-600 spectrometer (operating at 600 MHz for ^{1}H) interfaced with an Aspect-X32 computer. The sample tube is placed in the magnetic field ($\mathbf{B}_0$). It is important to use a high-sensitivity 5-mm probe for recording the ^{1}H NMR spectra of the oligosaccharide. The D signal of the solvent serves as a reference for the field-frequency lock.

The temperature of the sample in the probe can be selected between 20 and 30°, but must be kept constant during the NMR experiment. If constant room temperature is maintained in the environment where the NMR spectrometer is located, spectra can be acquired at ambient temperature without temperature control by the NMR spectrometer. However, when recording the spectrum of a weak sample over 4 to 6 hr, the temperature may have to be controlled by the spectrometer. In that case, the temperature of the sample is typically controlled at a few degrees above room temperature (e.g., 27°), because most commercially available variable temperature (VT) control units on NMR spectrometers cannot maintain constant sample temperature well if set at room temperature. The temperature fluctuations induced in the sample by heating and cooling of the VT unit are reflected in the resonance position of the D signal of D_2O and the ^1H signal of HOD. Because the position of the HOD signal is by far the most temperature sensitive in the ^1H spectrum, the only sharp peak resulting from lock compensation is the solvent peak, whereas the remainder of the signals shift back and forth with temperature fluctuations, resulting in broad lines.

If the spectrometer has problems maintaining stable sample temperature in the range needed to conduct the NMR experiment, use another deuterated solvent, mixable with D_2O but with a temperature-insensitive lock frequency, for the lock. One such solvent is acetone-d_6. As little as 50 μl must be added to the solution; this amount is sufficient for lock in the presence of 450 μl of D_2O.

After carefully adjusting the shims, that is, optimizing the magnetic field homogeneity over the sample, the scene is set for data collection. During the NMR experiment, the sample is spun at a constant rate of $\sim$15 to 20 Hz. If the gain in resolution does not outweigh the occurrence of spinning side bands (especially around the strong residual HOD signal), the NMR spectra are recorded without spinning the sample. In our experience, it is not necessary to spin a sample if the nonspinning shim gradients are carefully adjusted.

Recording an ^1H Nuclear Magnetic Resonance Spectrum

Nuclear magnetic resonance data acquisition involves pulsing radiofrequency radiation into the sample (typically for 5 to 15 μsec), and then measuring the response of the spin system (by the receiver coil) as a function of time during the acquisition period. The induced magnetization is digitized and stored into the memory of a computer. This process (a scan) is repeated many times, and the resulting responses are coadded in the computer memory, giving rise to a free induction decay (FID) (see Fig. 6). Standard acquisition parameters for routine ^1H NMR spectroscopic analysis of oligosaccharides or glycopeptides in D_2O are as follows. With the spectral width set at 10 to 12 ppm and using 16K or 32K data points in the computer memory, the acquisition time is 1.5 to 3.4 sec/scan. The flip angle of the pulse used is 70 to 75° (i.e., the magnetization **M** is tilted over 70 to 75° away from the $\mathbf{B}_0$ axis; see Fig. 2b). An additional relaxation delay between consecutive scans is not necessary under these conditions. (Typical T_1 values for ^1H in oligosaccharides are 0.1 to 0.5 sec).

Data collection is continued until the signal-to-noise (S/N) ratio in the anomeric proton region of the spectrum (obtained after FT of the FID; see Data Processing, below) is at least 3, but preferably 5 or better. Depending on the amount of carbohydrate material available for analysis, reaching this S/N value may require a few hundred up to several thousand scans. Thus, the total time to obtain the NMR spectrum is a few minutes (for ~1 μmol of carbohydrate) up to 6 hr (for 10 nmol of carbohydrate).

Despite all of these precautions when preparing the sample for NMR analysis, the residual HOD signal will still appear as the dominant peak in all spectra but those of the most concentrated samples. The HOD signal is found at δ 4.75–4.83 ppm; its exact position varies with temperature, pH, and concentration of the solution and, therefore, should not be used for calibration of the chemical shift scale. If the residual HOD signal obscures any signals of interest, we have two ways to make signals in the immediate vicinity of the HOD peak visible. We can either modify the routine single-pulse NMR experiment (into a "water suppression" experiment) while maintaining sample temperature, or we can repeat the standard NMR experiment at a higher temperature, because raising the sample temperature to 40 or 45° is usually sufficient to observe undisturbed the region between δ 4.7 and 4.9. However, solvent suppression is preferred over temperature elevation. Not only does the HOD signal shift when the sample temperature is changed, but some of the carbohydrate proton signals shift as well. Although the chemical shifts vary only slightly with temperature changes, such effects usually prevent unambiguous assignment of signals on the basis of ambient temperature data. There is also the risk of degrading the sample (e.g., desialylation) at high temperature.

The simplest way to suppress the solvent signal is by fast pulsing. That is the major reason for not using an additional relaxation delay in data collection when the acquisition time is already on the order of ~2 sec. Because carbohydrate protons have much shorter relaxation times than the HOD proton, the signals of the former will not be affected by "fast pulsing," and the sensitivity of the method is not degraded.

Alternatively, presaturation can be used to suppress the residual HOD signal.[23] Careful adjustment of the irradiation time and power level of the presaturation pulse, usually generated by the ^{1}H decoupler channel of the spectrometer, is required to obtain a spectrum that still holds information on signals close to the HOD peak.

A third water suppression technique is a multiple-pulse sequence called a WEFT (water-eliminated Fourier transform) experiment. It uses the difference in relaxation times between the HOD and carbohydrate protons.

[23] P. J. Hore, this series, Vol. 176, p. 64.

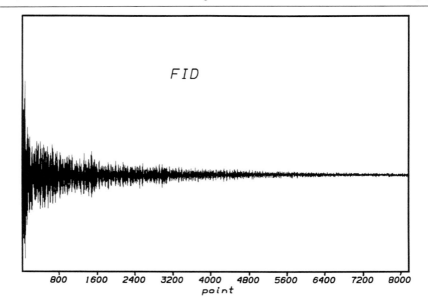

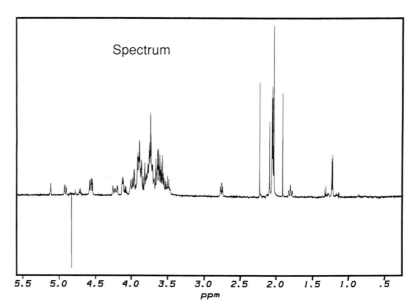

This experiment is based on the "inversion recovery" principle. Typically, a {180°–τ–90°–acquisition} sequence, in which the delay τ is empirically optimized, gives satisfactory results,[24] especially if the 180° pulse is a composite ($90°_x 180°_y 90°_x$). If, for sensitivity reasons, selective inversion of the HOD peak is desired, the 180° pulse in the above scheme may be replaced by a selective 180° pulse (usually a DANTE pulse train[25] or a shaped pulse). Both the nonselective and the selective WEFT experiment leave the regions immediately to the right and the left of the HOD signal unaffected.

Data Processing

The result of the data collection described earlier is called a free induction decay (FID), or an NMR spectrum in the time domain (see Fig. 6). To convert the FID into the NMR spectrum in the frequency domain, one routinely applies Fourier transformation (FT), followed by phase correction. The FID can be manipulated before FT, depending on the aspect of the resulting NMR spectrum to be emphasized. When multiplied by an exponential function, the result after FT is a sensitivity-enhanced spectrum at the cost of resolution (line broadening); however, when a sinusoidal or Gaussian function is used for window multiplication (signal-to-noise ratio permitting), the result is a resolution-enhanced spectrum. Often, after Gaussian multiplication, the number of data points is increased before FT ("zero filling") to ensure a sufficient number of data points to achieve a digital resolution of ~0.2 to 0.3 Hz/point. This technique is most useful in the methyl proton region of the ^{1}H NMR spectrum, where small differences in chemical shift between sharp methyl singlets (NAc signals) or doublets (Fuc CH$_3$ protons) are significant for structural analysis.

Spectral integration may be applied in the anomeric proton region of the spectrum, mainly to verify the purity of the sample by determining the ratio in which two or more components occur in the sample.[26]

[24] B. M. Nilsen, K. Sletten, B. Smestad Paulsen, M. O'Neill, and H. van Halbeek, *J. Biol. Chem.* **266,** 2660 (1991).

[25] C. A. G. Haasnoot, *J. Magn. Reson.* **52,** 153 (1983).

[26] C. C. Yu Ip, W. J. Miller, D. J. Kubek, A.-M. Strang, H. van Halbeek, S. J. Piesecki, and J. A. Alhadeff, *Biochemistry* **31,** 285 (1992).

FIG. 6. Data processing: from FID to spectrum (from time domain to frequency domain) by Fourier transformation.

Data processing can, of course, be carried out by the microcomputer that is part of the NMR spectrometer and controls the data acquisition and storage process. Alternatively, the data may be transferred to a workstation (a general-purpose computer without the data acquisition capability, such as a Sun SPARC station, a Silicon Graphics Personal Iris, or even an IBM PC or a Macintosh). Several companies (in addition to the NMR spectrometer manufacturers) market software packages for off-line data processing. For example, the spectra depicted in this chapter were processed on an IBM PC (with a 486 processor), using the Felix version 2.0 software (Hare Research, Inc.).

Spectral Interpretation

The ^{1}H NMR spectrum obtained represents a unique "identity card" of the carbohydrate; mere comparison of the ^{1}H NMR spectra of compounds (even without detailed interpretation) can demonstrate the identity of compounds. Thus, a one-dimensional ^{1}H NMR study may suffice for primary structure determination if the oligosaccharide has been characterized previously. Several glycoprotein carbohydrate ^{1}H NMR databases are available for N-type glycopeptides and oligosaccharides[27-29] and for O-type oligosaccharide alditols[30] in D_2O. When the spectrum does not match any of the spectra in existing databases, attempts can be made to interpret the ^{1}H NMR spectrum in terms of (partial elements of) the primary structure of the carbohydrate (including anomeric configurations and positions of glycosidic linkages) by using the structural reporter group concept.[2,27,31] As an example of this approach, the interpretation of the 500-MHz ^{1}H NMR spectra of two oligosaccharide samples released[5] from rEPO (Figs. 4 and 7), and of the spectrum of a mixture of glycopeptides generated[4] from HNE (Fig. 8), is discussed. In addition, the interpretation of two spectra (Fig. 9) of typical O-type oligosaccharide alditols also obtained[5] from rEPO is discussed. The actual values for all relevant acquisition parameters can be found in the captions to Figs. 7 to 9.

We ignore the crowded region in the center of the spectrum (between 3 and 4 ppm; see Fig. 4); only the positions and patterns of those signals that are more or less individually observable are examined. Particularly

[27] J. F. G. Vliegenthart, L. Dorland, and H. van Halbeek, *Adv. Carbohydr. Chem. Biochem.* **41,** 209 (1983).

[28] J. P. Carver and A. A. Grey, *Biochemistry* **20,** 6607 (1981).

[29] I. Brockhausen, A. A. Grey, H. Pang, H. Schachter, and J. P. Carver, *Glycoconjugate J.* **5,** 419 (1988).

[30] J. P. Kamerling and J. F. G. Vliegenthart, *Biol. Magn. Reson.* **10,** 1 (1992).

[31] H. van Halbeek, *Biochem. Soc. Trans.* **12,** 601 (1984).

useful structural reporter groups for ^{1}H NMR analyses are (1) the anomeric (H-1) protons, (2) the protons attached to the carbon atoms in the direct vicinity of a linkage position, (3) the protons attached to deoxy carbon atoms, and (4) methyl protons, for example, in N-acetyl groups. The structural reporter group regions of the spectrum in Fig. 4 are expanded in Fig. 7a. The chemical shifts of the structural reporter groups were measured and are compiled in Table II.

Figures 4 and 7 represent the ^{1}H NMR spectrum of a doubly charged oligosaccharide fraction [released from rEPO by N-glycanase, then reduced and purified by high-pH anion-exchange (HPAE) chromatography[16]]. The NMR spectrum indicates that the sample contains a diantennary N-acetyllactosamine-type oligosaccharide ending in reduced N,N'-diacetylchitobiose.[32] The diantennary type of branching is evident from the set of chemical shifts of the Man H-1 and H-2 atoms (Table II).[20,21,26,27] Both branches of the diantennary oligosaccharide terminate in NeuAc attached in $\alpha(2 \rightarrow 3)$ linkage to Gal, as seen by the pair of NeuAc H-3ax and H-3eq signals (δ 1.80 and 2.76, respectively; Table II). The precise position of the NeuAc H-3ax signal reflects the branch location of the NeuAc residue. The position of the H-3ax signal is different for NeuAc in the C–3-linked (δ 1.800) vs the C-6-linked branch (δ 1.805). These values, along with the positions of the Gal-6 and Gal-6' H-1 doublets and the GlcNAc-5 and GlcNAc-5' NAc signals, are valuable for determining both the type of linkage of the NeuAc residue to Gal and the branch location of the residue.[21,27] It is worth noting that $\alpha(2 \rightarrow 6)$-linked NeuAc residues (although not found in the carbohydrate chains of glycoproteins expressed in CHO cells) have different characteristic chemical shifts for their H-3ax and H-3eq signals. They also exert different effects on the chemical shifts of the aforementioned structural reporter groups of residues in the sialylated branch.[32,33]

The diantennary oligosaccharide alditol from rEPO has in addition Fuc $\alpha(1 \rightarrow 6)$-attached to GlcNAc-1-ol in the core. The presence of the Fuc residue manifests itself in the NMR spectrum of the oligosaccharide by (1) structural reporter group signals of the Fuc residue itself, and (2) chemical shifts induced on the reporter group signals of the GlcNAc-2 residue. Typical chemical shifts for Fuc H-1, H-5, and the CH_3 protons are δ 4.90, 4.075, and 1.225, respectively (cf. Table II). Extending the chitobiose unit by Fuc $\alpha(1 \rightarrow 6)$ at GlcNAc-1-ol affects the chemical shifts of H-1 ($\Delta\delta$ 0.075 ppm) and the NAc protons ($\Delta\delta$ 0.01 ppm) of GlcNAc-2

[32] E. D. Green, G. Adelt, J. U. Baenziger, S. Wilson, and H. van Halbeek, *J. Biol. Chem.* **263**, 18253 (1988).
[33] D. Fu and H. van Halbeek, *Anal. Biochem.* **206,** 53 (1992).

TABLE II

¹H CHEMICAL SHIFTS OF STRUCTURAL REPORTER GROUPS OF CONSTITUENT MONOSACCHARIDES FOR *N*-TYPE DI-, TRI-, AND TETRAANTENNARY OLIGOSACCHARIDE ALDITOLS

Reporter group	Residue[a]	Chemical shift[b] (ppm) in[c]:					
		Di + F	Di − F	Tri + F	Tri − F	Tri′ + F	Tetra + F
H-1	Fucα(1 → 6)	4.901	—	4.903	—	4.903	4.905
	GlcNAc-**2**	4.711	4.637	4.709	4.636	4.709	4.708
	Man-**3**	4.777	4.78	4.78	4.77	4.772	4.765
	Man-**4**	5.117	5.116	5.115	5.119	5.124	5.133
	Man-**4′**	4.925	4.926	4.909	4.911	4.872	4.858
	GlcNAc-**5**	4.573	4.572	4.575	4.572	4.578	4.563
	GlcNAc-**5′**	4.573	4.572	4.575	4.572	4.578	4.589
	GlcNAc-**7**	—	—	4.549	4.548	—	4.563
	GlcNAc-**7′**	—	—	—	—	4.562	4.563
	Gal-**6**	4.544	4.544	4.549	4.548	4.546	4.544
	Gal-**6′**	4.550	4.550	4.549	4.548	4.546	4.544
	Gal-**8**	—	—	4.549	4.548	—	4.544
	Gal-**8′**	—	—	—	—	4.562	4.544
H-2	GlcNAc-**1**-ol	4.227	4.237	4.225	4.239	4.219	4.22
	Man-**3**	4.255	4.250	4.217	4.216	4.257	4.213
	Man-**4**	4.194	4.190	4.217	4.216	4.198	4.218
	Man-**4′**	4.112	4.114	4.111	4.116	4.094	4.095
H-3	Gal-**6,6′**	4.118	4.115	4.120	4.116	4.118	4.119
	Gal-**8,8′**	—	—	4.120	4.116	4.118	4.119
H-3ax	NeuAc	1.800	1.798	1.804	1.802	1.804	1.805
	NeuAc′	1.805	1.798	1.804	1.802	1.804	1.805
	NeuAc*,*′	—	—	1.804	1.802	1.804	1.805
H-3eq	NeuAc	2.757	2.756	2.755	2.757	2.756	2.755
	NeuAc′	2.757	2.756	2.755	2.757	2.756	2.755
	NeuAc*,*′	—	—	2.755	2.757	2.756	2.755
H-5	Fucα(1 → 6)	4.077	—	4.075	—	4.075	4.077
CH₃	Fucα(1 → 6)	1.226	—	1.225	—	1.226	1.226
NAc	GlcNAc-**1**-ol	2.057	2.056	2.055	2.056	2.055	2.055
	GlcNAc-**2**	2.090	2.080	2.088	2.081	2.087	2.086
	GlcNAc-**5**	2.050	2.048	2.044	2.045	2.053	2.049
	GlcNAc-**5′**	2.043	2.042	2.042	2.043	2.040	2.040
	GlcNAc-**7**	—	—	2.075	2.075	—	2.077
	GlcNAc-**7′**	—	—	—	—	2.040	2.040

TABLE II (continued)

		Chemical shift[b] (ppm) in[c]:					
Reporter group	Residue[a]	Di + F	Di − F	Tri + F	Tri − F	Tri' + F	Tetra + F
	NeuAc	2.030	2.030	2.029	2.032	2.030	2.030
	NeuAc'	2.030	2.030	2.029	2.032	2.030	2.030
	NeuAc*,*'	—	—	2.029	2.032	2.030	2.030

[a] The numbering system used for denoting glycosyl residues in the diantennary oligosaccharide alditols is as follows:

N'   **6'**   **5'**   **4'**   **F**
NeuAcα(2 → 3)Galβ(1 → 4)GlcNAcβ(1 → 2)Manα(1 → 6)   Fucα(1 → 6)
                                                       Manβ(1 → 4)GlcNAcβ(1 → 4)GlcNAc-ol
                                                       3   **2**   **1**
NeuAcα(2 → 3)Galβ(1 → 4)GlcNAcβ(1 → 2)Manα(1 → 3)
N   **6**   **5**   **4**

The numbering system used for denoting glycosyl residues in the tri- and tetraantennary oligosaccharide alditols is as follows:

N*'   **8'**   **7'**
NeuAcα(2 → 3)Galβ(1 → 4)GlcNAcβ(1 → 6)

N'   **6'**   **5'**   **4'**   **F**
NeuAcα(2 → 3)Galβ(1 → 4)GlcNAcβ(1 → 2)Manα(1 → 6)   Fucα(1 → 6)
                                                       Manβ(1 → 4)GlcNAcβ(1 → 4)GlcNAc-ol
                                                       3   **2**   **1**
NeuAcα(2 → 3)Galβ(1 → 4)GlcNAcβ(1 → 2)Manα(1 → 3)
N   **6**   **5**   **4**
NeuAcα(2 → 3)Galβ(1 → 4)GlcNAcβ(1 → 4)
N*   **8**   **7**

[b] Data were acquired at 500 MHz for neutral solutions of the compounds in D_2O at 21° (for the fucosyl oligosaccharides) or 27° (for the nonfucosyl oligosaccharides).

[c] Fucosyl oligosaccharides were released from recombinant human erythropoietin [E. Watson, A. Bhide, and H. van Halbeek, submitted for publication (1993)]; nonfucosyl oligosaccharides were prepared from bovine fetuin [E. D. Green, G. Adelt, J. U. Baenziger, S. Wilson, and H. van Halbeek, J. Biol. Chem. **263**, 18253 (1988)]. Di, Diantennary; tri, tri(2,4)-antennary; tri', tri(2,6)-antennary; tetra, tetraantennary oligosaccharide; F, α(1 → 6)-fucosyl residue at GlcNAc-1-ol. Structures are schematically illustrated: ■, GlcNAc; ●, Gal; ○, Man; ▲, NeuAc; □, Fuc; ⊕, GlcNAc-ol. In the diantennary oligosaccharide alditols, the peripheral unit on the left corresponds to the glycosyl residues **5-6-N**, the unit on the right to the **5'-6'-N'** glycosyl residues. The peripheral units in the tri- and tetraantennary oligosaccharides, from left to right, correspond to the glycosyl residues **5-6-N**, **7-8-N***, **5'-6'-N'**, and **7'-8'-N*'**, respectively.

(a)

NeuAcα(2→3)Galβ(1→4)GlcNAcβ(1→2)Manα(1→6)\
      N'   6'   5'   4'

                                      Fucα(1→6)\
                                      F

                           Manβ(1→4)GlcNAcβ(1→4)GlcNAc-ol

NeuAcα(2→3)Galβ(1→4)GlcNAcβ(1→2)Manα(1→3)/ 3   2   1
     N   6   5   4

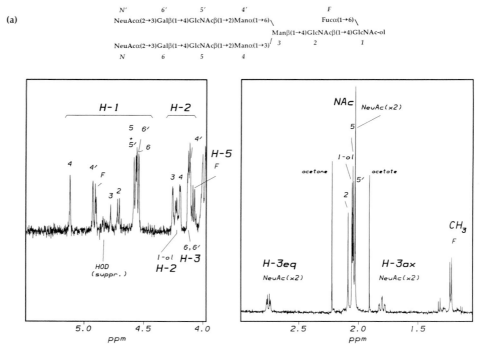

FIG. 7. Structural reporter group regions of the 500-MHz ^{1}H NMR spectra (D$_2$O, pD 6, 21°) of N-type oligosaccharide alditols prepared from recombinant human erythropoietin by N-glycanase digestion, subsequent reduction, and HPAE chromatography [E. Watson, A. Bhide, and H. van Halbeek, submitted for publication (1993)]. (a) Disialyl diantennary oligosaccharide alditol (250 μg). (b) Tetrasialyl tetraantennary oligosaccharide alditol (120 μg). Water suppression by low-power presaturation (1 sec). Acquisition parameters: number of data points, 16K in the time domain; spectral width, 5000 Hz; acquisition time, 1.64 sec; total interpulse delay, 2.64 sec; pulse width, 5.0 μsec (~75° flip angle); number of scans, 400 (a) and 2000 (b). The numbers in the spectra refer to the corresponding glycosyl residues in the structures (see also Table II).

(see Table II). The NMR appearance of Fuc residues α(1 → 6)-linked to reducing GlcNAc in N-glycanase-released oligosaccharides is slightly more complicated, because of the anomerization effect.[2,20,21]

    Di-, tri-, tri'-, and tetraantennary structures have distinctly different sets of Man H-1 and H-2 chemical shifts (see Table II), making determination of the degree of branching of an N-type glycopeptide/oligosaccharide possible on the basis of the NMR spectrum. For example, Figure 7 compares the spectra of a sialyl diantennary oligosaccharide alditol (Fig. 7a)

(b)

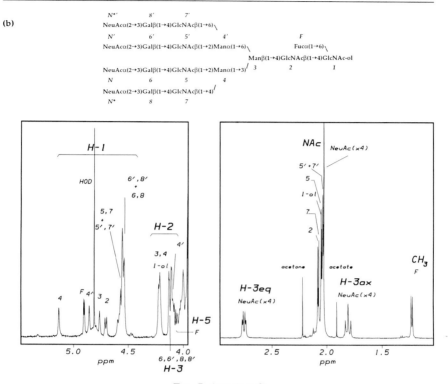

FIG. 7. (*continued*)

and a sialyl tetraantennary oligosaccharide alditol (Fig. 7b) obtained[5] from rEPO.

The ¹H NMR spectra of glycopeptides and those of reducing oligosaccharides (*N*-glycanase products) are slightly more difficult to unravel than those of oligosaccharide alditols. Figure 8 shows the 500-MHz ¹H NMR spectrum of a glycopeptide preparation obtained from glycosylation site Asn-95 in HNE isozyme E-3 through pronase digestion followed by purification on high-performance liquid chromatography (HPLC).[4,34] The chemical shifts of the structural reporter groups of the glycopeptide sample are compiled in Table III. Characteristically, the anomeric doublet of GlcNAc-**1** is found at δ 5.078 ($^3J_{12}$ ~10 Hz), because of its N-glycosidic β linkage to asparagine. The spectrum reveals that the structure of the

[34] H. van Halbeek, W. Watorek, and J. Travis, *Abstr. 17th Annu. Meet. Soc. Complex Carbohydr.*, San Antonio, TX, p. 97 (1988).

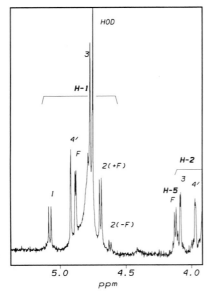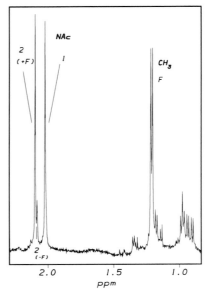

FIG. 8. Structural reporter group regions of the ^{1}H NMR spectrum (500 MHz, 27°, pD 6) of a mixture of two glycopeptides (200 μg) derived from isozyme E-3 of human neutrophil elastase.[34a] The glycopeptide(s) contain glycosylation site Asn-95. The numbers refer to the corresponding glycosyl residues in the structures. Any signals not marked in this spectrum are due to amino acid protons. Water suppression was achieved by a WEFT experiment. Other acquisition parameters are as in the legend to Fig. 7.

oligosaccharide portion of the E-3 Asn-95 glycopeptide is predominantly a pentasaccharide.

$$
\begin{array}{llll}
\mathbf{4'} & & \mathbf{F} & \\
\text{Man}\alpha(1 \to 6)\!\!\searrow & & \text{Fuc}\alpha(1 \to 6)\!\!\searrow & \\
& \text{Man}\beta(1 \to 4)\text{GlcNAc}\beta(1 \to 4)\text{GlcNAc}\beta(1 \to \text{N})\text{Asn} \\
& \quad\;\mathbf{3} & \quad\;\mathbf{2} & \quad\quad\;\mathbf{1}
\end{array}
$$

However, a minor component of the apparent mixture has the tetrasaccharide devoid of Fuc attached to asparagine, as evidenced by low-intensity signals observed for GlcNAc-**2** H-1 and NAc at δ 4.622 and 2.082, respectively (compare Fig. 8 and Table III). The intensity ratio of the

[34a] W. Watorek, H. van Halbeek, and J. Travis, *Biol. Chem. Hoppe Seyler* **374,** 385 (1993).

TABLE III

[1]H CHEMICAL SHIFTS OF STRUCTURAL REPORTER GROUPS OF CONSTITUENT MONOSACCHARIDES FOR N-TYPE GLYCOPEPTIDES DERIVED FROM HUMAN NEUTROPHIL ELASTASE (ISOZYMES E-1 AND E-3)

Reporter group	Residue[a]	Chemical shift[b] (ppm) in[c]:			
		E-1[d] (Asn-95)	E-1[d] (Asn-144)	E-3 + F (Asn-95)	E-3 − F (Asn-95)
H-1	GlcNAc-1	5.070	5.030	5.078	5.078
	Fucα(1 → 6)	4.878	4.875	4.883	—
	GlcNAc-2	4.683	4.683	4.690	4.622[e]
	Man-3	4.780	4.767	4.773	4.773
	Man-4	5.122	5.119	—	—
	Man-4′	4.917	4.919	4.920	4.920
	GlcNAc-5	4.579	4.581	—	—
	GlcNAc-5′	4.579	4.581	—	—
	Gal-6	4.470	4.469	—	—
	Gal-6′	4.472	4.472	—	—
H-2	Man-3	4.250	4.248	4.087	4.087
	Man-4	4.190	4.189	—	—
	Man-4′	4.108	4.110	3.973	3.973
H-5	Fucα(1 → 6)	4.12	4.118	4.119	—
CH$_3$	Fucα(1 → 6)	1.211	1.198	1.205	—
NAc	GlcNAc-1	2.013	2.017	2.020	2.020
	GlcNAc-2	2.092	2.090	2.097[e]	2.082[e]
	GlcNAc-5	2.051	2.051	—	—
	GlcNAc-5′	2.049	2.047	—	—

[a] For numbering system of glycosyl residues, see footnote a to Table II.

[b] Data were acquired at 500 MHz for neutral solutions of the compounds in D$_2$O at 27°.

[c] For complete structures of the glycopeptides, see text. Structures are schematically illustrated in the table heading, using a short-hand symbolic notation: ■, GlcNAc; ●, Gal; ○, Man; ▲, NeuAc; □, Fuc. The peripheral unit on the left corresponds to the glycosyl residues 5-6, the unit on the right to the 5′-6′ glycosyl residues.

[d] E-1, Isozyme HNE 1; E-3, isozyme HNE 3; Asn-95 and Asn-144 are the two glycosylation sites in both E-1 and E-3. The diantennary structure at both Asn-95 and Asn-144 in E-1 was found to be partially sialylated, as evidenced by the occurrence of NeuAc H-3ax signals at δ 1.716 and 1.796, H-3eq signals at δ 2.67 and 2.76, an NAc signal at δ 2.030, and small Man-4 H-1, GlcNAc H-1, NAc, and Gal H-1 signals at δ 5.134, 4.546, 2.069, and 4.440, respectively (cf. Table II).

[e] The truncated core structure was found to be partially (85%) fucosylated, that is, structures E-3 + F and E-3 − F constitute the mixture of glycopeptides in the ratio 85 : 15 (see Fig. 8).

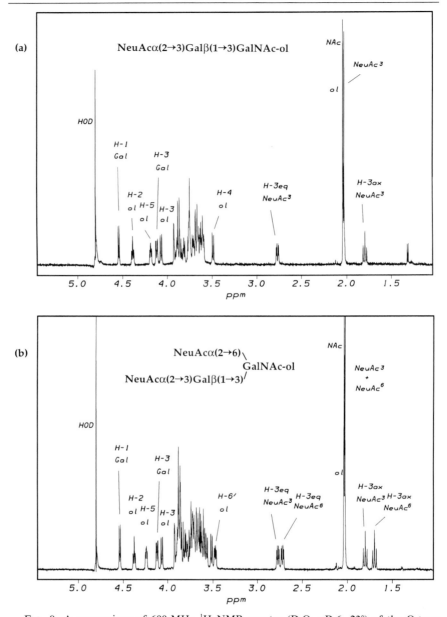

FIG. 9. A comparison of 600-MHz ^{1}H NMR spectra (D$_2$O, pD 6, 23°) of the O-type oligosaccharide alditols released from recombinant human erythropoietin (expressed in CHO cells) by reductive alkalinolysis [E. Watson, A. Bhide, and H. van Halbeek, submitted for

NAc signals of GlcNAc-**2**, at δ 2.097 for the fucosyl and δ 2.082 for the nonfucosyl compound, was used to determine that the ratio of penta- and tetrasaccharide in mixture E-3 was 85 : 15 (Fig. 8). This example illustrates the power of NMR spectroscopy in the analysis of a mixture of structurally related components.

The third example features the ^{1}H NMR spectra of two O-type oligosaccharide alditols, derived from oligosaccharide chains commonly found in natural (e.g., fetuin and glycocalicin) and recombinant glycoproteins. Figure 9 shows the 600-MHz ^{1}H NMR spectra of the linear trisaccharide alditol and the branched tetrasaccharide alditol obtained[5] from rEPO. The chemical shifts of the reporter groups of the O-type oligosaccharide alditols are listed in Table IV. Figure 9b clearly illustrates the difference in the NMR appearance of $\alpha(2 \rightarrow 3)$- vs $\alpha(2 \rightarrow 6)$-linked sialic acid (note the H-3ax and H-3eq signals).

Comments on Method

Advantages and Disadvantages of Nuclear Magnetic Resonance

Nuclear magnetic resonance spectroscopy is a powerful method for primary structural characterization of glycoprotein carbohydrates, but it has its limitations. Therefore NMR should be the first, but never the only, step in the structural analysis procedure. Partial or even complete primary structure determination is possible from the one-dimensional ^{1}H NMR spectrum as long as structurally related compounds have been previously characterized by ^{1}H NMR spectroscopy. It is recommended that the glycosyl residue composition be obtained independently by chemical analysis[35] and the molecular weight be verified by (electrospray ionization) mass spectrometry.[36]

The most important advantage of NMR spectroscopy over other techniques used for structural analysis of carbohydrates is its nondestructive

---

[35] R. K. Merkle and I. Poppe, this volume [1].
[36] A. Dell, A. J. Reason, K-H. Khoo, M. Panico, R. A. McDowell, and H. R. Morris, this volume [8].

---

publication (1993)]. (a) Trisaccharide alditol (100 μg). (b) Tetrasaccharide alditol (75 μg). Acquisition parameters: number of data points, 32K (both in the time domain and in the frequency domain); spectral width, 5435 Hz; acquisition time, 3.01 sec; pulse width, 9.0 μsec (~90° flip angle); number of scans, 1000 for each spectrum. The samples were not spun. The residual HOD signal was suppressed (1 sec) by low-power presaturation. Superscripts in NeuAc^3 and NeuAc^6 refer to the position of the linkage in which these residues are involved.

TABLE IV
¹H Chemical Shifts of Structural Reporter
Groups of Constituent Monosaccharides for
Typical O-Type Oligosaccharide Alditols

Residue[a]	Reporter group	Chemical shift[b] (ppm) in[c]:	
GalNAc-ol	H-2	4.389	4.380
	H-3	4.071	4.068
	H-4	3.492	3.523
	H-5	4.188	4.244
	H-6'	nd[d]	3.462
	NAc	2.042	2.041
Gal	H-1	4.545	4.543
	H-3	4.121	4.121
NeuAcα(2 → 3)	H-3ax	1.799	1.805
	H-3eq	2.769	2.772
	NAc	2.030	2.030
NeuAcα(2 → 6)	H-3ax	—	1.701
	H-3eq	—	2.722
	NAc	—	2.030

[a] Structures are schematically illustrated in the table heading, using a short-hand symbolic notation: ●, Gal; ▲, NeuAc; ⊕, GalNAc-ol. For complete structures of the compounds, see Fig. 9.

[b] Data were acquired at 600 MHz for neutral solutions of the compounds in D_2O at 23°.

[c] Oligosaccharide alditols were obtained from recombinant human erythropoietin [E. Watson, A. Bhide, and H. van Halbeek, submitted for publication (1993)].

[d] nd, Not determined.

nature. The oligosaccharide/glycopeptide sample, after NMR analysis, can be recovered 100% unimpaired and used for other analyses, biological activity tests, and so on. Also, mixtures of components with closely related structures can be analyzed successfully. The most important limitation of NMR spectroscopy is its sensitivity. Not only is at least 10 to 15 nmol of pure carbohydrate required to record an NMR spectrum, but even at 600 MHz, heterogeneity occurring in low abundance in the sample may escape attention. ^{1}H NMR spectroscopy may also fail to detect the presence of nonmagnetically active nuclei in the carbohydrate; although it is relatively simple to detect the presence of a phosphate[37] or O-acetate[38] group in an oligosaccharide, sulfate may escape detection by NMR.[39]

With regard to size limitations, glycoprotein oligosaccharides with 15 to 20 constituent glycosyl residues have been successfully characterized by 500-MHz ^{1}H NMR spectroscopy.[21,26] Degrees of branching as great as six (so-called "intersected pentaantennary" structures) have been identified by NMR.[40,41] However, with increasing size, it may be possible to define structural elements that extend the core and backbone of the common structures (Tables II and IV), but it is not always possible to delineate unambiguously their branch location by NMR alone.[42,43]

Expert Assistance

Primary structure determination of glycoprotein-derived oligosaccharides can be performed by ^{1}H NMR spectroscopy, using radio frequencies greater than 300 MHz. We strongly advise using a spectrometer operating at 500 or 600 MHz because of the advantages in sensitivity and spectral resolution. Although high-field (300 to 600 MHz) NMR spectrometers are widespread these days, mere access to such an instrument does not always guarantee the successful structural elucidation of the carbohydrate in question. Only a skillful NMR spectroscopist familiar with the peculiar aspects of carbohydrate NMR spectra will record spectra of the quality, and under the conditions, that allow reliable interpretation. An experienced specialist is almost always required to interpret the NMR data in terms of the structure of the carbohydrate. Efforts are beginning, however, to automate interpretation of ^{1}H NMR spectra of carbohydrates (see the next section). Thus, biological researchers often seek collaboration with laboratories specializing in NMR spectroscopy of carbohydrates.

[37] R. O. Couso, H. van Halbeek, V. N. Reinhold, and S. Kornfeld, *J. Biol. Chem.* **262,** 4521 (1987).
[38] J. B. L. Damm, K. Voshol, K. Hård, J. P. Kamerling, and J. F. G. Vliegenthart, *Eur. J. Biochem.* **180,** 101 (1989).
[39] P. de Waard, A. Koorevaar, J. P. Kamerling, and J. F. G. Vliegenthart, *J. Biol. Chem.* **266,** 4237 (1991).
[40] J. Paz Parente, J.-M. Wieruszeski, G. Strecker, J. Montreuil, B. Fournet, H. van Halbeek, L. Dorland, and J. F. G. Vliegenthart, *J. Biol. Chem.* **257,** 13173 (1982).
[41] J. Paz Parente, G. Strecker, Y. Leroy, J. Montreuil, B. Fournet, H. van Halbeek, L. Dorland, and J. F. G. Vliegenthart, *FEBS Lett.* **152,** 145 (1983).
[42] M. Fukuda, B. Bothner, P. Ramsamooj, A. Dell, P. R. Tiller, A. Varki, and J. C. Klock, *J. Biol. Chem.* **260,** 12957 (1985).
[43] J. Finne, M. E. Breimer, G. C. Hansson, K.-A. Karlsson, H. Leffler, J. F. G. Vliegenthart, and H. van Halbeek, *J. Biol. Chem.* **264,** 5720 (1989).

Automation of Method

With a dedicated microcomputer and an automatic sample changer connected to the NMR spectrometer, the method of recording ^{1}H NMR spectra of oligosaccharides can be automated. However, sample preparation will remain the responsibility of a researcher. The most time-consuming part of the structural analysis of carbohydrates by NMR has been spectral interpretation, until now. Efforts to automate this aspect of the method have begun along two different paths. The use of a computer search algorithm to compare a list of chemical shifts of structural reporter groups with all those in a database appears straightforward. Indeed, several such computer programs have been written to assist in the interpretation of glycoprotein oligosaccharide ^{1}H NMR spectra.[44,45] A much more elegant and potentially faster way is to use the entire spectrum (including the 3- to 4-ppm envelope region with its high information density) for pattern recognition. The NMR spectrum, already available in digital format, is not reduced to a list of chemical shifts, as is done for human interpretation. In our laboratories at the Complex Carbohydrate Research Center (University of Georgia, Athens, GA) artificial neural networks (ANN) have been successfully applied for automated NMR spectral interpretation.[46,47] This work has been extended to include the 500-MHz ^{1}H NMR spectra of glycoprotein carbohydrates.[48] In the foreseeable future, NMR spectral databases will be linked to the complex carbohydrate structure database (CCSD).[49] The neural network search algorithms, much like CarbBank, will be made available to the scientific community.

Perhaps the most exciting development in the application of the ANN technology for the analysis of ^{1}H NMR spectra is the increase in sensitivity of the method that may be achieved. For an S/N spectrum of a given oligosaccharide already in the database (i.e., when a neural net has been trained to recognize this spectrum in terms of the associated carbohydrate structure), this oligosaccharide can be recognized again from its ^{1}H NMR spectrum with an S/N ratio of a factor of 100 to 1000 less than in the original spectrum (such as would be obtainable for subnanomole amounts of carbohydrate). Once perfected, the ANN technology will make NMR

[44] D. S. M. Bot, P. Cleij, H. A. van't Klooster, H. van Halbeek, G. A. Veldink, and J. F. G. Vliegenthart, *J. Chemom.* **2**, 11 (1988).

[45] E. F. Hounsell and D. J. Wright, *Carbohydr. Res.* **205**, 19 (1990).

[46] J. U. Thomsen and B. Meyer, *J. Magn. Reson.* **84**, 212 (1989).

[47] B. Meyer, T. Hansen, D. Nute, P. Albersheim, A. Darvill, W. York, and J. Sellers, *Science* **251**, 542 (1991).

[48] J. P. Radomski, H. van Halbeek, and B. Meyer, submitted for publication (1993).

[49] S. Doubet, K. Bock, D. Smith, A. Darvill, and P. Albersheim, *Trends Biochem. Sci.* **14**, 475 (1989).

competitive in sensitivity with any other technique in primary structural determination of carbohydrates related to glycoproteins.

^1H Nuclear Magnetic Resonance of Nonglycoprotein Carbohydrates

The ^1H NMR spectroscopy principles and methodology discussed in this chapter are not valid only for glycoprotein-related carbohydrates. They apply equally well to glycolipids,[50] including glycosylphosphatidylinositol (GPI) anchors,[51] to polysaccharides,[52] including lipopolysaccharides and glycosaminoglycans,[53] and to proteoglycans. Obviously, if oligosaccharides generated from these glycoconjugates/polysaccharides are to be examined by ^1H NMR spectroscopy, there are essentially no differences with the procedures listed earlier. Sample preparation and, particularly, choice of solvent need to be modified accordingly when dealing with intact glycolipids and lipopolysaccharides. As stated before, the larger the polymer, the less resolved its ^1H NMR spectrum, and the more difficult it is to extract detailed structural information from the spectrum.

De Novo Structural Elucidation by One-, Two-, and Three-Dimensional, Homo- and Heteronuclear Nuclear Magnetic Resonance

In presenting NMR spectroscopy as a ''fingerprinting'' technique, we only discussed solution-state, single-pulse, one-dimensional ^1H NMR spectroscopy. Nuclear magnetic resonance of carbohydrates in the solid state is beyond the scope of this chapter. However, the application of multiple-pulse, multidimensional, multinuclear (^1H and ^{13}C) NMR techniques to carbohydrates should not be left unmentioned.

When the normal, so-called one-dimensional ^1H spectrum does not resemble that of a known oligosaccharide structure, the combination of multiple-pulse ^1H NMR spectroscopic techniques [chiefly, two-dimensional total correlation spectroscopy (TOCSY) and rotating-frame NOE spectroscopy (ROESY)] may be applied for the de novo sequencing of the carbohydrate,[54–56] provided that 1 to 3 μmol of pure substance (100

[50] R. K. Yu, T. A. W. Koerner, J. N. Scarsdale, and J. H. Prestegard, Chem. Phys. Lipids 42, 27 (1986).

[51] S. W. Homans, M. A. J. Ferguson, R. A. Dwek, T. W. Rademacher, R. Anand, and A. F. Williams, Nature (London) 333, 269 (1988).

[52] K. Bock, C. Pedersen, and H. Pedersen, Adv. Carbohydr. Chem. Biochem. 42, 193 (1984).

[53] A. S. Perlin, in ''NMR Applications in Biopolymers'' (J. W. Finley, W. Schmidt, and A. S. Serianni, eds.), p. 7. Plenum, New York, 1990.

[54] C. A. Bush, Bull. Magn. Reson. 10, 73 (1988).

[55] T. A. W. Koerner, J. H. Prestegard, and R. K. Yu, this series, Vol. 138, p. 38.

[56] J. Dabrowski, this series, Vol. 179, p. 122.

times the amount of sample mentioned for the one-dimensional analysis) is available for the analysis. Two-dimensional NMR spectra contain a wealth of information, spread into two dimensions (rather than along one frequency axis) and therefore show a dramatic improvement in resolution over one-dimensional spectra. The TOCSY technique permits subspectral editing of the ^{1}H spectrum for each constituting glycosyl residue and, consequently, the virtually complete assignment of all the multiplet patterns in the ^{1}H NMR spectrum (including those in the envelope region between 3 and 4 ppm). Subsequently, from the ROESY spectrum we can infer the sequence of the glycosyl residues, including the positions of glycosidic linkages, based on the detection of through-space connectivities. A discussion of these more sophisticated NMR techniques is beyond the scope of this chapter. The interested reader is referred to review articles.[54,56-59] Extensions of ^{1}H one- and two-dimensional techniques to three- and four-dimensional NMR techniques have been described for carbohydrates[60,61]; however, full-blown three- and four-dimensional data acquisition is time consuming and often not necessary to solve a carbohydrate structural problem. Selective (one- and two-dimensional) analogs of multidimensional NMR experiments[62] are far more efficient for obtaining the desired information from a carbohydrate ^{1}H NMR spectrum.

As mentioned under Principle of Method (above), NMR spectroscopy of carbohydrates is not limited to studying ^{1}H nuclei. If not for its low natural abundance, the ^{13}C nucleus might be even more popular for NMR studies of carbohydrates than it is today. ^{13}C NMR spectra can also be used for fingerprinting of glycoprotein carbohydrates[63]; these spectra, usually recorded under ^{1}H decoupling conditions, consist of singlets only. Favored further by the large chemical shift dispersion ($\Delta\delta$ ~100 ppm), ^{13}C-NMR spectra (even for glycopolymers) are often characterized by high resolution. When ^{13}C and ^{1}H NMR spectroscopy are used together, NMR demonstrates its real power in the *de novo* delineation of carbohydrate structures from first principles. Heteronuclear {^{13}C, ^{1}H} correlation NMR techniques are the most informative NMR experiments for carbohy-

[57] H. van Halbeek, *in* "Frontiers of NMR in Molecular Biology" (D. Live, I. M. Armitage, and D. Patel, eds.), UCLA Symp. Ser., Vol. 109, p. 195. Alan R. Liss, New York, 1990.
[58] A. Bax, *Curr. Opin. Struct. Biol.* **1**, 1030 (1991).
[59] H. Kessler, M. Gehrke, and C. Griesinger, *Angew. Chem., Int. Ed. Engl.* **27**, 490 (1988).
[60] G. W. Vuister, P. de Waard, R. Boelens, J. F. G. Vliegenthart, and R. Kaptein, *J. Am. Chem. Soc.* **111**, 772 (1989).
[61] T. J. Rutherford and S. W. Homans, *Glycobiology* **2**, 1 (1992).
[62] H. Kessler, S. Mronga, and G. Gemmecker, *Magn. Reson. Chem.* **29**, 527 (1991).
[63] D. A. Cumming, C. Hellerqvist, and O. Touster, *Carbohydr. Res.* **179**, 369 (1988).

drate molecules.[57,64] The sensitivity limitations for these types of experiments were overcome by the introduction of the ^{1}H-detected versions of many {^{13}C, ^{1}H} correlated spectroscopies. At the present time, isolating and purifying carbohydrates in sufficient amounts is the bottleneck to their structural characterization; the NMR methodology for (nondestructively) deciphering any unknown carbohydrate primary structure is available.

Solution Conformation Analysis by Nuclear Magnetic Resonance

^{1}H NMR spectroscopy is presented here as a method eminently suited for the elucidation of the primary structure of glycoprotein carbohydrates. It is also the method of choice for solution conformation analysis. Complete ^{1}H resonance assignments and primary structure determination are prerequisites for the analysis of the solution conformation based on quantitation of {^{1}H, ^{1}H} NOEs. Often assisted by other NMR parameters (^{13}C chemical shifts, heteronuclear coupling constants and NOE effects, isotope shift effects, etc.)[13] and evaluated by theoretical conformational analysis, that is, potential energy calculations of one sort or another (HSEA, AMBER, MM2, GEGOP/Monte Carlo, molecular dynamics, etc.[65–68]), two- and three-dimensional ^{1}H (and ^{13}C) NMR spectroscopy is the key experimental technique for solution conformation analysis of carbohydrates and glycoconjugates. Furthermore, the internal dynamics (segmental motions) of carbohydrate molecules in solution can be assessed by NMR spectroscopy if the rearrangements occur much faster (or much slower) than overall molecular tumbling.[69–71]

Ultimately, the knowledge of primary structures, three-dimensional conformations, and the dynamics/flexibility of glycoprotein oligosaccharides in the natural environment of their glycoprotein macromolecule will broaden our insights into their functioning as mediators of numerous biological cell–cell and cell–molecule interactions. Nuclear magnetic resonance spectroscopy is a most valuable contributor toward this understanding.

[64] C. Abeygunawardana and C. A. Bush, *Adv. Biophys. Chem.* **3** (1993) (in press).

[65] S. W. Homans, *Prog. NMR Spectrosc.* **22,** 55 (1990).

[66] B. Meyer, *Top. Curr. Chem.* **154,** 141 (1990).

[67] C. A. Bush and P. Cagas, *Adv. Biophys. Chem.* **2,** 149 (1992).

[68] A. D. French and J. W. Brady, eds., "Computer Modeling of Carbohydrate Molecules," ACS Symp. Ser. 430. Am. Chem. Soc., New York, 1990.

[69] L. Poppe and H. van Halbeek, *J. Am. Chem. Soc.* **114,** 1092 (1992).

[70] H. van Halbeek and L. Poppe, *Magn. Reson. Chem.* **30,** S74 (1992).

[71] L. Poppe and H. van Halbeek, *Adv. Carbohydr. Chem. Biochem.* (1993) (in press).

Acknowledgments

Research in the author's laboratory is supported by National Institutes of Health Grants
P41-RR-05351, P01-AI-27135, R01-AI-29751, R01-AI-32899, and S10-RR-04720. The author
thanks Dr. Michael W. Spellman (Genentech, Inc., San Francisco, CA), Dr. James Travis
(the University of Georgia, Athens, GA), and Dr. Eric Watson (Amgen, Inc., Thousand
Oaks, CA) for fruitful collaborations, Dr. John Glushka for recording most of the NMR
spectra discussed in this chapter, and Ms. Allyson Mann for editing the manuscript.

[10] Determination of Sialic Acids

By Gerd Reuter and Roland Schauer

Introduction

Sialic acids are constituents of glycoconjugates in some viruses, bacteria, protozoa, and a wide range of higher animals. They represent a family of about 30 similar sugars, most of them being O-acetylated (Table I).[1,2] There are a large and rapidly increasing number of examples based on their biological roles that often depend on the exact type of sialic acid present,[3-6] thus necessitating a discrimination between different sialic acid species. Excellent analytical systems are needed to allow qualitative and quantitative analyses of these sugars, which is complicated by the significantly greater lability of *O*-acetyl groups when compared to the non-*O*-acetylated parent sialic acids.

Additional compounds have been described in *Shigella boydii* and *Pseudomonas aeruginosa*[7-10] that may be considered as sialic acid derivatives; however, they have a different stereochemistry at C-5, C-7, and

[1] G. Reuter and R. Schauer, *Glycoconjugate J.* **5**, 133 (1988).
[2] A. P. Corfield and R. Schauer, *Cell Biol. Monogr.* **10**, 5 (1982).
[3] R. Schauer, *Trends Biochem. Sci.* **10**, 357 (1985).
[4] R. Schauer and G. Reuter, *NATO ASI Ser., Ser. H* **7**, 17 (1987).
[5] G. Reuter and R. Schauer, *NATO ASI Ser., Ser. H* **7**, 155 (1987).
[6] A. Varki, *Glycobiology* **2**, 25 (1992).
[7] Y. A. Knirel, E. V. Vinogradov, A. S. Shaskov, and N. K. Kochetkov, *Carbohydr. Res.* **141**, C1 (1985).
[8] Y. A. Knirel, N. A. Kocharova, A. S. Shashkov, and N. K. Kochetkov, *Carbohydr. Res.* **145**, C1 (1986).
[9] Y. A. Knirel, E. V. Vinogradov, A. S. Shashkov, B. A. Dimitriev, N. K. Kochetkov, E. S. Stanislavsky, and G. M. Mashilova, *Eur. J. Biochem.* **163**, 627 (1987).
[10] Y. A. Knirel, N. A. Kocharova, A. S. Shashkov, B. A. Dimitriev, N. K. Kochetkov, E. S. Stanislavsky, and G. M. Mashilova, *Eur. J. Biochem.* **163**, 639 (1987).

C-8 and a 3-hydroxybutyramido or formamido group at C-7. 3-Deoxy-D-*glycero*-D-*galacto*-nonulosonic acid (ketodeoxynonulosonic acid, KDN), which has an OH group in place of the acetamido group at C-5, was discovered in the eggs of rainbow trout as a terminal sugar on polysialic acid chains linked to glycoprotein,[11] as poly-KDN,[12] and as KDN-containing ganglioside[13] in this fish species. It has also been found in the egg jelly coat from the newt *Pleurodeles waltlii*.[14] In these examples, with the exception of poly-KDN, the modified sialic acid has not been found in the internal position. In the jelly coat of Mexican axolotl *Ambystoma mexicanum* eggs, it has also been found substituted by fucose.[15] A number of questions about the presence of this sugar in other animals, its biosynthesis, catabolism, and biological function remain unanswered.

This chapter adds to previous reviews on sialic acid analysis by Schauer[16,17] and gives examples of the procedures currently used by our group and of new developments from other laboratories. Isolation and purification of sialoglycoconjugates are also dealt with (see [4, 11, 15, and 22] in this volume). The following procedures describe isolation, purification, and characterization of monomeric sialic acids. Synthetic sialic acid derivatives will be mentioned only if there is special relevance to this chapter.

Isolation of Sialic Acids

Because of the possible heterogeneity of sialic acids in biological materials, special care must be taken to obtain the complete spectrum of these compounds, which includes release of all sialic acids from their glycosidic linkages and preservation of all substituents during release. This, however, cannot be achieved by the methods available up to now.

Before release of sialic acids from sialoglycoconjugates, it is advisable to dialyze the material in order to remove low molecular weight impurities that may interfere with the analytical procedures to be used.

[11] D. Nadano, M. Iwasaki, S. Endo, K. Kitajima, S. Inoue, and Y. Inoue, *J. Biol. Chem.* **261**, 11550 (1986).
[12] S. Kitazume, K. Kitajima, S. Inoue, and Y. Inoue, *Anal. Biochem.* **202**, 25 (1992).
[13] Y. Song, K. Kitajima, S. Inoue, and Y. Inoue, *Trends Glycosci. Glycotechnol.* **4**, 216 (1992).
[14] G. Strecker, J.-M. Wieruszeski, J.-C. Michalski, C. Alonso, B. Boilly, and J. Montreuil, *FEBS Lett.* **298**, 39 (1992).
[15] G. Strecker, J.-M. Wieruszeski, J.-C. Michalski, C. Alonso, Y. Leroi, B. Boilly, and J. Montreuil, *Eur. J. Biochem.* **207**, 995 (1992).
[16] R. Schauer, this series, Vol. 50, p. 64.
[17] R. Schauer, this series, Vol. 138, p. 132.

TABLE I

NATURALLY OCCURRING SIALIC ACIDS DERIVED FROM *N*-ACETYLNEURAMINIC ACID OR 2-DEOXY-2,3-DIDEHYDRO-*N*-ACETYLNEURAMINIC ACID AND FROM *N*-GLYCOLOYLNEURAMINIC ACID OR 2-DEOXY-2,3-DIDEHYDRO-*N*-GLYCOLOYLNEURAMINIC ACID

Acid	Abbreviated form of acid
N-Acetylneuraminic acid	2-Deoxy-2,3-didehydro-*N*-acetylneuraminic acid

Acid	Abbreviated form of acid
N-Acetylneuraminic acid	Neu5Ac
N-Acetyl-4-*O*-acetylneuraminic acid	Neu4,5Ac$_2$
N-Acetyl-7-*O*-acetylneuraminic acid	Neu5,7Ac$_2$
N-Acetyl-8-*O*-acetylneuraminic acid	Neu5,8Ac$_2$
N-Acetyl-9-*O*-acetylneuraminic acid	Neu5,9Ac$_2$
N-Acetyl-4,9-di-*O*-acetylneuraminic acid	Neu4,5,9Ac$_3$
N-Acetyl-7,9-di-*O*-acetylneuraminic acid	Neu5,7,9Ac$_3$
N-Acetyl-8,9-di-*O*-acetylneuraminic acid	Neu5,8,9Ac$_3$
N-Acetyl-7,8,9-tri-*O*-acetylneuraminic acid	Neu5,7,8,9Ac$_4$
N-Acetyl-9-*O*-acetyl-8-*O*-methylneuraminic acid	Neu5,9Ac$_2$8Me
N-Acetyl-9-*O*-lactoylneuraminic acid	Neu5Ac9Lt
N-Acetyl-4-*O*-acetyl-9-*O*-lactoylneuraminic acid	Neu4,5Ac$_2$9Lt
N-Acetyl-8-*O*-methylneuraminic acid	Neu5Ac8Me
N-Acetylneuraminic acid 9-phosphate	Neu5Ac9P
N-Acetylneuraminic acid 8-sulfate	Neu5Ac8S
2-Deoxy-2,3-didehydro-*N*-acetylneuraminic acid	Neu5Ac2en
2-Deoxy-2,3-didehydro-*N*-acetyl-9-*O*-acetylneuraminic acid	Neu5,9Ac$_2$en
2-Deoxy-2,3-didehydro-*N*-acetyl-9-*O*-lactoylneuraminic acid	Neu5Ac2en9Lt

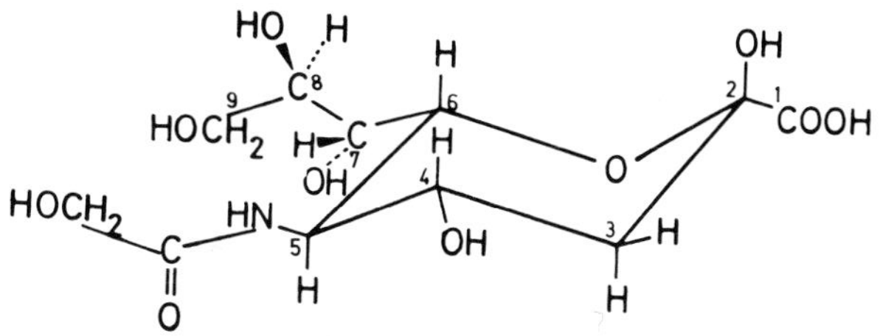

2-Deoxy-2,3-didehydro-N-glycoloylneuraminic acid

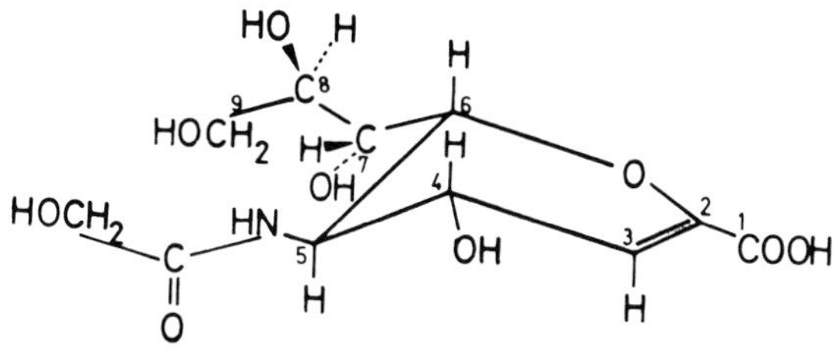

N-Glycoloylneuraminic acid

Neu5Gc
Neu4Ac5Gc
Neu7Ac5Gc
Neu9Ac5Gc
Neu7,9Ac$_2$5Gc
Neu8,9Ac$_2$5Gc
Neu7,8,9Ac$_3$5Gc
Neu9Ac5Gc8Me
Neu5GcAc
Neu5Gc9Lt
Neu5Gc8Me
Neu5Gc8S
Neu2en5Gc
Neu9Ac2en5Gc
Neu2en5Gc9Lt
Neu2en5Gc8Me

N-Glycoloylneuraminic acid
N-Glycoloyl-4-O-acetylneuraminic acid
N-Glycoloyl-7-O-acetylneuraminic acid
N-Glycoloyl-9-O-acetylneuraminic acid
N-Glycoloyl-7,9-di-O-acetylneuraminic acid
N-Glycoloyl-8,9-di-O-acetylneuraminic acid
N-Glycoloyl-7,8,9-tri-O-acetylneuraminic acid
N-Glycoloyl-9-O-acetyl-8-O-methylneuraminic acid
N-(O-Acetyl)glycoloylneuraminic acid
N-Glycoloyl-9-O-lactoylneuraminic acid
N-Glycoloyl-8-O-methylneuraminic acid
N-Glycoloylneuraminic acid 8-sulfate
2-Deoxy-2,3-didehydro-N-glycoloylneuraminic acid
2-Deoxy-2,3-didehydro-N-glycoloyl-9-O-acetylneuraminic acid
2-Deoxy-2,3-didehydro-N-glycoloyl-9-O-lactoylneuraminic acid
2-Deoxy-2,3-didehydro-N-glycoloyl-8-O-methylneuraminic acid

Care must be taken to prevent loss of sialic acids and/or O-acetyl groups due to the action of enzymes of sialic acid catabolism.[18] Therefore, freshly isolated and cooled materials should be kept in 50% (v/v) ethanol for 30–60 min at room temperature, or, especially in the case of tissues, sonicated in ethanol for about 15 min, concentrated by rotary evaporation of ethanol (bath temperature not higher than 30°) or heated to 96° for 3 min and dialyzed as described below. The diffusate must be tested for the presence of sialic acids by the methods described below. If only small amounts of sialic acids (<200 μg) are present, this predialysis should not be carried out, even if relatively large amounts of contaminants are present.

There are two basic procedures for liberating sialic acids from glycosidic linkages: enzymatic and acid hydrolysis. In the former case, a variety of sialidases (EC 3.2.1.18) may be used, which differ in specificity for sialic acid linkage or species. It should be noted that in most cases much lower amounts of sialic acids are released by sialidase than by acid,[19–21] which may be because of the different accessibility of the sialic acids in the materials to be analyzed. The ratio of cleavage rates for $\alpha(2 \rightarrow 3)$ versus $\alpha(2 \rightarrow 6)$ linkages depends on the type of sialidase used (Table II). The preferred cleavage of $\alpha(2 \rightarrow 3)$ linkages does not exclude hydrolysis of $\alpha(2 \rightarrow 6)$ linkages by the same enzyme. Thus, a discrimination of linkages by enzymatic methods alone is hardly possible, although hints for the prevalence of one type of linkage may be obtained under appropriate incubation conditions, for example, with Newcastle disease virus sialidase. From the different species, N-acetylneuraminic acid (Neu5Ac; for abbreviations of sialic acids, see Table I[1]) is released faster than N-glycoloylneuraminic acid (Neu5Gc). Substitution of these parent sialic acids by O-acetyl groups again reduces the hydrolysis rate.[22] In the case of 4-O-acetylated sialic acids, most enzymes do not work at all; a slow release has been observed only for fowl plague virus sialidase.[23] 9-O-Acetylation reduces the relative cleavage rates of the sialidases tested so far by more than 50%.[24]

[18] A. P. Corfield and R. Schauer, *Cell Biol. Monogr.* **10,** 195 (1982).
[19] C. Fischer, S. Kelm, B. Ruch, and R. Schauer, *Carbohydr. Res.* **213,** 263 (1991).
[20] E. Müller, C. Schröder, R. Schauer, and N. Sharon, *Hoppe-Seyler's Z. Physiol. Chem.* **364,** 1419 (1983).
[21] A. Kluge, G. Reuter, H. Lee, B. Ruch-Heeger, and R. Schauer, *Eur. J. Cell Biol.* **59,** 12 (1992).
[22] R. Schauer, *Adv. Carbohydr. Chem.* **40,** 131 (1982).
[23] R. G. Kleinedam, K. Furuhata, H. Ogura, and R. Schauer, *Biol. Chem. Hoppe-Seyler* **371,** 715 (1990).
[24] R. Schauer, this series, Vol. 138, p. 611.

TABLE II
SUBSTRATE SPECIFICITY OF VARIOUS SIALIDASES[a]

Source of sialidase	pH optimum	Neu5Acα2-3Lac	Neu5Acα2-6Lac	Neu5Acα2-8-Neu5Acα2-3Lac	Neu5Gcα2-3Lac
Vibrio cholerae	5.0–6.5	100	53	31	25
Clostridium perfringens	5.5–7.2	100	44	44	20
Arthrobacter ureafaciens	4.4–5.5	100	167	53	12
Fowl plague virus	5.5–6.0	100	2	14	37
Newcastle disease virus	5.0–6.0	100	0.2	78	11
Influenza A2 virus		100	0.4	3	15

[a] From Schauer.[22]

The incubation buffer used should not exceed pH 6 to prevent a possible migration or saponification of O-acetyl groups at higher pH values.[25] The amount of enzyme needed depends on the type and amount of sialic acids present and on the nature of the sialylated compound, that is, on whether it is a soluble complex carbohydrate or a tissue or cell. Typically, 1–10 mM sialic acid is incubated with 2–10 mU of sialidase in 500 μl of an appropriate buffer (e.g., 100 mM acetate buffer, pH 4.5–5.5, the pH optimum for most sialidases) for 1 hr at 37°. Attention must be paid to cation requirements, for example, Ca^{2+} is needed with *Vibrio cholerae* sialidase. To ensure that the release of sialic acids is due to enzyme action, a control experiment should be performed in the presence of 1 mM 2-deoxy-2,3-didehydro-N-acetylneuraminic acid (Neu5Ac2en; see Table I), the most potent sialidase inhibitor known so far[26] that is commercially available. A protocol for desialylation of mammalian erythrocytes is now given.

Enzymatic Desialylation of Rat Erythrocytes

Rat erythrocytes (10^{10} cells) are suspended in 1 ml of phosphate-buffered saline (PBS), pH 7.2, containing 1 mM Ca^{2+}, and 50 mU of *V. cholerae* sialidase is added. After incubation for 1 hr at 37°, the mixture is cooled on ice for 5 min and then centrifuged for 10 min at 900 g. The supernatant containing liberated sialic acids is removed and the cell pellet washed three times in 20 ml of PBS, pH 7.2, followed by centrifugation under the same conditions. The combined supernatants are further purified as described below.

Although the pH of this incubation mixture is higher than it should be for optimal preservation of O-acetyl groups and enzyme activity, the cells are more stable in this buffer and may be used as desialylated erythrocytes for further studies. If these modified cells are not needed and only sialic acid analysis is needed, acid hydrolysis of the cells or, better, of their membranes is preferred. Termination of the enzymatic reaction together with precipitation of the (asialo)glycoprotein can be achieved by the addition of 12% (w/v) trichloroacetic acid in 1% (w/v) phosphotungstic acid.[27] Finally, dialysis may be carried out to isolate the released sialic acids. This latter procedure is also applicable during enzymatic hydrolysis.[28]

[25] J. P. Kamerling, R. Schauer, A. K. Shukla, S. Stoll, H. van Halbeek, and J. F. G. Vliegenthart, *Eur. J. Biochem.* **162**, 601 (1987).

[26] R. Schauer and A. P. Corfield, *in* "Medicinal Chemistry Advances" (F. G. de las Heras and S. Vega, eds.), p. 423. Pergamon, Oxford and New York, 1981.

[27] J. M. Beau and R. Schauer, *Eur. J. Biochem.* **106**, 531 (1980).

[28] R. Schauer and H. Faillard, *Hoppe-Seyler's Z. Physiol. Chem.* **349**, 961 (1968).

Several methods are available for acid hydrolysis of sialic acid-containing complex carbohydrates. All these procedures suffer from being not optimal in giving the spectrum of the sialic acids originally present, but a complete release of sialic acids from their glycosidic linkages is possible and this is a valuable insight into their chemical composition. On treatment with aqueous acid, esters such as O-acetyl groups will at least be partially hydrolyzed. In addition, migration of these substituents, for example, from position 7 to 9, occurs in acidic as well as in alkaline solutions.[25] With milder conditions of hydrolysis, a greater portion of O-acetyl groups can be preserved, leading, however, to incomplete hydrolysis, that is, the total amount of sialic acids is too low and the relative amount of O-acetylated derivatives too high when compared with the results obtained after complete release of sialic acids by acid hydrolysis. Methanol–HCl treatment, often used for sugar analysis,[29] gives neuraminic acid methyl ester methylglycoside with loss of all ester groups and of the nitrogen substituent. Thus a discrimination between different species is not possible by this method, although the total amount and general presence of sialic acids can be deduced accurately after this type of hydrolysis.

In our experience, complete release of sialic acids from their glycosidic linkages together with a preservation of 50–70% of the O-acetyl groups is achieved by a two-step acid hydrolysis with formic acid, pH 2, for 1 hr at 70°, followed by HCl, pH 1, for 1 hr at 80°. After each step liberated sialic acids must be recovered either by centrifugation or by dialysis and in most cases the acid should be carefully removed before further processing (see the next section). The use of 2 M acetic acid for 3 hr at 80° as described in the literature[30] did not improve the yield of O-acylated sialic acids in our hands. The major disadvantage in using sulfuric acid for hydroysis[16,17] is that this acid is not removed as easily as HCl, formic, or acetic acid. Precipitation with barium hydroxide has been used, but this requires efficient filtration of barium sulfate, which is difficult to perform. As an example of the two-step acid hydrolysis, release of sialic acids from a mucin is now described.

Acid Hydrolysis of Bovine Submandibular Gland Mucin

Dialyzed and lyophilized bovine submandibular gland mucin (BSM)[31] (5 g) is suspended in about 20 ml of water in a glass-stoppered round-bottom flask. The pH is adjusted to 2 (± 0.2) with 50% (v/v) formic acid

[29] J. P. Kamerling and J. F. G. Vliegenthart, *Cell Biol. Monogr.* **10**, 95 (1982).
[30] A. Varki and S. Diaz, *Anal. Biochem.* **137**, 236 (1984).
[31] G. Reuter, R. Pfeil, S. Stoll, R. Schauer, J. P. Kamerling, C. Versluis, and J. F. G. Vliegenthart, *Eur. J. Biochem.* **134**, 139 (1983).

and the suspension incubated for 1 hr at 70°. After cooling on ice, the suspension is poured into a dialysis bag, which has been intensively washed to remove glycerol and other contaminants. The volume of the hydrolysate, including washings, should not exceed two-thirds of the volume of the dialysis bag. Dialysis is carried out against the 10- to 20-fold volume of the hydrolysate for about 16 hr at 4° with 3 changes of the diffusate. The content of the dialysis bag is then transferred to the round-bottom flask, the pH adjusted to 1–1.2 with 3 M HCl, and hydrolysis carried out for 1 hr at 80°, followed by dialysis as described. Much longer times and/or higher temperatures (room temperature) for dialysis will result in the loss of O-acetyl groups. All diffusates of one hydrolytic step are combined and preferably lyophilized; in the case of volumes higher than about 1 liter, diffusates may be concentrated to one-third by rotary evaporation under vacuum at 30° bath temperature (maximum). Concentration must not be carried out to dryness because this will lead to degradation of sialic acids. If the volume of the diffusate is small, the purification procedures described below may be carried out without the foregoing lyophilization. The content of the dialysis bag is also lyophilized and tested for any sialic acids that remain (see section on Quantitative Sialic Acid Analysis).

The reason for the separate processing of the HCOOH and HCl hydrolysates is that the diffusate from the first one contains the majority of O-acetylated sialic acids, whereas that from the second one contains mostly Neu5Ac and Neu5Gc. Thus, the HCOOH hydrolysis represents a fraction, in which O-acetylated species are enriched and therefore more easily recognized. Mixing of the HCOOH and HCl hydrolysates results in a decrease of pH in the former, which may lead to further loss of O-acetylation.

Purification of Sialic Acids

Sialic acids of the two fractions obtained after acid hydrolysis or the one from sialidase incubation are purified prior to analyses. Depending on the nature of the starting material different procedures should be followed.

Ether Extraction

Lyophilized sialic acid-containing samples isolated from tissues or cells are dissolved in a small volume of water and extracted three times with an equal volume of diethyl ether each time. The ether layers are discarded and the water phase carefully concentrated by rotary evaporation, as described above, to remove all of the ether. This procedure reduces the amount of lipophilic contaminants that may be present in the samples and that may hinder further analyses.

Ion-Exchange Chromatography

With or without the foregoing ether extraction, all samples are dissolved in a small volume of water and passed over cation-exchange resin (Dowex 50W-X8, 20–50 mesh, H⁺ form, about 2 ml of resin for up to 300 μg of sialic acids). After application of the sample, the column is washed with water (a three- to fivefold column volume). The eluate is collected together with the washings. Before anion-exchange chromatography on Dowex 2-X8 or 1-X8 (200–400 mesh, HCOO⁻ form), the pH of the solution is measured and, if it is below 2, carefully adjusted to about 4 under cooling on ice with diluted ammonia. Instead, the solution may be lyophilized before anion-exchange chromatography, which should be done in any case if the total volume exceeds about 20 ml. The sample is given on the anion-exchange resin (the same volume of resin as for cation-exchange chromatography) and the column washed with 3 to 5 vol of water. Adsorbed sialic acids are then eluted with three column volumes of 1.5–2 M formic acid. If many impurities are expected to be present, gradient elution may be performed instead, ranging from 0 to 2 M formic acid, with about 20 column volumes in total. This latter procedure will result in purer sialic acid preparations; however, the individual fractions must be tested for the presence of sialic acids. Therefore the elution volume of sialic acids should be determined on the same column with a known sialic acid standard. Alternatively, ³H or ¹⁴C-labeled Neu5Ac can be added to the sample and elution monitored by radioactivity. Gradient elution is an ideal method for the preparation of larger amounts of individual sialic acids.[31] Anion-exchange chromatography may be repeated if the purity does not seem to be appropriate after the first run. The sialic acid-containing fractions are diluted with water and lyophilized. Because adsorption of sialic acids to the anion-exchange resin may be incomplete, due to a low pH value of the solution, high salt concentrations, or application of too much sialic acid, the water fraction must also be tested for the presence of sialic acids.

Gel Filtration

Instead of or in addition to a second anion-exchange chromatographic separation, further purification steps may be required for optimal qualitative and quantitative sialic acid analysis (e.g., for sialic acid determination on erythrocytes); these steps often involve gel filtration on BioGel P-2 or P-4 (Bio-Rad, Richmond, CA). Column dimensions of 30–40 × 1–2 cm are appropriate for quantities up to about 300 μg of sialic acids. Elution is performed with a pump, using water as eluent; if possible, elution should be controlled by ultraviolet (UV) measurement at 200 to 214 nm, with decreasing sensitivity at higher wavelengths. Again, the column should be calibrated with standard sialic acids. Good results have also been obtained with Fraktogel TSK HW-40(S) (Merck, Rahway, NJ), as gel matrix, which can be used at higher flow rates than the BioGel material. Gel chromatographic media composed of saccharide polymers should not be used because the sample will often become contaminated, complicating further analysis.

Column Chromatography on Cellulose

Similar to the thin-layer chromatography (TLC) procedure described below, sialic acid purification is possible by column chromatography (30 × 1 cm for about 200 μg of sialic acids) on cellulose (MN 2100ff; Macherey & Nagel, Düren, Germany) with 1-propanol–1-butanol–water (2:1:1, v/v/v). In contrast to the TLC system, this solvent must not contain 0.1 M HCl, which may lead to partial hydrolysis of the cellulose matrix and consequently introduce impurities into the sample. Elution should be calibrated with a standard and can easily be monitored at 206 nm or with the orcinol/Fe^{3+}/HCl assay (see that section). This separation procedure is also valuable for preparation of individual sialic acids.[31]

Reversed-Phase Chromatography

In the presence of hydrophobic contaminants (e.g., colored substances) purification over cartridges containing RP-18 material is advisable. The cartridge [1 ml; e.g., Waters (Milford, MA) or Baker (Phillipsburg, NJ)] is washed successively with 5 ml of methanol, 5 ml of methanol–chloroform (1:1, v/v), 5 ml of methanol–chloroform (1:2, v/v), 5 ml of methanol, and finally twice with 5 ml of water. The aqueous sample is then applied and the cartridge washed with 2 ml of water. Elution is performed with 2 ml of water and 2 ml of water–methanol (2:1, v/v). All eluates are kept separate and tested for the presence of sialic acids.

Preparative (High-Performance) Thin-Layer Chromatography

Finally, purification can be achieved by preparative [high performance (HP)] TLC in the systems described below for analytical purposes. Indication of the migration positions of sialic acids is obtained by cutting small stripes on both sides of the plastic plate and staining in the conventional way or, in the case of glass plates, by spraying with water, which shows migrated material as opaque bands. The bands are scraped from the plate, the material collected, and sialic acids extracted with water, followed by water–methanol (1:1, v/v). Care must be taken in this procedure because sialic acids occurring in only small amounts may not be detected on the plates. Extraction of sialic acids from the TLC support (cellulose or silica gel) is not quantitative; in our experience losses of up to 40% are often observed.

Quantitative Sialic Acid Analysis

Quantitation of sialic acids usually requires purified samples because a number of substances are known to interfere with certain tests. Colorimetric sialic acid quantitation is still one of the most important methods to determine the amount of these sugars in a given sample accurately. Two basically different tests are used: the orcinol/Fe^{3+}/HCl[32] or resorcinol/Cu^{2+}/HCl[33] assay and the periodic acid/thiobarbituric acid test according to Warren[34] or Aminoff,[35] which have been described in detail earlier.[16,36] The procedures for a microadaptation of the orcinol/Fe^{3+}/HCl and the periodic acid/thiobarbituric acid assays as performed in our laboratory are given below.

Orcinol/Fe^{3+}/HCl Assay

Reagent. Concentrated (37%) HCl (81.4 ml), 0.2 g of orcinol, and 2 ml of 1% $FeCl_3$ solution (1.66 g $FeCl_3 \cdot 6H_2O$ in 100 ml of H_2O) are mixed and made up to 100 ml with H_2O. The solution should be stored at 4° and is stable at this temperature for several weeks. For spraying TLC plates, three parts of this solution are mixed with one part of water.

Test. Reagent (100 μl) is placed in 100 μl of solution containing up to 8 μg of sialic acids, and the solution is thoroughly mixed and incubated for 15 min at 96°. After cooling on ice for 5 min, 700 μl of isoamyl alcohol is added, and the mixture thoroughly vortexed and centrifuged for 5 min at 3000 g. The upper organic layer is read by photometry at 572 nm in appropriate cuvettes.

Comments. The HCl that must be used for preparing the reagent may contain compounds that interfere with the assay; this must be tested. Attention should be paid to take only the organic layer for photometry; traces of water result in turbid solutions that do not allow accurate photometric reading. The molar extinction coefficients of sialic acids [5300 liter/(mol × cm) for Neu5Ac] are listed in Schauer[22] and Schauer and Corfield.[36]

[32] P. Böhm, S. Dauber, and L. Baumeister, *Klin. Wochenschr.* **32**, 289 (1954).
[33] L. Svennerholm, *Biochim. Biophys. Acta* **24**, 604 (1957).
[34] L. Warren, *J. Biol. Chem.* **234**, 1971 (1959).
[35] D. Aminoff, *Virology* **7**, 355 (1959).
[36] R. Schauer and A. P. Corfield, *Cell Biol. Monogr.* **10**, 77 (1982).

Periodic Acid/Thiobarbituric Acid Assay according to Warren[34]

  Reagents

Reagent 1: Concentrated H_3PO_4 (75 ml) and 5.7 g of H_5IO_6 prepared
  uniformly are made up to 100 ml with H_2O
Reagent 2: $NaAsO_2$ (5 g) and 7.1 g of Na_2SO_4 dissolved in 100 ml H_2O
Reagent 3: Thiobarbituric acid (0.9 g) and 7.1 g of Na_2SO_4 dissolved in
  100 ml of H_2O. This solution should be stored at $-20°$ and thawed
  at about 30° before use
Reagent 4: Cyclohexanone

  Test. Reagent 1 (20 μl) is added to 100 μl of solution containing up to
3.5 μg of sialic acid, and the solution is mixed and incubated for 30 min
at 37°. Then 200 μl of reagent 2 is added and the solution vigorously
vortexed until the yellow-brown color formed disappears. Reagent 3 (200
μl) is added, the solution is mixed and incubated for 15 min at 96°, followed
by cooling on ice for 5 min. Reagent 4 (700 μl) is added, and the solution
is vigorously mixed and centrifuged for 5 min at 3000 g. The upper organic
layer is read photometrically at 549 and 532 nm in appropriate cuvettes.

  Comments. The sialic acid amount should be calculated as discussed
below. After addition of reagent 2, formation and disappearance of a
yellow-brown color must take place, otherwise the test must be repeated
and the reagents checked. This color is due to a reduction of IO_6^{5-} by
AsO_2^- to I^- via the colored I_2. Again, careful removal of only the organic
layer is necessary for reliable photometry. The molar extinction coefficients for sialic acids (61,000 liter/(mol $\times$ cm) for Neu5Ac) are listed in
Schauer[22] and Schauer and Corfield.[36]

  It should be mentioned that the orcinol/Fe^{3+}/HCl assay is suited to
determine the amount of sialic acids as free sugars or in glycosidic linkages,
whereas the periodic acid/thiobarbituric acid test requires free sialic acids.
Thus, when both tests are used in combination, a differentiation between
total and free sialic acid is possible, allowing calculation of the amount
of glycosidically bound sialic acids. On the other hand, the orcinol assay
may be used to roughly determine the sialic acid amount already in the
starting material by mixing equal volumes of sample and reagent followed
by incubation for 15 min at 96°; for this purpose, it is not necessary to
perform the whole procedure. It should be noted, however, that this value
will not be correct in most cases because of interfering substances, but
it allows a first estimation and appropriate setup of further purification
conditions. The presence of organic solvents prevents color formation in
this assay. In these cases, the occurrence of sialic-acid-containing substances is checked by spotting small amounts (up to 50 μl) of the solution

on silica gel TLC plates and, after drying, staining the plates with the orcinol/Fe^{3+}/HCl spray reagent.

A severe problem of the periodic acid/thiobarbituric acid assay is that the yield of chromophore is drastically reduced in the presence of substituents (e.g., O-acetyl groups) at the C-7/C-8/C-9 side chain of sialic acids that prevent periodate oxidation at this part,[37] a prerequisite of color formation.[34] A possible influence of ester substituents, which is the most frequently found type of sialic acid modification, can easily be recognized when this test is repeated after saponification (see Saponification of O-Acetylated Sialic Acids).

In the presence of contaminants additional chromophores may be formed and, in consequence, the typical color of the sialic acid derivatives may not appear. In other cases, the chromophore formed contributes to the extinction observed for sialic acid (e.g., deoxyribose in the periodic acid/thiobarbituric acid assay, with an extinction maximum at 532 nm). To estimate whether sialic acids are nevertheless present in the sample, it is advisable to record the spectrum of the chromophores between 400 and 800 nm. The corresponding spectra obtained with pure N-acetylneuraminic acid in the orcinol/Fe^{3+}/HCl and periodic acid/thiobarbituric acid tests are depicted in Fig. 1. Various methods have been used to calculate the amount of sialic acids in the latter assay. Neither reading at 549 nm alone nor calculation of the difference of OD_{549} − OD_{532} is satisfactory for samples from biological materials, although both procedures will give good results for purified preparations, the use of which is highly recommended for accurate colorimetric quantification. In our experience, it is best to calculate the sialic acid amount on the basis of the empirical formula given in the literature,[34] which must be corrected for the volume of 700 μl of cyclohexanone used in the modified procedure. It then reads as

$$\text{Sialic acid } (\mu\text{mol}) = 0.015 OD_{549} - 0.0054 OD_{532}$$

The detection limit in this assay is 0.5 μg (1.6 nmol); it is linear up to 3.5 μg of sialic acid in the test volume of 100 μl and should be used only in this range if the experimenter has no experience in using nonlinear calibration curves for accurate calculations. The orcinol/Fe^{3+}/HCl assay is linear up to 8 μg of sialic acid and has a detection limit of 2 μg (6.5 nmol).

Further methods for colorimetric sialic acid quantitation have been described in former reviews,[16,17] including a procedure that allows discrimination between C-8/C-9-substituted and unsubstituted sialic acids after

[37] J. Haverkamp, R. Schauer, M. Wember, J. P. Kamerling, and J. F. G. Vliegenthart, *Hoppe-Seyler's Z. Physiol. Chem.* **356,** 1575 (1975).

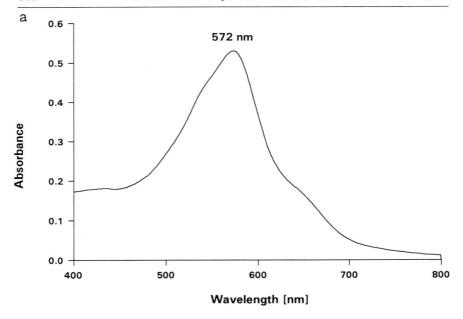

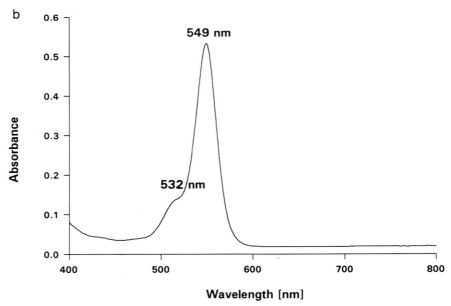

Fig. 1. Absorption spectrum of Neu5Ac in the orcinol/Fe^{3+}/HCl (a) and in the periodic acid/thiobarbituric acid (b) assay recorded from 400 to 800 nm.

periodate oxidation.[38] In a combined enzymatic assay using sialidase, β-galactosidase, and galactose dehydrogenase, a luminescence test has been developed for determination of sialic acids (as sialyllactose) at the nanogram level.[39] A further enzymatic test is available that needs sialidase followed by N-acylneuraminate lyase (EC 4.1.3.3), which cleaves N-acylneuraminic acids to the corresponding N-acylmannosamine derivatives and pyruvate.[40] The amount of pyruvate is determined by further enzymatic tests. Although this assay is sensitive, it will easily give too-low values due to incomplete enzymatic hydrolysis of sialic acids that may be caused by linkages that are not well hydrolyzed and/or by modifications, for example, O-acetylation, of N-acetylneuraminic acid, which in turn will also lead to slower action of the lyase.

The determination of sialic acids in serum, urine, or tissues or on cells is often used in clinical applications, because it seems to be a valuable marker for certain malignancies.[41-45] In some diseases, a change in the type of sialic acid has been observed. For human colon carcinoma, the occurrence of N-glycoloylneuraminic acid, which is normally absent in humans, has been found.[46] In mammary carcinoma, an increase in N-acetylneuraminic acid 9-O-acetylation of T cells has been described.[47] In human melanoma, a higher proportion of N-acetyl-9-O-acetylneuraminic acid in gangliosides was found.[6]

The other methods appropriate for qualitative sialic acid analysis as specified below may all be used for quantitation after calibration with known standards. Because these calibration procedures are well covered in a number of books dealing with the corresponding apparatuses [high-performance liquid chromatography (HPLC), gas-liquid chromatography (GLC), mass spectrometry (MS), densitometry, and radio-TLC] they will not be discussed here.

[38] A. K. Shukla and R. Schauer, *Hoppe-Seyler's Z. Physiol. Chem.* **363,** 255 (1982).
[39] J. A. Cabezas, N. Perez, M. Llanillo, A. Reglero, and P. Calvo, *Hoppe-Seyler's Z. Physiol. Chem.* **365,** 415 (1984).
[40] K. Sugahara, K. Sugimoto, O. Nomura, and T. Usui, *Clin. Chim. Acta* **108,** 493 (1980).
[41] E. M. J. Schutter, J. J. Visser, G. J. van Kamp, S. Mensdorff-Pouilly, W. van Dijk, J. Hilgers, and P. Kenemans, *Tumor Biol.* **13,** 121 (1992).
[42] G. Lindberg, L. Råstam, B. Gullberg, and G. A. Eklund, *Int. J. Epidemiol.* **21,** 253 (1992).
[43] C. H. Chien, Y. H. Wei, and J. F. Shaw, *Enzyme Microb. Technol.* **13,** 45 (1991).
[44] R. O'Kennedy, G. Berns, E. Moran, H. Smyth, K. Carroll, R. D. Thornes, and A. O'Brien, *Cancer Lett.* **58,** 91 (1991).
[45] C. Tautu, D. Pee, M. Dunsmore, J. J. Prorok, and J. A. Alhadeff, *J. Clin. Lab. Anal.* **5,** 247 (1991).
[46] T. Kawai, A. Kato, H. Higashi, S. Kato, and M. Naiki, *Cancer Res.* **51,** 1242 (1991).
[47] H. Stickl, W. Huber, H. Faillard, A. Becker, R. Holzhauser, and H. Graeff, *Klin. Wochenschr.* **69,** 5 (1991).

Qualitative Sialic Acid Analysis

Thin-Layer Chromatography

For TLC of sialic acids several systems have been described. Specific detection of sialic acid-containing bands is performed with the orcinol/ Fe^{3+}/HCl spray reagent described earlier. Development takes place for 15–20 min at 120° in a chamber containing the plate with the spray reagent. Sialic acids become visible as purple bands. Appearance of blue-gray bands points to the presence of other sugars; high amounts of salt become visible as white spots that in most cases prevent a normal separation of sialic acids on the corresponding lane. In the case of radioactive sialic acids obtained, for example, from metabolic labeling experiments, detection is performed with a TLC scanner. Nonradioactive reference sialic acids should also be applied on the same plate or lane to allow colorimetric detection, and thus to identify the bands of radioactivity. The most versatile system for the analysis of sialic acids, including their O-acetylated derivatives, in our hands consists of plastic plates precoated with cellulose (Merck) and 1-propanol–1-butanol– 0.1 *M* HCl (2 : 1 : 1, v/v/v) as solvent system running over a distance of about 8 cm for normal analysis; if better separations are needed (e.g., for radio-TLC or densitometric quantitation), the running distance can be extended to 18 cm without unacceptably severe band broadening. Substitution of 1-propanol by 2-propanol has hardly any effect on the separation. The use of HPTLC plates gives significantly better separations than the conventional layers.

Cellulose always contains relatively high amounts of impurities. Therefore the plates are twice prerun in the solvent, resulting in a yellow-brown band with an R_f value of about 0.6 after the first run. Prerun plates should not be kept for long periods of time (> 4 weeks), as separation becomes worse with storage, probably because of residual traces of acid present in the layer after the prerun, which may affect the cellulose matrix. Because separation may vary considerably among different cellulose batches, analysis of a sialic acid standard is highly recommended to test new plates. If sialic acids from BSM are used as standard (Fig. 2), the bands of Neu5Gc, Neu5Ac, Neu9Ac5Gc (which also contains Neu5,7Ac$_2$), Neu5,9Ac$_2$, and, depending on the preparation, sometimes also a band corresponding to higher-O-acetylated sialic acids should be visible. Because of the often observed variations in R_f values even on plates that show good separation of individual sialic acids, analyses should be carried out in the presence of appropriate reference sialic acids on a separate

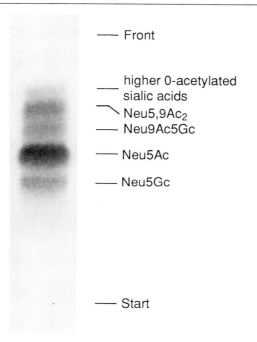

——— Front

___ higher 0-acetylated
    sialic acids
⟍ Neu5,9Ac₂
——— Neu9Ac5Gc

——— Neu5Ac

——— Neu5Gc

——— Start

FIG. 2. Thin-layer chromatographic analysis of sialic acids from bovine submandibular gland glycoproteins on cellulose with 1-propanol–1-butanol–0.1 M HCl (2 : 1 : 1, v/v/v) as solvent system and indication by the orcinol/Fe^{3+}/HCl spray reagent.

lane. The use of cellulose with fluorescence indicator is not recommended; we observed darkening of the whole plate after spraying with the orcinol/Fe^{3+}/HCl reagent. Aluminum-backed TLC plates are difficult to handle in this test because of the acidic spray reagent.

Two-dimensional TLC with intermediate ammonia treatment for saponification of sialic acid O-acetyl groups on cellulose plates gives information about parent sialic acids. By this method a discrimination, for example, between Neu5,7Ac₂ and Neu9Ac5Gc, which comigrate in the cellulose system described, is possible.[48] This test, however, is difficult because it usually takes a number of experiments before satisfactory results are obtained, and cannot be considered a routine method. In the case of O-acetylated sialic acid-containing gangliosides that are separated on silica

---

[48] H.-P. Buscher, J. Casals-Stenzel, and R. Schauer, *Eur. J. Biochem.* **50,** 71 (1974).

gel HPTLC plates, two-dimensional separation including saponification has been used successfully.[49,50]

Confirmation of the results obtained here can be achieved in other TLC systems.[16,17,37] The use of HPTLC silica gel with fluorescence indicator and ethanol–pyridine–1-butanol–water–acetic acid (50 : 5 : 5 : 15 : 1.5, v/v/v/v/v) as solvent system is also possible for sialic acid analysis.[17] Separation in this system is less effective than in the cellulose system described, but the sensitivity is higher.

Comparison of the results obtained in different systems allows reliable detection of the presence of certain sialic acids, which is further facilitated by the specific spray reagent or, in the case of metabolic labeling experiments,[51,52] by the specificity of incorporation of the precursors added.

Nondestructive sensitive detection of sialic acid-containing glycosphingolipids on HPTLC plates has been described by using primuline[53] or fluorochromes,[54,55] making both types of assays useful for preparative thin-layer chromatography.

Saponification of O-Acetylated Sialic Acids

An indication of the presence of O-acetylated sialic acids is obtained after saponification of a part of the sample as described in the following.

Sialic acids (up to 10 μg/μl) are mixed with a fivefold volume of 4 M NH$_3$ and kept for 4 hr at room temperature. The sample is dried by lyophilization and resuspended in water to the original concentration. Alternatively, an equal volume of 0.1 M NaOH is added to the sialic acids and the solution kept for 30–45 min at 4°, followed by addition of an equal volume of 0.1 M HCl.[24] Although saponification with ammonia may not be complete, this sample can easily be concentrated and is free of salt. The procedure using NaOH should be used when sufficient material is available and NaCl does not affect further analysis. Removal of NaCl by ion-exchange or gel chromatography as described above is also possible.

[49] D. C. Gowda, G. Reuter, A. K. Shukla, and R. Schauer, *Hoppe-Seyler's Z. Physiol. Chem.* **365,** 1247 (1984).

[50] S. Sonnino, R. Ghidoni, V. Chigorno, M. Masserini, and G. Tettamanti, *Anal. Biochem.* **128,** 104 (1983).

[51] H. J. Schoop, R. Schauer, and H. Faillard, *Hoppe-Seyler's Z. Physiol. Chem.* **350,** 155 (1969).

[52] H. C. Yohe, K. Ueno, N.-C. Chang, G. H. Glaser, and R. K. Yu, *J. Neurochem.* **34,** 560 (1980).

[53] S. B. Levery, E. D. Nudelman, M. E. K. Salyan, and S.-I. Hakomori, *Biochemistry* **28,** 7772 (1989).

[54] J. Müthing and D. Heitmann, *Anal. Biochem.* **208,** 121 (1993).

[55] J. Müthing and F. Unland, *Biomed. Chromatogr.* **6,** 227 (1992).

High-Performance Liquid Chromatography

Separation of native, that is, underivatized sialic acids was first performed on Aminex A-28 or A-29 anion-exchange resin (40 × 4.6 mm; Bio-Rad), using 0.75 mM Na$_2$SO$_4$ for isocratic elution and UV detection at 200 nm.[56] The detection limit is at 30 ng (0.1 nmol). Higher wavelengths may be used instead but only with considerable loss of intensity. A major disadvantage of this method is that detection is unspecific and separation of the frequently occurring sialic acids Neu5,9Ac$_2$ and Neu5Gc is not possible. This can, however, be overcome by saponification or by use of enzymes,[57] as will be described in the section Esterase Incubation.

A different HPLC assay has been developed in which fluorigenic derivatives of free sialic acids are formed, allowing relatively specific and highly sensitive detection. From the various protocols given in the literature,[58–60] the procedure adopted in our laboratory is now detailed.

Derivatization. 1,2-Diamino-4,5-methylenedioxybenzene (DMB) (15.75 mg) is dissolved in 5 ml of water. 2-Mercaptoethanol (680 μl) and 39 μl of 37% NaHSO$_3$ are added and the volume made up to 10 ml with water. Aliquots of 100 μl are stored in caps at $-20°$ in the dark until use. Up to 20 μl of the sialic acid-containing sample is added to one cap and the solution incubated for 2.5 hr at 56° in the dark. The sample is cooled on ice and is then ready for analysis. The derivatized samples may be kept at $-20°$ in the dark for several weeks without significant loss of peak intensity.

High-Performance Liquid Chromatography Analysis. HPLC is performed in the isocratic mode on an RP-18 cartridge (4 × 250 mm, 5-μm particle size; Merck) with methanol–acetonitrile–water (7 : 9 : 84, v/v/v) as eluent at a flow rate of 1 ml/min and fluorimetric detection at 373 nm excitation and 448 nm emission wavelengths. An appropriate cutoff filter may be used instead. Up to 20 μl of sample solution is injected. The retention time of Neu5Ac is at about 10 min in this system (Fig. 3).

The detection limit of this assay is in the femtomole range. It is specific for α-keto acids. 3-Deoxy-D-*manno*-octulosonic acid (KDO) and 3-deoxy-D-*glycero*-D-*galacto*-nonulosonic acid (KDN) can also be derivatized and

[56] A. K. Shukla and R. Schauer, *J. Chromatogr.* **244**, 81 (1982).
[57] A. K. Shukla and R. Schauer, *Anal. Biochem.* **158**, 158 (1986).
[58] S. Hara, M. Yamaguchi, Y. Takemori, and M. Nakamura, *J. Chromatogr.* **377**, 111 (1986).
[59] S. Hara, Y. Takemori, M. Yamaguchi, M. Nakamura, and Y. Ohkura, *Anal. Biochem.* **164**, 138 (1987).
[60] S. Hara, M. Yamaguchi, Y. Takemori, K. Furuhata, H. Ogura, and M. Nakamura, *Anal. Biochem.* **179**, 162 (1989).

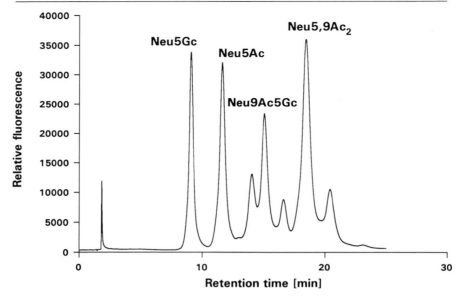

FIG. 3. HPLC analysis of sialic acids from BSM, using fluorimetric detection with the DMB reagent under the conditions described in text.

analyzed in this HPLC system. The retention times of these compounds and various sialic acids are listed in Table III.

Separation of N-acetyl- and N-glycoloylneuraminic acid as per-O-benzoylated derivatives has been described, using RP-18 as stationary phase,

TABLE III

RETENTION TIMES OF SIALIC ACIDS AND OTHER ULOSONIC ACIDS IN
FLUORIMETRIC HIGH-PERFORMANCE LIQUID CHROMATOGRAPHY ASSAY[a]

Sialic acid	Relative retention time
Neu5Ac	1.00
Neu4,5Ac$_2$	1.71
Neu5,9Ac$_2$	1.62
Neu5Gc	0.78
Neu9Ac5Gc	1.30
Neu5Gc8Me	0.98
3-Deoxy-D-*manno*-octulosonic acid (KDO)	0.76
3-Deoxy-D-*glycero*-D-*galacto*-nonulosonic acid (KDN)	0.73

[a] Relative to N-acetylneuraminic acid. For further details, see text.

isocratic elution, and UV detection at 230 nm.[61] The detection limit in this assay is 30 pmol. Measuring of chemiluminescent N-acetylneuraminic acid derivatives by isocratic reversed-phase HPLC has been reported with a detection limit of 9 fmol.[62]

Another sensitive HPLC method has been established, in which mono- or oligosaccharides are eluted from the HPLC column with strong alkali followed by pulsed amperometric detection.[63] Although this method is sensitive and useful in carbohydrate analysis (see [12] in this volume), it is not suited for general sialic acid analysis because O-acetyl groups are lost due to the alkaline treatment. The use of eluents with a pH value of 4.65 has also been described[63] for acidic compounds, but to our knowledge it has not yet been applied to the separation of alkali-sensitive oligosaccharides.

A system for testing the activity of sialate 9-O-acetylesterase (EC 3.1.1.53) has been set up with the synthetic sialic acid derivative 2-α-(N-dansyl-4-aminophenylthio)-N-acetyl-9-O-acetylneuraminic acid, which serves as substrate for this enzyme. The 9-de-O-acetylated reaction product can easily be separated from the substrate by HPLC and detected by a fluorimetric reading that allows determinations in the picomole range.[64]

Similarly, a sialyltransferase assay has been developed in which the fluorescent sialic acid N-acetyl-9-O-fluoresceinylneuraminic acid is transferred onto appropriate acceptors by sialyltransferases.[65] In addition to having a sensitive nonradioactive sialyltransferase test, it is also possible to introduce a fluorescence label, for example, on cells by this method.

Gas-Liquid Chromatography

Gas-liquid chromatography (GLC) of sialic acids requires derivatization prior to analysis. This has often been done by transformation of these sugars into per-O-trimethylsilyl (Me$_3$Si) ether methyl ester derivatives,[66] which separate on packed columns better than the corresponding Me$_3$Si esters.[67] Introduction of WCOT capillary columns for GLC analysis of sialic acids also allows, however, the use of the latter derivatives, which can be prepared more easily than the methyl esters. Several deriva-

---

[61] F. Unland and J. Müthing, Biomed. Chromatogr. **6**, 155 (1992).

[62] J. Ishida, T. Nakahara, and M. Yamaguchi, Biomed. Chromatogr. **6**, 135 (1992).

[63] R. R. Townsend, M. R. Hardy, and Y. C. Lee, this series, Vol. 179, p. 65.

[64] B. Reinhard, A. Becker, J. Rothermel, and H. Faillard, Biol. Chem. Hoppe-Seyler **373**, 63 (1992).

[65] H. J. Gross, U. Sticher, and R. Brossmer, Anal. Biochem. **186**, 127 (1990).

[66] J. P. Kamerling, J. F. G. Vliegenthart, C. Versluis, and R. Schauer, Carbohydr. Res. **41**, 7 (1975).

[67] G. Reuter and R. Schauer, Anal. Biochem. **157**, 39 (1986).

TABLE IV

GAS-LIQUID CHROMATOGRAPHIC RETENTION TIMES OF SIALIC ACIDS RELATIVE
TO N-ACETYLNEURAMINIC ACID AND CHARACTERISTIC MASS SPECTROMETRIC
FRAGMENT IONS[a]

Sialic acid	Relative retention time	Fragment						
		A	B	C	D	E	F	G
Neu5Ac	1.00	726	624	536	356	375	205	173
Neu4,5Ac$_2$	1.05	696	594	506	356	—	205	143
Neu5,9Ac$_2$	1.02	696	594	536	356	375	175	173
Neu5,8,9Ac$_3$	1.04	666	564	536	356	375	—	173
Neu5Ac2en	1.01	636	—	446	356	285	205	—
Neu5Gc	1.19	814	712	624	444	375	205	261
Neu9Ac5Gc	1.21	784	682	624	444	375	175	261
Neu5Ac8Me	0.98	668	566	536	356	375	147	173
Neu5,9Ac$_2$8Me	1.00	638	536	536	356	375	117	173
Neu5Gc8Me	1.14	756	654	624	444	375	147	261
Neu9Ac5Gc8Me	1.17	726	624	624	444	375	117	261

[a] Sialic acids were measured as per-O-trimethylsilyl ether trimethylsilyl esters on a 25-m WCOT capillary column coated with CP-Sil 5 and electron impact mass spectra recorded at 70 eV.[67,69]

tization reagents are available, all of which have certain disadvantages. Pyridine–1,1,1,3,3,3-hexamethyldisilazane–chlorotrimethylsilane (5 : 1 : 1, v/v/v) and N-trimethylsilylimidazole are widely used, but they show severe tailing of the solvent peak. The residual derivatization reagent, especially in combination with mass spectrometric analysis, is still visible after 1 hr at about 200° oven temperature. N-Methyl-N-trimethylsilyl-2,2,2-trifluoroacetamide (MSTFA)–pyridine shows significantly less tailing but leads to formation of N-Me$_3$Si derivatives;[68] thus two different peaks are found for each sialic acid.

The retention times of the most widespread sialic acids as per-O-trimethylsilyl ether trimethylsilyl esters are given in Reuter and Schauer[67] and listed in Table IV, together with several sialic acids discovered in the starfish Asterias rubens[69]; the m/z values of characteristic fragment ions observed in mass spectrometry (see the next section) are also indicated.

Mass Spectrometry

The methods described so far do not allow identification of the individual sialic acids, but correlate sample data with known standards. Depend-

[68] C. Schröder and R. Schauer, Fresenius' Z. Anal. Chem. **311**, 385 (1982).
[69] A. A. Bergwerff, S. H. D. Hulleman, J. P. Kamerling, J. F. G. Vliegenthart, L. Shaw, G. Reuter, and R. Schauer, Biochimie **74**, 25 (1992).

ing on the selectivity of detection, this correlation may easily be wrong. Mass spectrometry, in contrast, allows identification of the individual compounds. If this method is used in the electron-impact mode, the fragmentation ions yield complete information about the type of sialic acid present and possible substitutions within this molecule.[31,70] The fragmentation scheme that has been set up on the basis of a few per-O-SiMe$_3$ ether methyl esters of sialic acids[66] has been found to be applicable to SiMe$_3$ ester derivatives (Fig. 4a and b) and to all sialic acids found so far.[67] However, before mass spectrometric analysis separation of sialic acids is necessary, which is often performed by GLC.

The information obtained in the chemical ionization (CI) mode (Fig. 4c) greatly depends on the parameters used for measurement. If peaks are formed that correspond to $[M + H]^+$ without significant further fragmentation, the exact structure of the corresponding sialic acid cannot be given. Adjustment of the conditions in a way that both typical CI and electron impact (EI) patterns are formed in one spectrum may give all the information needed for unequivocal identification; in our experience this may be achieved by the use of methane as the reactant gas (Fig. 4d), but not with isobutane or ammonia. If a sialic acid spectrum is overlaid with one derived from another compound, assignment may become impossible. In these cases recording of daughter ions derived from characteristic fragments including the $[M + H]^+$ ion will allow interpretation.

Fast atom bombardment mass spectrometry (FAB-MS) has been established as a useful method in the analysis of complex carbohydrates, including sialoglycoconjugates.[71-73] With this method, it is possible to obtain information about the presence of sialic acids and their substitution; however, localization of specific substituents within a sialic acid residue cannot be achieved.

Nuclear Magnetic Resonance Spectroscopy

A number of sialic acids have been analyzed by ^{1}H nuclear magnetic resonance (NMR) spectroscopy.[74] This method shows the migration of sialic acid O-acetyl groups from position 7 to 9, presumably via C-8.[25] A major drawback is the need to use pure sialic acids, that is, mixtures must first be separated into individual species. The advantage is the possibility

[70] J. P. Kamerling and J. F. G. Vliegenthart, Cell Biol. Monogr. 10, 95 (1982).

[71] J. Peter-Katalinić and H. Egge, this series, Vol. 193, p. 713.

[72] A. Dell, this series, Vol. 193, p. 647.

[73] G. Muralikrishna, G. Reuter, J. Peter-Katalinić, H. Egge, F.-G. Hanisch, H.-C. Siebert, and R. Schauer, Carbohydr. Res. 236, 321 (1992).

[74] J. F. G. Vliegenthart, L. Dorland, H. van Halbeek, and J. Haverkamp, Cell Biol. Monogr. 10, 127 (1982).

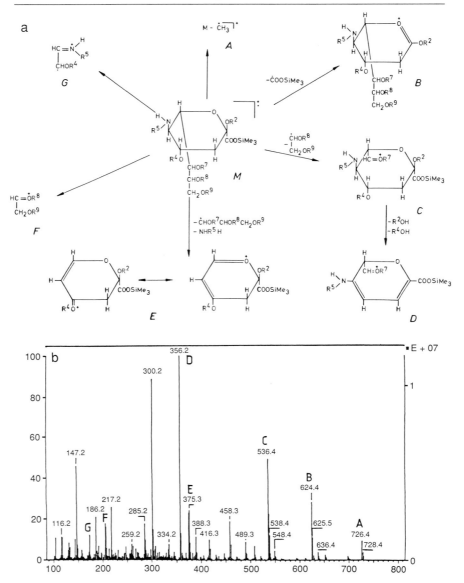

FIG. 4. (a) Fragmentation scheme of per-*O*-trimethylsilylsialic acids in electron-impact mass spectrometry.[49] (b) Electron-impact mass spectrum of per-*O*-trimethylsilyl-Neu5Ac. The major fragments are indicated. (c) Fragmentation scheme or per-*O*-trimethylsilylsialic acids in chemical ionization mass spectrometry using isobutane as reactant gas.[50] (d) mass spectrum of per-*O*-trimethylsilyl-Neu5Ac obtained by chemical ionization with methane. Fragments according to (a) and (c) are indicated.

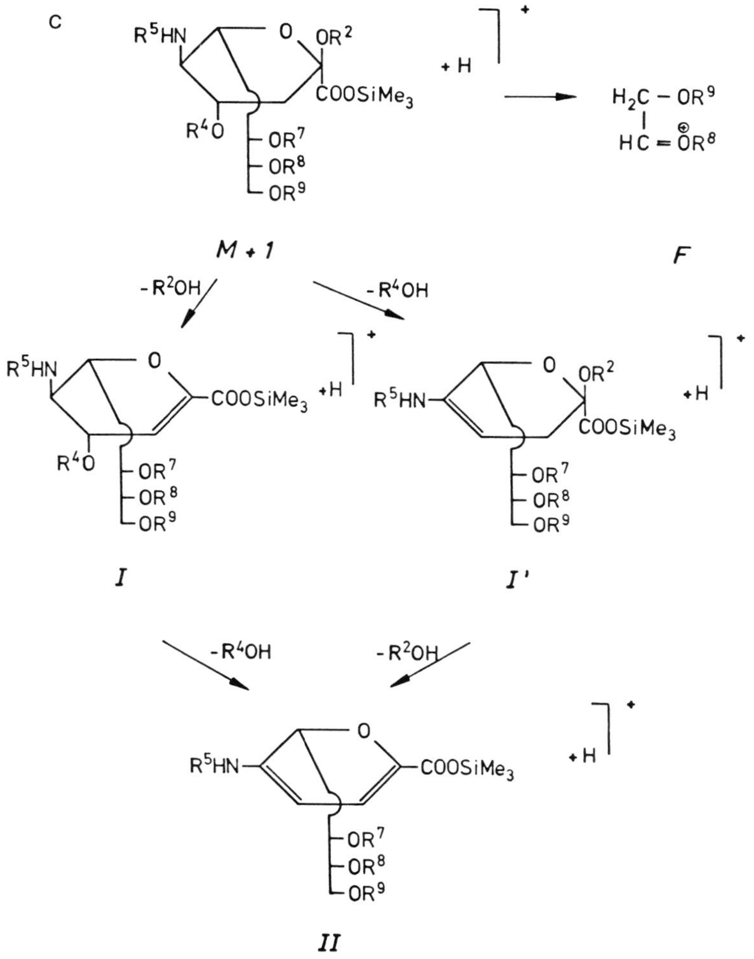

C

R⁵HN — O OR²
COOSiMe₃   + H   ⟶   H₂C – OR⁹
R⁴O — OR⁷   HC = ÖR⁸
OR⁸
OR⁹

M + 1   F

– R²OH   – R⁴OH

R⁵HN — O
COOSiMe₃ + H   R⁵HN — O OR²   + H
R⁴O — OR⁷   COOSiMe₃
OR⁸   OR⁷
OR⁹   OR⁸
OR⁹

I   I'

– R⁴OH   – R²OH

R⁵HN — O
COOSiMe₃   + H
OR⁷
OR⁸
OR⁹

II

FIG. 4. (*continued*)

to analyze even highly complex sialyloligosaccharides and to obtain data about the structure of sialic acids, sequence, anomeric configuration, type of linkage, and overall conformation of the whole molecule. In contrast to mass spectrometry, sialic acid substituents can be localized within a sialoglycoconjugate, for example, in gangliosides containing O-acetylated sialic acids.[75] Unusual sialic acid linkages as found in 8-O-methyl-N-glycol-

[75] H.-C. Siebert, G. Reuter, R. Schauer, C.-W. von der Lieth, and J. Dabrowski, *Biochemistry* **31,** 6962 (1992).

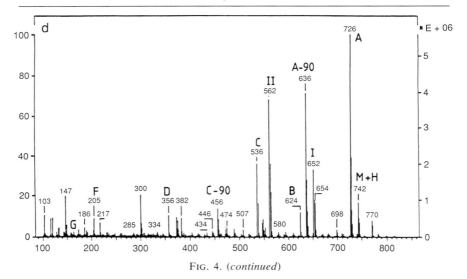

FIG. 4. (*continued*)

oylneuraminic acid oligomers from the starfish *A. rubens,* that is, α-glycosidic linkage to the OH of the *N*-glycoloyl group of the next sialic acid, have been established successfully by this method.[69] A detailed discussion of this method is given (see [9] in this volume).

Enzymatic Methods for Sialic Acid Analysis

Whenever unequivocal identification of sialic acids is not possible on the basis of the analytical data obtained, it may be helpful to confirm assignment by additional methods. This approach has been described in a previous volume of this series.[17] In the case of O-acetylated sialic acids additional information can be obtained by alkali treatment as discussed above. Because identical behavior of different sialic acids cannot be excluded in the various procedures, additional methods are needed to support the assignments. Besides the use of sialidases (EC 3.2.1.18) for release of sialic acids from their glycosidic linkages (see above, and Table II), two enzymes may be used for this purpose: sialate 9-*O*-acetylesterase (EC 3.1.1.53), hydrolyzing *O*-acetyl groups from the 9-position of sialic acids, and *N*-acylneuraminate lyase (EC 4.1.3.3), cleaving sialic acids to *N*-acylmannosamine derivatives and pyruvate. This latter enzyme works best with Neu5Ac, slower with Neu5Gc and O-acetylated sialic acids; it does not cleave 4-O-acetylated or glycosidically linked sialic acids.[22,24] The occurrence of sialate *O*-acetylesterases has been described in various sources, including bovine brain,[76] horse liver,[77] and influenza C virus.[78] The latter enzyme works equally well on free and glycosidically bound sialic acids, the others only on free sialic acids. From horse liver two different enzymes were isolated, one acting only on 9-O-acetylated sialic acids and a second one working equally well on both 4- and 9-O-acetylated sialic acids. The sialate O-acetylesterases from bovine brain and influenza C virus hydrolyze preferably 9-O-acetyl groups. Sialic acid 7-O-acetyl groups are resistant to all these esterases.

When an "enzyme-assisted" sialic acid assignment needs to be carried out due to indications of the presence of, for example, 9-O-acetylated sialic acids together with non-O-acetylated species, the following procedure can be used.

Esterase Incubation

The sample containing free or glycosidically bound sialic acids is dissolved in PBS, pH 7.2, to yield a concentration of 1–10 mM sialic acid, and is incubated with influenza C virus[78] (1–5 mU; activity adjusted with methylumbelliferyl acetate) for 30 min at 37°. The reaction is stopped by heating to 96° for 3 min. In the case of glycosidically linked sialic acids, hydrolysis with sialidase or acid must be performed after centrifugation, as described above. A part of the incubation mixture is analyzed. Shifts of peaks or bands from positions of O-acetylated sialic acids to Neu5Ac or Neu5Gc indicate the presence of O-acetylated sialic acids in the original mixture. If only free sialic acids need to be analyzed, other sialate 9-O-acetylesterases may also be used.[76,77]

N-Acylneuraminate Pyruvate Lyase Incubation

To the rest of the mixture from the esterase incubation of free sialic acids, 2–10 mU of N-acylneuraminate lyase is added, and incubation is carried out for 1 hr at 37°. Glycosidically linked sialic acids must be liberated by sialidase or acid treatment before lyase incubation. The reaction is again stopped by heating as described above, followed by analysis in the given system. The presence of Neu5Ac and/or Neu5Gc in the mixture is recognized by a decrease in peak/band intensity. Although 9-O-acetylated sialic acids are cleaved at a significantly slower rate by lyase, formation of 6-O-acetylated N-acylmannosamine can be observed by TLC or GLC analysis.

[76] R. Schauer, G. Reuter, S. Stoll, and A. K. Shukla, J. Biochem. (Tokyo) **106,** 143 (1992).
[77] R. Schauer, G. Reuter, and S. Stoll, Biochimie **70,** 1511 (1988).
[78] R. Schauer, G. Reuter, S. Stoll, F. Posadas del Rio, G. Herrler, and H.-D. Klenk, Biol. Chem. Hoppe-Seyler **369,** 1121 (1988).

Because of the specificity of the enzymes, the assignments made after these incubations are reliable. This procedure may be included in HPLC analysis by UV[57] or fluorimetric detection and in TLC analysis. For GLC, salts should be removed before derivatization, making this type of analysis less attractive. In cases in which the amounts of sialic acids are too small to allow intensive purification for GLC-MS, fluorimetric HPLC analysis of free sialic acids, followed by (1) de-O-acetylation by ammonia or sialate 9-O-acetylesterase and (2) degradation by N-acylneuraminate pyruvate lyase, is helpful in the analysis of these sugars.

Histochemistry

Classic histochemistry of sialic acid-containing glycoconjugates makes use of either binding of cationized dyes or selective periodate oxidation of sialic acids[79] that is strongly dependent on side-chain modifications.[37] These procedures have been summarized [17,80] and applied to the histochemical identification of 9-O-acylated sialic acids.[81]

The presence of glycosidically linked sialic acids in complex carbohydrates can be detected by specific lectins, viruses, or antibodies; however, structural requirements exceed by far those of sialic acid alone. Although this approach cannot be considered as a general method, it gives information about detailed structural features, but appropriate controls must be used to eliminate misinterpretation. Any positive result that is assumed to be due to the presence of terminal sialic acid must become negative after sialidase incubation, and remain positive if sialidase incubation is carried out in the presence of the natural inhibitor 2-deoxy-2,3-didehydro-N-acetylneuraminic acid. A number of papers have been published since the last review in this series[17] describing the specificity of the various lectins for individual sialic acids and their type of binding[82–94] (Table V).

[79] G. Reuter, R. Schauer, C. Szeiki, J. P. Kamerling, and J. F. G. Vliegenthart, *Glycoconjugate J.* **6**, 35 (1989).

[80] C. F. A. Culling and P. E. Reid, *Cell Biol. Monogr.* **10**, 173 (1982).

[81] P. E. Reid, J. Needham, and D. A. Owen, *Histochem. J.* **23**, 149 (1991).

[82] C. Mandal and C. Mandal, *Experientia* **46**, 433 (1990).

[83] W. C. Wang and R. D. Cummings, *J. Biol. Chem.* **263**, 4576 (1988).

[84] T. Sata, J. Roth, C. Zuber, B. Stamm, and P. U. Heitz, *Am. J. Pathol.* **139**, 1435 (1991).

[85] R. N. Knibbs, I. J. Goldstein, R. M. Ratcliffe, and N. Shibuya, *J. Biol. Chem.* **266**, 83 (1991).

[86] A. N. Yatskovskii and A. D. Lutsik, *Bull. Exp. Biol. Med. (Engl. Transl.)* **111**, 96 (1991).

[87] N. Shibuya, I. J. Goldstein, W. F. Broekaert, M. Nsimba-Lubaki, B. Peeters, and W. J. Peumans, *J. Biol. Chem.* **262**, 1596 (1987).

[88] H. Harada, M. Kondo, M. Yanagisawa, and S. Sunada, *Anal. Biochem.* **189**, 262 (1990).

[89] S. Ratanapo and M. Chulavatnatol, *Comp. Biochem. Physiol. B* **97B**, 515 (1990).

[90] A. Ouadia, Y. Karamanos, and R. Julien, *Glycoconjugate J.* **9**, 21 (1992).

TABLE V
SIALIC ACID-SPECIFIC LECTINS AND SPECIFICITIES[a]

Lectin from:	Source	Specificity	Ref.
Maackia amurensis	Seeds, bark	Neu5Acα2-3glycan	83–85
Sambucus nigra	Elderberry (bark)	Neu5Acα2-6glycan	84–87
Sambucus sieboldiana	Elderberry bark	Sialic acidα2-glycan	88
Peneaus monodon	Black tiger prawn	Neu5Ac	89
Escherichia coli K99		Neu5Gcα2-glycan	90
Pila globosa	Snail	Neu5Gcα2-glycan	91
Cepaea hortensis	Snail	Neu5Ac/Neu5,9Ac$_2$α2-	92
Marine crab	Hemolymph	Neu5Gc	93
Rat brain		Neu5Ac/Neu5Gc	94

[a] Structural requirements outside of the sialic acid molecule or linkage are not indicated. For further sialic acid-recognizing lectins, see Schauer[17] and Mandal and Mandal.[82]

This specificity makes them ideal tools for the isolation of compounds with this given structure.

With the selectins[95–97] and sialoadhesin,[98] new classes of sialic acid-binding proteins have been discovered that have great biological importance and exhibit high specificities for certain sialooligosaccharide structures.

Hemagglutinins of viruses are also proteins with specific carbohydrate-binding specificities. The sialic acid-binding properties of a number of viruses have been established.[99] Influenza C virus,[78,100,101] bovine corona

[91] S. Swarnakar, P. S. Chowdhury, and M. Sarkar, *Biochem. Biophys. Res. Commun.* **178,** 85 (1991).

[92] R. Brossmer, M. Wagner, and E. Fischer, *J. Biol. Chem.* **267,** 8752 (1992).

[93] S. P. D. Mercy and M. H. Ravindranath, *Experientia* **48,** 498 (1992).

[94] M. Popoli and A. Mengano, *Neurochem. Res.* **13,** 63 (1988).

[95] M. Bevilacqua, E. Butcher, B. Furie, B. Furie, M. Gallatin, M. Gimbrone, J. Harlan, K. Kishimoto, L. Lasky, R. McEver, J. Paulson, S. Rosen, B. Seed, M. Siegelman, T. Springer, L. Stoolman, T. Tedder, A. Varki, D. Wagner, I. Weissman, and G. Zimmerman, *Cell (Cambridge, Mass.)* **67,** 233 (1991).

[96] A. Varki, *Trends Glycosci. Glycotechnol.* **3,** 10 (1991).

[97] A. Aruffo, *Trends Glycosci. Glycotechnol.* **4,** 146 (1992).

[98] P. R. Crocker, S. Kelm, C. Dubois, B. Martin, A. S. McWilliam, D. M. Shotton, J. C. Paulson, and S. Gordon, *EMBO J.* **10,** 1661 (1991).

[99] M. A. K. Markwell *in* "Microbial Lectins and Agglutinins: Properties and Biological Activity" (D. Mirelman, ed.), p. 51. Wiley, New York, 1986.

[100] G. Zimmer, G. Reuter, and R. Schauer, *Eur. J. Biochem.* **204,** 209 (1992).

[101] B. Schultze, H.-J. Gross, R. Brossmer, H.-D. Klenk, and G. Herrler, *Virus Res.* **16,** 185 (1990).

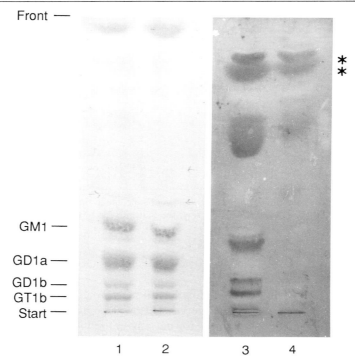

FIG. 5. Analysis of bovine grain gangliosides by thin-layer chromatography [HPTLC silica gel; chloroform–methanol–H_2O, 60:40:9 (v/v/v) as solvent, containing 0.02% $CaCl_2 \cdot 2H_2O$] and staining with orcinol/Fe^{3+}/HCl (lanes 1 and 2) or TLC overlay with influenza C virus (lanes 3 and 4). The samples applied on lanes 2 and 4 were saponified before development by exposure of the TLC plate to ammonia vapor overnight at room temperature. Gangliosides carrying 9-O-acetylated sialic acids become visible on lane 3 and do not comigrate with the orcinol-stained reference gangliosides on lanes 1 and 2. The fast-migrating bands indicated with an asterisk that are seen on lanes 3 and 4 are caused by unspecific staining, because the color is different from the other bands on lane 3 and they do not disappear after saponification (lane 4).

virus,[101–103] and encephalomyelitis virus[101] recognize 9-O-acetylated sialic acids. A shift in specificity of the receptor-destroying enzyme of influenza A virus for sialic acid linkages from $\alpha(2 \rightarrow 3)$ to $\alpha(2 \rightarrow 6)$ has occurred over the last two decades.[104]

A test has been developed that uses the hemagglutinin and the receptor-

[102] B. Schultze, H. J. Gross, R. Brossmer, and G. Herrler, *J. Virol.* **65**, 6232 (1991).
[103] B. Schultze and G. Herrler, *J. Gen. Virol.* **73**, 901 (1992).
[104] L. G. Baum and J. C. Paulson, *Virology* **180**, 10 (1991).

destroying enzyme activities of the influenza C virus to specifically detect 9-O-acetylated sialic acids in complex carbohydrates.[100] Either developing the sialic acid-containing compounds on TLC plates (gangliosides) or immobilizing them on microtiter plates, followed by incubation with influenza C virus at 4°, allows binding of the virus to 9-O-acetylated sialic acids via the hemagglutinin. At this temperature, the esterase does not cleave the ester groups at significant rates. Excess virus is removed by washing. The following incubation step is carried out at 37° with an esterase substrate that is hydrolyzed much faster than sialic acids, for example, methylumbelliferyl acetate in the microtiter assay or α-naphthyl acetate in the TLC test (Fig. 5). A positive reaction indicates bound virus and thus the presence of 9-O-acetylated sialic acids. A control for unspecific binding must be performed with the de-O-acetylated substrate.

A sialic acid modified at C-9, 9-acetamido-N-acetylneuraminic acid, has been synthesized[105] that allows binding of influenza C virus but not release of the C-9 part by the viral esterase. This sialic acid can be activated to the CMP-glycoside and transferred to cells with sialyltransferases and is thus an excellent tool to study the role of 9-substituted sialic acids on infection by this virus. Additional tests are available commercially that allow detection of sialic acids and other carbohydrates on blots using specific digoxigenin-labeled lectins.[106]

In summary, sialic acid analysis has been improved to allow detection of small amounts of free sugars and even analysis of highly complex structures in intact sialoglycoconjugates. With modern analytical tools and the methods available the biological function of these molecules can be studied in much more detail and it may be possible to recognize additional specific functions of individual sialic acids.

Acknowledgments

Thanks are due to Dr. Wolfgang Wagner-Redeker (Finnigan-MAT, Bremen, Germany) for recording the mass spectra, Dr. Jean-Claude Michalski (University of Lille, France) for the kind gift of 3-deoxy-D-*glycero*-D-*galacto*-nonulosonic acid (KDN), Toshiaki Suguri for analysis of bovine brain gangliosides, Margret Wember and Sabine Stoll for excellent technical assistance, and the Deutsche Forschungsgemeinschaft, Fonds der Chemischen Industrie, and Sialic Acids Society for financial support.

[105] G. Herrler, H. J. Gross, A. Imhof, R. Brossmer, G. Milks, and J. C. Paulson, *J. Biol. Chem.* **267**, 12501 (1992).
[106] A. Haselbeck and W. Hösel, *Glycoconjugate J.* **7**, 63 (1990).

[11] Size Fractionation of Oligosaccharides

By Akira Kobata

Introduction of endo-β-N-acetylglucosaminidases[1] as tools to release the N-linked sugar chains quantitatively as oligosaccharides has opened a new age for the study of these sugar chains. These enzymes, however, have strict substrate specificities. Although the specificities are useful in determining the structures of some sugar chains, they limit the use of the enzymes for the study of high mannose-type and hybrid-type sugar chains.[2-5] Development of hydrazinolysis[6] (see [5] in this volume) and the finding of glycopeptidases[7,8] have removed this barrier, and all asparagine-linked sugar chains can now be released quantitatively as oligosaccharides. Because all these oligosaccharides contain N-acetylglucosamine at their reducing termini, they can be labeled by either NaB^3H_4 reduction[9] or by pyridylamination.[10] The ^3H-labeled oligosaccharides can be fractionated by using the appropriate methods. Gel-permeation chromatography using ultrafine (under 400 mesh) BioGel P-4 (Bio-Rad, Richmond, CA) columns has been developed as one of the effective tools to fractionate oligosaccharides by their effective sizes.[11] Although the size fractionation of radioactive oligosaccharides can also be performed by paper chromatography, the gel-permeation column chromatography is more efficient because it can fractionate a wide range of oligosaccharides in a short time. By using appropriate internal standards, the effective size of an oligosaccharide can be precisely determined. Because the value can be used as a constant that is specific for the oligosaccharide, and because the number of oligosac-

[1] A. Kobata, *Anal. Biochem.* **100**, 1 (1979).

[2] T. Tai, K. Yamashita, M.-A. Ogata, N. Koide, T. Muramatsu, S. Iwashita, Y. Inoue, and A. Kobata, *J. Biol. Chem.* **250**, 8569 (1975).

[3] S. Ito, K. Yamashita, R. G. Spiro, and A. Kobata, *J. Biochem.* (*Tokyo*) **81**, 1621 (1977).

[4] T. Tai, K. Yamashita, S. Ito, and A. Kobata, *J. Biol. Chem.* **252**, 6687 (1977).

[5] K. Yamashita, Y. Tachibana, and A. Kobata, *J. Biol. Chem.* **253**, 3862 (1978).

[6] S. Takasaki, T. Mizuochi, and A. Kobata, this series, Vol. 83, p. 263.

[7] K. Sugiyama, H. Ishihara, S. Tejima, and N. Takahashi, *Biochem. Biophys. Res. Commun.* **112**, 155 (1983).

[8] T. H. Plummer, J. H. Edler, Jr., S. Alexander, A. W. Phelan, and A. L. Tarentino, *J. Biol. Chem.* **259**, 10700 (1984).

[9] S. Takasaki and A. Kobata, this series, Vol. 50, p. 50.

[10] H. Takemoto, S. Hase, and T. Ikenaka, *Anal. Biochem.* **145**, 245 (1985).

[11] K. Yamashita, T. Mizuochi, and A. Kobata, this series, Vol. 83, p. 105.

charides with similar effective sizes is limited, they can be discriminated easily by simple auxiliary methods such as exoglycosidase digestion.[1]

Selection of Gels

A series of polyacrylamide gels (BioGel P-2, P-4 and P-6; Bio-Rad) as well as a series of dextran gels (Sephadex G-15 and G-25; Pharmacia, Piscataway, NJ) are available as the molecular sieve gel suitable for fractionation of oligosaccharides. Columns containing these gels have been used for the rough fractionation of oligosaccharides and glycopeptides. However, exact separation of glucose oligomers larger than octasaccharide could not be performed in these experiments. Furthermore, the elution position of an oligosaccharide changes with changes in column temperature, probably because of the interaction of gel with sugar chains.

We examined the behavior of dextran hydrolysates and found that an effective and reproducible separation of glucose oligomers can be obtained by operating the chromatography at 55° with use of a column containing ultrafine BioGel beads. The BioGel P-2 column is suitable for the separation of glucose oligomers up to 10 glucose units. Successful separation of oligosaccharides from 1 to 30 glucose units can be obtained by using a BioGel P-4 column. In the case of a BioGel P-6 column, oligosaccharides smaller than five glucose units cannot be mutually separated, but will be effective in separating larger oligosaccharides. Because the effective sizes of most of the N-linked sugar chains fall into the range of 10 to ~20 glucose units, a BioGel P-4 column has been widely used for the study of these oligosaccharides.

It must be remembered that the separation characteristics of commercial ultrafine BioGel P-4 vary by the lot of preparation. Therefore it is important to preexamine the effectiveness of separation of the column by using a dextran hydrolysate as a probe.

Reagents

Water for running chromatography: Deaerate distilled water in a suction bottle for 60 min at 60° by suction with an aspirator

Gel slurry: Suspend ultrafine BioGel P-4 (under 400 mesh) in 5 vol of distilled water; allow to swell in a suction bottle. Extremely small particles, which are included in commercial ultrafine BioGel P-4, cause clogging of the column and should be removed by the following procedures. The swollen gel suspension is thoroughly mixed and left still for 3 hr. The small particles, which do not precipitate, are removed by decantation. The precipitated gel is mixed thoroughly with 3 vol of distilled water. After standing for 3 hr, the supernatant containing small particles is decanted. This procedure is repeated eight more times to remove as many excessively small particles as possible. Final suspension of BioGel P-4, thus obtained, is deaerated by suction with an aspirator at 55° for 60 min.

Standard oligosaccharide mixtures: The mixtures are important for examining the performance of a BioGel P-4 column as well as an internal standard for actual analysis of radioactive oligosaccharides. A mixture of glucose oligomers and a mixture of N-acetylglucosamine oligomers are used for this purpose. One N-acetylglucosamine residue behaves as two glucose residues. Accordingly, glucose oligomers afford more precise value for calculating the effective sizes of sample oligosaccharides. Furthermore, use of N-acetylglucosamine oligomers is not suitable for the sequential exoglycosidase digestion of sample oligosaccharides because they inhibit the action of β-N-acetylhexosaminidase. The benefit of using N-acetylglucosamine oligomers is that they can be detected by an ultraviolet (UV) monitor, which is less expensive than a refractive index monitor

Preparation of Mixture of Glucose Oligomers. One gram of dextran with an average molecular weight of 200,000 is dissolved in 10 ml of 0.1 N HCl, and heated in boiling water for 4 hr in a sealed tube. The reaction mixture is freed from HCl by passing it through a column containing 1 ml of Bio-Rad AG 3-X4A (100–200 mesh, OH⁻ form). To the neutral hydrolysate, 20 mg of dextran is added for the determination of void volume of the BioGel P-4 column and diluted to 20 ml by adding distilled water.

Preparation of Mixture of N-Acetylglucosamine Oligomers. Chitin (0.5 g) is suspended in 25 ml of ice-cold concentrated HCl, and dissolved by mixing on a magnetic stirrer. The solution is then warmed at 40° for 2 hr. After cooling the reaction mixture with ice, it is freed from HCl completely by passing the mixture through a column containing 190 ml of Bio-Rad AG 3-X4A (100–200 mesh, OH⁻ form). After washing the column thoroughly with distilled water, the mixture of effluent and washings is passed through a column containing 3 ml of Bio-Rad AG 50W-X12 (50–100 mesh, H⁺ form) to remove de-N-acetylated oligosaccharides. The effluent is concentrated *in vacuo* and the volume of the solution is adjusted to 50 ml by adding distilled water. Any precipitation formed after this procedure is removed by centrifugation, and the supernatant is used as a mixture of N-acetylglucosamine oligomers.

Construction of BioGel P-4 Column Chromatography System

A schematic diagram of the gel-permeation chromatography apparatus is shown in Fig. 1. For elution, distilled water is supplied by a precision

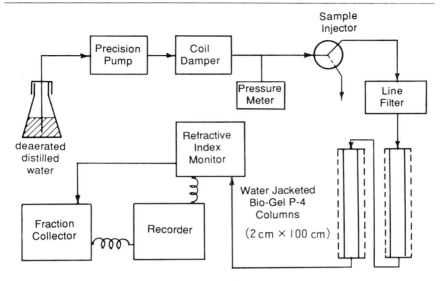

FIG. 1. Schematic diagram of BioGel P-4 column chromatography.

pump. Because small particles of gel are used, use of a positive displacement pump is required. We use a low-cost single-plunger pump for this purpose. The weak pulsation, which cannot be avoided by using such an inexpensive pump, can be eliminated by inserting a coil damper between the pump and the sample injector. Sugar samples are injected and pass through a line filter to reach the column. This line filter contains a Teflon filter with a pore size of 10 μm, which removes the small particles accidentally passed into the sample solution and distilled water, and protects the BioGel P-4 column against deterioration. The dirt accumulated on the filter can be monitored by the indicator of the pressure meter, and the filter replaced appropriately.

A thick-walled column with a water jacket, 100 × 2 cm i.d., can be purchased from Umetani Precision Co., Ltd., Tokyo (GCV-20 glass column) or from other suitable sources. Hot water (55 ± 0.5°) is circulated through the jacket of the column from a constant-temperature circulating water bath (Eyela, HS-1; Tokyo Rikakikai Co., Ltd.).

Two to four columns packed with ultrafine BioGel P-4 are connected by tubes as shown in Fig. 1. The flow rate is less than 0.5 ml/min. The effluent from the column is passed through a refractive index monitor to detect glucose oligomers and then collected by a fraction collector. The emission signal of the fractionation from the collector is recorded by the same automatic recorder, which also records the signal from the refractive index monitor. It is convenient if the dead volume within the tube connecting the monitor to the fraction collector can be compensated as delay time at the event marker of the collector. The inexpensive fraction collector model FRAC 100, equipped with an electromagnetic valve (Pharmacia LKB Biotechnology, Ltd.) is suitable for that purpose. A Teflon connecting tube with 0.5 mm i.d. and 2.0 mm o.d. is used.

When using N-acetylglucosamine oligomers as internal standards, the UV monitor replaces the refractive index monitor. This monitor can also be used for the detection of oligosaccharides containing N-acetylamino sugar residues.

Column Preparation

The gel slurry is poured into the column prewarmed by circulating warm (55°) water in the jacket and packed at a flow rate of 0.5–1.0 ml/min. It is important to watch the pressure meter during this column packing procedure. Restart packing of the column if an abrupt rise in the column pressure is observed during the packing process, because such columns tend to result in the leading phenomena of sample elution, and disconnection of the flow line during the fractionation of samples. The packed column should be examined for the satisfactory fractionation of glucose oligomers before use.

Fractionation and Identification of Radioactive Oligosaccharides

Studies of N-linked sugar chains have been described.[11,12] Here we will introduce the investigation of O-linked sugar chains. A method for labeling quantitatively the oligosaccharide alcohols obtained by the alkaline borohydride treatment of O-linked sugar chains has been reported.[13] In short, the oligosaccharide alcohols are completely de-N-acetylated by exhaustive hydrazinolysis, and the amino sugar residues in the oligosaccharide alcohols are re-N-acetylated with either [14]C- or ^{3}H-labeled acetic anhydride. This method makes the use of the BioGel P-4 column chromatography for the study of O-linked sugar chains possible.

Sialylated and sulfated oligosaccharides are converted to neutral oligosaccharides by sialidase digestion and by methanolysis followed by N-acetylation, respectively. Salts in the sample solution can deteriorate a BioGel P-4 column by assisting the growth of microorganisms. Therefore a radioactive neutral oligosaccharide solution is best desalted beforehand by passing it through a small mixed-bed column containing Bio-Rad AG 50W-X12 (H[+] form) in the upper layer and Bio-Rad AG3-X4A (OH[−] form)

---

[12] A. Kobata, K. Yamashita, and S. Takasaki, this series, Vol. 138, p. 84.
[13] J. Amano and A. Kobata, this series, Vol. 179, p. 261.

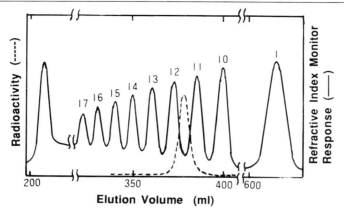

FIG. 2. Analysis of an oligosaccharide by BioGel P-4 column chromatography. —, Elution pattern of glucose oligomers; – – –, Fucα(1 → 2)Galβ(1 →4)GlcN[³H]Acβ(1 → 6)[Fucα(1 → 2)Galβ(1 → 4)GlcN[³H]Acβ(1 → 3)Galβ(1 → 3)]GalN[³H]Ac$_{OH}$. V_0 is the same as in Fig. 1. Analysis is performed by using two × 100 cm columns.

in the lower layer. The effluent and washings (five bed volumes) are mixed and evaporated to dryness. The residue is mixed with 0.2–0.4 ml of distilled water containing about 4–8 mg of a mixture of glucose oligomers, and applied to a BioGel P-4 column through the sample injector. Distilled water is passed through the column at a flow rate of 0.3–0.5 ml/min. Glucose oligomers are recorded by refractive index monitor, and the radioactivity is determined on an aliquot of effluent fractionated into about 2–3 ml/tube. Because the actual amounts of radioactive oligosaccharide alcohols are much smaller than the glucose oligomers, they are not detected by the refractive index monitor. The exact elution volume of each radioactive oligosaccharide alcohol can be determined by tracing its elution curve over those of glucose oligomers, as shown in Fig. 2. In Table I, the effective sizes of several oligosaccharide alcohols are summarized in glucose units. The value of a particular oligosaccharide does not change because of the conditions of analysis and these glucose units can be used for identification.

Comments

As in the case of an N-acetylglucosaminitol residue, the N-acetyl-galactosaminitol residue behaves as 2.5 glucose units. Furthermore, the

TABLE I
EFFECTIVE SIZES OF OLIGOSACCHARIDE ALCOHOLS RELEASED FROM O-LINKED
SUGAR CHAINS OF GLYCOPROTEINS

Structures	Glucose units
Galβ1 → 3GalNAc$_{OH}$	3.5
Fucα1 → 2Galβ1 → 3GalNAc$_{OH}$	4.0
GlcNAcβ1 $\searrow$6 $\quad$GalNAc$_{OH}$ Galβ1 $\nearrow$3	5.3
GlcNAcβ1 → 3Galβ1 → 3GalNAc$_{OH}$	5.6
GalNAcα1 $\searrow$3 $\quad$Galβ1 → 3GalNAc$_{OH}$ Fucα1 $\nearrow$2	6.0
Galβ1 → 4GlcNAcβ1 $\searrow$6 $\quad$GalNAc$_{OH}$ Galβ1 $\nearrow$3	6.3
Galβ1 → 4GlcNAcβ1 → 3Galβ1 → 3GalNAc$_{OH}$	6.7
Fucα1 $\searrow$3 Galβ1 → 4GlcNAcβ1 $\searrow$6 $\quad$GalNAc$_{OH}$ Galβ1 $\nearrow$3	7.0
Fucα1 → 2Galβ1 → 4GlcNAcβ1 $\searrow$6 $\quad$GalNAc$_{OH}$ Galβ1 $\nearrow$3	7.3

TABLE I (*continued*)

Structures	Glucose units
Galα1 $\rightarrow$ 3Galβ1 $\rightarrow$ 4GlcNAcβ1 $\rightarrow$ 3GalNAc$_{OH}$	7.7
GalNAcβ1 $\rightarrow$ 3 or 4GlcNAcβ1 $\rightarrow$ 3Galβ1 $\rightarrow$ 3GalNAc$_{OH}$	7.7

Fucα1

3

Fucα1 $\rightarrow$ 2Galβ1 $\rightarrow$ 4GlcNAcβ1

6
3 GalNAc$_{OH}$ 7.8

Galβ1

Galα1

3

Fucα1 $\rightarrow$ 2Galβ1 $\rightarrow$ 4GlcNAcβ1

6
3 GalNAc$_{OH}$ 7.8

Galβ1

| Galβ1 $\rightarrow$ 3GlcNAcβ1 $\rightarrow$ 6Galβ1 $\rightarrow$ 4GlcNAcβ1 $\rightarrow$ 3Galβ1 $\rightarrow$ 3GalNAc$_{OH}$ | 9.6 |

Galβ1 $\rightarrow$ 4GlcNAcβ1

Fucα1 $\rightarrow$ 2   6
3 GalNAc$_{OH}$ 10.5

Galβ1 $\rightarrow$ 4GlcNAcβ1 $\rightarrow$ 3Galβ1

| Fucα1 $\rightarrow$ 2Galβ1 $\rightarrow$ 3GlcNAcβ1 $\rightarrow$ 6Galβ1 $\rightarrow$ 4GlcNAcβ1 $\rightarrow$ 3Galβ1 $\rightarrow$ 3GalNAc$_{OH}$ | 10.6 |

Galα1

3

| Fucα1 $\rightarrow$ 2Galβ1 $\rightarrow$ 4GlcNAcβ1 $\rightarrow$ 6Galβ1 $\rightarrow$ 4GlcNAcβ1 $\rightarrow$ 3Galβ1 $\rightarrow$ 3GalNAc$_{OH}$ | 11.3 |

Fucα1 $\rightarrow$ 2Galβ1 $\rightarrow$ 4GlcNAcβ1

6
3 GalNAc$_{OH}$ 11.5

Fucα1 $\rightarrow$ 2Galβ1 $\rightarrow$ 4GlcNAcβ1 $\rightarrow$ 3Galβ1

[a] GalNAc$_{OH}$, *N*-Acetylgalactosaminitol.

effective sizes of *N*-acetylhexosamines, galactose, and various fucosyl residues are almost the same as those of N-linked sugar chains.[11] Therefore, the data from studies of N-linked sugar chains can be applied to the preliminary estimation of sugar chain structures.

[12] High-pH Anion-Exchange Chromatography of Glycoprotein-Derived Carbohydrates

By MARK R. HARDY and R. REID TOWNSEND

Introduction

High-pH anion-exchange chromatography (HPAEC) has been used successfully to analyze covalently linked carbohydrates from natural and recombinant glycoproteins.[1,2] Sensitivities in the low picomole range have been routinely achieved with either pulsed amperometric or radiometric detection. A major advantage of HPAEC, when coupled with pulsed amperometric detection (PAD), is its utility for analyzing both monosaccharides and all classes of oligosaccharides without derivatization. In a single separation, neutral sugars [mannose (Man), glucose (Glc), fucose (Fuc), and galactose (Gal)] and amino sugars [galactosamine (GalN) and glucosamine (GlcN)] from acid hydrolysates of glycoproteins can be separated isocratically and quantified.[3] In another chromatographic run, using the same column, sialic acids without base-labile groups [*N*-acetylneuraminic acid (Neu5Ac) and *N*-glycolylneuraminic acid (NeuGc)] are well resolved with an acetate gradient.[4] Sialic acids with *O*-acetyl groups can be analyzed at lower pH and detected with PAD after addition of postcolumn base.[4] High-pH anion-exchange chromatography has been an effective method for separating N-linked hybrid, oligomannosidic and lactosamine-type oligosaccharides from glycoproteins.[5–7] Isomeric forms differing in only a single linkage or in the branch position of one residue have been re-

[1] R. R. Townsend and M. R. Hardy, *Glycobiology* **1**, 139 (1991).

[2] M. W. Spellman, *Anal. Chem.* **62**, 1714 (1990).

[3] M. R. Hardy, R. R. Townsend, and Y. C. Lee, *Anal. Biochem.* **170**, 54 (1988).

[4] A. E. Manzi, S. Diaz, and A. Varki, *Anal. Biochem.* **188**, 20 (1990).

[5] M. W. Spellman, L. J. Basa, C. K. Leonard, J. A. Chakel, J. V. O'Connor, S. Wilson, and H. van Halbeek, *J. Biol. Chem.* **264**, 14100 (1989).

[6] G. Pfeiffer, H. Geyer, R. Geyer, I. Kalsner, and Wendorf, *Biomed. Chromatogr.* **4**, 183 (1990).

[7] R. R. Townsend, P. H. Atkinson, and R. B. Trimble, *Carbohydr. Res.* **215**, 211 (1991).

solved.[6,7] High-pH anion-exchange chromatography has been particularly useful for analyzing sialylated lactosamine-type oligosaccharides.[8] These separations are marked by large differences (~15 min) in retention times among mono-, di-, tri-, and tetrasialylated species. Within each isocharged complex, resolution has been achieved on the basis of sialic acid linkage [$\alpha(2 \rightarrow 3)$ versus $\alpha(2 \rightarrow 6)$] and/or branch position of these residues.[8] Importantly, HPAEC with PAD obviates the requirement for derivatization under acidic conditions that can hydrolyze sialic residues. Reduced, neutral O-linked oligosaccharides have been separated according to size, branching, and, in particular, according to fucosyl linkages.[9] High-pH anion-exchange chromatography with radiometric detection has been used not only to characterize the carbohydrate core of glycosylphosphatidyl-inositol membrane anchors, but also for the elucidation of their biosynthetic pathways.[1] This chapter details basic HPAEC methods for monosaccharide and oligosaccharide analysis of glycoproteins and a sample preparation approach that extends these methods to the analysis of glycoproteins requiring complicated sample matrices (e.g., salts and detergents) for their isolation.

Principles of Method

Carbohydrates, as polyhydric compounds, are weak acids with pK_a values of 12–14.[10] Thus, in sufficiently basic solutions all monosaccharides and oligosaccharides are negatively charged from oxyanion formation—the basis for their retention on strong anion-exchange columns. The elution order of monosaccharides correlates well with their pK_a values. For example, Glc and Man have pK_a values of 12.28 and 12.08, respectively, and accordingly, Man elutes later than Glc (Fig. 1). The retention times of approximately 100 monosaccharides, using HPAEC, have been reported.[11] Because of the ring oxygen, the acidity of reducing monosaccharides is largely from ionization at the anomeric hydroxyl.[10] Polar, steric, and hydrogen-bonding effects produce a hierarchy of acidity for the other ring hydroxyls (2-OH > 6-OH > 3-OH > 4-OH). For acetamido sugars, the 3-OH is considerably more acidic than the 3-OH of neutral sugars.[12] Monosaccharide alditols are too poorly retained on available quaternary ammonium columns to enable their chromatography, but oligo-

---

[8] R. R. Townsend, M. R. Hardy, D. A. Cumming, J. P. Carver, and B. Bendiak, *Anal. Biochem.* **182,** 1 (1989).

[9] G. P. Reddy and C. A. Bush, *Anal. Biochem.* **198,** 278 (1991).

[10] J. A. Rendleman, *Adv. Chem. Ser.* **117,** 51 (1971).

[11] T. J. Paskach, H.-P. Lieker, P. J. Reilly, and K. Thielecke, *Carbohydr. Res.* (in press).

[12] A. Neuberger and B. M. Wilson, *Carbohydr. Res.* **17,** 89 (1971).

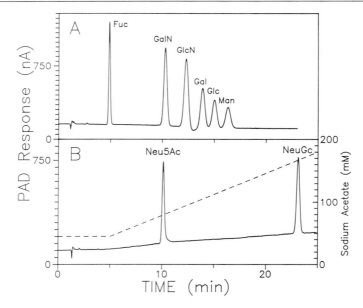

FIG. 1. HPAEC with PAD of standard neutral and amino monosaccharides (A) and sialic acids (B).

saccharide alditols can be separated with comparable resolution, as will be shown in Preparation of Oligosaccharide Alditols. Thus, oxyanions at nonanomeric centers contribute to the chromatographic resolution.

Predicting elution order of oligosaccharides is largely empirical. In addition to the considerations discussed earlier for monosaccharides, there are statistical and linkage effects. A larger oligosaccharide with similar structure should be more retained than smaller ones on an adsorption-type column and this result has been observed, as will be discussed further in High-pH Anion-Exchange Chromatography of High-Mannose Oligosaccharides. Glycosidic bond formation to a highly acidic hydroxyl group can dramatically shift retention times, as has been observed with fucosylated oligosaccharides.[13] Finally, the solution conformation may be an important factor in determining which oxyanionic centers interact with the quaternary ammonium groups on the column resin. For example, the separation of isomers, which differ only in the branch location of residues (no linkage, size, or composition differences), is best explained by a steric argument. Oligosaccharides with formal negative charges (e.g., sialic acids and phos-

[13] M. R. Hardy and R. R. Townsend, *Carbohydr. Res.* **188,** 1 (1989).

phate groups) are predictably more retained on strong anion-exchange columns; however, separation attributable to the neutral portion of the oligosaccharide is not lost.

Although sufficient data for a theoretical prediction of retention times from oligosaccharide structure are not extant, reports from a number of laboratories have identified consistent trends. Oligosaccharides with an intact chitobiosyl core, although one monosaccharide residue larger, elute ~1 min earlier than those with one core GlcNAc.[7] Oligosaccharides with a fucose $\alpha(1 \rightarrow 6)$-linked to the innermost GlcNAc of the chitobiosyl core elute earlier than nonfucosylated counterparts.[2,6] Sialylated oligosaccharides with $\alpha(2 \rightarrow 3)$-linked Neu5Ac residues are more retained than those with $\alpha(2 \rightarrow 6)$-linked Neu5Ac.[8] For multisialylated oligosaccharides with both of these linkages, the order of elution is dependent on the proportion of $\alpha(2 \rightarrow 3)$- and $\alpha(2 \rightarrow 6)$-linked residues.[8] For example, a trisialylated triantennary oligosaccharide with two $\alpha(2 \rightarrow 6)$- and one $\alpha(2 \rightarrow 3)$-linked Neu5Ac residues elutes prior to its structural homolog with one $\alpha(2 \rightarrow 6)$- and two $\alpha(2 \rightarrow 3)$-linked Neu5Ac residues. Elongation of sialylated oligosaccharides with one or two $Gal\beta(1 \rightarrow 4)GlcNAc$ units decreases retention time.[14] Many of these elution order trends are global in the sense that they apply regardless of size or charge state. For example, neutral or sialylated oligosaccharides, with a $Gal\beta(1 \rightarrow 3)GlcNAc$ unit, elute later than their counterparts with a $Gal\beta(1 \rightarrow 4)GlcNAc$ sequence in structures with 4 to 14 monosaccharide units or with 0 to 4 negative charges.[8,15]

Materials and Solutions

Monosaccharide Analysis

Glass-distilled water (Corning MegaPure System, Corning, NY)
NaOH solution: 50% (w/w) (Fisher, Pittsburgh, PA)
Anhydrous sodium acetate (Fluka, Ronkonkoma, NY)
Powderless gloves (Tagg Industries, Laguna Hills, CA)
Plastic pipettes (25 ml)
Polypropylene microcentrifuge tubes (1.5 ml), caps, and O rings (Cat. No. 72-692; Sarstedt, Newton, NC)
Autosampler vials (Sun Brokers, Wilmington, NC): 12 × 32 mm disposable, limited-volume sample vials (Cat. No. 500-116)
Caps for disposable 12 × 32-mm sample vials (Cat. No. 1251; Sun Brokers)

[14] P. Hermentin, R. Witzel, J. F. G. Vliegenthart, J. P. Kamerling, M. Nimtz, and H. S. Conradt, *Anal. Biochem.* **203,** 281 (1992).
[15] M. R. Hardy and R. R. Townsend, *Proc. Natl. Acad. Sci. U.S.A.* **85,** 3289 (1988).

Teflon/silicone septa (Cat. No. 1258; Sun Brokers)

Reference-grade monosaccharides (Pfanstiehl Laboratories, Wauke-gan, IL)

Concentrated (13 N) trifluoroacetic acid (TFA), in sealed 1-ml ampoules (Pierce, Rockford, IL)

Sequencer-grade constant boiling HCl (Applied Biosystems, Foster City, GA)

Heating block for 12-mm tubes (Pierce)

Heating module for block (Reacti-Therm; Pierce)

Nylon filters (Gelman Sciences, Ann Arbor, MI)

MonoStandards (Dionex, Sunnyvale, CA)

Neu5Ac MonoStandard (Dionex)

NeuGc MonoStandard (Dionex)

Neuraminidase from *Arthrobacter ureafaciens*

Bovine fetuin, Spiro method (GIBCO, Grand Island, NY)

Dilute HCl (0.1 N) for sialic release

Sodium acetate buffer for neuraminidase digestion (0.1 M sodium ace-tate buffer, pH 5.0): Add 5.7 ml of glacial acetic acid to 994 ml of water. Adjust the pH to 5.0 with 50% NaOH solution (Fisher). Filter through a 0.2-μm pore size filter. Store at 4° when not in use

Oligosaccharide Analysis

Vanguard Microspin cartridges (Cat. No. 8504-06; Vanguard VG, Nep-tune, NJ)

Endoglucosaminidase H

PNGase [peptide-N^4-(N-acetyl β-D-glucosaminyl)asparagine amidase]

Ribonuclease B (Sigma, St. Louis, MO)

Recycling Dowex 50W-X8 (H$^+$) (100–200 mesh) (Sigma): Add 90 ml of water to a new bottle (100 g) of Dowex resin. Add 10 ml of concen-trated HCl. Cap the bottle and shake vigorously for about 30 sec. Let the bottle sit until the Dowex resin has settled to the bottom. Decant the liquid from the bottle. Add 100 ml of water and repeat the process of shaking, sitting, and decanting. Repeat the water wash five times

Sodium acetate buffer (25 mM), pH 5.5

Sodium phosphate buffer (25 mM), pH 7.6

Sodium borohydride solution (1 M): Weigh 190 mg of sodium borohy-dride into a clean scintillation vial. Add 5 ml of water and dissolve. Use the resulting solution immediately and store the solid sodium borohydride over a desiccant

Carbohydrate Analysis on Polyvinylidene Difluoride Membranes

See the two previous sections for materials and solutions for mono- and oligosaccharide analysis.

Polyvinylidene difluoride (PVDF) sheets [Immobilon PSQ (15 × 15 cm),
0.1-μm pore size (Millipore, Bedford, MA)]
Reduced Triton X-100 (Calbiochem, La Jolla, CA)

Instrumentation

If the time between injections is not equal (as is often the case for manual injections), significant changes (1–3 min) in retention times of monosaccharides will occur. Further, three or four chromatographic runs with standards are usually required for consistent retention times and peak areas ($\pm 1.5\%$). Thus, an automated system is recommended.

Instrument Configuration

The chromatograph (Dionex) consists of a gradient pump, a PAD-II, and an eluent degas module (EDM). The EDM is used to sparge and pressurize the eluents with helium. The system is controlled and data are collected with Dionex AI450 software. Sample injection is with a Spectra-Physics SP8880 autosampler (Fremont, CA) equipped with a 200-μl sample loop. The Rheodyne (Cotati, CA) injection valve is fitted with a Tefzel rotor seal to withstand the alkalinity of the eluents. The pulse potentials for the PAD-II are $E_1 = 0.05$ V, $t_1 = 480$ msec (range 2, position 5): $E_2 = 0.60$ V, $t_2 = 120$ msec (position 2); and $E_3 = -0.60$ V, $t_3 = 60$ msec (position 1). The time constant is set to 3 sec. The Dionex basic PAD cell, equipped with a thin gasket (0.005 inch), is used. Monosaccharides are separated with a CarboPac PA1 column (4 × 250 mm) (Dionex). A Carbo-Pac PA100 is used for oligosaccharide analyses. Eluent 1 (E1) is water; eluent 2 (E2) is 200 mM NaOH; eluent 3 (E3) is 100 mM NaOH; and eluent 4 (E4) is 1 M sodium acetate in 100 mM NaOH. Eluent 4 is filtered through a 0.2-μm pore size nylon membrane before use.

Preparation of Eluents

Rinse each of the bottles with deionized water and with a final rinse of 200 ml of glass-distilled water. Make sure that each bottle is equipped with a clean polyethylene inlet frit filter. An inlet filter needs to be replaced when it becomes discolored. Monthly examination is recommended. In-stall E1 as water in a 2- or 4-liter bottle (4 liters of water provides sufficient eluent for performing monosaccharide analysis unattended for 48 hr). Prepare E2 as 200 mM NaOH. Because of the ion selectivity of the column, NaOH-containing eluents for HPAEC must be prepared to minimize contamination with carbonate. Commercial NaOH pellets are usually coated with a film of sodium carbonate produced by adsorption of CO_2 from the air. To prepare E2, measure 1980 ml of glass-distilled water into a 2-liter plastic, graduated cylinder. Using a disposable 25-ml plastic pipette, transfer 20.8 ml of 50% (w/w) NaOH (19.3 M) from the stock bottle. Any CO_2 adsorbed by the NaOH will precipitate in 50% NaOH. Therefore, draw the liquid from the middle of the solution, rather than from the sides or edges. Transfer the NaOH solution to the graduate; avoid bubbling air into the eluent. Mix the resulting solution thoroughly. Pour the fresh 200 mM NaOH into its bottle (E2). Prepare E3 as 100 mM NaOH from 50% NaOH, as described above for E2. Prepare a 1 M sodium acetate solution in 100 mM NaOH, as E4. It is important to filter this eluent before installation, so that the valves of the high-performance liquid chromatography (HPLC) pump are protected from particulate matter that may be present in commercial preparations of salts. Measure approximately 800 ml of water into a 1-liter graduated cylinder and begin stirring. Weigh out 82 g of anhydrous, crystalline sodium acetate. Add the solid sodium acetate steadily to the briskly stirred water, to avoid the formation of clumps. Add water to a final level of 990 ml. Vacuum filter through a 0.45-μm pore size nylon filter. After filtration, add 5 ml of 50% NaOH for a final concentration of 100 mM NaOH. Install the solution as eluent 4 on the chromatograph. Keep all eluents blanketed under helium (5 to 7 psi) at all times, unless the instrument will not be used for more than a week. Freshly prepared, filtered eluent should be useable for 1 to 2 weeks. The water for all eluents should be glass-distilled, using a Corning (Corning, NY) Mega Pure system. Water should be collected and stored in a glass reservoir without a plastic tubing extension, which can be a site of microbial contamination.

Monosaccharide Analysis

Analogous to amino acid analysis of proteins, hydrolysis with heat and acid is used to release monosaccharides from glycoproteins. The susceptibility of different types of monosaccharides to these conditions varies considerably. For example, the amino sugars, GalN and GlcN, are stable for hours in 6 N HCl at 100°, whereas the neutral sugar Man is quickly destroyed under these conditions. Ketosides, such as sialic acids, are the least stable of glycoprotein monosaccharides. Thus, different conditions are needed for quantitation of glycoprotein monosaccharides, as is detailed below.

Preparation of Monosaccharide Standard Solution

Because monosaccharides are hygroscopic, a significant portion of their weight may be from retained water. To prepare a standard monosaccharide solution, the sample must be dried to a constant weight under vacuum. Prepare a vacuum desiccator containing approximately 100 g of NaOH pellets. Into separate, tared glass scintillation vials, weigh approximately 200–300 mg of each monosaccharide standard (Fuc, GalN, GlcN, Gal, Glc, or Man). (*Note:* Always handle the vials with dry, powderless gloves.) Obtain the gross weight to ± 10 μg. Loosely place a clean cap on each vial and place in the desiccator under continuous vacuum for 24 hr. Include six additional, labeled vials in the desiccator. Determine the gross weight of the standards daily until a change of $<1\%$ is observed ($\sim$3–4 days). Transfer each dry standard into a dry, tared vial. Determine the exact weight of the monosaccharide added and add sufficient glass-distilled water to the vial in the balance chamber to give a 0.1 M solution of each standard. Swirl and invert to completely dissolve the standard. The primary standards should be divided into ten 1-ml aliquots and kept at $-20°$ until they are needed to prepare a standard mixture. To prepare a working solution of all six standards, make a 1 : 1000 dilution of each of the primary standard solutions by pipetting 100 μl of each primary standard solution into a 100-ml volumetric flask. Confirm the volume by noting the weight after the addition of each standard solution to the flask. After thorough mixing, prepare aliquot sizes that are sufficient for 1 day of analyses ($\sim$1 ml). Store the aliquots at $-20°$. The secondary standard solutions should be used within 1–2 days and not refrozen. A mixture of Fuc, GalN, GlcN, Gal, Glc, and Man (100 nmol each) is commercially available (Dionex MonoStandards).

Remove a 200-μl aliquot of the standard solution (0.1 nmol/μl) and transfer to an autosampler vial. Place a Teflon/silicone septum (Teflon face down) in a screw cap, and cap the vial. Often, a small air bubble forms at the bottom of the autosampler vial. Because the SP8880 autosampler withdraws sample from the bottom of the vial, this bubble must be removed. Tap the vial firmly on a solid surface. Samples that have been refrigerated or frozen will usually form bubbles as they warm to room temperature. Allow ample time for the samples to come to room temperature before clearing the bubbles.

Neutral and Amino Sugar Analysis

Whenever glycoprotein sample is not limiting, duplicate HCl and TFA hydrolyses should be performed. Glycoprotein samples are pipetted into a 1.5-ml microfuge tube, placed into a SpeedVac (Savant, Hicksville, NY), and evaporated to dryness. Then either add either 2 M TFA (60 μl of concentrated TFA, 340 μl of water) or 400 μl of constant boiling HCl. The tubes are placed in the heating block for 4 hr at 100°.

Trifluoroacetic acid and HCl are strong, volatile acids. Add the TFA and HCl to samples in a fume hood. Be careful to avoid skin contact with TFA or HCl, and avoid breathing the vapor. Unused TFA should not be saved in ampoules for later use. A digital microliter pipettor can be used to dispense TFA or HCl, but care should be taken not to "splash" TFA inside the tip while pipetting. If this occurs, eject the acid into a suitable waste receptacle, unscrew the end of the pipettor, and clean with a Kimwipe. Always store the pipettor in an upright position, because liquid will enter the pipettor and corrode the plunger.

After the tubes cool, centrifuge for 5 min in a microcentrifuge to return the condensate to the liquid in the tubes. Remove the caps, and place them into a SpeedVac. Evaporate to dryness, without heat. Dissolve each hydrolyzed, dried sample with 200 μl of water and transfer to an autosampler vial. Program the autosampler and HPLC pump according to manufacturer recommendations. The gradient program used for monosaccharide analysis is as follows: $t = 0$ min, E1 = 92%, E2 = 8%; $t =$ 25 min, E1 = 92%, E2 = 8%; $t = 27$ min, E1 = 0%, E2 = 100%; $t =$ 37 min, E1 = 0, E2 = 100%; and $t = 39$ min, E1 = 92%, E2 = 8%. The cycle time for the autosampler (time between each injection) should be set to 50 min. For samples containing large amounts of interfering substances (e.g., salts, protein, and detergents) a more vigorous column cleaning portion of the gradient may be advantageous, using 100% of E4 (1 M acetate, 100 mM NaOH). A separation of these six monosaccharide standards is shown in Fig. 1A. During acid hydrolysis, GalNAc and GlcNAc are quantitatively de-N-acetylated within 3 hr and therefore GalN and GlcN are detected in the acid hydrolysates of glycoproteins.

Preparation of Standard Sialic Acid Solution

Standard solutions of Neu5Ac and NeuGc can be prepared as described earlier for monosaccharides (Preparation of Monosaccharide Standard Solution). Prepare 500 μl of a 50 μM solution of Neu5Ac and NeuGc. A mixed sialic acid standard (25 μM) should be prepared by pipetting 100

μl of the Neu5Ac standard and 100 μl of the NeuGc standard into an autosampler vial. Mix, clear the vial of bubbles, and place it into the autosampler tray. Standard Neu5Ac and NeuGc are also available commercially.

Analysis of Neu5Ac and NeuGc

Pipette glycoprotein samples into 1.5-ml microfuge tubes and dry in a SpeedVac, without heat. Add 200 μl of 0.1 N HCl to each hydrolysis tube, mix by vortexing, and centrifuge for 5 min. Place in the heating block for 60 min at 80°. Dry in a SpeedVac, without heat, add 200 μl of water, and transfer each hydrolysate into a like-labeled autosampler vial. Program the autosampler and analyze, using the following gradient: t = 0 min, E3 = 95%, E4 = 5%; t = 5 min, E3 = 95%, E4 = 5%; t = 25 min, E3 = 82%, E4 = 18%; t = 30 min, E3 = 82%, E4 = 18%; t = 32 min, E3 = 95%, E4 = 5%. The cycle time should be set for 60 min. A separation of Neu5Ac and NeuGc with this gradient is shown in Fig. 1B.

Digestion with a nonspecific neuraminidase is another way to release sialic acids quantitatively. For example, the neuraminidase from *A. ureafaciens* cleaves both $\alpha(2 \rightarrow 3)$- and $\alpha(2 \rightarrow 6)$-linked sialic acids and is not inhibited by most substitutions (4-O-acetylation is one exception).[16] This enzyme is sufficiently active toward most substrates that sialic acids can be released at room temperature. Thus, the kinetics of desialylation can be directly analyzed with the autosampler. Prepare a "positive enzyme control" by pipetting 20 μl of fetuin (1 mg/ml) and 175 μl of 0.1 M sodium acetate buffer (pH 5.0) into an autosampler vial. The neuraminidase will be added later. Program the autosampler for a 60-min cycle time and for two injections of the sialic acid standard mixture (20 μl), five injections of the neuraminidase digest (30 μl), and two injections of sialic acid standard (20 μl). This schedule will run two chromatographies of standards, and then inject aliquots of the neuraminidase digest at 1-hr intervals for 4 hr. Program the HPLC pump to analyze the sialic acids as described above for the analysis of mild acid hydrolysates. Examine the results of the first injection to ensure that the chromatography is comparable to that obtained previously for sialic acid standards (Fig. 1B). If so, allow the schedule of injections to continue. Immediately after the first injection from vial 2 (enzyme blank) add 5 μl of a neuraminidase dilution (0.1 mU/μl), and allow the schedule of injections to continue. The amount of Neu5Ac released should reach a maximum in 2–3 hr. A similar procedure

---

[16] A. Varki and S. Diaz, *J. Biol. Chem.* **258**, 12465 (1983).

can be employed with unknown samples. Quantitation of sialic acid after enzymatic release usually gives a 15–20% greater value than after mild acid hydrolysis of fetuin, presumably due to partial destruction of Neu5Ac with acid. Mild acid hydrolysis of an appropriate sialic acid standard can be used to account for this loss.

Oligosaccharide Analysis

Because HPAEC gives sufficient chromatographic resolution for all classes of oligosaccharides found on glycoproteins there is a high probability that each peak represents a single oligosaccharide structure; thus, HPAEC is a useful "oligosaccharide mapping" method. Further retention times have been shown to vary by less than 1% when internal standards are employed.[14] This procedure details protocols for analyzing oligosaccharides released from glycoproteins by endo-β-N-acetylglucosaminidase H (Endo H) (specific for oligomannosidic and hybrid structures) and the amidase, peptide-N^4-(N-acetyl-β-D-glucosaminyl)asparagine amidase (PNGase) (see Ref. 17 and [4] in this volume for a review of endoglycosidases). Endo H cleaves between the two GlcNAc residues of the chitobiosyl core, which results in one GlcNAc remaining with the protein and the other GlcNAc with the free oligosaccharide. Endo H requires a Manα(1 → 6)Manα(1 → 3)Manβ(1 → 4) consensus sequence.[17] PNGase has a much broader oligosaccharide specificity and releases the oligosaccharide with the amino group of the asparagine residue attached to the oligosaccharide. The resulting glycosylamine is unstable, and eventually converts to the reducing oligosaccharide. Both forms can be observed during HPAEC unless conversion is facilitated by mildly acidic conditions, as described below.

High-pH Anion-Exchange Chromatography Analysis
  of High-Mannose Oligosaccharides

The content of the incubations and their controls for an Endo H digest are given in Table I. Ribonuclease B, a model glycoprotein that contains only high-mannose oligosaccharides[18] is used. Because the proportion of glycosylated ribonuclease is low in commercial preparations, 500 μg of protein is needed to obtain an oligosaccharide map at a detector sensitivity of 300 nA (full scale). The digest buffer used is 25 mM sodium acetate, pH 5.5. The Endo H digestions should be performed in 1.5-ml microfuge tubes. Incubate all tubes at 37° for 18 hr.

[17] F. Maley, R. B. Trimble, A. L. Tarentino, and T. H. Plummer, Jr., *Anal. Biochem.* **180**, 195 (1989).
[18] C.-J. Liang, K. Yamashita, and A. Kobata, *J. Biochem.* (*Tokyo*) **88**, 51 (1980).

TABLE I
ENDO H DIGESTION OF RIBONUCLEASE B

Tube no.	Ribonuclease B[a] (μl)	Sodium acetate buffer (μl)	Water (μl)	Endo H[b] (μl)	Comments
1	0	145	55	0	Buffer blank
2	50	145	5	0	Substrate only
3	0	145	50	5	Enzyme only
4	50	145	0	5	Digest

[a] A 10 mg/ml solution.
[b] A 1 mU/μl solution.

Preparation of Microspin Cartridges for Decationization of Endo H Digest. After starting the enzyme incubations, prepare four Dowex microspin cartridges by pipetting 200 μl of washed Dowex 50W-X8 (H⁺ form) resin into each of four Vanguard microspin cartridges. Cap the cartridges and place them into their microfuge tube holders. Place the tube/cartridge assembly into a microcentrifuge and spin it for 30 sec. Remove the cartridges from the tubes. Using a Pasteur pipette, remove the liquid from the bottom of the tubes. Return the cartridges to the tubes and add 100 μl of water. Repeat the centrifugation and liquid removal steps two times, as described above. After the overnight incubation with Endo H, pipette each sample into the appropriate cartridge (while inserted in the tube) and centrifuge for 30 sec. Remove the cartridges and dry the samples in a SpeedVac to remove water and the acetic acid formed during the desalting process. Dissolve the samples in 200 μl of water and transfer to autosampler vials. Program the HPLC pump as follows: $t = 0$, E3 = 98%, E4 = 2%; $t = 5$ min, E3 = 98%, E4 = 2%; $t = 60$ min, E3 = 80%, E4 = 20%; $t = 62$ min, E3 = 98%, E4 = 2%. Set the autosampler for a cycle time of 90 min.

A separation of the oligosaccharides from ribonuclease B is shown in Fig. 2. The numbered peaks indicate the elution positions of Man_5Glc NAc–$Man_9GlcNAc$. Their structures (**1–3**) are shown in Table II. Except for $Man_7GlcNAc$, all species appear as single peaks. The doublets (labeled as peak 7 in Fig. 2) are likely isomeric forms that differ only in the position of a terminal mannose residue.[18]

PNGase Digestion

Fetuin is used as a model for oligosaccharide mapping of sialylated lactosamine structures after their release by PNGase. Table III gives the content of the incubations for the enzymatic release of N-linked oligosac-

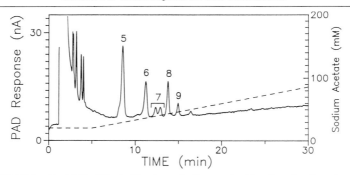

FIG. 2. HPAEC with PAD of Endo H-released oligosaccharides from bovine ribonuclease.

charides from bovine fetuin. Two hundred micrograms of fetuin will give detectable signals at 300 nA (full scale). To avoid loss of resolution after direct injection of the digest, the final buffer concentration is 12.5 mM sodium phosphate buffer. It is recommended that the buffer be filtered through a 0.22-μm pore size membrane prior to use, because the incubation will take place at 37° for 18 hr at near-neutral pH. Perform the digestions in 1.5-ml microfuge tubes and incubate for 18 hr at 37°.

Conversion of Glycosylamine in PNGase Digestions. As discussed earlier, PNGase is an amidase that initially produces an unstable glyco-sylamine. After an overnight enzyme digestion, as much as 30% of this form of the oligosaccharides has been observed. Glycosylamine oligosac-charides are much less retained during HPAEC and thus, in a mixture of these two forms, each oligosaccharide structure appears as two peaks. For oligosaccharide mapping, the glycosylamine must be converted quan-titatively to its single reducing form by lowering the pH of the digest as described below.

After the overnight incubation, add 10 μl of 174 mM acetic acid (1 : 100 dilution of glacial acetic acid) to each sample vial. Incubate the tubes for 2 hr at room temperature. Dry all samples in a SpeedVac at ambient temperature. Then dissolve each sample in 200 μl of water and analyze a 50-μl aliquot. Use the same gradient program as described for separating the high-mannose oligosaccharides from ribonuclease B. A separation of PNGase-released oligosaccharides is shown in Fig. 3A. Three areas, indicated by the brackets, correspond to di-, tri-, and tetrasialylated oligo-saccharides. The structures of the major trisialylated oligosaccharides (peaks 1 and 2) are shown as **4** and **5** in Table II. The only difference between these two structures is the linkage of sialic acid on the uppermost

TABLE II
Structures of N-Linked Oligosaccharides[a]

Oligosaccharide	Structure

Oligomannosides

1

$\text{Man}\alpha(1 \to 6)$
$\text{Man}\alpha(1 \to 6)$
$\text{Man}\alpha(1 \to 3)$
$\text{Man}\beta(1 \to 4)\text{GlcNAc}\alpha,\beta$
$\text{Man}\alpha(1 \to 3)$

2

$\text{Man}_{1-3}\alpha(1 \to 2) \to$ [$\text{Man}\alpha(1 \to 6)$
$\text{Man}\alpha(1 \to 6)$
$\text{Man}\alpha(1 \to 3)$
$\text{Man}\beta(1 \to 4)\text{GlcNAc}\alpha,\beta$
$\text{Man}\alpha(1 \to 3)$]

3

$\text{Man}\alpha(1 \to 2)\text{Man}\alpha(1 \to 6)$
$\text{Man}\alpha(1 \to 6)$
$\text{Man}\alpha(1 \to 2)\text{Man}\alpha(1 \to 3)$
$\text{Man}\alpha(1 \to 4)\text{GlcNAc}\alpha,\beta$
$\text{Man}\alpha(1 \to 2)\text{Man}\alpha(1 \to 2)\text{Man}\alpha(1 \to 3)$

Sialylated lactosamines

4

$\text{Neu5Ac}\alpha(2 \to 6)\text{Gal}\beta(1 \to 4)\text{GlcNAc}\beta(1 \to 2)\text{Man}\alpha(1 \to 6)$
$\text{Man}\beta(1 \to 4) \to R$
$\text{Neu5Ac}\alpha(2 \to 6)\text{Gal}\beta(1 \to 4)\text{GlcNAc}\beta(1 \to 2)\text{Man}\alpha(1 \to 3)$
$\text{Neu5Ac}\alpha(2 \to 3)\text{Gal}\beta(1 \to 4)\text{GlcNAc}\beta(1 \to 4)$

5

$\text{Neu5Ac}\alpha(2 \to 3)\text{Gal}\beta(1 \to 4)\text{GlcNAc}\beta(1 \to 2)\text{Man}\alpha(1 \to 6)$
$\text{Man}\beta(1 \to 4) \to R$
$\text{Neu5Ac}\alpha(2 \to 6)\text{Gal}\beta(1 \to 4)\text{GlcNAc}\beta(1 \to 2)\text{Man}\alpha(1 \to 3)$
$\text{Neu5Ac}\alpha(2 \to 3)\text{Gal}\beta(1 \to 4)\text{GlcNAc}\beta(1 \to 4)$

[a] R, GlcNAcβ(1 → 4)GlcNAcα,β.

branch [$\alpha(2 \to 6)$ versus $\alpha(2 \to 3)$]. The other peaks within each charged class represent separation on the basis of sialic linkage, branch position, and the linkage of Gal in one branch of the triantennary oligosaccharides.[8]

Preparation of Oligosaccharide Alditols

Separations of oligosaccharides require longer column residence times and higher base concentrations than does monosaccharide analysis. Con-

TABLE III
PNGASE DIGESTION OF BOVINE FETUIN

Tube no.	Fetuin[a] (μl)	Buffer (μl)	Water (μl)	Enzyme[b] (μl)	Comments
1	0	60	60	0	Buffer blank
2	20	60	40	0	Substrate only
3	0	60	55	5	Enzyme only
4	20	60	35	5	Digest

[a] A 10 mg/ml solution.
[b] A 0.2 mU/μl solution.

sequently, under these conditions epimerization of GlcNAc to ManNAc has been observed to occur with a $t_{1/2} \approx 20$ min.[19] At equilibrium 80% of the starting GlcNAc remains, whereas 20% is converted to ManNAc. Oligosaccharides that differ only in a GlcNAc or ManNAc at the reducing terminal are separable, appearing as two peaks during HPAEC,[19] and thus a single oligosaccharide species may appear as two peaks with relative ratios of 8 : 2. A method to avoid this complication, by reduction of the terminal GlcNAc residues, is described below.

Dissolve oligosaccharides (~5 nmol) in 500 μl of 1 M sodium borohydride solution in a 20-ml scintillation vial. Add the borohydride solution slowly, as it may bubble and foam after addition to the oligosaccharides. Cap the reaction mixture and incubate at room temperature for 1 hr. *Caution:* This reaction releases hydrogen gas. Periodically loosen the cap to vent. Place the reduction mixture into an ice–water bath. Destroy the excess borohydride by the careful addition of a slight excess of acetic acid. To the chilled reaction mixture, add dropwise 35 μl of glacial acetic acid. The reaction mixture will bubble vigorously from released hydrogen as the acid is added. Carefully add the acetic acid a small drop at a time, and swirl the scintillation vial as the acid is added to minimize the foaming. As the last few droplets of acid are added, the foaming should decrease in intensity, and eventually stop when all the borohydride is converted to borate.

Sodium ion is removed by a strong cation-exchange resin. Prepare a Pasteur pipette column (5 cm in length) of recycled Dowex 50W-X8 (H⁺ form), 100–200 mesh. Position a clean 20-ml scintillation vial under the column outlet. Pipette the entire reduction mixture onto the Dowex column and collect the effluent in the vial. Rinse the Dowex column with

[19] M. R. Hardy, R. R. Townsend, and Y. C. Lee, *Proc. 16th Annu. Meet., Soc. Complex Carbohydr.*, Washington DC, Abstr. No. 12 (1987).

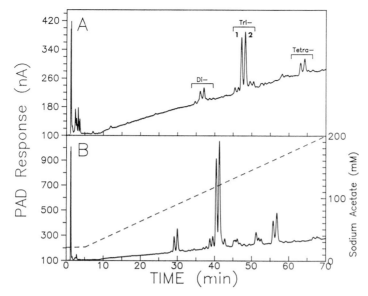

FIG. 3. HPAEC with PAD of PNGase-released oligosaccharides from bovine fetuin.

three 3-ml portions of distilled water, collecting all rinses in the vial. The boric acid is now removed by three evaporations with methanol. Transfer the samples to autosampler vials and inject 20% of the sample at a detection setting of 300 nA (full scale). Figure 3B shows the separation of sialylated fetuin oligosaccharide alditols. All oligosaccharides elute approximately 10 min earlier after conversion of the reducing end from the hemiacetal to the alcohol form. Yet there is no loss of chromatographic resolution. Note the disappearance and reduction of small trailing peaks from the oligosaccharide map of reducing oligosaccharides (Fig. 3A).

Carbohydrate Analysis of Stained, Electroblotted Protein Bands on Polyvinylidene Difluoride Membranes

Sodium dodecyl sulfate–polyacrylamide gel electrophoresis is a commonly employed technique to analyze proteins by removing interfering substances, minimizing preparative losses, and facilitating sample handling. This procedure describes both monosaccharide and oligosaccharide analysis of Coomassie-stained protein bonds on PVDF membranes, using HPAEC with PAD. This method has been found to be particularly useful for the carbohydrate analysis of membrane glycoproteins, which require detergents for isolation and characterization.[20]

Monosaccharide Analysis

To determine the monosaccharide composition of a stained, electroblotted band on PVDF membrane, neutral and amino monosaccharides can be released in a single TFA hydrolysis, as described in Neutral and Amino Sugar Analysis and analyzed at 100 nA. These hydrolysis conditions release 80% of the amino sugars from N-linked structures.[3] Stained bands are excised from methanol–wetted PVDF membranes, placed into 1.5-ml microfuge tubes, and submerged with 340 μl of distilled, deionized water. Sixty microliters of TFA is then added (see precautions for handling TFA in Neutral and Amino Sugar Analysis). Care must be taken to ensure that the stained band is completely submerged. A comparable-sized strip of PVDF from a parallel lane should be excised, hydrolyzed, and analyzed as a blank. All tubes are capped, and subjected to hydrolysis at 100° for 4 hr. The hydrolysates are centrifuged for 2 min in a bench-top microcentrifuge to unite the condensate with the bulk fluid. The hydrolysates are then dried in a SpeedVac centrifuge. The dried samples are reconstituted and transferred with 200 μl of water in autosampler vials.

Sialic acids are released in a separate, milder hydrolysis as described in Analysis of Neu5Ac and NeuGc. Stained, excised bands are prewetted in methanol and submerged in 400 μl of 0.1 N HCl. Hydrolysis is performed at 80° for 60 min. The samples are then processed for HPAEC/PAD as described earlier, and analyzed as given in Analysis of Neu5Ac and NeuGc, at 100 nA (full scale). If sufficient sample is available a 6 N HCl hydrolysis can be performed for quantitative release of amino sugars.

Oligosaccharide Analysis

Because oligosaccharides do not bind to PVDF membranes and all glycoproteins examined remained immobilized under the incubation conditions as required for enzymatic release of oligosaccharides,[20] sequential treatment with a battery of endoglycosidases is a facile approach for releasing and classifying oligosaccharides from glycoprotein bands.

Described here is the sequential treatment of an electroblotted band with Endo H, first to release hybrid and high-mannose oligosaccharides, and then treatment of the same washed blot with PNGase to cleave the

---

[20] M. Weitzhandler, D. Kadlecek, N. Avdolavic, J. G. Forte, D. Chow, and R. R. Townsend, *J. Biol. Chem.* **268,** 13914 (1993).

remaining oligosaccharides. After excising and wetting in methanol, stained bands are immersed in 150 μl of sodium acetate buffer (5 mM, pH 5.5) containing 0.1% reduced Triton X-100 in a 1.5-ml microfuge tube. First, Endo H (1 mU) is added, followed by incubation at 37° for 16 hr. The PVDF membrane is removed and the supernatant is analyzed for neutral oligosaccharides as described in High-pH Anion-Exchange Chromatography Analysis of High-Mannose Oligosaccharides at 100 nA (full scale). The PVDF membrane is rinsed with water and then placed in a tube containing 150 μl of 12.5 mM sodium phosphate buffer, pH 7.6, with 0.1% reduced Triton X-100. One milliunit of PNGase is added and incubated for 16 hr at 37°. These oligosaccharides are then analyzed as described earlier at 100 nA (full scale).

Acknowledgment

The authors thank Richard Lahti for assistance in preparing this manuscript.

[13] High-Performance Liquid Chromatography of Pyridylaminated Saccharides

By Sumihiro Hase

Introduction

When the structures of a minute amount of saccharides are to be analyzed, derivatization of the saccharides is the first step. The label must be detectable at a low level by high-performance liquid chromatography (HPLC), and must be stable during the purification and structure determination. The label should also be removable if the purified saccharides are to be used after characterization. Pyridylamination is one method of labeling that fulfills these requirements.

Pyridylamination involves reductive amination (Fig. 1, reaction 1)[1] and removal of the pyridyl group (Fig. 1, reaction 2).[2] These reactions have been used mostly for asparagine-linked saccharides.[3]

[1] S. Hase, T. Ikenaka, and Y. Matsushima, *Biochem. Biophys. Res. Commun.* **85,** 257 (1978).
[2] S. Hase, *J. Biochem. (Tokyo)* **112,** 266 (1992).
[3] S. Hase, T. Ikenaka, and Y. Matsushima, *J. Biochem. (Tokyo)* **90,** 407 (1981).

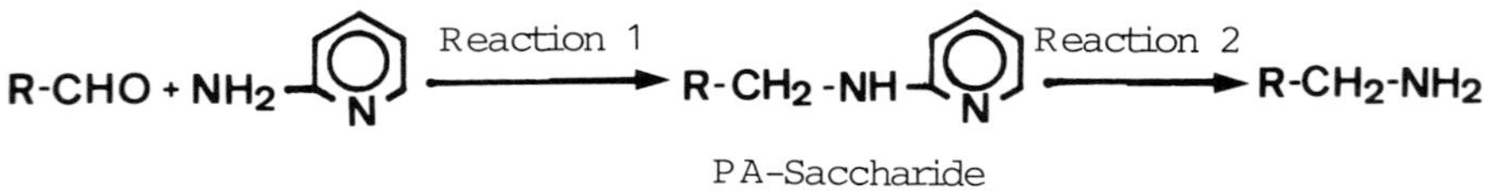

FIG. 1. Chemical reactions for pyridylamination. R-CHO represents the aldehyde form of a saccharide.

Saccharides are analyzed as follows.

1. liberation of saccharides from glycoconjugates[4]
2. pyridylamination of saccharides (Fig. 1, reaction 1)[1,5]
3. separation of pyridylaminated (PA) saccharides by anion-exchange HPLC,[6] size-fractionation HPLC,[7,8] and reversed-phase HPLC[3,8]
4. structural analysis of purified PA saccharides by conventional analytical methods
5. use of a combination of reversed-phase chromatography and size fractionation (giving a two-dimensional mapping),[3] exoglycosidase digestion, and the additivity rule for reversed-phase HPLC,[9] assisting the assessment of the structures of PA saccharides
6. conversion of PA saccharides to 1-amino-1-deoxy derivatives[2] (Fig. 1, reaction 2), which are further coupled with fluorescein, biotin, and so on

In this chapter, pyridylamination[5] and separation of asparagine-linked PA saccharides[6,8] are the main focus, with further applications for analysis of O-linked sugar chains,[5] glycosphingolipids,[10] enzymes that hydrolyze glycosaminoglycans,[11] specificity of lectins,[12] glycosyltransferases,[13,14]

[4] A. L. Tarentino and T. H. Plummer, Jr., this volume [4] and T. Patel and R. Parekh this volume [5].
[5] N. Kuraya and S. Hase, *J. Biochem.* (*Tokyo*) **112**, 122 (1992).
[6] S. Yamamoto, S. Hase, S. Fukuda, O. Sano, and T. Ikenaka, *J. Biochem.* (*Tokyo*) **105**, 547 (1989).
[7] S. J. Mellis and J. U. Baenziger, *Anal. Biochem.* **134**, 442 (1983).
[8] S. Hase, S. Koyama, H. Daiyasu, H. Takemoto, S. Hara, Y. Kobayashi, Y. Kyogoku, and T. Ikenaka, *J. Biochem.* (*Tokyo*) **100**, 1 (1986).
[9] S. Hase, S. Natsuka, H. Oku, and T. Ikenaka, *Anal. Biochem.* **167**, 321 (1987).
[10] K. Ohara, M. Sano, A. Kondo, and I. Kato, *J. Chromatogr.* **586**, 35 (1991).
[11] K. Takagaki, T. Nakamura, H. Kawasaki, A. Kon, S. Ohishi, and M. Endo, *J. Biochem. Biophys. Methods* **21**, 209 (1990).
[12] T. Mega and S. Hase, *J. Biochem.* (*Tokyo*) **109**, 600 (1991).
[13] N. Morita, S. Hase, K. Ikenaka, K. Mikoshiba, and T. Ikenaka, *J. Biochem.* (*Tokyo*) **103**, 332 (1988).
[14] A. Nishikawa, S. Fujii, T. Sugiyama, and N. Taniguchi, *Anal. Biochem.* **170**, 349 (1988).

products by partial acetolysis,[15] and structures by nuclear magnetic resonance (NMR).[16]

Pyridylamination of Saccharides

Glassware and Equipment

Glass test tubes tapered at the bottom (10 × 100 mm) or Reacti-Vials
    (1 ml; Pierce, Rockford, IL)
Water bath at 90° and 80°
Centrifuge with a swinging bucket rotor (about 3000 rpm or 1000 × g)
Gel-filtration column (1.0 × 66 cm) packed with TSK gel HW-40F
    (TosoHaas, Philadelphia, PA)
Lyophilizing apparatus and an ultrasonic washer

Chemicals

Distilled water
2-Aminopyridine (colorless, leaflet-like crystals) twice crystallized from
    1-hexane and stored desiccated at −15° or below
Other chemicals: Of the highest grade commercially available.

Caution: 2-Aminopyridine and borane–dimethylamine complex (flash
point, 43°) are toxic; injurious to the eyes, skin, and mucosa. Avoid
inhalation and store them desiccated at −15° or below in small amounts.

Stock Solutions

Coupling reagent: 2-Aminopyridine (552 mg) is dissolved in 200 μl of
    acetic acid (when the reagent is diluted with 9 vol of water, the pH
    of the solution should be 6.8). The solution is stored at −15° or below
Reducing reagent: Borane–dimethylamine complex (200 mg; Fluka
    Chemie AG, Buchs, Switzerland) is dissolved in a mixture of 50
    μl of water and 80 μl of acetic acid. This reagent is prepared just
    before use
Ammonium acetate solution (0.01 M), pH 6.0: Aqueous ammonia (1.5
    M) is added to 0.01 M acetic acid to bring the solution to pH 6.0.
    The solution is prepared just before use

Procedure for Pyridylamination

A sample from glycoconjugates obtained by hydrazinolysis–N-acetylation or by an enzyme reaction is pyridylaminated without purification of

[15] S. Natsuka, S. Hase, and T. Ikenaka, *Anal. Biochem.* **167,** 154 (1987).
[16] S. Koyama, H. Daiyasu, S. Hase, Y. Kobayashi, Y. Kyogoku, and T. Ikenaka, *FEBS Lett.* **209,** 265 (1986).

saccharides. When amounts of saccharides or glycoconjugates larger than those given below are to be used, the sample is divided into portions small enough for use in the procedure described below. The portions are pyridylaminated, the reagents are evaporated, and the residues are combined.

Saccharides (0.05–100 nmol) or saccharides from 1.5 mg (or less) of glycoconjugates are lyophilized in a glass test tube tapered at the bottom. Coupling reagent (20 μl) is added to the residue and the solution is thoroughly mixed (reagents and pipettes are warmed before use). The sealed tube is spun to bring the solution to the bottom. The whole tube is heated at 90° for 60 min, cooled, and opened.

Reducing reagent (70 μl) is added to the solution. The solution is mixed and spun. The tube is resealed and the whole tube is heated at 80° for 35 min.

Caution: Hydrogen is released during the reaction.

Removal of Excess Reagents

Most excess reagents are removed first by evaporation or by extraction with chloroform. The small excess that remains is removed by gel filtration.

The tube is opened (do not heat sealed glass tubes to open them, due to the presence of hydrogen in the tube), and 40 μl of a 3 : 1 (v/v) mixture of methanol and triethylamine is added. To the solution, 40 μl of toluene is added, and excess reagents are removed by flushing with nitrogen gas at about 300 ml/min at 50° for 10 min under reduced pressure (about 150 mmHg). This evaporation procedure can be done by commercially available apparatus (Palstation Model 1000; Takara Biomedicals, Kyoto, Japan). The evaporation procedure is repeated a total of three times with 60 μl of a 2 : 1 (v/v) mixture of toluene and methanol.

The residue is dissolved in 0.5 ml of water, and the solution is placed on a TSK gel HW-40F column (1 × 66 cm). PA saccharides are eluted with 0.01 M ammonium acetate solution, pH 6.0. A typical chromatogram is shown in Fig. 2. PA N-linked saccharides elute in fractions A and B (Fig. 2, PA sialosaccharides in fraction A and PA asialosaccharides in fraction B), Gal-GalNAc-PA elutes in fraction C, and PA glucose elutes in fraction D. Excess 2-aminopyridine appears first at about the elution volume of 100 ml. The main peak of contamination often appears at 41 ml. After use, the column is washed with 10 or more bed volumes of the eluent before being used again.

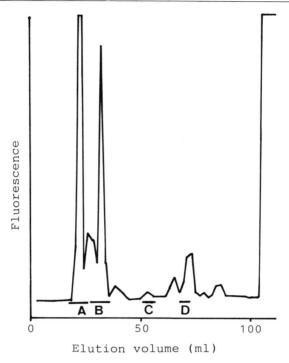

FIG. 2. Gel chromatography of PA saccharides. A sample obtained from galactoglycopro-tein (1.4 mg; kindly supplied by K. Schmid, Boston University) by hydrazinolysis and N-acetylation was pyridylaminated, and excess reagents were removed by evaporation. PA saccharides were then chromatographed on a TSK gel HW-40F column (1.0 × 66 cm) with 0.01 M ammonium acetate solution, pH 6.0, at a flow rate of 7.6 ml/hr. Fractions of 1.9 ml were collected.

Alternative Method for Removal of Excess Reagents

Instead of evaporation, most of the excess reagents can be removed by extraction with chloroform before gel filtration. On a preparative scale, the amount of reagents is increased in proportion to the amount of saccha-rides or glycoconjugates. An example follows: Glycoconjugates (10 mg) are hydrazinolyzed, followed by N-acetylation. The product is pyridylami-nated with 100 μl of the coupling reagent and 350 μl of the reducing reagent as described above. Aqueous ammonia (2.8 M) is added to the reaction mixture to bring the solution to pH 10. Excess reagents are extracted three times or more with 2 vol of chloroform. The water phase is lyophilized and the residue is dissolved in 0.5 ml of water. The solution is placed on a TSK gel HW-40F column (1.5 × 40 cm) and PA saccharides are eluted as described earlier. The PA saccharides first appear at the elution volume of 23 ml, PA Glc appears at 96 ml, and 2-aminopyridine first appears at 130 ml.

Separation and Purification of PA Saccharides by High-Performance Liquid Chromatography

Before HPLC is carried out, 10 μl of water is injected to check that no peaks appear on the chromatogram, except for a few small peaks due to contamination at an elution time of 2–5 min. A few picomoles of PA saccharides are usually injected.

Equipment

The HPLC apparatus has two pumps for gradient elution, a Rheodyne Model 7125 injector, and an in-line fluorescence spectrophotometer or an HPLC monitor with a flow cell of 10–20 μl, a 150-W xenon lamp, and two monochromaters. PA GlcNAc has a UV_{max} at 236 and 309 nm at pH 4.0 and at 239 and 302 nm at pH 7.3. The excitation and emission wavelengths are 315 and 400 nm, respectively, at pH < 6.5; they are at 310 and 380 nm at pH > 7.3. A water jacket is used to maintain the HPLC column at 25°.

Anion-Exchange High-Performance Liquid Chromatography

Eluent and Column

Column: Mono Q HR 5/5 (0.5 × 5 cm; Pharmacia, Uppsala, Sweden)
Eluent A: Aqueous ammonia, pH 9 (about 0.7 mM)
Eluent B: 0.5 M acetic acid adjusted to pH 9.0 with 4 M aqueous ammonia

Eluents are degassed by sonication under reduced pressure just before use.

Column Washing. Before injection of about 10 samples, the column is washed first with 6% (v/v) acetic acid for 10 min and then with 0.4 M aqueous ammonia for 10 min at a flow rate of 1 ml/min. The column is equilibrated with eluent A.

Procedure for Gradient. The pH of a sample to be injected is adjusted to pH 9 with diluted aqueous ammonia. After injection of the sample, elution is done with a linear gradient of eluent B from 0 to 12% for the first 3 min, a linear gradient from 12 to 40% for the next 14 min, a linear gradient from 40 to 100% for the next 5 min, and 0% eluent B for the last 13 min. The flow rate is 1.0 ml/min. Injections can be made every 35 min.

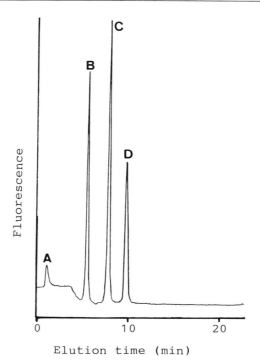

FIG. 3. Elution profile of PA saccharides from a Mono Q column. Peak A, Gal-GalNAc-PA; peak B, PA-monosialobiantenna; peak C, PA-disialobiantenna; peak D, PA-trisialotriantenna.

Relationship of Elution Time and Structure. The separation of PA saccharides is mostly dependent on the number of negative charges (e.g., sialic acid), but some effects of gel permeation occur (Fig. 3).

Size-Fractionation High-Performance Liquid Chromatography

Eluent and Column

Column: MicroPak AX-5 (0.46 × 15 cm; Varian Aerograph, Walnut Creek, CA). A precolumn (any silica gel; 0.75 × 7.5 cm) is placed between the injector and the gradient mixer to prevent damage to the MicroPak AX-5 column. The volume of sample injected is less than 20 µl

Eluent C: Triethylamine is added to 3% acetic acid in a 4 : 1 (v/v) mixture of acetonitrile and water to bring the solution to pH 7.3

Eluent D: Triethylamine is added to 3% acetic acid in water to bring the solution to pH 7.3

Eluents are degassed by sonication under reduced pressure just before use.

Column Washing. Before the injection of about 20 samples, the column is washed with methanol for 30 min or more at a flow rate of 1 ml/min and then equilibrated with the initial eluent.

Procedure for Gradient. After injection of the sample, elution is done with a linear gradient of eluent D from 0 to 10% for the first 2 min, a linear gradient from 10 to 30% for the next 28 min, and 0% eluent B for the last 25 min. The flow rate is 1.0 ml/min. Injections can be made every 55 min.

Relationship of Elution Time and Structure. When PA isomaltooligosaccharides are used as a scale for elution, the interval between two adjacent peaks is defined as 1 glucose unit. A hexose residue equals 1.0 glucose unit, an *N*-acetylhexosamine residue equals 0.6 glucose unit, a bisecting *N*-acetylglucosamine residue equals 0.2 glucose unit, a fucosyl-$\alpha(1 \rightarrow 6)$ residue equals 0.5 glucose unit, and a PA glycose residue equals 1.0 glucose unit, as shown in Fig. 4.

Reversed-Phase High-Performance Liquid Chromatography

Eluents E and F can also be used for PA sialosaccharides, but separation from PA asialosaccharides is better with eluents G and H. Resolution

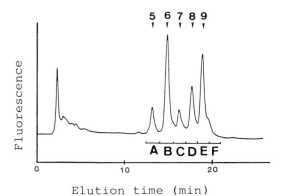

FIG. 4. Elution profile of PA saccharides from the third component of human complement (human C3). The PA saccharide fraction obtained by gel filtration was chromatographed on a MicroPak AX-5 column. Arrowheads 5 to 9 indicate the elution positions of $Man_5GlcNAc_2$-PA to $Man_9GlcNAc_2$-PA, respectively. [Data from Hase *et al., Anal. Biochem.* **167,** 321 (1987), with permission.]

of small PA saccharides is improved by adjustment of the pH of the eluents to between pH 4 and pH 6.5. When the pH of the eluent is increased, the elution times increase.

Eluent and Column

Column: Cosmosil 5C18-P (0.46 × 15 cm; JM Science, Inc., Buffalo, NY)

Eluent E: Aqueous ammonia (4 M) is added to 0.1 M acetic acid to bring the solution to pH 4.0

Eluent F: 1-Butanol is added to eluent E to a final concentration of 1.0%

The eluents are degassed by sonication under reduced pressure just before use.

Eluents for PA sialosaccharides:

Eluent G: 0.1 M acetic acid

Eluent H: 1-Butanol is added to eluent G to the final concentration of 1.0%

Column Washing. Before the injection of about 20 samples, the column is washed with methanol for 30 min or more at the flow rate of 1 ml/min and equilibrated with the initial eluent.

Procedure for Gradient. After injection of the sample, elution is carried out with a linear gradient of eluent F (or H) from 2.5 to 50% for the first 55 min and 2.5% eluent F (or H) for the last 20 min. The flow rate is 1.5 ml/min. Injections can be made every 75 min. An example of the use of eluents E and F is shown in Fig. 5, and one with eluents G and H is shown in Fig. 6.[17]

Rule for Elution Time and Structure. The additivity rule is observed[9]: A sugar residue in a PA saccharide makes an intrinsic contribution to the elution time of the PA saccharide with a certain HPLC apparatus, column, column temperature, and mode of gradient elution. Elution time (E) and the contribution (partial elution time, E_i) are related by the equation: $E = \Sigma E_i$.[18] Elution times of PA saccharides not in hand can be calculated; this rule is used for the assessment of the structures of PA saccharides without the need for the use of many standard PA saccharides. A similar rule has been reported elsewhere.[19]

[17] S. Yamamoto, S. Hase, H. Yamauchi, T. Tanimoto, and T. Ikenaka, *J. Biochem.* (*Tokyo*) **105**, 1034 (1989).

[18] S. Hase and T. Ikenaka, *Anal. Biochem.* **184**, 135 (1990).

[19] Y. C. Lee, B. I. Lee, N. Tomiya, and N. Takahashi, *Anal. Biochem.* **188**, 259 (1990).

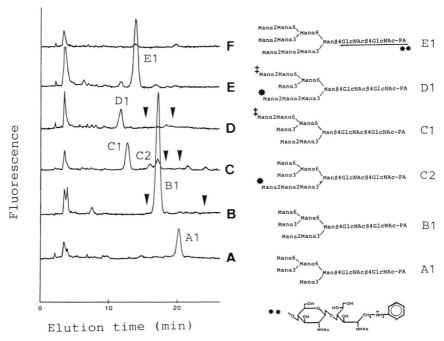

FIG. 5. Reversed-phase HPLC of fractions A to F from Fig. 4 and structures of the fractions. Each fraction shown in Fig. 4 was chromatographed on a Cosmosil 5C$_{18}$-P column with eluents E and F. Arrowheads indicate the elution positions of other isomers of high mannose-type PA saccharides. [Data from Hase *et al., Anal. Biochem.* **167**, 321 (1987), with permission.]

Two-Dimensional Mapping. The principles of two-dimensional mapping [20] can be applied to PA saccharides.[1,21] When the elution times as determined by size-fractionation HPLC (Fig. 4) and by reversed-phase HPLC (Fig. 5) are used as *y*-axis and *x*-axis coordinates, respectively, a pseudo-two-dimensional map can be generated (Fig. 7). The ensemble of saccharide structures of a glycoprotein is represented in Fig. 7 as a set of closed circles. Two-dimensional mapping, combined with the use of many standard PA saccharides, together with the additivity rule and exoglycosidase digestion, allows pattern analysis for structure of saccharides, which can reduce the amount of glycoprotein needed for structure assessment.[21] Not all C$_{18}$ reversed-phase columns can be used for this purpose. Thus far, Cosmosil 5C$_{18}$-P and μBondasphere 5 μm

---

[20] S. A. Barker, E. J. Bourne, P. M. Grant, and M. Stacey, *Nature* (*London*) **177**, 1125 (1956).

[21] S. Hase, T. Sugimoto, H. Takemoto, T. Ikenaka, and K. Schmid, *J. Biochem.* (*Tokyo*) **99**, 1725 (1986).

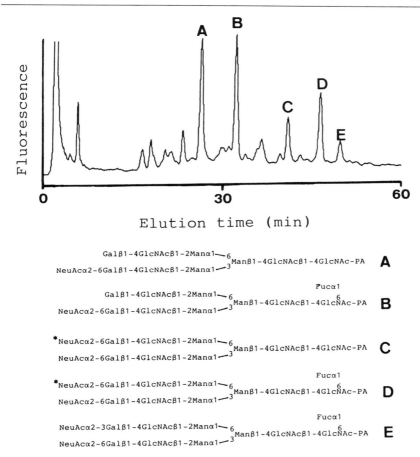

FIG. 6. Reversed-phase HPLC profile of PA saccharides of interferon γ (0.016 mg) from peripheral blood lymphocytes and structures of the peaks. The PA saccharide fraction was chromatographed on a Cosmosil 5C$_{18}$-P column with eluents G and H. The peaks eluting between 2 and 4 min arose from contaminating materials. The additivity rule is illustrated by peaks A and B, and peaks C and D, which differ by one Fucα1-6 residue; and by peaks A and C, and peaks B and D, which differ by one Neu5Acα2-6 residue with an asterisk. (Data from Yamamoto *et al.*[17])

C$_{18}$-300A (Nihon Waters, Ltd., Tokyo, Japan) have been used. Two-dimensional mapping has been done by the use of a larger number of PA saccharides.[22,23]

[22] S. Hase, K. Ikenaka, K. Mikoshiba, and T. Ikenaka, *J. Chromatogr. Biomed. Appl.* **434,** 51 (1988).
[23] N. Tomiya, J. Awaya, M. Kurono, S. Endo, Y. Arata, and N. Takahashi, *Anal. Biochem.* **171,** 73 (1988).

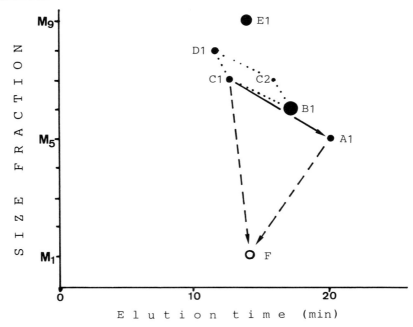

FIG. 7. Two-dimensional HPLC mapping of human C3. Positions of PA saccharides (solid circles) are taken from Figs. 4 and 5. The area ratios of closed circles correspond to those of the peaks (i.e., molar ratio). The solid arrow shows the digestion by α-1,2-mannosidase (*Aspergillus oryzae*) and the dashed arrows the digestion by α-mannosidase (jack bean). The dotted lines connect pairs of C1 and D1, and B1 and C2, that differ by one Manα1-2 residue with an asterisk (Fig. 5) (additivity rule). The additivity rule is also illustrated by B1 and C1, and C2 and D1, which differ by one Manα1-2 residue with a double dagger (Fig. 5). M_1 to M_9, Man$_1$GlcNAc$_2$-PA to Man$_9$GlcNAc$_2$-PA, respectively. F, Manβ1-4GlcNAcβ1-4GlcNAc-PA. For the other keys, see legend to Fig. 5.

Comments

When first using the pyridylamination technique large amounts of the saccharides or glycoconjugates (50–100 nmol or about 1.5 mg, respectively) should be employed together with a known PA saccharide as a standard for chromatography.

The peaks obtained are not always from saccharides (Fig. 2), although most peaks seen in Figs. 4–6 are from PA saccharides (except for peaks at 2–4 min). This can be confirmed by exoglycosidase digestion or component sugar analysis. Solutions of PA saccharides should be stored at $-15°$ or below and those of standard PA saccharides for quantitative purposes should be used within 2 years.

Instead of the columns mentioned earlier, other resins can be used: for example, Sephadex G-25 and BioGel P-2 for gel filtration. Other C_{18} reversed-phases may give slightly different elution profiles. If the separation is not satisfactory, the use of a column with a larger diameter or other C_{18} reversed-phase column is recommended. TSK gel DEAE-5PW can be used instead of Mono Q HR5/5, and TSK gel amide-80 and other amine-bearing bonded phases can be used for size fractionation. Columns with large diameters (e.g., 0.6–1.0 cm) give better separation. The elution gradient (mostly the rate of increase in the percentage of the second eluent) can be changed if desired.

The fluorescence intensity of PA saccharides allows relative molar quantitation of individual saccharides labeled by PA that is virtually independent of the sugar sequence. The deviation of the molar fluorescence intensities of seven PA N-linked saccharides was less than ±7%.[24]

[24] S. Hase, N. Kuraya, and A. Kondo, unpublished data (1990).

[14] High-Performance Liquid Chromatography of Oligosaccharides

By Jacques U. Baenziger

Introduction

Enzymatic[1-3] and chemical methods[4] for the release of N-linked and O-linked oligosaccharides have permitted the development of strategies providing much greater resolution for the separation of oligosaccharides. These play an important role both in purification of individual structures from highly complex and heterogeneous mixtures and in structural analysis. A number of approaches have been devised, offering specific advantages and disadvantages related to the method used for oligosaccharide release, the method of detection, and the purpose of analysis. No single approach is universally effective and various combinations

[1] T. H. Plummer, J. H. Elder, S. Alexander, A. W. Phelan, and A. L. Tarentino, *J. Biol. Chem.* **259,** 10700 (1984).

[2] A. Kobata, *Anal. Biochem.* **100,** 1 (1979).

[3] N. Takahashi and H. Nishibe, *Biochim. Biophys. Acta, Gen. Subj.* **657,** 457 (1981).

[4] S. Takasaki, T. Mizuochi, and A. Kobata, *in* "Methods in Enzymology" (V. Ginsburg, ed.), Vol. 83, p. 263. Academic Press, Orlando, FL, 1982.

of the methods described in the current publication may be required for a specific problem.

The separations described in this chapter have been devised primarily for oligosaccharides that have been radiolabeled by metabolic incorporation of ^{3}H- or ^{14}C-labeled sugar, metabolic incorporation of ^{35}SO$_4$ or ^{32}PO$_4$, or chemical incorporation of ^{3}H at the reducing terminus by reduction with NaB^{3}H$_4$. The most common strategy is to first separate the radiolabeled oligosaccharides into different groups on the basis of their anionic character, using anion-exchange high-performance liquid chromatography (HPLC).[5,6] Oligosaccharides differing in the number and/or character of their anionic group(s) are then further fractionated on the basis of size, using ion suppression–amine adsorption (ISAA) HPLC.[7] The effects of chemical modifications and enzymatic digestions on individual species are subsequently monitored by anion-exchange HPLC and ISAA-HPLC along with other methods such as lectin affinity HPLC.[8]

Anion-Exchange High-Performance Liquid Chromatography

Four methods for separation of anionic oligosaccharides (Fig. 1)[9–12] are illustrated. Fractionation of oligosaccharides bearing zero to four anionic moieties, using MicroPak AX-10 or AX-5 columns (Varian, Palo Alto, CA) and a phosphate buffer (pH 4.0), is effective for oligosaccharides bearing terminal sialic acid.[6] Oligosaccharides-bearing sulfate or phosphate can be fractionated under the same conditions; however, the concentration of the initial eluant is reduced to 2.5 mM in order to retain monosulfated species.[5] The linkage of the sialic acid [$\alpha(2 \rightarrow 3)$ vs $\alpha(2 \rightarrow 6)$] does not significantly affect separation under these conditions. Oligosaccharides (see Fig. 1) with two sialic acids (N-2), two sulfates (S-2), and one sialic acid and one sulfate (S-N) are not resolved at pH 4.0 (Fig. 2). Resolution of N-2, S-2, and S-N

[5] E. D. Green and J. U. Baenziger, *Anal. Biochem.* **158**, 42 (1986).

[6] J. U. Baenziger and M. Natowicz, *Anal. Biochem.* **112**, 357 (1981).

[7] S. J. Mellis and J. U. Baenziger, *Anal. Biochem.* **134**, 442 (1983).

[8] E. D. Green and J. U. Baenziger, *Trends Biochem. Sci.* **14**, 168 (1989).

[9] E. D. Green and J. U. Baenziger, *J. Biol. Chem.* **263**, 25 (1988).

[10] E. D. Green and J. U. Baenziger, *J. Biol. Chem.* **263**, 36 (1988).

[11] Y. Endo, K. Yamashita, Y. Tachibana, S. Tojo, and A. Kobata, *J. Biochem.* (*Tokyo*) **85**, 669 (1979).

[12] E. D. Green, G. Adelt, J. U. Baenziger, S. Wilson, and H. van Halbeek, *J. Biol. Chem.* **263**, 18253 (1988).

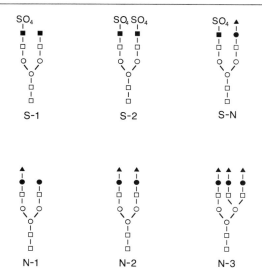

FIG. 1. Oligosaccharides used as standards. The sulfated and sialylated oligosaccharides shown, each with a ^{3}H-labeled N-acetylglucosaminitol at its reducing terminus, were used as standards for the separations shown in Figs. 2–7. S-1, S-2, and S-N were obtained from the glycoprotein hormones,[9,10] N-1 and N-2 from human chorionic gonadotropin,[11] and N-3 from fetuin.[12] ■, GalNAc; □, GlcNAc; ○, Man; ●, Gal; ▲, sialic acid.

oligosaccharides is significantly improved by using phosphoric acid as the eluant (Fig 3). The pH of the eluant changes as the concentration of phosphoric acid increases, but is indicated to be pH 1.7 to distinguish this condition from elution at pH 4.0. The rate of desialylation is significantly increased at lower pH values and fractions should be neutralized immediately if oligosaccharides are being isolated for further analysis.

A guard column packed with pellicular C_8 silica (Whatman, Clifton, NJ) is used rather than a guard column packed with an anion-exchange medium for analysis on MicroPak AX-5 and AX-10 columns. The guard column removes the remaining extraneous ^{3}H following labeling by reduction with NaB^{3}H$_4$ and any remaining peptide following release of metabolically labeled oligosaccharides with endoglycosidases. Relatively large amounts, 1–10 μmol, of oligosaccharide can be isolated with a 4.6 × 300 mm column without deterioration of resolution. This is a particularly critical issue for large, multibranched oligosaccharides bearing a single anionic moiety, as they are weakly bound and can easily be shifted into the nonretained fraction by overloading.

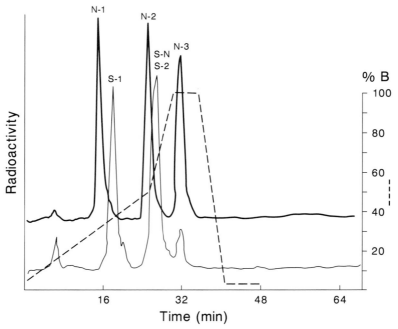

FIG. 2. Anion exchange on MicroPak AX-10 at pH 4.0. Sulfated (S-1, S-2, and S-N) and sialylated (N-1, N-2, and N-3) standards were analyzed separately. For the separations shown, 1500–2000 dpm of each ^{3}H-labeled oligosaccharide was analyzed. The gradient profile is indicated by the dashed line.

Fractions are collected at 0.3-min intervals at a flow rate of 1 ml/min and an appropriate scintillation cocktail added. Alternatively, radioactivity can be monitored continuously with a flow monitor. The flow rate for the eluant is reduced to 0.5 ml/min for continuous monitoring. The reduced flow rate has little effect on elution times but reduces the amount of scintillation cocktail required and increases sensitivity. The scintillation cocktail is combined with the eluant through a T connection, using a metering pump delivering cocktail at a flow rate of 2.0 ml/min. Alternative scintillation cocktails can be substituted for those listed; however, the uniformity and efficiency of counting must be established over the entire range of conditions encountered during the elution. Although as little as 300–500 disintegrations per minute (dpm) of ^{3}H label can be detected as a peak during continuous monitoring, we prefer to use 1000–2000 dpm of ^{3}H label.

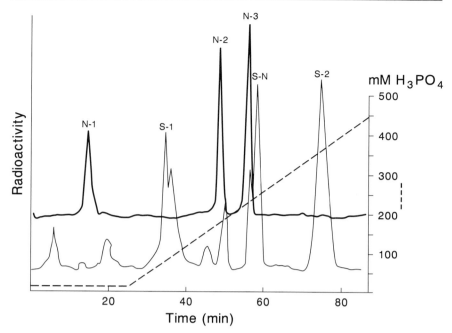

FIG. 3. Anion exchange on MicroPak AX-10 in phosphoric acid. Sulfated (S-1, S-2, and S-N) and sialylated (N-1, N-2, and N-3) standards were analyzed separately. For the separations shown, 1500–2000 dpm of each ^{3}H-labeled oligosaccharide was analyzed. The gradient profile is indicated by the dashed line.

Anion-exchange separations on MicroPak AX-10 and AX-5 columns are able to tolerate considerable variation in the sample preparation. The samples are loaded in 200–400 μl of the initial eluant, using a 500-μl loop. Because oligosaccharides bearing one or more anionic groups are retained and eluted by increasing the ionic strength of the eluant, the large sample volumes do not cause a discernible decrease in resolution. Removal of sialic acid by treatment with 2 N acetic acid at 100° for 15 min can be monitored by drying the sample and resuspending it in the initial eluant without further preparation. Neuraminidase digests in volumes of 50–100 μl can be directly applied to the column after dilution in the initial eluant if they are carried out in 50–100 mM cacodylate, pH 6.0. Digests using other buffers may in some instances require desalting by gel filtration prior to analysis.

Mono Q (Pharmacia, Piscataway, NJ) and GlycoPak (Waters/Milli-pore, Milford, MA) offer better resolution of sialylated and sulfated

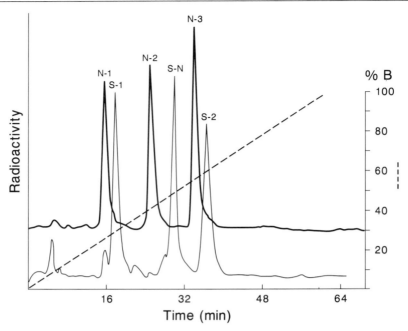

FIG. 4. Anion exchange on Mono Q. Sulfated (S-1, S-2, and S-N) and sialylated (N-1, N-2, and N-3) standards were analyzed separately. For the separations shown, 1500–2000 dpm of each ^{3}H-labeled oligosaccharide was analyzed. The gradient profile is indicated by the dashed line.

oligosaccharides (Figs. 4 and 5). However, we have not found either column to be satisfactory for preparative isolation of anionic oligosaccharides, due to limited capacity and deterioration of resolution when increasing amounts of oligosaccharide are loaded onto the column. For analytic purposes they are, however, particularly useful because we are able to separate virtually all the sulfated and sialylated oligosaccharides we have encountered under the same conditions. Samples must be essentially salt free for analysis with the GlycoPak column.

Conditions for Separation on MicroPak AX-10 or AX-5, 4.6 × 300 mm (Varian Associates) at pH 4.0

Guard column: Pellicular C_8 silica, dry packed
Eluants: 250 mM KH_2PO_4 titrated to pH 4.0 with phosphoric acid
    Reservoir A: 2.5 mM KH_2PO_4, pH 4.0
    Reservoir B: 250 mM KH_2PO_4, pH 4.0

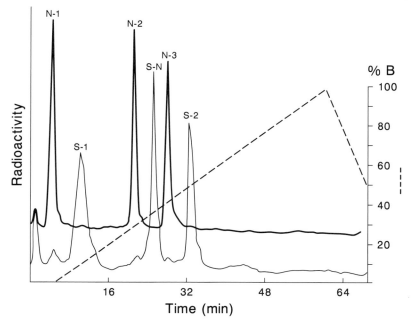

FIG. 5. Anion exchange on GlycoPak DEAE. Sulfated (S-1, S-2, and S-N) and sialylated (N-1, N-2, and N-3) standards were analyzed separately. For the separations shown, 1500–2000 dpm of each ^{3}H-labeled oligosaccharide was analyzed. The gradient profile is indicated by the dashed line.

Gradient:

Time	Reservoir	
	A (%)	B (%)
0 min	100	0
25 min	50	50
30 min	0	100
35 min	0	100
40 min	100	0

Flow rate for collecting fractions: 1.0 ml/min (0.5 min/fraction)
Flow rate for radioactivity monitor: 0.5 ml/min with a 2.0-ml flow cell.
     Scintillation cocktail 3a70B (Research Products International, Mount Prospect, IL) is added at 2.0 ml/min through a T connection

Conditions for Separation on MicroPak AX-10 or AX-5, 4.6 × 300 mm (Varian Associates) at pH 1.7[12a]

Guard column: Pellicular C_8 silica, dry packed
Eluants:
   Reservoir A: 10 mM H_3PO_4
   Reservoir B: 460 mM H_3PO_4
Gradient:

	Reservoir	
Time	A (%)	B (%)
0 min	100	0
25 min	100	0
85 min	0	100
90 min	0	100
95 min	100	0

Flow rate for collecting fractions: 1.0 ml/min (0.5 min/fraction)
Flow rate for radioactivity monitor: 0.5 ml/min with a 2.0-ml flow cell. Scintillation cocktail 3a70B (Research Products International) is added at 2.0 ml/min through a T connection

Conditions for Separation on Mono Q (Pharmacia)

Guard column: None
Eluants:
   Reservoir A: 2.5 mM Tris-HCl (pH 7.4)
   Reservoir B: 2.5 mM Tris-HCl (pH 7.4), 100 mM NaCl
Gradient:

	Reservoir	
Time	A (%)	B (%)
0 min	100	0
5 min	100	0
60 min	0	100
75 min	100	0

[12a] Phosphoric acid is prepared at the concentrations indicated and is not brought to a specific pH. The pH of the eluant will change during the course of the elution and is only nominally pH 1.7.

Flow rate for radioactivity monitor: 0.5 ml/min with a 2.0-ml admix cell. Scintillation cocktail 3a70B (Research Products International) is added at 2.0 ml/min through a T connection

Conditions for Separation on GlycoPak (Waters/Millipore)

Guard column: Pellicular C_8 silica, dry packed
Eluants:
     Reservoir A: Water
     Reservoir B: 100 mM NaCl
Gradient:

	Reservoir	
Time	A (%)	B (%)
0 min	100	0
60 min	0	100
70 min	100	0

Flow rate for radioactivity monitor: 0.5 ml/min with a 2.0-ml flow cell. Scintillation cocktail 3a70B (Research Products International) is added at 2.0 ml/min through a T connection

Size Fractionation of Neutral and Anionic Oligosaccharides

Neutral and anionic oligosaccharides that have been fractionated into groups on the basis of the number and type of anionic moieties can be further separated on the basis of size, that is, number of sugars, using normal-phase chromatography.[7] The column we have most frequently used is the MicroPak AX-5 (Varian Associates); however, other columns such as the TSKgel amide-80 (TosoHaas, Philadelphia, PA) will give similar separations with the same mobile phase. The initial separations of neutral oligosaccharides were performed with acetonitrile–H_2O, increasing the proportion of H_2O.[13] Anionic oligosaccharides cannot be eluted under these conditions. The ionic interactions of oligosaccharides bearing sialic acid, sulfate, and/or phosphate can be largely suppressed by including 3% (v/v) acetic acid titrated to pH 5.5 with triethylamine in the mobile phase. This modification of the mobile phase allows oligosaccharides bearing the same number and type of anionic moieties to be eluted in order of size. Fractionation of neutral oligosaccharides is largely unchanged by

[13] S. J. Mellis and J. U. Baenziger, *Anal. Biochem.* **114,** 276 (1981).

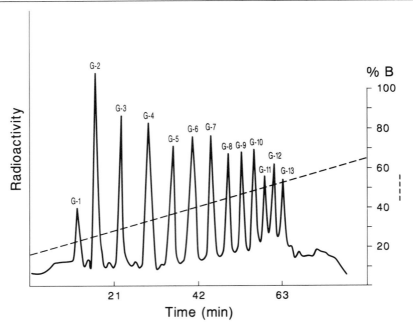

FIG. 6. Ion-suppression amine adsorption HPLC on MicroPak AX-5. Glucose oligomers with 1–13 glucose moieties were prepared from dextran and labeled by reduction with NaB^{3}H$_4$.[9] They were separated under the conditions described in text. The amounts of each oligomer differ and the decrease in peak area with increasing size is not a reflection of any change in the counting efficiency, which remains constant.

the addition of acetic acid and triethylamine to the mobile phase (Fig. 6). Addition of acetic acid and triethylamine to the mobile phase improves the reproducibility and life time of the AX-5 columns markedly. Acetic acid and triethylamine were chosen as ion-suppression agents because they will not precipitate as the proportion of acetonitrile is increased and they do not interfere with scintillation counting. For continuous flow counting, Pico-Fluor 40 (United Technologies Packard, Meriden, CT) has proved to be the most effective scintillation cocktail for our applications. Water is added to the Pico-Fluor 40 at a concentration of 10% (v/v). The Pico-Fluor 40 containing 10% water is then mixed with the eluant at a 6:1 ratio. Addition of water is essential to obtain uniform counting efficiency for oligosaccharides throughout the gradient.

Separations by ISAA-HPLC are sensitive to the identity of the anionic moiety. For example, oligosaccharides terminating with $\alpha(2 \rightarrow 3)$-linked sialic acid elute earlier than the identical oligosaccharides bearing $\alpha(2 \rightarrow$

6)-linked sialic acid.[5] Differences in location and branching patterns can also produce species that separate under the conditions used for ISAA-HPLC.[12] Each sialic acid contributes roughly 6 min to the retention time, whereas each sulfate contributes roughly 8 min to the retention time. Separations of oligosaccharides bearing a single GalNAc-4-SO$_4$ but differing in the number of hexose and hexosamine residues are illustrated in Fig. 7. Separations by ISAA-HPLC are relatively insensitive to structural features of the oligosaccharide such as linkage and branching patterns. This is useful when performing sequential digestions with exoglycosidases, such as those shown in Fig. 7, because the elution time will decrease by ~3 min for each hexose and ~1.5 min for each hexosamine released regardless of location. Resolution of structural isoforms is better accomplished by other columns or by lectin affinity HPLC,[8] methods that are highly sensitive to structural features not resolvable by ISAA-HPLC.

Samples are loaded in a volume of 200–400 μl, using a 500-μl sample loop. Samples of 50–100 μl are diluted with 300–400 μl of the initial eluant for injection. Samples that have been taken to dryness should be dissolved in water and then diluted with the initial eluant. This is necessary because oligosaccharides are only minimally soluble in the initial eluant, which consists of 65% acetonitrile and 35% water (v/v). Enzyme digests can be examined without removing salts or proteins prior to analysis of ISAA-HPLC.

Conditions for Separation by Amine Adsorption HPLC: MicroPak AX-5, 4.6 × 300 mm (Varian Associates)

Guard column: Pellicular C$_8$ silica, dry packed
Mobile phases:
   Reservoir A: Acetonitrile
   Reservoir B: H$_2$O
Gradient:

	Reservoir	
Time	A (%)	B (%)
0 min	65	35
60 min	35	65
70 min	65	35

Flow rate for collecting fractions: 1.0 ml/min (0.3 min/fraction)
Flow rate for radioactivity monitor: 0.5 ml/min with a 2.0-ml flow cell.

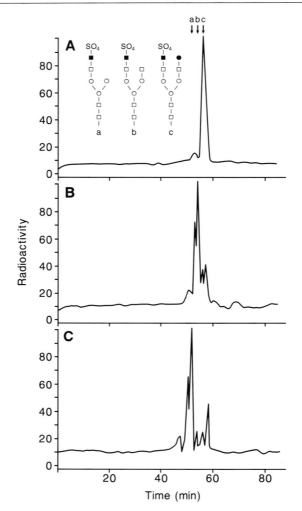

FIG. 7. Separation of monosulfated oligosaccharides differing in their underlying structures. The S-1 structure shown in (A), having the terminal sequence Galβ1-4GlcNAcβ1-2Man on one branch, was sequentially digested with diplococcal β-galactosidase (B) and diplococcal β-hexosaminidase (C). The resulting products were analyzed individually by ISAA-HPLC under the conditions described in text. The elution positions are indicated by a, b, and c. The smaller leading peak seen in (B) and (C) reflects the presence of an α(1 → 6)-linked fucose on 30% of the S-1 oligosaccharide. ■, GalNAc; □, GlcNAc; ○, Man; ●, Gal; ▲, sialic acid.

Pico-Fluor 40 scintillation cocktail (United Technologies Packard) containing 10% (v/v) H_2O is added at a flow rate of 3.0 ml/min through a T connection

Conditions for Separation by Ion Suppression–Amine Adsorption HPLC: MicroPak Ax-5, 4.6 × 300 mm (Varian Associates)

Guard column: Pellicular C_8 silica dry packed
Mobile phases:
  Reservoir A: Glacial acetic acid (3%, v/v) in an 80 : 20 (v/v) mixture of acetonitrile–H_2O titrated to pH 5.5 with triethylamine
  Reservoir B: Glacial acetic acid (3%, v/v) titrated to pH 5.5 with triethylamine
Gradient:

	Reservoir	
Time	A (%)	B (%)
0 min	85	15
100 min	35	65
110 min	85	15

Flow rate for collecting fractions: 1.0 ml/min (0.3 min/fraction)
Flow rate for radioactivity monitor: 0.5 ml/min with a 2.0-ml flow cell
  Pico-Fluor 40 scintillation cocktail (United Technologies Packard) containing 10% (v/v) H_2O is added at a flow rate of 3.0 ml/min through a T connection.

Conclusions

The HPLC separations described for neutral and anionic oligosaccharides have proved to be highly useful for fractionation and analysis of radiolabeled oligosaccharides. We have not found variations in conditions to enhance or improve the separations significantly. The reproducibility, long-term stability, and insensitivity to contaminants of these separations are important factors when considering approaches to oligosaccharide analysis. Also of key importance is the ability to use the same columns and procedures for preparative and analytical separations. These basic procedures and variations have been used by many laboratories with similar results. As with all such methods, it is essential to maintain a library of standards such as glucose oligomers and anionic oligosaccharides to establish the quality and reproducibility of separations.

[15] High-Resolution Polyacrylamide Gel Electrophoresis of Fluorophore-Labeled Reducing Saccharides

By PETER JACKSON

Introduction

A wide variety of standard analytical methods have been used for the separation of carbohydrates. These include chromatographic methods such as low pressure gel permeation chromatography,[1] high-performance liquid chromatography (HPLC),[2,3] GLC, and thin-layer and paper chromatography.[4,5] Some electrophoretic methods have also been used widely, particularly those employing paper and cellulose acetate as matrices.[6] More recently, capillary electrophoresis has begun to be applied.[7] In contrast, polyacrylamide gel electrophoresis (PAGE), which has become the standard and ubiquitous method for the analytical separation of large molecules such as proteins and for DNA sequencing, has been used in limited, although important, ways for the separation of carbohydrates. Existing examples are such as those for the separation of glycosaminoglycans,[8] acidic polysaccharides,[9] and small tritiated neutral saccharides in buffers containing borate.[10] PAGE has also been applied to the separation of fluorophore-labeled glycopeptides.[11] A novel addition to this panoply of techniques is described here whereby saccharides, which contain a reducing (aldose) terminal, are separated by PAGE after they have been derivatized by a suitable fluorophore.[12–16] This new method is called fluor-

[1] A. Kobata, K. Yamashita, and S. Takasaki, this series, Vol. 138, p. 84.
[2] R. R. Townsend, M. R. Hardy, and Y. C. Lee, this series, Vol. 179, p. 65.
[3] S. Honda, *Anal. Biochem.* **140**, 1 (1984).
[4] C. A. White and J. F. Kennedy, in "Carbohydrate Analysis: A Practical Approach" (M. F. Chaplin and J. F. Kennedy, eds.), p. 1. IRL Press, Oxford, 1986.
[5] C. A. White and J. F. Kennedy, in "Carbohydrate Analysis: A Practical Approach" (M. F. Chaplin and J. F. Kennedy, eds.), p. 37. IRL Press, Oxford, 1986.
[6] H. Weigel, *Adv. Carbohydr. Chem. Biochem.* **16**, 61 (1963).
[7] M. Taverna, A. Baillet, D. Biou, M. Schluter, R. Werner, and H. Schroder, *Electrophoresis* **13**, 359 (1992).
[8] I. N. Hampson and J. T. Gallagher, *Biochem. J.* **221**, 697 (1984).
[9] S. Pelkonen and J. Finne, this series, Vol. 179, p. 104.
[10] S. Weitzman, V. Scott, and K. Keegstra, *Anal. Biochem.* **97**, 438 (1979).
[11] R. D. Poretz and G. Pieczenik, *Anal. Biochem.* **115**, 170 (1983).
[12] P. Jackson, *Biochem. J.* **270**, 705 (1990).
[13] P. Jackson and G. R. Williams, *Electrophoresis* **12**, 94 (1990).
[14] P. Jackson, *Anal. Biochem.* **196**, 238 (1991).

METHODS IN ENZYMOLOGY, VOL. 230

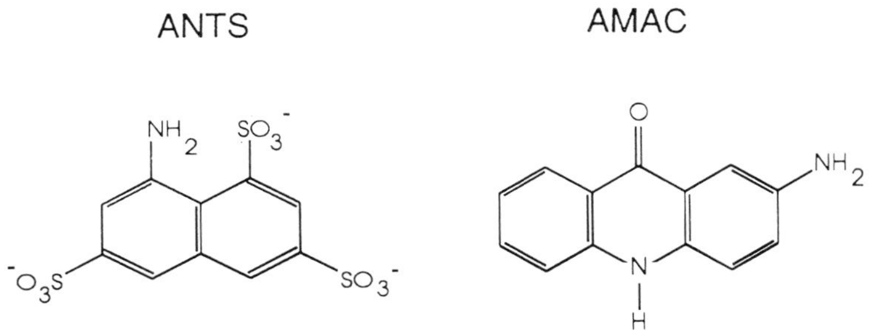

ANTS   AMAC

FIG. 1. Structural formulas of ANTS (8-aminonaphthalene-1,3,6-trisulfonate) and AMAC (2-aminoacridone).

ophore-assisted carbohydrate electrophoresis (FACE) and has a combination of high resolution, high sensitivity, and ease of use that is comparable with the other analytical methods previously listed. It has been found to be particularly useful for revealing profiles of asparagine-linked glycans after enzymatic release from glycoproteins, as described below.

Principle

Saccharides that have a reducing (aldose) terminus are reacted with a fluorophore that contains a primary amino group. This group forms a Schiff base with the carbonyl carbon of the saccharide and the bond formed is stabilized through reduction to a secondary amine, using sodium cyanoborohydride (reductive amination). As a result of the fluorescence conferred onto the saccharide by the reaction with the fluorophore the derivatives can be detected with great sensitivity. The fluorophore that has been used most often is 8-aminonaphthalene-1,3,6-trisulfonic acid (ANTS),[12,13,15,16] (Fig. 1). The three sulfonic acid groups impart to the saccharide–ANTS derivatives a net negative charge that enables both acidic and neutral saccharides to be electrophoresed. Thus the saccharide–ANTS derivatives can be both separated by electrophoresis and detected by their fluorescence. A second fluorophore, 2-aminoacridone (AMAC)[14] (Fig. 1), which confers no charge onto the saccharide derivatives, has also been used. The electrophoresis of derivatives of this latter fluorophore requires that either the saccharides themselves be charged or that an electrophoretic buffer system containing borate ions be used.

[15] P. Jackson, *Biochem. Soc. Trans.* **21**, 121 (1993).
[16] R. J. Stack and M. T. Sullivan, *Glycobiology* **2**, 85 (1992).

Because neutral saccharides derivatized with AMAC can only be electro-phoresed in a buffer containing borate ions, acidic and neutral saccharides can be distinguished readily by electrophoresis in a nonborate buffer when only the derivatives of the acidic saccharides will move. An additional property of AMAC, in contrast to ANTS, is that, it can be used to deriva-tize N-acetylneuraminic acid (see below).

The fluorophore-derivatized saccharides are analyzed by PAGE. The principle of PAGE has been described in numerous reviews. A useful description can be found in Garfin.[17] The most important elements of the method will be outlined here. The key to the high resolution, which is obtained with ANTS and for the separation of charged saccharides that have been derivatized by 2-aminoacridone, lies in the electrophoretic buffer systems used. These are so-called stacking systems that are also termed discontinuous, moving boundary, multizone, or multiphasic. Briefly, the saccharide–fluorophore derivatives are introduced into the electrophoretic system at an interface between a cationic buffer containing an anionic counterion that has a low mobility at the pH of the buffer and a second cationic buffer that contains an anionic counterion with a high mobility. The mobilities of the saccharide derivatives lie between those of the two buffer ions and, during the early part of the electrophoresis, they are swept ahead of the slow anions to become highly concentrated at the sharp boundary that forms between the fastest and slowest. During this process the individual saccharide–fluorophore derivatives sort them-selves into a series of thin adjacent zones in order of their mobilities, which are determined by their charge-to-mass ratios. When the moving boundary between the fastest and slowest buffer anions reaches the high-concentration polyacrylamide of the resolving gel the slow buffer anions increase in relative mobility owing to a change in pH in that gel that increases their net charge. At the same time the saccharide derivatives, being somewhat larger than the slow buffer anions, are retarded by the small pore size of the dense polyacrylamide gel (PAG) and, after being concentrated in an ultrathin zone, are overtaken by the slow buffer anions; the saccharide derivatives become unstacked while maintaining their sepa-ration and narrow banding. They continue to move into the resolving gel, spreading further apart as they move down the gel until the current is turned off. The PAG inhibits convection and diffusion and thus preserves the separation. In the case of the separation of neutral saccharides after labeling with AMAC it is necessary to use a continuous buffer containing borate ions; that is, one that is the same throughout the gel and the electrode buffer reservoirs. The PAG and the buffers have low intrinsic

[17] D. E. Garfin, this series, Vol. 182, p. 425.

fluorescence, which enables high sensitivity to be obtained. The resulting electrofluorograms can be viewed on a graphics display unit after imaging with a suitable electronic camera. An imaging system based on a camera that uses a cooled charge-coupled device (CCD) is available commercially (Millipore Corporation, Bedford, MA) and this enables high sensitivity, quantitation, convenience, and electronic data storage.

A wide variety of monosaccharides, oligosaccharides, and small polysaccharides have been separated by PAGE. In particular, the ANTS derivatives have high resolution that enables the separation of some pairs of isomers, anomers, and epimers. For instance, glucose and galactose are separated, as are maltose, isomaltose, and cellobiose. The range of separation for ANTS is shown in Fig. 2 (refer to the saccharides listed in Table I). The method has found particular use for the analysis of glycans released from glycoproteins by peptide N-glycosidase F (PNGase F) (see [15] in this volume) and protocols for quantities of protein in the range from 1 μg to 200 μg will be described. Typical separations of glycans obtained from some readily available glycoproteins are shown in Fig. 3, together with a maltooligosaccharide ladder standard, derived from wheat starch; typical separations of AMAC-labeled oligosaccharides are shown in Fig. 4.

Procedures

Using PNGase F to Release Asparagine-Linked Oligosaccharides from Glycoproteins

Reagents

All reagents should be of analytical grade.

Phosphate-buffered saline (PBS): 10 mM sodium phosphate buffer (pH 7.0), 0.15 M NaCl

Ethanol: Containing no denaturants

Denaturing solution: 1.0% (w/v) sodium dodecyl sulfate (SDS), 0.5 M 2-mercaptoethanol, 0.1 M disodium ethylenediaminetetraacetic acid (EDTA)

Incubation buffer: 0.2 mM sodium phosphate buffer, pH 8.6, at 37°

Nonidet P-40 (7.5%, v/v) (Calbiochem, La Jolla, CA)

PNGase F (EC 3.5.1.52) from *Flavobacterium meningosepticum*, recombinant form expressed in *Escherichia coli*. Obtained from Boehringer Mannheim (Indianapolis, IN) in solution [1.0 unit (supplier definition)/5 μl]

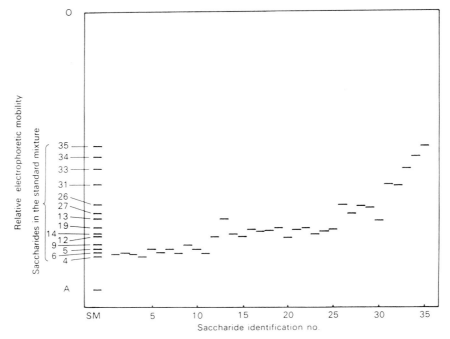

FIG. 2. Diagrammatic representation of an electrofluorogram of ANTS-derivatized saccharides, showing their relative electrophoretic mobilities. Each band represents the position of an individual saccharide derivative relative to unchanged ANTS (labeled A), which moved at the buffer front. Each gel lane number corresponds to the number of the saccharide shown in Table I. The lane labeled SM shows the separation of 14 saccharides that have been found to be useful as a standard mixture to monitor the quality of electrophoreses and as mobility markers. In order of decreasing mobilities the identities of the bands are as follow: **4**, 6-deoxyglucose; **6**, glucose; **5**, galactose; **9**, *N*-acetylgalactosamine; **12**, galactosylgalactose; **14**, lactose; **19**, maltose; **13**, galactobiose; **27**, cellotriose; **26**, maltotriose; **31**, maltotetraose; **33**, maltopentaose; **34**, maltohexaose; **35**, maltoheptaose. The top of the resolving gel is marked by an O.

Procedure for 50 to 200 μg of Glycoprotein

1. Take a suitable volume of solution containing between 50 and 500 μg of the glycoprotein to be analyzed.

2. Dry the sample in a microcentrifuge tube (volume, 1.5 ml), using a centrifugal vacuum evaporator (CVE) [e.g., Savant (Hicksville, NY) SpeedVac].

3. Add 50 μl of PBS and dissolve or suspend the protein.

4. Add 500 μl of cold ethanol, mix, and incubate on ice for at least 1 hr.

5. Centrifuge at 10,000 g for 2 min at room temperature.

TABLE I

SACCHARIDES USED TO CHARACTERIZE ELECTROPHORESIS[a]

No.	Abbreviated formula	Trivial name
1	2-Deoxy-D-Gal	2-Deoxygalactose
2	6-Deoxy-L-Gal	L-Fucose
3	2-Deoxy-D-Glc	2-Deoxyglucose
4	6-Deoxy-D-Glc	6-Deoxyglucose
5	D-Gal	Galactose
6	α-D-Glc	Glucose
7	D-Man	Mannose
8	3-O-Methyl-α-D-Glc	3-O-Methylglucose
9	D-GalNAc	N-Acetylgalactosamine
10	α-D-GlcNAc	N-Acetylglucosamine
11	D-GlcNAc6SO$_3$	N-Acetylglucosamine 6-sulfate
12	α-D-Gal-(1 → 4)-D-Gal	Galactosylgalactose
13	β-D-Gal-(1 → 6)-D-Gal	Galactobiose
14	β-D-Gal-(1 → 4)-D-Glc	Lactose
15	α-D-Gal-(1 → 6)-D-Glc	Mellibiose
16	β-D-Gal-(1 → 4)-D-Man	Galactosylmannose
17	α-D-Glc-(1 → 3)-D-Glc	Nigerose
18	β-D-Glc-(1 → 3)-D-Glc	Laminaribiose
19	α-D-Glc-(1 → 4)-D-Glc	Maltose
20	β-D-Glc-(1 → 4)-D-Glc	Cellobiose
21	α-D-Glc-(1 → 6)-D-Glc	Isomaltose
22	β-D-Glc-(1 → 6)-D-Glc	Gentiobiose
23	α-D-Man-(1 → 3)-D-Man	Mannobiose
24	β-D-Gal-(1 → 4)-D-GlcNAc	N-Acetyllactosamine
25	β-D-GlcNAc-(1 → 4)-D-GlcNAc	Diacetylchitobiose
26	α-D-Glc-(1 → 4)-α-D-Glc-(1 → 4)-α-D-Glc	Maltotriose
27	β-D-Glc-(1 → 4)-β-D-Glc-(1 → 4)-β-D-Glc	Cellotriose
28	α-D-Glc-(1 → 6)-α-D-Glc-(1 → 4)-α-D-Glc	Panose
29	α-D-Glc-(1 → 6)-α-D-Glc-(1 → 6)-α-D-Glc	Isomaltotriose
30	α-Neu5Ac-(2 → 3)-β-D-Gal-(1 → 4)-D-Glc	N-Acetylneuramin-lactose
31	[α-D-Glc-(1 → 4)-]$_3\alpha$-D-Glc	Maltotetraose
32	α-D-Glc-(1 → 6)-[α-D-Glc-(1 → 4)-]$_2\alpha$-D-Glc	
33	[α-D-Glc-(1 → 4)-]$_4\alpha$-D-Glc	Maltopentaose
34	[α-D-Glc-(1 → 4)-]$_5\alpha$-D-Glc	Maltohexaose
35	[α-D-Glc-(1 → 4)-]$_6\alpha$-D-Glc	Maltoheptaose

[a] From Ref. 12.

6. Remove the supernatant and discard.

7. Dry the pellet for 15 min in a CVE.

8. Add to the dry pellet 5.0 μl of glycoprotein denaturing solution and vortex to mix well.

9. Centrifuge briefly to bring the solution to the tip of the tube.

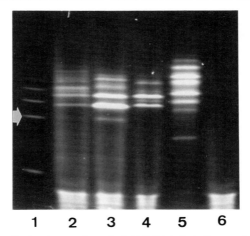

1   2   3   4   5   6

FIG. 3. Graphics display image of a gel of ANTS-labeled oligosaccharides cleaved from various glycoproteins, using PNGase F. The gel was imaged for 10 sec, using a Millipore (Bedford, MA) imager. The length of the resolving gel was approximately 85 mm and the position of maltopentaose in the maltooligosaccharide standard ladder is indicated by an arrow. Lane 1, maltooligosaccharide standard derived from hydrolyzed wheat starch; lane 2, ovalbumin; lane 3, ribonuclease B (bovine pancreas); lane 4, transferrin (human); lane 5, trypsin inhibitor (chicken egg white); lane 6, control. Each lane contains oligosaccharides derived from 10 μg of glycoprotein, except for trypsin inhibitor (in which 4 μg was analyzed) and the control (in which no glycoprotein was added). A greater number of bands can be revealed in each sample and in the maltooligosaccharide standard by adjustment of the gray scaling of the image on the graphics display.

10. Incubate at room temperature for 30 min to wet and denature the dry protein.

11. Add 40 μl of incubation buffer and mix.

12. Heat the sample in a boiling water bath for 5.0 min.

13. Allow the sample to cool to room temperature.

14. Centrifuge briefly to bring the suspension to the tip of the tube.

15. Add 5.0 μl of Nonidet P-40 solution and mix. Ignore the fact that the protein may not be in solution.

16. Add 5.0 μl [1.0 unit (supplier definition)] of PNGase F solution and mix with a minimum of frothing.

17. Centrifuge briefly to bring all the suspension to the tip of the tube.

18. Tap the tube gently to resuspend the protein fully.

19. Incubate at 37° for 18 to 20 hr.

20. Add 165 μl (3 vol) of cold ethanol.

21. Incubate on ice for at least 1 hr.

22. Centrifuge at 10,000 g for 2 min at room temperature.

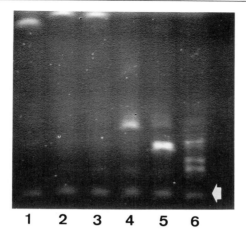

FIG. 4. Graphics display image of a gel of AMAC-labeled asparagine-linked oligosaccharides that were obtained in purified form from Oxford Glycosystems, Ltd., and separated by PAGE in a borate buffer. Twenty picomoles of each sample was loaded per lane. Their compositional formulas are as follows: lane 1, $GlcNAc_4Man_5$ (biantennary); lane 2, $GlcNAc_6$-Man_3 (tetraantennary); lane 3, $GlcNAc_5Man_3$ (triantennary); lane 4, $Neu5Ac_2Gal_2GlcNAc_4$-Man_3 (biantennary); lane 5, $Neu5Ac_3Gal_3GlcNAc_5Man_3$ (triantennary); and lane 6, $Neu5Ac_4$-$Gal_4GlcNAc_6Man_3$ (tetraantennary). The sample in lane 6 is known to consist of several isomers (supplier data). The band marked by an arrow is an artifact.

23. Remove the supernatant, which contains the released glycans, and dry in a CVE. Do not heat above 45° while drying or degradation of the saccharides can occur. The glycans are now ready for derivatization with fluorophore, using the standard method (see below).

24. Save the pellet for further analysis if required. Dry in a CVE and store frozen.

Procedure for 1 to 50 μg of Glycoprotein

The procedure for 1 to 50 μg of glycoprotein is the same as that for larger amounts, as described above, except for the following changes: Dissolve the protein in 20 μl of PBS and precipitate with 200 μl of ethanol. After precipitation dissolve/suspend the protein in 2.0 μl of 0.5% (w/v) SDS, 0.25 M 2-mercaptoethanol, 0.05 M EDTA and, after a 30-min incubation at room temperature, add 8.0 μl of 0.2 M sodium phosphate buffer, pH 8.6. Heat on a boiling water bath for 5 min and, after cooling, add 1.0 μl of 7.5% (v/v) Nonidet P-40 followed by 1.0 μl of the PNGase F. After incubation at 37° for 18 to 20 hr add 36 μl (3 vol) of ethanol to precipitate the deglycosylated protein. Continue as described above. Use the reduced-volume method for the fluorophore derivatization (see below).

Comments on Procedure. The glycoproteins are precipitated (if necessary) to concentrate them and to remove them from any solution that may inhibit the activity of PNGase F. This may be done by a variety of methods. If the method described above is ineffective care should be taken, when using any alternative (such as precipitation with acid), to ensure that this step has no subsequent effect on the activity of the enzyme. It should be noted that many proteins will not be precipitated efficiently by ethanol if they are dissolved in either water or low-salt solution. Precipitation from PBS has been found to be effective for a variety of glycoproteins that are available commercially. Bovine serum albumin (BSA) can be added as a carrier (1 mg/ml) to aid precipitation of glycoproteins that are in low concentrations. The particular batch of BSA used should be tested alone for the presence of any material that may introduce artifactual bands into the electrofluorogram. Glycopeptides can also be analyzed. It is advisable to purify them in a salt-free medium such as by HPLC in a volatile eluant before treatment with PNGase F.

Preparation of Wheat Starch Digest Standard

Reagents

Heat-hydrolyzed wheat starch (Sigma, St. Louis, MO)
α-Amylase (EC 3.2.1.1) from *Bacillus amyloliquefaciens* (Boehringer Mannheim, Indianapolis, IN)
Enzyme incubation buffer: 0.1 M ammonium acetate buffer, pH 5.5, at 37°
Ethanol

Procedure

1. Suspend the starch in the enzyme incubation buffer, using vigorous mixing so that the final concentration of starch is approximately 10 mg/ml.
2. Dissolve the amylase in enzyme incubation buffer to give a protein concentration of 0.75 μg/ml.
3. To 50 μl of starch suspension add 5.0 μl of amylase solution and mix.
4. Incubate for 30 min at 37°.
5. Add 1.0 ml of cold ethanol to stop the reaction and dry the mixture in a CVE.
6. React with ANTS, using the standard conditions.
7. Dry the reaction mixture and dissolve in a suitable volume (e.g., 250 μl) of glycerol–water solution (1 : 4, v/v) and load 2.0 μl/gel lane. If necessary the concentration of the standard can be adjusted to match the

starch digest band intensities to those of the oligosaccharides being analyzed.

Comments on Procedure. It has been found that the extent of digestion
is variable and it may be necessary to carry out a series of digestions,
using either different amounts of amylase or incubation times, to obtain
a mixture of oligosaccharides with a useful range of degrees of polymerization (DP). A typical ladder of bands will range from glucose to maltooligosaccharides with a DP of approximately 20. The identities (i.e., the DP
number) for individual bands should be determined by preparing and
electrophoresing the ANTS-labeled derivatives of the series of maltooligosaccharide standards ranging from maltose, maltotriose, and so on, to
maltoheptaose. The individual pure maltooligosaccharides in this series
are obtained from Sigma. Additional faint bands between the main bands
of the maltooligosaccharide series derived from the starch are sometimes
observed. These are probably due to oligosaccharides containing
branching points of the starch molecules. They can be identified readily
and ignored.

Derivatization of Oligosaccharides with ANTS: Standard Procedure

Reagents

ANTS, 0.15 *M* in acetic acid–water solution (3 : 17, v/v): Gentle warming
(60°) is required to dissolve the ANTS. This solution may be stored
at −70°

Sodium cyanoborohydride (1.0 *M*) in dimethyl sulfoxide (DMSO): Make
this solution fresh

Electrophoresis sample solution for ANTS derivatives: Glycerol–water
(1 : 4, v/v)

Procedure

1. Dry the saccharide(s) that are to be derivatized in a microcentrifuge
tube (volume, 1.5 ml), using a CVE. Do not heat above 45°.
2. Add 5.0 μl of ANTS solution and mix.
3. Add 5.0 μl of sodium cyanoborohydride solution and mix.
4. Centrifuge briefly to bring the reactants to the tip of the tube.
5. Incubate for 16 hr at 37°.
6. Dry the reaction mix in a CVE for about 1 hr. Some heating will
be required to vaporize the DMSO but avoid temperatures greater than
45°. Either store at −70° until required or dissolve in a suitable volume
of glycerol–water (1 : 4, v/v) sample solution for electrophoresis. Store
at −70°.

Derivatization of Oligosaccharides with ANTS:
  Reduced-Volume Method

The method can be scaled down for the analysis of low picomolar quantities of saccharides. In this case the reaction tube should be a small microcentrifuge tube (volume, 0.4 ml) or other low-volume tube with a conical tip. The volumes of ANTS solution and the sodium cyanoborohydride are reduced to 1.0 μl each. Care should be taken to agitate the mixture to dissolve the saccharides at the start of the reaction. Centrifuge briefly to bring all the reactants to the tip of the tube and incubate as described above.

Comments on Procedure. Complete derivatization can be obtained for quantities of saccharides up to 100 nmol/tube. The volume of sample solution will vary with the amount of saccharides derivatized, 10 pmol of saccharide per microliter is typical.

Derivatization of Oligosaccharides with AMAC[10]

  Reagents

  AMAC (Lambda Fluoreszenztechnologie GmbH, Graz, Austria), 0.1 *M*, in solution in acetic acid–DMSO (3 : 17, v/v). This solution may be stored at −70°
  Sodium cyanoborohydride (Aldrich, Milwaukee, WI), 1.0 *M* in solution in water. This solution must be made fresh and used within a few minutes
  Electrophoresis sample buffer: DMSO–glycerol–water (2 : 1 : 7, v/v)
  Procedure. The procedures are the same as those for ANTS, described above.

Polyacrylamide Gel Electrophoresis Separation of ANTS-Derivatized
  Oligosaccharides and for Acidic Oligosaccharides Derivatized
  with AMAC

A PAGE system designed specifically for the separation of ANTS-derivatized oligosaccharides is available commercially from Millipore. The system includes precast gels, prepacked buffers, an electrophoretic apparatus, and a power supply unit. The system is recommended for its convenience and resolving power and because the gel cassettes are designed to be used with the gel-imaging system. Full instructions for the use of the system are supplied by the manufacturer. However, gels can be made in the laboratory if desired. The method described below is one that uses one of several possible buffer systems[12-14]; it is based on the

SDS–PAGE method of Laemmli,[18] but with the SDS omitted through-out.

Reagents and Equipment

Gel electrophoresis apparatus (SE600; Hoefer Scientific Instruments, San Francisco, CA)

Recirculating water cooler (RCB 300; Hoefer Scientific Instruments)

DC power supply unit capable of delivering at least 1000 V and 250 mA and capable of operating at either constant voltage or constant current

Stock acrylamide solution: An aqueous solution of 60% (w/v) acryl-amide (Millipore), 1.6% (w/v) N,N'-methylenebisacrylamide (Millipore)

An aqueous solution of 10% (w/v) ammonium persulfate (Millipore) either made fresh or from a frozen stock

$N,N,N'N'$-Tetramethylenediamine (TEMED) (Millipore)

Stock gel buffer (4× concentrated): 1.5 M Tris-HCl buffer, pH 8.5, made by mixing 1.5 M Tris base and 1.5 M Tris-HCl

Stock electrode buffer (10× concentrated): 1.92 M glycine, 0.25 M Tris base, pH 8.5

Marker dyes (Millipore)

ANTS-labeled wheat starch digest standard

Store all the reagent solutions in the dark at 4°.

Procedure

1. Dilute the stock electrode buffer 10-fold and place the required volume in the anode compartment of the apparatus; allow it to cool to 5°. The gels should be surrounded by well-stirred buffer over as much of their length as is compatible with the equipment.

2. Assemble the gel-casting apparatus according to manufacturer in-structions. Use spacers that are 0.5 mm thick.

3. For each 10 ml of the resolving gel solution mix 5.0 ml of stock acrylamide solution, 2.5 ml of stock gel buffer (4× concentrated), 50 μl (10%, w/v) of ammonium persulfate solution, 2.5 ml of water, and 10 μl of TEMED.

4. Pour the resolving gel solution into the gel mold without delay after adding the TEMED. The height of the gel will depend on the size of the gel apparatus. Any height from 8 to 14 cm is suitable—the longer the gel the greater the resolution.

---

[18] U. K. Laemmli, *Nature (London)* **227**, 680 (1970).

5. Ensure that there are no bubbles in the gel solution before it sets.

6. Overlay the gel solution with water immediately to a depth of at least 1 cm.

7. Allow the acrylamide to polymerize; this takes about 10 min.

8. Make the stacking gel solution by mixing 0.63 ml of stock acrylamide solution, 2.5 ml of stock gel buffer ($4\times$ concentrated) (a special stacking gel buffer is not required), 100 μl (10%, w/v) of ammonium persulfate solution, water to a volume of 10 ml, and 10 μl of TEMED.

9. Pour off the water overlay and fill all the space in the mold above the resolving gel with stacking gel solution.

10. Insert a comb containing the required number of teeth to generate the sample wells. Typically 8 to 12 wells are used for each 8 cm of gel width.

11. Ensure there are no bubbles in the gel solution before it sets.

12. Allow the acrylamide to polymerize for at least 30 min before removing the comb.

13. Remove the comb with care immediately prior to sample loading.

14. Fill the wells with electrode reservoir buffer that has been diluted from $10\times$ stock concentrate or with gel buffer diluted $4\times$ from stock concentrate.

15. Load the samples by layering underneath the buffer in the wells.

16. Assemble the gel cassette into the electrophoretic apparatus according to supplier instructions.

17. Add 10-fold diluted stock electrode buffer to the cathode compartment.

18. Turn on the current. For a gel with the dimensions $0.5 \times 80 \times 80$ mm (thickness $\times$ height $\times$ width) use a current of 15 mA (constant) for 1 hr, then increase to 30 mA (constant) for 2 hr.

19. At the end of the electrophoresis turn off the current and view the electrofluorogram (see below).

20. For gels used for the analysis of AMAC derivatives rinse the wells with water, using a syringe and needle, to reduce the excess fluorescence owing to the unreacted AMAC that remains in the sample wells during the electrophoresis.

Comments on Procedure. The glass of the gel molds must be scrupulously clean before assembly. Any dust particles will be fluorescent and will be detected by the imager. For small gels ($0.5 \times 80 \times 80$ mm) the solutions need not be degassed before polymerization. For large gels ($0.5 \times 140 \times 140$ mm) degassing improves the adhesion of the gel to the glass. Gel may peel away from the glass at the edges of the cassette to a small extent but this can be ignored and the central unblemished area used. Pour the stacking gel onto the resolving gel within 20 min of the setting of the resolving gel. Use a Hamilton syringe (5-μl volume) for loading samples if high accuracy is needed for quantitative work. When using a pipettor, use a tip with a flat end (Cat. No. T 0906; Sigma). Typical sample volumes are between 1 and 5 μl in a well 8 mm wide.

Marker dye can be added to the outer lanes loaded also with the starch digest standard. Bromphenol blue dye has a mobility similar to that of maltotetraose. Gels may be viewed during the electrophoresis by illuminating them in the dark with a hand-held ultraviolet (UV) lamp. Battery-operated lamps are convenient for this and available from UVP, Inc. (San Gabriel, CA). Viewing will be aided if the gel cassette plates are made of low-fluorescence glass such as Pyrex. Only the most intense bands can be seen in this way: the band of unreacted ANTS moves rapidly through the gel ahead of the saccharide derivatives. The duration of the electrophoresis can be adjusted to suit each experiment. In general, the further down the gel each band moves, then the greater will be the separation from neighboring bands. It is most important to maintain good cooling of the gels during the electrophoresis. This is best achieved by surrounding the gels with a stirred liquid. In the case of the recommended equipment this is achieved by a design in which the gels are immersed in the surrounding anode reservoir buffer, which is maintained at 5° by a recirculating water cooler. If cooling is uneven between opposite sites of the gel cassette, then resolution will be lost and some bands may even appear as artifactual doublets.

Gels can be used preparatively to isolate individual fluorophore-labeled saccharides. The band fluorescence tends to fade rapidly when exposed to high- or even moderate-intensity UV light. Therefore for preparative work place the gel over a low-intensity UV light (e.g., Mineralite; UVP Inc.) in a dark room and excise the bands of interest with a scalpel as rapidly as possible. Place the gel strip in a microcentrifuge tube (2.0-ml volume), smash the gel with a spatula, and elute with 1.5 ml of water for 3 hr while mixing either on a rocking table or rotating end over end in the dark. Repeat the elution, combine the eluents, and dry in a CVE.

Polyacrylamide Gel Electrophoresis for Separation of Neutral and Acidic AMAC-Derivatized Saccharides

Reagents and Equipment

Stock gel and electrode buffer (10× concentrated): 1.0 M Tris base, boric acid, pH 8.3. This buffer is made by adjusting the pH of an approximately 1.5 M Tris base solution by adding solid boric acid (approximately 60 g/liter) and diluting the solution to give a final concentration of 1.0 M Tris. Store at 4°

All the other reagents and equipment (except the buffers) are as described in the previous section.

Procedure. The procedure is as described for ANTS gels (see earlier). The final buffer concentration in both electrode compartments and in both the resolving and stacking gels is the same, 0.1 *M* Tris base–borate/boric acid, pH 8.3. The concentration of acrylamide in the resolving gel solution should be 20% (w/v), and 4% (w/v) in the stacking gel. When using an 80-mm long resolving gel electrophorese the samples at 250 V for 30 min, followed by 500 V for 120 min. After this time the bromophenol blue dye will be at the base of the gel. At the end of the electrophoresis rinse the sample wells with water to wash away unreacted AMAC before viewing.

Comments on Procedure. This is not a stacking electrophoretic system. There is little concentration of sample during the electrophoresis and the final bandwidth and therefore the resolution is influenced greatly by the depth of the samples in the wells. The smaller the sample the better the resolution. Samples should be preferably no greater than 2.0 μl in an 8-mm wide well. The so-called "stacking" gel in this system is a low-concentration gel that reduces band distortion. The mobility of neutral saccharides is much lower than that of acidic ones. Run times as long as 4 hr may be required to obtain adequate migration of neutral saccharides. The mobilities of acidic N-linked glycans are usually less than that of the faster moving blue dye in the marker dye mixture. The reaction of *N*-acetylneuraminic acid with AMAC will give two major yellow fluorescent bands in this method. These products have not yet been fully investigated. The faster band moves near the other acidic monosaccharides and the slower band near the neutral monosaccharides.

Viewing and Imaging Electrofluorograms

The electrophoretic band patterns (electrofluorograms) may be viewed in any of three ways: by visual inspection, by photography, and by electronic imaging, using a suitable camera system. In all these cases the gels must be illuminated by UV light. For visual inspection the gels are usually removed from their glass molds and placed on a UV transilluminator (UVP, Inc.) with maximum emissions at wavelengths of either 254 or 365 nm. Gels illuminated in this way can be photographed through a yellow filter (Wratten 8; Kodak, Rochester, NY), using standard black-and-white film. A Polaroid type 55 film with a speed of ISO50 and an aperture of f4,5 works well and may require a 60-sec exposure. Workers must be properly masked during the photography. Great care should be taken not to expose the eyes or any skin to any UV light. A much more advantageous method of viewing the gels is by using an imaging system based on a cooled charge-coupled device (CCD). The cooling enables these cameras to have low intrinsic noise levels, which allows them to be used for relatively long exposures. Photons detected by the camera are converted to electrons and digitized. Images are viewed on a graphics display associated with the computer that controls the whole system. The electronic images can be stored on disk. This method of imaging has severalfold higher sensitivity than film. It is also more convenient as gels are viewed *in vitro,* that is, without being removed from their glass molds. The outside of the glass is rinsed and wiped dry and clean before imaging and then simply slotted into the gel holder of the imager. Exposure times are in a range from 0.5 to 20 sec and as little as 0.2 pmol of ANTS-labeled saccharide per band can be detected. The computer system has a gray scale from 0 to 65,000 and it has been shown that it is possible to achieve a linear response from 10 to 500 pmol.[10] The system is therefore highly suited to quantitative analysis of electrofluorograms. An imaging system based on this technology and designed for use with the associated electrophoretic equipment is available from Millipore. Typical electrofluorograms obtained with this system are shown in Figs. 3 and 4.

[16] Capillary Electrophoresis of Oligosaccharides

By Robert J. Linhardt

Introduction

Electrophoresis is a method of separation that relies on the migration of charged substances in a conducting solution under the influence of an applied electrical field. It is the principal tool for the analysis of peptides and nucleic acids and has been applied to the analysis of oligosaccharides as well.

Charged molecules migrate in an applied electric field at neutral pH. There are many charged carbohydrates that can be directly analyzed by electrophoresis. Ionic carbohydrates include (1) acidic polysaccharides such as pectin and alginic acid, (2) glycosaminoglycans, including hyaluronic acid, chondroitin sulfates, dermatan sulfates, heparan sulfate, heparin, and keratan sulfate, and (3) acidic oligosaccharides, including sialylated, sulfated, and phosphorylated oligosaccharides.

Because electrophoresis generally involves migration of charged species in an electric field, it is not immediately apparent how this technique applies to neutral carbohydrates. Neutral sugars can be given charge by simply adjusting the pH of their environment, thus most show mobility at high pH. Alternatively, neutral sugars can be given a charge through complexation. Borate ions complex with the vicinal hydroxyl groups of sugars, forming anionic species that can be separated by electrophoresis.[1,2] Steric hindrance by carbohydrate side chains and the competitive interaction with other ligands may alter the formation of complexes and thereby provide a wide range of saccharide mobilities. Neutral carbohydrates can also be given charge by their conjugation to a charged species. In addition to ensuring their mobility in an electric field, such derivatization reactions can enhance carbohydrate detection.

Despite its relative ease of use and widespread availability, gel electrophoresis has certain limitations, particularly in sample quantitation and ease of automation. Capillary electrophoresis (CE)[3-5] can overcome these limitations but has only recently been used for analyzing carbohydrates. The most common type of CE is capillary zone electrophoresis (CZE). This rapid, high-resolution, sensitive method is based primarily on the differential migration of solutes through a narrow-bore fused silica capillary.[5] This chapter focuses on the analysis of both underivatized and derivatized oligosaccharides by CZE in unmodified, unfilled fused silica capillaries. Capillary zone electrophoresis is not generally useful for preparative applications.

Background

Capillary zone electrophoresis is a fast and simple method that represents a powerful tool for analysis and separation of acidic saccharides, as first demonstrated independently by Al-Hakim and Linhardt[6] and Carney and Osborne.[7] Capillary zone electrophoresis was used to analyze disaccharides derived from chondroitin sulfates, dermatan sulfate, and hyaluronic acid (Fig. 1). These disaccharides were prepared through the action of a lyase that introduces an unsaturated uronic acid having an ultraviolet absorbance at the nonreducing end of the oligosaccharide. These disaccharides had a net charge of -1 to -4 and were resolved primarily on the basis of net charge and secondarily on the basis of charge distribution. The sensitivity of detection of this method surpassed all other

[1] H. Weigel, *Adv. Carbohydr. Chem.* **18,** 61 (1963).
[2] A. A. Foster, *Adv. Carbohydr. Chem.* **12,** 81 (1957).
[3] W. G. Kuhr, *Anal. Chem.* **62,** 403R (1990).
[4] B. L. Karger, A. S. Cohen, and A. Guttman, *J. Chromatogr.* **492,** 585 (1989).
[5] J. W. Jorgenson and K. D. Lukas, *Science* **222,** 266 (1983).
[6] A. Al-Hakim and R. L. Linhardt, *Anal. Biochem.* **195,** 68 (1991).
[7] S. L. Carney and D. J. Osborne, *Anal. Biochem.* **195,** 132 (1991).

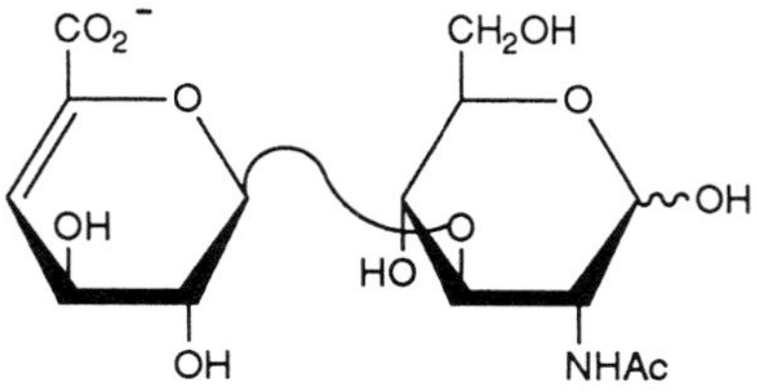

CO_2^- CH_2OH

OH

HO

OH   NHAc

1

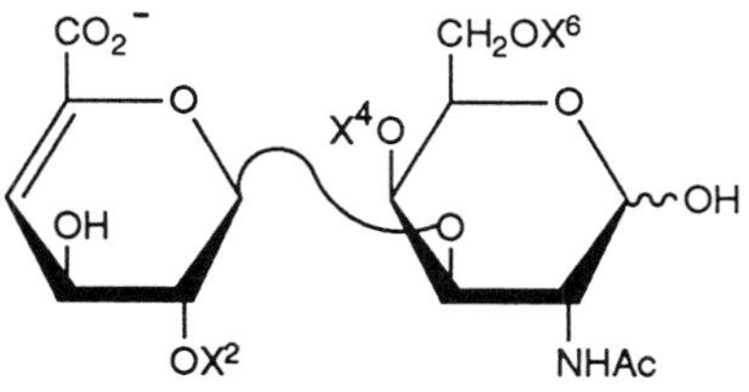

CO_2^- CH_2OX^6

X^4O

OH

OX^2 NHAc

2 $X^2=X^4=X^6=H$ 6 $X^4=X^6=SO_3^-$, $X^2=H$
3 $X^2=X^4=H$, $X^6=SO_3^-$ 7 $X^2=X^6=SO_3^-$, $X^4=H$
4 $X^2=X^6=H$, $X^4=SO_3^-$ 8 $X^2=X^4=SO_3^-$, $X^6=H$
5 $X^4=X^6=H$, $X^2=SO_3^-$ 9 $X^2=X^4=X^6=SO_3^-$

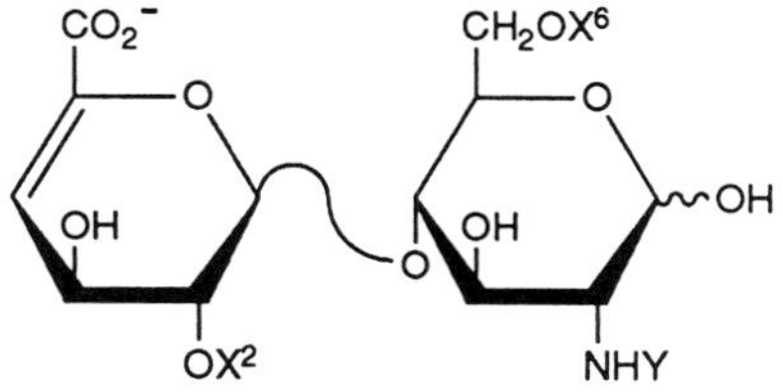

CO_2^- CH_2OX^6

OH   OH

OX^2 NHY

10 $X^2=X^6=H$, Y=$COCH_3$ 14 $X^2=X^6=SO_3^-$, Y=$COCH_3$
11 $X^2=H$, $X^6=SO_3^-$, Y=$COCH_3$ 15 $X^2=H$, $X^6=Y=SO_3^-$
12 $X^2=SO_3^-$, $X^6=H$, Y=$COCH_3$ 16 $X^2=SO_3^-$, $X^6=H$, Y=SO_3^-
13 $X^2=X^6=H$, Y=SO_3^- 17 $X^2=X^6=Y=SO_3^-$

FIG. 1. Structures of acidic, glycosaminoglycan-derived disaccharides analyzed by CZE.

analytical methods for these compounds, permitting the detection of 500 amol.[6] Separation of chondroitin disaccharides in the presence of sodium dodecyl sulfate (SDS) gave improved peak shape and resolution.[7] Ampofo et al.[8] extended this method by separating eight disaccharides prepared from heparin, heparan sulfate, and derivatized heparins (Fig. 1). The effect of capillary length, buffer prepared in ^2H_2O, and SDS in the absence of buffer on peak resolution was also examined.

Sialylated oligosaccharides released from α-acid glycoprotein by hydrazinolysis have been analyzed by CZE by Hermentin et al.[9,10] After

[8] S. A. Ampofo, H. M. Wang, and R. J. Linhardt, *Anal. Biochem.* **199**, 249 (1991).
[9] P. Hermentin, R. Witzel, R. Doenges, H. Haupt, T. Patel, and D. Brazel, *J. Cell. Biochem.*, *Suppl.* **16D**, 146 (1992).
[10] P. Hermentin, R. Witzel, R. Doenges, R. Bauer, H. Haupt, T. Patel, R. B. Parekh, and D. Brazel, *Anal. Biochem.* **206**, 419 (1992).

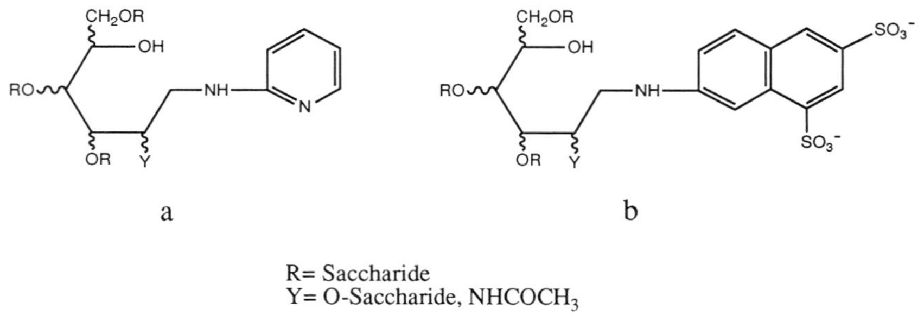

R= Saccharide
Y= O-Saccharide, NHCOCH₃

FIG. 2. Structures of oligosaccharide derivatives analyzed by CZE. (a) N-2-Pyridylgly-camine derivatives; (b) AGA derivatives.

release, the N-glycans were fractionated by chromatography on the basis of charge. The total N-glycans as well as fractions containing a net charge of −1 to −5 were analyzed in buffer at neutral pH. These oligosaccharides were detected without derivatization by monitoring absorbance at 190 nm.

Most of carbohydrate analysis using CE has relied on separation of derivatized sugars in unfilled, underivatized, fused silica capillaries, and detection has been by ultraviolet absorbance. Honda et al.[11] first demonstrated that carbohydrates could be analyzed by CZE. Reducing monosaccharides, converted to N-2-pyridylglycamines by reductive amination (Fig. 2a), were separated by using a capillary tube of fused silica containing basic borate buffer as carrier. The anionic borate complexes facilitated separation and the ultraviolet absorbance of the derivatized sugars permitted the detection of picomole quantities. This method was applied to the determination of the monosaccharide composition of various oligosaccharides and polysaccharides, including lactose, melibiose, rutin, digitonin, and arabic gum.

Nashabeh and El Rassi[12] later reported conditions for the separation of pyridylamino derivatives of maltooligosaccharides in the absence of borate ion. Nashabeh and El Rassi[13] have used this method to separate sialylated oligosaccharides in a derivatized capillary. High separation efficiencies were obtained in phosphate buffer at acidic pH (3.0–4.5). Under these conditions the N-2-pyridylglycamine derivatives (Fig. 2a) are protonated and carry a positive charge. The inclusion of small amounts of tetrabutylammonium bromide in the electrolyte facilitates this separation at pH 5.0 and gives higher separation efficiency. The oligosaccharides eluted in

[11] S. Honda, S. Iwase, A. Makino, and S. Fujiwara, Anal. Biochem. 176, 72 (1989).
[12] W. Nashabeh and Z. El Rassi, J. Chromatogr. 514, 57 (1990).
[13] W. Nashabeh and Z. El Rassi, J. Chromatogr. 536, 31 (1991).

the order of increasing size, with mobility being a linear function of the number of glucose residues in the homologous series. These results were consistent with the independent study of Honda *et al.*,[14] which extended application of CZE to oligosaccharides derived from glycoproteins. Pyridylaminated derivatives of oligosaccharides released from ovalbumin with hydrazine were analyzed by CZE with fluorescence detection. By using polyacrylamide-coated capillary tubes and phosphate buffer (pH 2.5), both ovalbumin oligosaccharide derivatives and isomaltooligosaccharide derivatives [up to a degree of polymerization (DP) of 20] were separated in order of decreasing size. Each oligosaccharide derivative had a single protonated amino group at pH 2.5 that facilitated the separation.

A major limitation of using CZE for carbohydrate analysis has been the absence of sensitive detection methods. Ultraviolet detection has been used for both underivatized and derivatized carbohydrates, but it has limited detection sensitivity. In contrast, fluorescence detection is easily adapted for use in CE and a number of fluorophores have been introduced into the carbohydrate molecules for on-column detection. Lee *et al.*[15] first described the use of a charged fluorescent tag, 7-amino-1,3-naphthalenedisulfonic acid (Fig. 2b), for the analysis of carbohydrates by gel electrophoresis. This method was extended to CZE when an approach was developed to examine the action pattern of chitinase on fluorescent conjugates of chitooligosaccharides.[16] Lee *et al.*[17] also demonstrated the usefulness of CZE for measuring fucosyltransferase activity. Capillary zone electrophoresis measures these fluorescent conjugates with a sensitivity of 80 fmol.[17]

Laser sources are capable of focusing light in small capillaries and thus give increased sensitivity of detection. Liu *et al.*[18] have employed fluorogenic reagents for laser-induced fluorescence measurements of carbohydates. By using 3-(4-carboxybenzoyl)-2-quinoline carboxaldehyde as a precolumn derivatization agent for various amino sugars (degradation products of chitosan and a glycoprotein), highly fluorescent isoindole derivatives were determined by CZE at attomole levels.

A wide variety of compounds has been determined by indirect detection without the need for chemical derivatization.[19] This method relies on the physical displacement of a chromophore or fluorophore in the electro-

[14] S. Honda, A. Makino, S. Suzuki, and K. Kakehi, *Anal. Biochem.* **191**, 228 (1990).

[15] K. B. Lee, A. Al-Hakim, D. Loganathan, and R. J. Linhardt, *Carbohydr. Res.* **214**, 155 (1991).

[16] K. B. Lee, Y. S. Kim, and R. J. Linhardt, *Electrophoresis* **12**, 636 (1991).

[17] K. B. Lee, U. R. Desai, M. M. Palcic, O. Hindsgaul, and R. J. Linhardt, *Anal. Biochem.* **205**, 108 (1992).

[18] J. Liu, O. Shirota, and M. Novotny, *Anal. Chem.* **63**, 413 (1991).

[19] B. L. Hogan and E. S. Yeung, *J. Chromatogr. Sci.* **28**, 15 (1990).

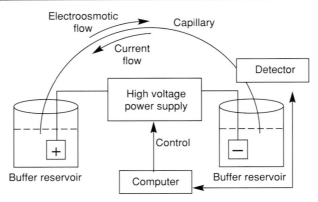

FIG. 3. Schematic of instrumentation used for CE.

lyte by an analyte of similar charge. This results in negative peaks in the electropherogram, due to the absence of fluorophore, and is independent of the spectral characteristics of the analyte. Indirect detection with fluorescein and coumarin fluorophores allows the detection of attomole amounts. Garner and Yeung[20] extended the use of indirect fluorescence detection for CZE to the visible region, permitting the use of visible optics and more powerful visible-light sources. Glucose, fructose, and sucrose were separated by CZE without prior derivatization at a high pH (resulting in the ionization of the sugar hydroxyl groups). Coumarin, which works well with a 442-nm helium–cadmium laser, was used in the detection of this sugar mixture.

Principle of Method

Capillary zone electrophoresis relies on simple instrumentation that includes a high-voltage supply, a capillary column, and a detector (Fig. 3). The capillary column invariably carries a negative charge on its inner surface because the silanol groups of the uncoated capillary ionize above pH 3. Positive ions in the buffer electrolyte are attracted to the negatively charged surface of the capillary. Under high voltage, the positive buffer ions migrate toward the cathode, creating a bulk flow of electrolyte called the electroosmotic flow. If a sample containing both ionic and nonionic solutes is introduced into the capillary at the anode side, all the components are carried in the electroosmotic flow to the cathode. The positive species elute first, followed by the neutral, and finally by the negative

[20] T. W. Garner and E. S. Yeung, *J. Chromatogr.* **515**, 639 (1990).

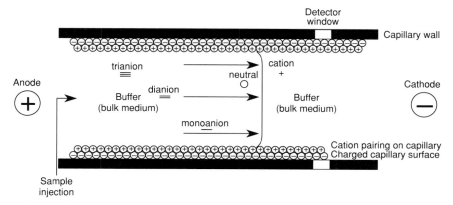

FIG. 4. Separation of charged and neutral species by CZE. Resolution of these species depends on both electrophoretic mobility and electroosmotic flow. The electroosmotic flow is shown by arrows.

species, thus effecting a separation (Fig. 4). To optimize separation, the electroosmotic flow is adjusted either by altering buffer composition and pH or by chemically derivatizing the capillary wall. Micellar electrokinetic capillary electrophoresis (MECE) exploits a different separation principle. The addition of sodium dodecyl sulfate (SDS) to the electrolyte buffer results in partitioning of the analyte between the micelle and the surrounding aqueous phase. Separation is accomplished through differential migration of the phases.

Procedure

Instrumentation

There are a variety of different instruments that are commercially available. In addition, it is possible to build an instrument from the component parts. All CE instruments are composed of a high-voltage power supply, buffer reservoirs, a capillary, and a detector (Fig. 3). Most commercial instruments also include a sampler/injector and a programmable computer for automated analysis and data processing.

Preparing Capillary

The capillary column is typically made of fused silica, externally coated with a 20- to 75-μm i.d. and a 300- to 500-μm o.d., and is usually available in long rolls. It is first cut to the desired length (usually between 0.2 to

1 m), and the external coating is removed from the area where the capillary passes through the detector. The capillary tube is activated by washing extensively with 5 vol each of 0.1 M phosphoric acid, 0.5 M sodium hydroxide, deionized water, and operating buffer and inserted into the machine.

Preparing Buffer

The buffer to be used must first be carefully selected. Not all buffers support electroosmotic flow in a capillary. The instruction manual accompanying each instrument usually suggests buffers that can be used for different CZE applications. The buffer should be selected, carefully prepared, and its pH adjusted. Finally, the buffer must be filtered (through a 0.2-μm pore size syringe filter) to remove particles that can clog the narrow capillary and then thoroughly degassed. It is best to prepare fresh buffer each day.

Preparing Sample

The sample must be filtered (through a 0.2-μm pore size syringe filter) or centrifuged before analysis to remove particulates. The sample should be at a high concentration, typically between 100 μg/ml and 10 mg/ml (for a carbohydrate of M_r 500–2500). The sample should be free of salt and other contaminants that could interfere with either electrophoresis or sample detection.

Injecting Sample

There are a number of different methods that can be used to inject a sample into the capillary column for analysis. Gravity injection is an unbiased method of introducing both charged and neutral analytes. Typically, gravity injection that relies on capillary action is used. The amount of sample injected can be calculated from the difference in height between the ends of the capillary, the internal diameter of the capillary, and the length of time the elevated end of the capillary is immersed in the sample. This injection method is also referred to as vacuum injection or injection by hydrostatic pressure. Other methods of injection include electroinjection and syringe injection. Electroinjection is biased toward either anionic or cationic species (depending on the direction of current flow), making quantitation difficult. Typically only 1 to 100 nl of sample is injected for each analysis. The volume of analyte injected depends on the capillary volume, the sample concentration, and the detector sensitivity. Typically, an injection volume of >10% of the capillary volume leads to a loss of resolution.

Electrophoresis Conditions

A high voltage, typically ranging from 10 to 20 kV, is used in CZE. The length of time required for each analysis is usually from 2 to 90 min, depending on capillary length, voltage, buffer, and the net charge of the sample. A constant voltage is most commonly used; however, voltage gradients can be used to improve resolution or to minimize analysis times. Electrophoresis can also be done under constant power or constant current.

Detection

Only absorbance and fluorescence detectors are commercially available. The absorbance detectors are either fixed wavelength detectors or variable wavelength detectors. The fluorescence detectors require filters of the appropriate wavelength. A single-wavelength laser-induced fluorescence detector has become available.

Specific Applications

Analysis of Underivatized Carbohydrates

Glycosaminoglycans and Derivatives

Glycosaminoglycans (GAGs) are highly charged, linear, sulfated polysaccharides that can be broken down with polysaccharide lyases[21] into acidic oligosaccharides that absorb at 232 nm.[6–8] Disaccharides **1–17** (Fig. 1) can be prepared directly from glycosaminoglycans, including hyaluronic acid, chondroitin sulfate and dermatan sulfate, heparin, heparan sulfate, or chemically modified heparin, with polysaccharide lyases [from Seikagaku America, Inc. (Rockville, MD), Grampian Enzymes (Aberdeen, Scotland), and Sigma Chemical (St. Louis, MO)][21] and purified by strong anion-exchange high-performance liquid chromatography (HPLC).[22] Alternatively, these disaccharide standards are now available from these same companies.

1. Disaccharide standards are accurately weighed, and a 10-mg/ml stock solution of each is prepared in distilled, deionized water. Portions of these stock solutions are mixed together to prepare standard mixtures

[21] R. J. Linhardt, C. L. Cooney, and P. M. Galliher, *Appl. Biochem. Biotechnol.* **12,** 135 (1986).
[22] K. G. Rice and R. J. Linhardt, *Carbohydr. Res.* **190,** 219 (1989).

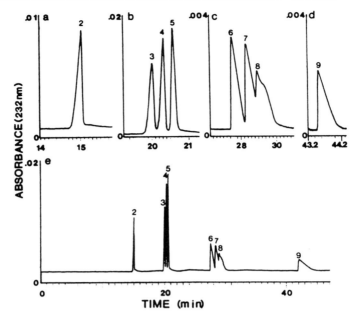

FIG. 5. A typical electropherogram showing the separation of acidic, glycosaminoglycan-derived disaccharides (Fig. 1). A sample containing 1 ng of each disaccharide is applied to a 68-cm capillary by gravity injection. The buffer is 10 mM sodium borate containing 50 mM boric acid at pH 8.8, and electrophoresis is performed at 10 kV. [Figure taken from A. Al-Hakim and R. J. Linhardt, *Anal. Biochem.* **195**, 68 (1991), with permission.]

and for coinjection studies. These standard solutions are stored frozen at $-70°$.

2. Glycosaminoglycan (10 μg/μl) is treated with the appropriate lyase (2 mU of enzyme/100 μg of substrate) in 5 mM sodium phosphate buffer at pH 7.0 for 12 hr at $30°$.

3. A CE system equipped with an ultraviolet detector set at 232 nm is used for this analysis. Separation uses a fused silica capillary column having an effective length of 68 cm (distance between injector and detector). The buffer used is 10 mM sodium borate containing 50 mM boric acid, having a pH of 8.5, or 10 mM sodium borate and 50 mM SDS, having a pH of 8.5.

4. The sample is injected by gravity injection (1 nl) at the anodic side (positive electrode), and electrophoresis is performed for 0.5 to 1 hr at 10 kV, using the operating buffer. The temperature is maintained at $30°$. A typical electropherogram is shown in Fig. 5.

Comments on Method. A simple mixture containing single disaccharides having a net charge of -1 to -4 can be used to examine the effect of voltage, capillary, buffer, and pH on migration time. A doubling of the voltage, from 10 to 20 kV, will reduce the migration time by 50% but decreases resolution. The replacement of one capillary tube for a second, identically prepared capillary will only slightly ($\pm 5\%$) affect the migration times. The use of phosphate buffer (pH 8.8), in place of borate buffer, will increase migration time and decrease resolution. Acidification of the borate buffer to pH 5, to decrease complexation, surprisingly increases disaccharide migration times.

Lengthening the capillary above 70 cm results in only a modest improvement in resolution but increases analysis time. The introduction of 50 mM SDS[8] in the presence or absence of borate buffer can improve resolution but results in a slightly decreased sensitivity. Under optimum conditions, the detection sensitivity of this method is 500 amol to 1 fmol.

The disaccharide composition of a glycosaminoglycan can be analyzed by this method. Oligosaccharides of net charge > -4 have similar migration times under the conditions described, thus glycosaminoglycan should be completely converted to disaccharides with a mixture of lyases.[8] These analyses required 15 ng of polysaccharide and a 20-min analysis time. Full quantitation of glycosaminoglycan may require the use of multiple sets of buffer conditions.

Sialylated Oligosaccharides

α_1-Acid glycoprotein contains a number of different N-linked sialylated oligosaccharides. These oligosaccharides can be released by hydrazinolysis or by using PNGase F to obtain an *N*-glycan pool. These acidic oligosaccharides can be fractionated by Mono Q FPLC (fast-protein liquid chromatography, Pharmacia, Piscataway, NJ) ion-exchange chromatography to obtain five fractions containing oligosaccharides with net charge of -1 to -5.[9,10]

1. Oligosaccharide Mono Q fractions prepared from α_1-acid glycoprotein are dissolved in distilled water at approximately 2 mg/ml and stored frozen.

2. A CE system equipped with an ultraviolet detector set at 190 nm is used for the analysis. Separation is in a 100-cm fused silica capillary column. The buffer s 80 mM ammonium sulfate, 20 mM sodium phosphate, 2 mM diaminobutane, adjusted with phosphoric acid to pH 7.

3. The sample is injected by vacuum injection (5 nl) at the anodic side (positive electrode), and electrophoresis is performed for 1 hr at 20 kV. The temperature is maintained at 30°.

Comments on Method. The method of Hermentin and co-workers[9,10] described earlier is one of the most general methods for the analysis of sialylated, glycoprotein-derived oligosaccharides. The most important feature of this method is that no derivatization is required for the detection of femtomole amounts of *N*-glycan by absorbance at 190 nm. The analysis of α_1-acid glycoprotein-derived oligosaccharides as pyridylaminated derivatives by Nashabeh and El Rassi[13] resulted in only 6 major structures, whereas the analysis described above facilitated the detection of 40 different structures. Whether differences in the observed complexity of the analyte are due to the superior resolution of this method or simply to differences in the heterogeneity of the starting glycoprotein remains to be established.

Indirect Fluorescence Detection

Simple underivatized monosaccharides, disaccharides, and trisaccharides can be analyzed at elevated pH and detected by indirect fluorescence detection.[20]

1. Saccharide samples are analyzed in 10 m*M* sodium bicarbonate buffer containing 1 m*M* coumarin adjusted to pH 11.5 with sodium hydroxide.[20]

2. Separations is in a 90-cm fused silica capillary column driven by the ionization of the sugar hydroxyl groups at high pH.

3. Injection of 640 fmol at the anodic side is followed by indirect fluorescence detection with a helium–cadmium laser operating at 442 nm.

Comments on Method. This method avoids the requirement of derivatization by using a high pH to drive the separation and an indirect method of detection. As more complex mixtures of sugars are analyzed, the resolution achieved at high pH may be insufficient for adequate separation. Additionally, high pH is known to result in oligosaccharide peeling[23] and other reactions that damage sugar fine structure. Indirect detection of oligosaccharides can also be performed in a standard fluorescence detector.[24]

Analysis of Derivatized Carbohydrates

Conjugation with Neutral Tag

Oligosaccharides that are tagged with 2-aminopyridine (Fig. 2a) are neutral derivatives and can be effectively analyzed by CZE only if they are first given a charge. Two approaches are typically used to accomplish

---

[23] R. L. Whistler and J. N. BeMiller, *Adv. Carbohydr. Chem.* **13,** 289 (1958).
[24] S. A. Ampofo and R. J. Linhardt, unpublished results, 1992.

this. In the first, the pyridylamino derivatives of neutral oligosaccharides are converted *in situ* to anionic borate complexes by using borate buffer at high pH. In the second, these pyridylamino derivatives are given a positive charge by protonation in an acidic (low-pH) buffer.

2-Aminopyridine Reagent Solution. To a mixture of methanol (350 μl) and glacial acetic acid (80 μl), 2-aminopyridine (184 mg) and sodium cyanoborohydride (35 mg) are added. The solution is mixed in vortexer and used immediately.

1. Preparation of oligosaccharide conjugates is accomplished by adding 40 μl of the 2-aminopyridine reagent solution to 0.1–1 mg of oligosaccharide in 10 μl of water.[14,25] The mixture is heated for 15 hr at 80°.

2. To purify the oligosaccharide derivatives, the reaction is first cooled and then applied to a Sephadex G-15 column (1 × 50 cm), followed by elution with 15 mM ammonium acetate in 25% aqueous methanol. The column is monitored by fluorescence (λ_{ex} = 316 nm, λ_{em} = 395 nm). The void and included volumes, containing derivatized oligosaccharide, are combined and dried *in vacuo*. The partially purified product is dissolved in water and applied to a Dowex 50-X2H$^+$ form (5 × 25 mm). The column is thoroughly washed with water and the sample eluted with 200 mM ammonium hydroxide, neutralized with acetic acid, and freeze-dried. The product may contain residual 2-aminopyridine (particularly if the oligosaccharide being derivatized contains only one or two saccharide residues) and can be further purified if necessary.[25]

3. The 2-aminopyridine derivatives of monosaccharides[11] and neutral oligosaccharides[14] (such as those prepared from ovalbumin by hydrazinolysis) can be analyzed in 200 mM potassium borate buffer at pH 10.5.[11] Electrophoresis is performed in a 65- to 95-cm fused silica capillary at 15–20 kV. Injection is at the anodic end and detection is at the cathodic end with ultraviolet at 240 nm or by fluorescence (λ_{ex} = 316 nm, λ_{em} = 395 nm).

4. Maltooligosaccharides derivatives (tetrasaccharides through heptasaccharides) can be analyzed in 0.1 M sodium phosphate containing 50 mM tetrabutylammonium bromide at pH 5.0.[12] Electrophoresis is performed in a 50-cm fused silica capillary at 15 kV. Injection and detection is as described in step 3.

Conjugation with Charged Tag

High-resolution separation and high-sensitivity fluorescence detection of sugars can be achieved by using saccharide conjugation to a fluorescent tag containing a fixed charge (Fig. 2b).

[25] S. Hase, S. Hara, and Y. Matsushima, *J. Biochem.* (*Tokyo*) **85**, 217 (1979).

1. The monopotassium salt of 7-amino-1,3-naphthalenedisulfonic acid [amido-G acid (AGA)] is used after recrystallization from deionized water. Oligosaccharide (3.5 μmol) is dissolved in 750 μl of AGA solution [50% (w/v) in water adjusted with sodium hydroxide to pH 6.2]. After heating at 80° for 60 min, sodium cyanoborohydride (16 μmol) is added. The mixture is then heated for 24 hr at 70°. After the reaction is complete the products are fractionated on a BioGel P-4 (Bio-Rad, Richmond, CA) or Sephadex G-25 column (2.5 × 50 cm) eluted with distilled water to remove salt and excess fluorescent tag. The sugar conjugate (Fig. 2b) is recovered in the void or included volume of the column and is detected either by using a hand-held ultraviolet light (long wavelength) or in a spectrometer by absorbance at 314 nm or by fluorescence (λ_{ex} = 314 nm and λ_{em} = 452 nm).

2. Oligosaccharide–AGA conjugates can be purified to homogeneity, if desired, by gradient preparative polyacrylamide gel electrophoresis, using a linear gradient from 12 to 22% (w/v) acrylamide that contains 0.5 to 2% (w/v) N,N'-bisacrylamide.[26] Electrophoresis is done on a gel (32 × 16 × 0.75 cm) for 18 hr at 400 V (constant voltage) with cooling. Following electrotransfer onto positively charged nylon membranes, oligosaccharide–AGA conjugates are recovered by washing the membranes with 2 M sodium chloride.[26] The salt is removed, using a BioGel P-2 column eluted with distilled water, and the sample is freeze-dried.

3. Electrophoresis is performed with a CE system in a fused silica capillary tube (68 cm long) by gravity injection (10 nl) at 20 kV, using 10 mM sodium borate and 50 mM boric acid, having a pH of 8.8

4. Fluorescence detection requires the installation of a long-pass optical emission filter in the cell. To optimize detection, an emission filter of 420 nm is used. The excitation wavelength is varied at 50-nm intervals to obtain the optimum excitation wavelength. With the optimum excitation wavelength various emission filters are tested to determine which gives the greatest sensitivity. Optimum detection sensitivity by fluorescence [on a Dionex (Sunnyvale, CA) capillary electrophoresis system] is obtained at λ_{ex} = 250 nm and λ_{em} = 420 nm.

Comments on Method. Reductive amination is a useful method for labeling sugars with probes to permit their detection and facilitates their separation. Amido-G acid contains two fixed negative charges used to drive a separation by electrophoresis.[15] The greater the number of sugar residues in a particular oligosaccharide, the more vigorous are the conditions (i.e., longer reaction times, higher temperatures) required to ensure

[26] A. Al-Hakim and R. J. Linhardt, *Electrophoresis* **11**, 23 (1990).

complete derivatization.[15] For oligosaccharides having a terminal hexose, a yield of 90% is possible, whereas for N-acetylhexosamine reducing ends recovered yields of 60–80% are obtained. Neutral pH is used when derivatizing sugars containing acid-labile functionalities such as sialic acids.[15] Detailed studies may be required to optimize yield for each type of oligosaccharide being labeled.

Oligosaccharide-fluorescent conjugates prepared are typically contaminated with excess fluorescent tag (AGA) required to drive the reaction to completion. It is often possible to remove this excess reagent by gel-filtration chromatography. Often, however, the oligosaccharide being derivatized is not 100% pure. These impurities can lead to the formation of additional minor products during reductive amination. Preparative gradient polyacrylamide gel electrophoresis can be used to resolve, identify, and recover the desired product(s) in the crude reaction mixture.

Fluorescence detection of AGA–oligosaccharide conjugates is sensitive. A linear calibration curve constructed for the AGA derivative of chitotriose gave a linear response from 42 fmol to 1 pmol.

Capillary Zone Electrophoresis for Assaying Glycosyltransferases

Glycosyltransferases are an important group of enzymes that transfer sugar residues from specific activated sugar nucleotide donors to suitable oligosaccharide acceptors. A method that relies on capillary electrophoresis for the assay of glycosyltransferases has been reported that uses a sugar–AGA conjugate as acceptor molecules. In glycosyltransferase assays both sugar–conjugate acceptor and sugar–conjugate product are rapidly quantified by CZE.

1. β-D-Gal-(1 → 4)-β-D-GlcNAc-(1 → 6)-β-D-Gal is conjugated to AGA by reductive amination in the presence of sodium cyanoborohydride (see above). The products are desalted on a 2.5 × 50 cm BioGel P-2 column and purified by preparative PAGE.

2. The AGA–acceptor (10 μg) is incubated at 37° for 26 hr with α(1 → 3/4)-fucosyltransferase (13 μU) and GDPfucose (420 μmol) in 150 μl of 16 mM sodium cacodylate buffer at pH 6.8, containing 3 mM manganese chloride. The mixture is freeze-dried at the end of the reaction.

3. The product mixture (0.5 mg/ml) is reconstituted in distilled, deionized water. Analysis is performed on a CE system with fluorescence detection (as described above).

Comments on Method. Capillary zone electrophoresis analysis of the purified oligosaccharide–AGA acceptor, using fluorescent detection, showed a single major peak. The α(1 → 3/4)-fucosyltransferase reaction showed an additional peak that migrates faster than the acceptor, corresponding to product. These results are consistent with CZE studies that show that larger sugar–AGA conjugates migrate faster than smaller sugar–AGA conjugates. The presence of α-linked fucose at the nonreducing end of the product is confirmed by conversion of product to acceptor on treatment with α-L-fucosidase.

This approach may represent a general method to assay various glycosyltransferase activities, including N-acetylglucosaminyltransferases, fucosyltransferases, sialyltransferases, and mannosyltransferases, provided proper sugar–AGA acceptors are prepared for each. The sensitivity of this CZE analysis is 80 fmol. Time-course studies[16] that measure transfer of the activated sugar to oligosaccharide–AGA acceptor are also possible.

Acknowledgments

The author thanks Dr. Peter Hermentin at Behringwerke AG (Marburg, Germany); Dr. Stephen Ampofo at Hazelton Wisconsin (Madison, WI); and Dr. Umesh Desai at the University of Iowa for their valuable assistance in the preparation of this chapter.

[17] Glycosidases in Structural Analysis

By GARY S. JACOB and PETER SCUDDER

Introduction

The role of glycosylation in biology is defined by the relationship between structure and function, and with it the requirement for detailed structural information on the carbohydrate component of macromolecules.[1–3] Unfortunately, compared with the analysis of proteins and nucleic acids, characterization of the glycan component of glycoconjugates has lagged, due to the high degree of complexity that they display. The features that complicate structural analysis include the variety of positional glycosidic bonds that link individual monosaccharide residues, the anomeric configuration of each sugar, and the presence of branched structures and

[1] E. G. Berger, E. Buddecke, J. P. Kamerling, A. Kobata, J. C. Paulson, J. F. G. Vliegenthart, *Experimentia* **38**, 1129 (1982).

[2] T. W. Rademacher, R. B. Parekh, and R. A. Dwek, *Annu. Rev. Biochem.* **57**, 785 (1988).

[3] A. Kobata, *in* "Biology of Carbohydrates" (V. Ginsburg and P. W. Robbins, eds.), Vol. 2, p. 87. Wiley, New York, 1984.

substituents such as sulfate and phosphate sugars. Cumulatively, this has led to no simple recipe for determining the structures of these compounds. Rather, a variety of chemical and analytical methods have been used to obtain definitive structural information, the particular procedures being used depending substantially on the nature and type of starting material. In many cases, the material to be analyzed has been a glycoprotein or glycolipid, but this has broadened to include new classes of glycan-containing materials such as glycosylphosphatidylinositol-anchored proteins.

In most cases, structural analysis of the carbohydrate components of these various molecules begins with release of the intact carbohydrate, either by chemical or enzymatic means, followed by purification and structural determination. Glycosidases have enjoyed prominence as reagents for the sequencing and structural characterization of the released oligosaccharides,[4,5] in particular with respect to asparagine-linked glycans, which generally show greater uniformity in their monosaccharide composition and structural features than related serine/threonine-linked compounds. The use of enzymes for structural determinations is likely to diminish in the future with the increased sophistication of analytical methods for carbohydrate analysis such as multidimensional proton nuclear magnetic resonance (NMR)[6,7] and fast atom bombardment and electrospray-mass spectrometry.[8-10] However, in laboratories where these techniques are not available or when the material to be analyzed is in low abundance (in the picomole to low nanomole range), enzyme sequencing will continue to be an extremely powerful tool for structural elucidation. The reader should be aware, however, that the use of enzymes for determining structure requires great care on the part of the researcher, and should involve a large dose of healthy skepticism about any deduced structure until sufficient corroborative evidence has been obtained.

The principle of oligosaccharide sequencing relies on the ability of specific exoglycosidases to remove terminal monosaccharides from the nonreducing end of oligosaccharides. To achieve complete removal of the terminal monosaccharide residue(s), a relatively high concentration of

[4] A. Kobata, *Anal. Biochem.* **100**, 1 (1979).

[5] F. Maley, R. B. Trimble, A. L. Tarentino, and T. H. Plummer, *Anal. Biochem.* **180**, 195 (1989).

[6] J. Dabrowski, this series, Vol. 179, p. 122.

[7] J. F. G. Vliegenthart, H. van Halbeek, and L. Dorland, *Pure Appl. Chem.* **53**, 45 (1981).

[8] A. Dell, H. R. Morris, H. Egge, H. von Nicolai, and G. Strecker, *Carbohydr. Res.* **115**, 41 (1983).

[9] A. Dell, this series, Vol. 193, p. 647.

[10] D. Garozza, G. Impallomeni, E. Spina, B. N. Green, and T. Hutton, *Carbohydr. Res.* **221**, 253 (1991).

enzyme is used together with an extended incubation time. By using the information obtained from a series of digestion experiments involving different exoglycosidases, a structure can be deduced. In this chapter, we review the basic requirements for this work, including (1) discussion of the kind and type of enzymes needed, (2) methods for sample preparation and standard digestion protocols, (3) interpretation of digestion data, and (4) observations and pitfalls in the use of exoglycosidases as probes of oligosaccharide structure. In addition, the issue of enzyme purity will be addressed in some depth, with details given regarding chromatographic steps that can be used to remove contaminants commonly found in commercial preparations.

Exoglycosidase Library

To sequence a glycan of unknown structure completely could potentially require an unrealistically large array of exoglycosidases. However, in reality most analyses can be performed with about 12 enzymes, most of which can be obtained from commerical suppliers, including Boehringer Mannheim (Indianapolis, IN), Calbiochem (La Jolla, CA), Genzyme (Boston, MA), ICN Biomedical (Costa Mesa, CA), Oxford GlycoSystems (Abingdon, UK), Seikagaku (St. Petersburg, FL), and Sigma (St. Louis, MO). The basic enzymes required for sequencing are listed in Table I, together with details of their specificity and incubation conditions used in sequencing experiments. Many of these enzyme preparations contain unacceptably high levels of contaminant glycosidases, which severely compromise their use as sequencing reagents. Described below, therefore, are simple chromatographic protocols that can be used to remove the commonly found contaminants and to provide enzymes that are operationally pure.

Jack Bean β-Galactosidase.[11] Preparations of this enzyme typically contain low levels of β-hexosaminidase, which can be removed by hydrophobic interaction and cation-exchange fast protein liquid chromatography (FPLC). Samples are adjusted to 0.85 M ammonium sulfate, applied (flow rate, 0.5 ml/min) to a phenyl-Superose HR 5/5 column (Pharmacia, Piscataway, NJ) equilibrated with 50 mM KH$_2$PO$_4$/NaOH, pH 7.0, containing 0.85 M ammonium sulfate, and the enzyme eluted with a descending linear gradient of 0.85–0.0 M ammonium sulfate (β-galactosidase elutes at about 0.6 M ammonium sulfate). After dialysis against 20 mM sodium acetate buffer, pH 5.0, the enzyme is applied (1 ml/min) to a Pharmacia Mono S HR 5/5 column (Fig. 1) equilibrated with the same buffer and is eluted

[11] S.-C. Li, M. Y. Mazzotta, S.-F. Chien, and Y.-T. Li, *J. Biol. Chem.* **250,** 6786 (1975).

with a 40-min linear gradient of 0.0–0.125 M NaCl. The β-galactosidase elutes at about 0.05 M NaCl.

Jack Bean α-Mannosidase.[12] β-Hexosaminidase is a common contaminant that can be removed by hydrophobic interaction FPLC. The sample is adjusted to 0.6 M ammonium sulfate and applied (1.5 ml/min) to a phenyl-Superose HR 10/10 column equilibrated with 50 mM KH$_2$PO$_4$/NaOH, pH 6.5, containing 0.6 M ammonium sulfate (Fig. 2). The column is eluted for 15 min with starting buffer, 10 min with buffer containing 0.25 M ammonium sulfate, and 40 min with a descending linear gradient of 0.25–0.0 M ammonium sulfate. The β-hexosaminidase elutes between 0.6 and 0.25 M ammonium sulfate, whereas the α-mannosidase elutes between 0.15 and 0.2 M ammonium sulfate.

Jack Bean β-Hexosaminidase.[13] Contaminating activities of β-galactosidase and α-mannosidase can be removed by a combination of the two methods described above.

Coffee Bean α-Galactosidase.[14] Preparations of this enzyme invariably contain relatively high levels of β-galactosidase, which can be removed in a single step by hydrophobic interaction FPLC (Fig. 3). The enzyme is adjusted to 2.0 M ammonium sulfate and applied (0.5 ml/min) to a phenyl-Superose HR 5/5 column equilibrated with 50 mM citric acid/Na$_2$HPO$_4$, pH 6.0, containing 2.0 M ammonium sulfate. The column is eluted for 10 min with a descending linear gradient of 2.0–0.8 M ammonium sulfate, 4 min with buffer containing 0.8 M ammonium sulfate, 24 min with a descending linear gradient of 0.8–0.0 M ammonium sulfate, and 5 min with 50 mM citric acid/Na$_2$HPO$_4$, pH 6.0. The α-galactosidase elutes at about 0.5 M ammonium sulfate whereas the β-galactosidase elutes at 0.25 M ammonium sulfate.

Aspergillus saitoi α-1,2-Mannosidase (α-Mannosidase I). α-Mannosidase I is not commercially available. The purification procedure described by Ichishima *et al.*[15] gives a preparation that contains measurable α-mannosidase II and α- and β-galactosidase activities, but all of these activities can be removed by FPLC on columns of Mono Q and phenyl-Superose. The partially purified enzyme is equilibrated with 20 mM piperazine/hydrochloride, pH 5.2, and applied (1 ml/min) to a Mono Q HR 5/5 column. Under these conditions the α-mannosidase does not bind and is recovered from the void volume of the column. This eluate is adjusted to 2.0 M ammonium sulfate and then applied (0.5 ml/min) to a phenyl-

[12] Y.-T. Li, *J. Biol. Chem.* **242**, 5474 (1967).
[13] S.-C. Li and Y.-T. Li, *J. Biol. Chem.* **245**, 5153 (1970).
[14] J. E. Courtois and F. Petek, this series Vol. 8, p. 565.
[15] E. Ichishima, M. Arai, Y. Shigmatsu, H. Kumagai, and R. Sumida-Tanaka, *Biochim. Biophys. Acta* **658**, 45 (1981).

TABLE I
EXOGLYCOSIDASES COMMONLY USED IN OLIGOSACCHARIDE SEQUENCING

Enzyme	Source	Linkage hydrolyzed	Digestion condition[a]
α-Mannosidase I	*A. saitoi*	Manα1-2Man	0.5 mU/ml in 50 mM sodium acetate (pH 5.0), BSA (1 mg/ml), assayed vs reduced mannan[b]
α-Mannosidase	Jack bean	Manα1-2,3,6	10 U/ml in 50 mM citric acid/sodium citrate (pH 4.5), 1 mM zinc acetate
		Manα1-6(GlcNAcβ1-2Manα1-3)Manβ1-4GlcNAc	50 U/ml (assayed vs 3 mM pNP-α-D-Man)
β-Hexosaminidase	Jack bean	GalNAc/GlcNAcβ1-2,3,4,6	6 U/ml in 50 mM citric acid/sodium citrate (pH 3.5), BSA (1 mg/ml) (assayed vs 3 mM pNP-β-D-GlcNAc)
β-Hexosaminidase	*D. pneumoniae*	GlcNAcβ1-3/6Gal GlcNAcβ1-2Manα1-6(GlcNAcβ1-2Manα1-3)(GlcNAcβ1-4)Manβ1-4GlcNAc	0.4 U/ml in 50 mM citric acid/Na$_2$HPO$_4$ (pH 6.0), BSA (1 mg/ml); all linkages shown except bisecting GlcNAcβ1-4Man are cleaved at 0.01 U/ml[c] (assayed vs 4 mM pNP-β-D-GlcNAc)
β-Galactosidase	Jack bean	Galβ1-3,4,6	10 U/ml in 50 mM citric acid/sodium citrate (pH 3.5), BSA (1 mg/ml), assayed vs 2 mM pNP-β-D-Gal
β-Galactosidase	*D. pneumoniae*	Galβ1-4GlcNAc	0.4 U/ml in 50 mM citric acid/Na$_2$HPO$_4$, (pH 6.0), BSA (1 mg/ml), assayed vs 4 mM pNP-β-D-Gal
α-Fucosidase	*C. lampas*	Fucα1-2Gal Galβ1-3(Fucα1-4)GlcNAc GlcNAcβ1-4(Fucα1-6)GlcNAc	9 U/ml in 50 mM acetic acid/sodium acetate (pH 4.5), 0.15 M NaCl, BSA (1 mg/ml) assayed vs 3 mM pNP-α-L-Fuc
		Galβ1-4(Fucα1-3)GlcNAc is resistant but can be cleaved with almond α-fucosidase I or III	

TABLE I (*continued*)

Enzyme	Source	Linkage hydrolyzed	Digestion condition[a]
α-Fucosidase I[d] and III	Almond	Galβ1-3(Fucα1-4)GlcNAc Galβ1-4(Fucα1-3)GlcNAc/ Glc	12.5 mU/ml in 50 mM acetic acid sodium acetate (pH 5.0), BSA (3 mg/ml), assayed vs 1.6 mM lacto-N-fucopentaose II[e]
α-Galactosidase	Coffee bean	Galα1-3,4,6	5 U/ml in 50 mM citric acid/Na$_2$HPO$_4$ (pH 6.0), BSA (1 mg/ml), assayed vs 4 mM pNP-α-D-Gal
α-Neuraminidase[f]	Newcastle disease virus	NeuAc/NeuGcα2-3Gal NeuAcα2-8NeuAc	1 U/ml in 50 mM acetic acid/sodium acetate (pH 5.5)
α-Neuraminidase[g]	A. ureafaciens	NeuAc/NeuGcα2-3/6Gal NeuAc/NeuGcα2-6GalNAc NeuAcα2-8NeuAc	1 U/ml in 50 mM acetic acid/sodium acetate (pH 4.8)
α-Glucosidase I[h]	Porcine liver	Glcα1-2Glc	0.1 M sodium phosphate (pH 7.0), containing 10% glycerol, 0.1 M NaCl, and 25 units[i] of enzyme
α-Glucosidase II[j]	Porcine liver	Glcα1-3Glc, Glcα1-3Man	0.1 M sodium phosphate (pH 7.0), containing 10% glycerol, 0.1 M NaCl, and 25 units[k] of enzyme

[a] All digestions carried out for 16 hr at 37°.

[b] Prepared by treatment of 0.5 g of yeast mannan (Sigma) with 0.1 M sodium borohydride in 0.1 M Na$_2$CO$_3$, pH 11.3. Enzyme assayed by incubation with 2 mg of reduced mannan/ml. Released mannose is determined by measuring increase in reducing power [J. T. Park and M. J. Johnson, *J. Biol. Chem.* **28**, 110 (1949)].

[c] S. Amatayakul, Ph.D. thesis, University of Oxford (1989).

[d] M. J. Imber, L. R. Glasgow, and S. V. Pizzo, *J. Biol. Chem.* **257**, 8205 (1982).

[e] Released fucose determined by HPAEC/triple-pulsed amperometry[19] or by coupled enzyme assay [P. R. Finch, Yuen, H. Schachter, and M. A. Moscarello, *Anal. Biochem.* **31**, 296 (1969)].

[f] A. Garcia-Sastre, I. A. Cabezas, and E. Villar, *Biochim. Biophys. Acta* **999**, 171 (1989).

[g] Y. Uchida, Y. Tsukada, and T. Sugimori, *J. Biochem* (Tokyo) **86**, 1573 (1979).

[h] E. Bause, J. Schweden, A. Gross, and B. Orthen, *Eur. J. Biochem.* **183**, 661 (1989).

[i] Assayed [M. S. Kang, J. H. Zwolshen, B. S. Harry, and P. S. Sunkara, *Anal. Biochem.* **181**, 109 (1989)] against 45 μM [^{14}C-*Glc*]Glc$_3$Man$_9$GlcNAc$_2$ (5000 dpm), prepared as described by I. K. Vijay and G. H. Perdew [*Eur. J. Biochem.* **126**, 167 (1982)]. One unit of enzyme liberates 10 dpm of [^{14}C]Glc/min at 37°.

[j] D. Brada and U. C. Dubach, *Eur. J. Biochem.* **141**, 149 (1984).

[k] Assayed [M. S. Kang, J. H. Zwolshen, B. S. Harry, and P. S. Sunkara, *Anal. Biochem.* **181**, 109 (1989)] against 45 μM [^{14}C-*Glc*]Glc$_2$Man$_9$GlcNAc$_2$ (5000 dpm).

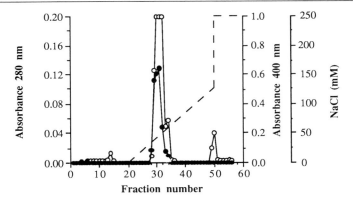

FIG. 1. FPLC Mono S chromatography of jack bean β-galactosidase. The sample is applied to the column and, after a 10-min wash period, the enzyme is eluted with a linear NaCl gradient, which is shown by the broken line. The eluate is monitored for protein by absorbance at 280 nm (○) and fractions assayed for enzyme activity (●).

Superose HR 5/5 column equilibrated with 50 m*M* potassium phosphate buffer, pH 6.5, containing 2.0 *M* ammonium sulfate (Fig. 4). The column is eluted for 25 min with starting buffer, 5 min with buffer containing 1.2 *M* ammonium sulfate, 30 min with a descending linear gradient of 1.2–0.7

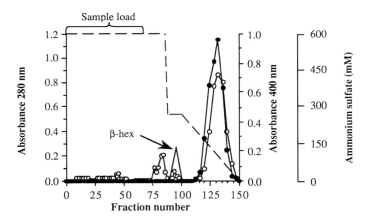

FIG. 2. FPLC phenyl-Superose chromatography of jack bean α-mannosidase. The sample is applied to the column and eluted with a discontinuous linear gradient of 0.6–0.25 and 0.25–0.0 *M* ammonium sulfate, as indicated by the broken line. The column eluate is monitored for protein by absorbance at 280 nm (○) and fractions are assayed for enzyme activity (●). A peak of contaminating β-*N*-acetylhexosaminidase activity (shown by the solid line and arrowed as "β-hex") elutes prior to the α-mannosidase.

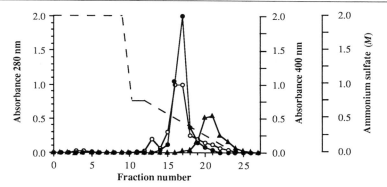

FIG. 3. FPLC phenyl-Superose chromatography of commercial coffee bean α-galactosidase. The sample is applied to the column and eluted with a discontinuous descending gradient (indicated by the broken line) of 2.0–0.0 M ammonium sulfate in 50 mM citric acid/Na$_2$HPO$_4$ buffer, pH 6.0. The column eluate is monitored for protein by absorbance at 280 nm (○) and fractions assayed for α-galactosidase (●) and β-galactosidase (▲).

M ammonium sulfate, and 5 min with buffer alone. The enzyme elutes at about 0.85 M ammonium sulfate. It is important to note that fractions should be desalted (using a Pharmacia fast-desalting column equilibrated with 50 mM sodium acetate, pH 5.0) to remove the ammonium sulfate, which interferes with the assay for α-mannosidase (see Table I, footnote b).

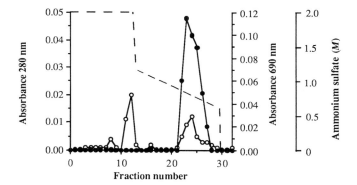

FIG. 4. FPLC phenyl-Superose chromatography of *A. saitoi* α-mannosidase I. The sample is applied to the column and eluted stepwise with a sharp, descending linear gradient (indicated by the broken line) of buffer containing 2.0–1.2 M ammonium sulfate followed by a shallow gradient of 1.2–0.7 M ammonium sulfate. The eluate is monitored for protein by absorbance at 280 nm (○) and fractions assayed for α-mannosidase (●).

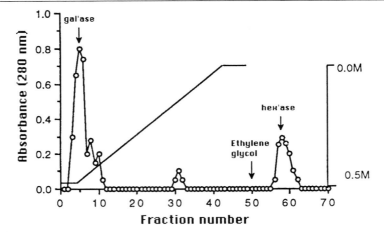

Fig. 5. FPLC phenyl-Superose chromatography of *D. pneumoniae* β-galactosidase and β-hexosaminidase. The sample is applied to the column and the β-galactosidase is eluted with a descending, linear gradient (indicated by the solid line) of buffer containing 0.5–0.0 *M* ammonium sulfate. The β-hexosaminidase is recovered by elution with 50% ethylene glycol. Fractions are monitored for protein by absorbance at 280 nm (○). The elution positions of β-galactosidase and β-hexosaminidase are indicated by gal'ase and hex'ase, respectively.

Diplococcus pneumoniae β-Galactosidase[16] *and β-Hexosaminidase.*[17] Preparations of these enzymes are invariably contaminated with one another but can be separated in a single step, using hydrophobic interaction FPLC (Fig. 5). The enzyme sample is adjusted to 0.5 *M* ammonium sulfate and applied (0.5 ml/min) to a phenyl-Superose HR 5/5 column equilibrated with 50 m*M* KH$_2$PO$_4$/NaOH, pH 7.0, containing 0.5 *M* ammonium sulfate. The column is eluted for 5 min with starting buffer, 40 min with a descending linear gradient of 0.5–0.0 *M* ammonium sulfate, 10 min with buffer minus ammonium sulfate, and 20 min with buffer containing 50% ethylene glycol. The β-galactosidase elutes at about 0.5 *M* ammonium sulfate, whereas the β-hexosaminidase is eluted with 50% ethylene glycol.

Charonia lampas α-Fucosidase. Commercial preparations of this enzyme are contaminated with a variety of other exoglycosidases that cannot easily be removed by FPLC. To date, the only satisfactory procedure to purify the α-fucosidase is to use ligand affinity chromatography using

---

[16] R. C. Hughes and R. W. Jeanloz, *Biochemistry* **3**, 1535 (1964).

[17] K. Yamashita, T. Ohkura, H. Yoshima, and A. Kobata, *Biochem. Biophys. Res. Commun.* **100**, 226 (1981).

carboxypentyldeoxyfuconojirimycin–Affi-Gel. Details of this procedure can be found in Butters *et al.*[18]

Imino Sugars as Ligands for Purification of Exoglycosidases

Carboxypentyldeoxyfuconojirimycin, an imino sugar analog of α-L-fucose, was originally used to purify α-fucosidase III from almond,[19] whereas deoxynojirimycin and deoxymannojirimycin (imino sugar analogs of Glc and Man, respectively) have been used as ligands for the isolation of processing α-glucosidases[20] and α-mannosidase,[21] respectively (see [19] in this volume). These sugar analogs (which have the ring oxygen replaced by a nitrogen) typically display K_i values in the low micromolar range against their respective glycosidases and thus show great promise as affinity ligands for the isolation of this class of enzyme (for a review, see Winchester and Fleet[22]). A large variety of imino sugars have been synthesized and several of them (analogs of Glc, Man, Gal, and Fuc) are now commercially available (Cambridge Research Biochemicals, Valley Stream, NY; Oxford GlycoSystems, Abingdon, UK; Toronto Research Chemicals, Inc., Toronto, Ontario, Canada). Although somewhat expensive, once immobilized the imino sugars can be reutilized for successive purifications or as a means of removing contaminant exoglycosidase activities. These ligands must first be converted to the corresponding carboxypentyl derivatives[23] in which form they can be covalently coupled (using the carbodiimide procedure) to matrices such as Affi-Gel or amino-Sepharose.

Assays for Contaminating Exoglycosidase Activities

Before a particular enzyme can be used for sequencing, it is screened against a panel of oligosaccharides and *p*-nitrophenyl-glycopyranosides (pNP-glycosides) at an enzyme concentration 5- to 10-fold above its operational concentration (i.e., the concentration to be used in a sequencing experiment; see Table I) and using saturating concentrations of substrate (usually in the low millimolar range). The choice of oligosaccharides is somewhat arbitrary. Sugars isolated from human milk[24] and mannosidosis

[18] T. D. Butters, P. Scudder, J. Rotsaert, S. Petursson, G. W. J. Fleet, F. W. Willenbrock, and G. S. Jacob, *Biochem. J.* **279**, 189 (1991).
[19] P. Scudder, D. C. A. Neville, T. D. Butters, G. W. J. Fleet, R. A. Dwek, T. W. Rademacher, and G. S. Jacob, *J. Biol. Chem.* **265**, 16472 (1990).
[20] H. Hettkamp, G. Legler, and E. Bause, *Eur. J. Biochem.* **142**, 85 (1984).
[21] J. Schweden, G. Legler, and E. Bause, *Eur. J. Biochem.* **157**, 563 (1986).
[22] B. Winchester and G. W. J. Fleet, *Glycobiology* **2**, 199 (1992).
[23] R. C. Bernotas and B. Ganen, *Biochem. J.* **270**, 539 (1990).
[24] A. S. R. Donald and J. Feeney, *Carbohydr. Res.* **178**, 79 (1988).

urine[25] provide the spectrum of substrates required to assay for most of the anticipated contaminants and can also be used to verify the substrate specificity of certain of the glycosidases such as the *A. saitoi* α-mannosidase I and almond fucosidase III. Alternatively, oligosaccharides can be generated by large-scale hydrazinolysis of plasma glycoproteins such as transferrin, which is a convenient source of a sialylated, biantennary complex sugar. In this case, substrates for β-galactosidase, β-hexosaminidase, and α-mannosidase can be generated by sequential exoglycosidase treatment of the purified oligosaccharide, utilizing neuraminidase, followed by β-galactosidase, and finally β-hexosaminidase. Oligosaccharide substrates can be used as the corresponding tritiated alditols, pyridylamino derivatives (the ³H labeling and pyridylamino derivatization of sugars is discussed below), or as the native sugars. Hydrolysis is monitored by medium- or high-pressure gel filtration chromatography,[26,27] reversed-phase high-performance liquid chromatography (HPLC)[28] and high-performance anion-exchange chromatography[29] (HPAEC), respectively. Both HPLC methods are extremely sensitive but HPAEC, using triple-pulsed amperometic detection, offers the added advantages of being able to use underivatized substrates and of detecting all of the enzyme products.

Experimental Procedures

Oligosaccharides to be analyzed may be native, free sugars or the result of cleavage from biomacromolecules such as a glycoprotein. In both cases, the starting oligosaccharide material will likely be heterogeneous and must first be separated into its individual components before structural analysis of any particular component can begin. The amount of any pure oligosaccharide that is needed for enzymatic analysis largely depends on the labeling and detection methods used, but is usually on the order of 10–100 pmol. A generalized scheme for the analysis of a purified glycoconjugate is given in Fig. 6 and comments about each individual step are given next.

[25] F. Matsuura, H. A. Nunez, G. A. Grabowski, and C. C. Sweeley, *Arch. Biochem. Biophys.* **207**, 337 (1981).
[26] K. Yamashita, T. Mizuochi, and A. Kobata, this series, Vol. 83, p. 105.
[27] A. Kobata, K. Yamashita, and S. Takasaki, this series, Vol. 138, p. 84.
[28] N. Tanaka, H. Nakada, N. Itoh, Y. Mizuno, M. Takanishi, T. Kawasaki, S. Tate, F. Inagaki, and I. Yamashina, *J. Biochem.* (*Tokyo*) **112**, 68 (1992).
[29] R. R. Townsend, M. R. Hardy, and Y. C. Lee, this series, Vol. 179, p. 65.

Biological Material

↓

Chemical or Enzymatic Release of Oligosaccharides

↓

Derivatization/Radiolabeling of Oligosaccharides

↓

Purification of Tagged Oligosaccharides

↓

Exoglycosidase Digestion

↓↑

Chromatographic Analysis of Enzyme Products

↓

Structural Assignment

FIG. 6. Generalized scheme for the analysis of a purified glycoconjugate.

Release of Oligosaccharides

Chemical Methods. Hydrazinolysis has generally been used to release N-linked glycans under conditions in which O-linked saccharides are destroyed.[30] Recently, however, the technique has been optimized such that it can be used to release first O-linked and then N-linked oligosaccharides selectively,[31,32] and an instrument is commercially available that automates this procedure.[33] Traditionally, O-linked sugars are released and simultaneously converted to their corresponding alditols by treatment of the

---

[30] D. Ashford, R. A. Dwek, J. K. Welply, S. Amatayakul, S. W. Homans, H. Lis, G. N. Taylor, N. Sharon, and T. W. Rademacher, *Eur. J. Biochem.* **166,** 311 (1987).

[31] R. Parekh, this volume [5].

[32] N. Kuraya and S. J. Hase, *J. Biochem. (Tokyo)* **112,** 122 (1992).

[33] E.-L. Allan and T. Merry, *Glycobiology* **2,** 267 (1992).

glycoprotein with 0.05 M NaOH (the β-elimination reaction) containing 1.0 M sodium borohydride.[34]

Enzymatic Methods. Deglycosylation of N-linked glycans can be achieved using glycoprotein-N^4-(*N*-acetyl-β-glucosaminyl)Asn amidase[35] (referred to as PNGase or *N*-glycanase). This enzyme is extremely versatile and will hydolyze all classes of N-linked oligosaccharides. However, the enzyme cannot be guaranteed to release quantitatively all N-linked glycans and conditions for hydrolysis of a particular glycoprotein must be determined empirically. Details of the use of this enzyme are discussed by Tarentino and Plummer.[36]

The endo-α-N-acetylgalactosaminidase (*O*-glycanase) from *D. pneumoniae* can be used to release O-linked glycans but its value is limited because of the narrow substrate specificity of the enzyme, which is only capable of releasing the disaccharide, Galβ1-3GalNAc, from glycopeptides or glycoproteins.[37] An endo-α-N-acetylgalactosaminidase that appears to have a broader substrate specificity has been identified in culture supernatants of *Streptomyces*.[38] Further work is needed to establish the value of this enzyme as a sequencing reagent.

Intact oligosaccharides can be released from neutral and sialylated glycosphingolipids, using *Rhodococcus* or *Corynebacterium* glycosyl-N-acylsphingosine 1,1-β-D-glucanohydrolase,[39,40] generally referred to as endoglycoceramidase. These enzymes have a broad substrate specificity and can hydrolyze Galβ1- and Glcβ1-ceramide linkages of neutral, sialylated, and sulfated glycosphingolipids. The utility of the enzyme from *Rhodococcus* for oligosaccharide sequencing has already been well demonstrated.[41,42] Released oligosaccharides are recovered in the upper, aqueous phase after partitioning with 6 vol of chloroform–methanol (2 : 1 v/v) and can subsequently be labeled with either sodium borotritide or 2-aminopyridine (see Derivatization of Oligosaccharides).

[34] J. Amano and A. Kobata, this series, Vol. 179, p. 261.
[35] S. Alexander and J. H. Elder, this series, Vol. 179, p. 505.
[36] A. Tarentino and T. Plummer, Jr., this volume [4].
[37] Y. Endo and A. Kobata, *J. Biochem. (Tokyo)* **80**, 1 (1976).
[38] H. Iwase, I. Ishii, K. Ishihara, Y. Tanaka, S. Omura, and K. Hotta, *Biochem. Biophys. Res. Commun.* **151**, 422 (1988).
[39] M. Ito and T. Yamagata, this series, Vol. 179, 488.
[40] H. Ashida, K. Yamamoto, H. Kumagai, and T. Tochikura, *Eur. J. Biochem.* **205**, 729 (1992).
[41] M. Shimamura, T. Hayase, M. Ito, M.-L. Rasilo, and T. Yamagata, *J. Biol. Chem.* **263**, 12124 (1988).
[42] S. Dasgupta, H. van Halbeek, and E. L. Hogan, *FEBS Lett.* **301**, 141 (1992).

Derivatization of Oligosaccharides

³H-Radiolabeling of Reducing Sugars. Oligosaccharides released by treatment with hydrazine, PNGase, or endoglycoceramidase are reduced with sodium borotritide, using an adaptation of the method described by Tomana *et al.*[43] The oligosaccharide or oligosaccharide mixture is taken to dryness, dissolved in 50 μl of 10 mM NaB^{3}H$_4$ (10–50 Ci/mol) in 0.1 M Na$_2$CO$_3$, and incubated for 16 hr at 4°. Excess NaBH$_4$ is destroyed by the addition of 50 μl of glacial acetic acid and the Na$^+$ removed with Dowex AG 50W-X12 (H$^+$, 100–200 mesh). The solution is dried under N$_2$ and borate removed by repeated (five) evaporations with 5% acetic acid in methanol (v/v). Finally, unincorporated tritium "blank" is removed by descending paper chromatography in 1-butanol–ethanol–water (4 : 1 : 1, v/v).[30] Approximately 5 × 10^6 disintegrations per minute (dpm) of an individual oligosaccharide is obtained from 0.5 nmol of starting material.

¹⁴C-Radiolabeling of Oligosaccharide Alditols. Purified O-linked oligosaccharide alditols are de-N-acetylated with anhydrous hydrazine and hydrazine sulfate, and then re-N-acetylated with [¹⁴C]acetic anhydride.[34] In principle, ³H labeling can be achieved with sodium borotritide in place of sodium borohydride in the original β-elimination reaction (see above), but this has not been practical because of the prohibitive expense of the radioisotope.

Derivatization of Reducing Sugars with 2-Aminopyridine. Chemically or enzymatically released reducing sugars can be pyridylaminated by reductive amination with 2-aminopyridine and sodium cyanoborohydride.[44,45]

Purification of Tagged Oligosaccharides

³H-Labeled Oligosaccharide Alditols. The material recovered from paper chromatography is now fractionated by ion-exchange chromatography to obtain pools of neutral and anionic sugars. Various methods, including QAE-Sephadex and DEAE-TSK (Pharmacia, Piscataway, NJ) have been described[46] that will separate mono-, di-, and trisialylated sugars. Another method for rapid fractionation of anionic oligosaccharides containing both sulfate and/or sialic acid moieties has been developed by

[43] M. Tomana, W. Niedermeyer, C. Spivey, and J. Gerard, *Michrochem. J.* **23,** 93 (1978).

[44] S. Hase, T. Ibuku, and T. Ikenaka, *J. Biochem. (Tokyo)* **95,** 197 (1984).

[45] A. Kondo, J. Suzuki, N. Kuraya, S. Hase, I. Kato, and T. Ikenaka, *Agric. Biol. Chem.* **54,** 2169 (1990).

[46] R. D. Cummings, R. K. Merkle, and N. L. Stults, *Methods Cell Biol.* **32,** 141 (1979).

Green and Baenziger[47] and uses HPLC separation on amine-bearing columns (Micropak AX-10 and AX-5) (see [14] in this volume). At this stage each fraction obtained from these procedures needs to be further purified to obtain a homogeneous preparation of each oligosaccharide. This will normally entail the use of methods such as gel filtration, lectin affinity chromatography,[48] HPLC,[49,50] HPAEC,[29] paper chromatography,[51] and paper electrophoresis.[52]

Neuraminidases (Table I) are frequently used to desialylate oligosaccharide alditols prior to BioGel P-4 (Bio-Rad, Richmond, CA) analysis and to establish the number and, in some cases, the linkage of terminal sialic acid residues in these materials. Resistance to treatment with neuraminidase would indicate the presence of sulfation or phosphorylation. Phosphate groups can readily be removed by treatment with alkaline phosphatase but there is no broad specificity sulfatase that can be used diagnostically to identify the presence of sulfate groups (sulfatases demonstrate activity toward monosaccharide rather than oligosaccharide sulfates). However, a hepatic sulfatase described by Baenziger et al.[53] may prove to be useful in some cases. Chemical methods of desulfation, such as solvolysis,[54] have been used but can be misleading because of accompanying glycosidic cleavage.[55] Where appropriate, the only reliable way to identify sulfation (at the picomole level) is to use metabolic ^{35}S labeling.

Pyridylamino Derivatives. Glycans derivatized in this way can be fractionated by any of the procedures mentioned above, but are typically purified by reversed-phase HPLC[28] with detection by fluorescence (see [13] in this volume).

Oligosaccharide alditols and derivatized oligosaccharides should be

[47] E. D. Green and J. U. Baenziger, *Anal. Biochem.* **158,** 42 (1986).
[48] T. Osawa and T. Tsuji, *Annu. Rev. Biochem.* **56,** 21 (1987).
[49] S. J. Turco, *Anal. Biochem.* **118,** 278 (1981).
[50] S. J. Mellis and J. U. Baenziger, *Anal. Biochem.* **114,** 276 (1981).
[51] A. Kobata, *in* "Methods in Enzymology" (V. Ginsburg, ed.), Vol. 28, p. 262. Academic Press, New York, 1972.
[52] K. Yamashita, A. Hitoi, N. Tateishi, T. Hihashi, Y. Sakamoto, and A. Kobata, *Arch. Biochem. Biophys.* **240,** 573 (1985).
[53] J. U. Baenziger, K. Swatantar, R. M. Brodbeck, P. L. Smith, and M. C. Beranek, *Proc. Natl. Acad. Sci. U.S.A.* **89,** 334 (1992).
[54] A. I. Usov, K. S. Adamyants, L. I. Miroshnikova, A. A. Shaposhnikova, and N. K. Kochetkov, *Carbohydr. Res.* **18,** 336 (1971).
[55] P. Scudder, P. W. Tang, E. F. Hounsell, A. M. Lawson, H. Mehmet, and T. Feizi, *Eur. J. Biochem.* **157,** 365 (1986).

judged pure by a minimum of two independent chromatographic techniques before being subjected to enzyme sequencing.

Exoglycosidase Sequencing

The sequencing of an individual sugar involves a series of sequential glycosidase digestions performed under conditions similar to those described in Table I. The oligosaccharide, typically 10–100 pmol, is incubated overnight with a relatively high concentration of the enzyme (to allow the reaction to go to completion) and the reaction products then subjected to chromatographic analysis. The choice of chromatography system will depend on the nature of the sugar, that is, radiolabeled or pyridylaminated oligosaccharide.

^{3}H-Labeled and ^{14}C-Labeled Oligosaccharide Alditols. Radiolabeled reaction products are usually fractionated on a column of BioGel P-4, using distilled water as the eluant (see [11] in this volume). This method, which has been extensively discussed by Yamashita et al.,[26] separates oligosaccharide alditols on the basis of their hydrodynamic volume. The relative elution volume for a particular alditol is generally expressed in terms of glucose units by comparison with an acid hydrolysate of dextran (see Fig. 7). Because of the regularity of structural features found in N-linked glycans and the empirical observation that the overall hydrodynamic volume is the sum of the volumes of those individual features, the elution position for a particular glycan can also be calculated from a set of basic observed rules described in Yamashita et al.[26] These calculated glucose unit values are generally in good agreement with those experimentally determined and tabulated by Kobata and colleagues.[26,27]

Analysis usually begins on a neutral substrate because sialylated and other anionic sugars are excluded from the BioGel P-4 beads, and consequently display ambiguous elution volumes when water is used as the eluant. A volatile buffer such as 0.1 M ammonium bicarbonate can be used as an eluant but, because a single sialic acid residue elutes with an hydrodynamic volume of 6.0 glucose units, is of limited practical use (the useful fractionation range of BioGel P-4 is about 1–24 glucose units). Exoglycosidase sequencing of a particular oligosaccharide involves stepwise cleavage from the nonreducing end. Figure 7 shows this process performed on a biantennary, core-fucosylated oligosaccharide alditol. The sample initially elutes from a BioGel P-4 column at a value of 14.5 glucose units (Fig. 7A). Incubation of the sample with individual exoglycosidases from the enzyme library produces successful cleavage only at the nonreducing end (as illustrated by the shift in elution to 12.3 glucose units; Fig.

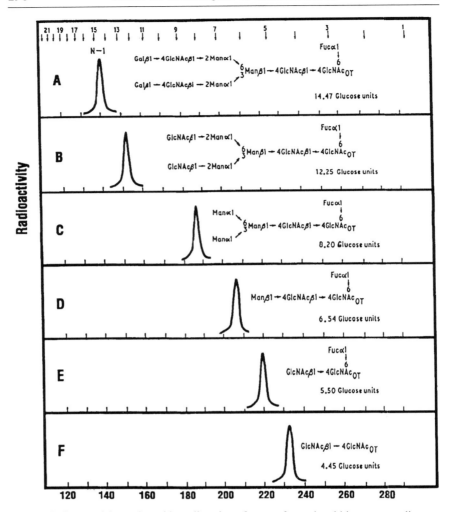

FIG. 7. Sequential exoglycosidase digestion of a core-fucosylated biantennary oligosaccharide isolated from transferrin. (A) Undigested oligosaccharide; (B) oligosaccharide treated with jack bean β-galactosidase; (C) treated with jack bean β-hexosaminidase; (D) treated with jack bean α-mannosidase; (E) treated with *Aspergillus fulica* β-mannosidase; (F) treated with *C. lampas* α-fucosidase. The X axis represents fraction number.

7B) when the sample is incubated in the presence of a β-galactosidase such as the jack bean enzyme. The shift is characteristic of the number of residues released by the enzyme, in this case demonstrating the release of two galactose residues (2.2 glucose unit shift). The resulting material is incubated again with the various members of the enzyme library; in this case cleavage is successful only with a β-hexosaminidase such as the jack bean enzyme, giving rise to a peak at 8.2 glucose units (Fig. 7C). The 4-glucose unit shift is indicative of the release of two N-acetylglucosamine residues. The process is repeated until the oligosaccharide has been successfully cleaved as shown in Fig. 7D–F, and at this point a tentative proposal of the structure of the starting material can be made on the basis of the sum total of all the data—successful cleavages as well as those cases in which incubation with an enzyme gave rise to no change in elution position. The proposed oligosaccharide structure must obviously be consistent with all of the digestion data generated on the material.

It should be emphasized here that for accurate and reliable structural assignments to be made, the purity of both the exoglycosidase reagents and of the oligosaccharide starting material must be high or even absolute. It is likely that many false assignments have been made owing to these criteria not being met. The presence of contaminating glycosidases in the enzyme preparation can give rise to misleading "positive" digestions due to the high concentrations of enzymes and extended digestion times oftentimes used in an effort to ensure that reactions go to completion.

A second problem is the inherent poor resolution of low-pressure gel-filtration methods, which cannot separate structurally related or isomeric mixtures of oligosaccharides; hence what appears to be a single peak can often be shown to represent a mixture of sugars.

Enzyme Array Technique. Edge *et al.*[56] describe an enzyme-sequencing strategy called reagent–array analysis, which utilizes enzyme digestions analyzed by BioGel P-4 to sequence an oligosaccharide, but differs from the method described earlier in that the whole procedure requires only a single BioGel P-4 analysis. The oligosaccharide sample is subdivided into aliquots that are incubated with precisely defined mixtures of exoglycosidases. The enzyme array used for analysis of a particular oligosaccharide should contain all of the exoglycosidases needed to digest the oligosaccharide into its individual monosaccharide components. One mixture of the array contains all of the exoglycosidases whereas the remaining mixtures are each missing a single exoglycosidase. After the digestions are complete, the individual samples plus the original oligosaccharide substrate are combined and subjected to a single analysis by BioGel P-4 chromatography. Each peak on the BioGel P-4, apart from the completely digested material, represents the result of omission of a particular glycosidase. The complex profile that is obtained is compared with a database of computer-

[56] C. J. Edge, T. W. Rademacher, M. R. Wormald, R. B. Parekh, T. D. Butters, D. R. Wing, and R. A. Dwek, *Proc. Natl. Acad. Sci. U.S.A.* **89,** 6338 (1992).

generated simulations of digestions of oligosaccharides generated with the particular enzyme array used for the digestion experiment to obtain a "fingerprint match." Advantages of this method are its speed and reduced sample requirement (lack of sample loss associated with repeated BioGel P-4 analyses, repurifications, and negative digestion incubations), whereas the major drawback is the absolute requirement for noncontaminating exoglycosidases in the enzyme array, as the determination is made on a single BioGel P-4 profile.

Pyridylamino Derivatives. Exoglycosidase sequencing of pyridylaminated oligosaccharides is similar in spirit to that described above for the tritiated alditols, but the enzyme products are usually analyzed by reversed-phase HPLC on an AQ-303 column with fluorescence detection. The sequencing of a biantennary disialylated oligosaccharide is found in Tanaka *et al.*[28] One advantage of the pyridylamino technique over that involving separation of ^{3}H-labeled alditols by BioGel P-4 chromatography is that sialylated and nonsialylated sugars need not be separated into individual pools prior to the sequential exoglycosidase-sequencing procedure, as the reversed-phase column is operational over the complete range of neutral and sialylated sugars.[28] In addition, the HPLC method can handle a higher throughput but suffers from the fact that, unlike BioGel P-4, a shift in relative elution position cannot always be used to deduce the number of residues of a particular monosaccharide that have been removed by treatment with the glycosidase.

Structural Assignment

Sequencing of an oligosaccharide relies on the selectivity of the enzymes used for analysis. Structural assignments are made on the basis of the known glycone specificities of the enzymes that make up the particular enzyme bank. As shown in Table I, some of these exoglycosidases, such as jack bean β-galactosidase and β-N-acetylhexosaminidase, have a broad specificity and are capable of cleaving terminal residues without much regard to the specific linkage. Other enzymes, such as *A. saitoi* α-mannosidase I (which cleaves only terminal $\alpha(1 \rightarrow 2)$-linked mannose residues), display a high degree of specificity and can thus be used to give specific linkage information. One application of exoglycosidase sequencing that is likely to increase during the next few years is the use of the processing α-glucosidases I and II to detect glycans terminating with glucose residues (Table I). This can result from the use of iminosugar inhibitors such as castanospermine and deoxynojirimycin to inhibit cellular glycan processing of glycoproteins. Examples of this will become particularly useful in the analysis of viral envelope glycoproteins such as influenza hemagglutinin and retroviral glycoproteins such as human immunodeficiency virus (HIV) gp 120.

A major issue in sequencing an unknown sugar concerns how one handles and interprets unsuccessful digestions. The standard exoglycosidases used for these studies all display some degree of aglycon specificity. Although they have been used operationally by various workers against a diverse array of oligosaccharide alditols, few in-depth kinetic studies have been performed, and K_m and V_{max} data against different related and unrelated sugars are virtually nonexistent for most of these enzymes. When one considers that in many instances of sequencing the amount of oligosaccharide is small (10–100 pmol) and the reaction rates are therefore likely to be first order, it is not unexpected to observe digestions that do not go to completion. This would be reflected in a BioGel P-4 analysis that does not give a single clean peak for the reaction product. A typical followup response is to treat the sample with a second incubation, using the same enzyme, in an attempt to drive the reaction to completion. Should this still be unsuccessful, the investigator is faced with a dilemma, and must decide how to proceed. The first thing to consider is whether the sample is indeed pure. BioGel P-4 chromatography is not capable of cleanly resolving linkage isomers or glycans having similar glucose unit values, and unless an independent analytical method such as HPAEC has been performed (see [12] in this volume), it is likely that glycan purity is the culprit. The other conclusion that is generally drawn is that the particular oligosaccharide alditol is a poor substrate for the enzyme and the reaction simply cannot be driven to completion. To illustrate the problem, consider the use of a *D. pneumoniae* β-galactosidase (which specifically cleaves Galβ1-4 termini) to release terminal galactose residues from a complex sugar such as a tri- or tetraantennary sugar terminating with Galβ1-4GlcNAc functionalities on the various arms. It is to be expected that the rates of galactose cleavage from different arms will vary considerably. Moreover, depending on which cleavage occurs first, different intermediate products can be formed that will display different rates of subsequent galactose cleavage. It is therefore conceivable that a situation can arise in which complete cleavage of all terminal galactose residues simply does not occur. When extended to an unknown oligosaccharide, a situation such as this illustrates where exoglycosidase-sequencing results alone should not be overinterpreted, and where additional analyses are needed for structural determinations.

[18] Glycosyltransferases in Glycobiology

By MONICA M. PALCIC

Introduction

Glycosyltransferases catalyze the transfer of a single monosaccharide unit from a nucleotide donor to the hydroxyl group of an acceptor saccharide, either a glycoprotein, glycolipid, or polysaccharide [Eq. (1)].[1] Mam-

$$\text{Sugar-O-}\overset{\overset{\text{O}}{\|}}{\text{P}}\text{(OH)-OR} + \text{HO-acceptor} \xrightarrow{\text{glycosyltransferase}}$$

$$\text{sugar-O-acceptor} + (\text{HO})_2\overset{\overset{\text{O}}{\|}}{\text{P}}\text{-OR} \quad (1)$$

malian oligosaccharide chains are largely composed of only seven monosaccharide units that are transferred by a glycosyltransferase to the appropriate acceptor. With few exceptions, there is at least one distinct transferase enzyme for every glycosidic linkage found in nature, so that well over 100 different enzymes must exist to account for all documented oligosaccharide structures.[2] The seven most common sugar units and their corresponding nucleotide donors are as follow: glucose (UDPGlc), galactose (UDPGal), N-acetylglucosamine (UDPGlcNAc), N-acetylgalactosamine (UDPGalNAc), mannose (GDPMan), N-acetylneuraminic acid/sialic acid (CMPNeuAc), and fucose (GDPFuc). All of the sugars in this series have a D configuration except for fucose, which has an L configuration.

The major use of glycosyltransferases in glycobiology is to establish structure–function relationships; that is, to add a monosaccharide to an acceptor, either soluble or on a cell surface, and then determine the effect of this addition. The effects of enzymatic glycosylation or reglycosylation for whole-cell systems have been well covered in this series[3-5] and will

---

[1] T. A. Beyer, J. E. Sadler, J. I. Rearick, J. C. Paulson, and R. L. Hill, *Adv. Enzymol.* **52,** 23 (1981).

[2] S. Roseman, *Chem. Phys. Lipids* **5,** 270 (1970).

[3] S. W. Whiteheart, A. Pasanti, J. S. Reichner, G. D. Holt, R. S. Haltiwanger, and G. W. Hart, *in* "Methods in Enzymology" (V. Ginsburg, ed.), Vol. 179, p. 83. Academic Press, San Diego, 1989.

[4] D. Spillman and J. Finne, *in* "Methods in Enzymology" (V. Ginsburg, ed.), Vol. 179, p. 270. Academic Press, San Diego, 1989.

not be reviewed in this chapter. Such additions can also provide the basis for enzyme assays in which levels of a particular transferase are correlated with a physiological or pathological state. Assays for glycosyltransferases have also been well covered.[6-9] The focus of this chapter is to tabulate the glycosyltransferases that have been isolated and characterized sufficiently for use in the synthesis of a defined oligosaccharide structure. As well, four detailed examples, representative of the use of different glycosyltransferases for the preparation of Galβ1 → 4GlcNAcβ-OR, Gal-NAcβ1 → 4GlcNAcβ-OR, NeuAcα2 → 3Galβ1 → 4[Fucα1 → 3]GlcNAcβ-OR (sialyl Lewis^x), and GlcNAcβ1 → 2Manα(1 → 3[Manα1 → 6]Manβ-OR will be given.

In Table I are listed the glycosyltransferases that can be obtained from readily available tissue sources for preparative applications, tabulated by the type of sugar transferred. These enzymes are present in at least milliunit amounts, which is sufficient for milligram scale synthesis. Only three of these enzymes are currently commercially available, the β(1 → 4)-galactosyltransferase (EC 2.4.1.38/2.4.1.90) from bovine and human milk, the α(2 → 6)-sialyltransferase (EC 2.4.99.1) from rat liver, and the α(2 → 3)-sialyltransferase (EC 2.4.99.4) from porcine liver. All seven of the nucleotide donors are commercially available, both radiolabeled and unlabeled, and most can be enzymatically or chemically synthesized.[10,11] No high-expression vectors for cloned transferases[12,13] are widely available, so for most applications the enzymes must be isolated from mammalian tissue sources. Isolations are complicated by the low natural abundance of the glycosyltransferases and the need for affinity matrices. As a result, functional enzyme purity is usually a sufficient criterion for use in preparative synthetic applications, rather than protein homogeneity. Latitudes

[5] J. C. Paulson and G. N. Rogers, *in* "Methods in Enzymology" (V. Ginsburg, ed.), Vol. 138, p. 162. Academic Press, Orlando, FL, 1987.

[6] B. E. Samuelsson, *in* "Methods in Enzymology" (V. Ginsburg, ed.), Vol. 138, p. 567. Academic Press, Orlando, FL, 1987.

[7] H. Schachter, I. Brockhausen, and E. Hull, *in* "Methods in Enzymology" (V. Ginsburg, ed.), Vol. 179, p. 351. Academic Press, San Diego, 1989.

[8] N. Taniguchi, A. Nishikawa, S. Fujii, and J. Gu, *in* "Methods in Enzymology" (V. Ginsburg, ed.), Vol. 179, p. 397. Academic Press, San Diego, 1989.

[9] J. E. Sadler, T. A. Beyer, C. L. Oppenheimer, J. C. Paulson, J. P. Prieels, J. I. Rearick, and R. L. Hill, *in* "Methods in Enzymology" (V. Ginsburg, ed.), Vol. 83, p. 458. Academic Press, New York, 1982.

[10] Y. Ichikawa, G. C. Look, and C.-H. Wong, *Anal. Biochem.* **202**, 215 (1992).

[11] E. J. Toone, E. S. Simon, M. D. Bednarski, and G. M. Whitesides, *Tetrahedron* **45**, 5365 (1989).

[12] J. C. Paulson and K. J. Colley, *J. Biol. Chem.* **264**, 17615 (1989).

[13] J. B. Lowe, *Semin. Cell Biol.* **2**, 289 (1991).

TABLE I
GLYCOSYLTRANSFERASES FOR PREPARATIVE SYNTHESIS CLASSIFIED BY TYPE OF MONOSACCHARIDE TRANSFERRED

Glycosyltransferase	EC number	Linkage synthesized[a]	Refs.[b]
UDPGal			
GlcNAc β1 → 4Gal-transferase	2.4.1.38	**Gal**β1 → **4**GlcNAcβ-R	1, 2
Galα1 → 3Gal-transferase	2.4.1.151	**Gal**α1 → **3**Galβ1 → 4GlcNAc-R	3
	2.4.1.124		
O-Glycan core 1 β1 → 3Gal-transferase	2.4.1.122	**Gal**β1 → **3**GalNAcα-Ser/Thr	4, 5
Blood group B α1 → 3Gal-transferase	2.4.1.37	**Gal**α1 → **3**Gal-R, with Fucα1 → 2	6
GlcNAcβ1 → 3Gal-transferase	None	**Gal**β1 → **3**GlcNAcβ-R	7
Lactosylceramideα1 → 4Gal-transferase	None	**Gal**α1 → **4**Galβ1 → 4Glc-R	8
UDPGalNAc			
Peptide αGalNAc-transferase	2.4.1.41	**GalNAc**α-Ser/Thr	9, 10
Blood group A α1 → 3GalNAc-transferase	2.4.1.40	**Gal**α1 → **3**Gal-R, with Fucα1 → 2	11, 12
Sd[a] antigen β1 → 4GalNAc-transferase	None	**GalNAc**β1 → **4**Galβ1 → 4GlcNAc-R, with NeuAc2 → 3	13
UDPGlcNAc			
β1 → 6GlcNAc-transferase V	2.4.1.155	**GlcNAc**β1 → **6**	14
β1 → 4GlcNAc-transferase VI	None	**GlcNAc**β1 → **4**-Manα1 → 6	15
β1 → 2GlcNAc-transferase II	2.4.1.143	**GlcNAc**β1 → **2**	16
β1 → 4GlcNAc-transferase III	2.4.1.144	**GlcNAc**β1 → **4**-Manβ1 → 4R	17
β1 → 2GlcNAc-transferase I	2.4.1.101	**GlcNAc**β1 → **2** Manα1 → 3	18, 19
β1 → 4GlcNAc-transferase IV	2.4.1.145	**GlcNAc**β1 → **4**	20

Enzyme	EC number	Structure	Ref.
O-Glycan core 2 β1 → 6GlcNAc-transferase	2.4.1.102	GlcNAcβ1 → **6** Galβ1 → 3GalNAcα-R	21, 22
O-Glycan core 3 β1 → 3GlcNAc-transferase	2.4.1.147	**GlcNAcβ1 → 3**GalNAcα-R	23
O-Glycan core 4 β1 → 6GlcNAc-transferase	2.4.1.148	**GlcNAcβ1 → 6**	23, 24
Core 1 or core 2 elongation β1 → 3GlcNAc-transferase	2.4.1.146	GlcNAcβ1 → 3GalNAcα-R	22, 25
Blood group i β1 → 3GlcNAc-transferase	2.4.1.149	**GlcNAcβ1 → 3Galβ1 → 3**GalNAcα-R	26, 27
Blood group I β1 → 6GlcNAc-transferase	None	**GlcNAcβ1 → 3Galβ-R** **GlcNAcβ1 → 6**	24, 28
		GlcNAcβ1 → 3Galβ-R	
GDPfucose			
GlcNAcα(1 → 3/4)fucosyltransferase	2.4.1.65	**Fucα1 → 3** Galβ1 → 4GlcNAc-R or **Fucα1 → 4** Galβ1 → 3GlcNAc-R	29, 30
GlcNAcα(1 → 3)fucosyltransferase	2.4.1.152	**Fucα1 → 3** Galβ1 → 4GlcNAc-R	30, 31

(continued)

TABLE I (continued)

Glycosyltransferase	EC number	Linkage synthesized[a]	Refs.[b]
Galα(1 → 2)fucosyltransferase	2.4.1.169	**Fucα1 → 2**Gal-R	32
Glycoprotein α(1 → 6)fucosyltransferase	2.4.1.68	Fucα1 → 6 | Manβ1 → 4GlcNAcβ1 → 4-GlcNAcβ-Asn	33
CMPNeuAc			
Galα(2 → 6)-sialyltransferase	2.4.99.1	**NeuAcα2 → 6**Galβ1 → 4GlcNAc-R	34–36
Galα(2 → 3)-sialyltransferase	2.4.99.6	**NeuAcα2 → 3**Galβ1 → 3/4GlcNAc-R	34
Galα(2 → 3)-sialyltransferase	2.4.99.4	**NeuAcα2 → 3**Galβ1 → 3GalNAcα-R	37
GalNAcα(2 → 6)-sialyltransferase	2.4.99.3	**NeuAcα2 → 6**GalNAcα-R	37, 38
GalNAcα(2 → 6)-sialyltransferase	2.4.99.7	**NeuAcα2 → 6** |	39
		NeuAcα2 → 3GalNAcβ1 → 3GalNAc-R	
N-Acetylneuramideα(2 → 8)-sialyltransferase	2.4.99.8	**NeuAcα2 → 8**NeuAcα2 → 3Galβ-R	40

[a] For linkage synthesized the sugar transferred is shown in bold; R, the rest of the glycoprotein, glycolipid, or oligosaccharide chain.

[b] Key to references: (1) R. Barker, K. W. Olsen, J. H. Shaper, and R. L. Hill, *J. Biol. Chem.* **247**, 7135 (1972); (2) A. C. Gerber, I. Kozdrowski, S. R. Wyss, and E. G. Berger, *Eur. J. Biochem.* **93**, 453 (1979); (3) W. M. Blanken and D. H. van den Eijnden, *J. Biol. Chem.* **260**, 12927 (1985); (4) J. Mendicino, S. Sivakami, M. Davila, and E. V. Chandrasekaran, *J. Biol. Chem.* **257**, 3987 (1982); (5) I. Brockhausen, G. Möller, A. Pollex-Krüger, V. Rutz, H. Paulsen, and K. Matta, *Biochem. Cell Biol.* **70**, 99 (1992); (6) M. Nagai, V. Davè, H. Muensch, and A. Yoshida, *J. Biol. Chem.* **253**, 380 (1978); (7) B. T. Sheares and D. M. Carlson,

J. Biol. Chem. **258**, 9893 (1983); (8) N. Taniguchi, K. Yanagisawa, A. Makita, and M. Naiki, *J. Biol. Chem.* **260**, 4908 (1985); (9) Y. Wang, J. L. Abenathy, A. E. Eckhardt, and R. L. Hill, *J. Biol. Chem.* **267**, 12709 (1992); (10) A. Elhammer and S. Kornfeld, *J. Biol. Chem.* **261**, 5249 (1986); (11) M. Schwyzer and R. L. Hill, *J. Biol. Chem.* **252**, 2338 (1977); (12) H. Tuppy and H. Schenkel-Brunner, *Eur. J. Biochem.* **10**, 152 (1969); (13) F. Serafini-Cessi and F. Dall'Olio, *Biochem. J.* **215**, 483 (1983); (14) M. G. Shoreibah, O. Hindsgaul, and M. Pierce, *J. Biol. Chem.* **267**, 2920 (1992); (15) I. Brockhausen, E. Hull, O. Hindsgaul, H. Schachter, R. N. Shah, S. W. Michnick, and J. P. Carver, *J. Biol. Chem.* **264**, 11211 (1989); (16) B. Bendiak and H. Schachter, *J. Biol. Chem.* **262**, 5775 (1987); (17) S. Narasimhan, *J. Biol. Chem.* **257**, 10235 (1982); (18) C. L. Oppenheimer and R. L. Hill, *J. Biol. Chem.* **256**, 799 (1981); (19) Y. Nishikawa, W. Pegg, H. Paulsen, and H. Schachter, *J. Biol. Chem.* **263**, 8270 (1988); (20) P. A. Gleeson and H. Schachter, *J. Biol. Chem.* **258**, 6162 (1983); (21) P. A. Ropp, M. R. Little, and P.-W. Cheng, *J. Biol. Chem.* **266**, 23863 (1991); (22) S. Sangadala, S. Sivakami, and J. Mendicino, *Mol. Cell. Biochem.* **101**, 125 (1991); (23) I. Brockhausen, K. L. Matta, J. Orr, and H. Schachter, *Biochemistry* **24**, 1866 (1985); (24) I. Brockhausen, K. L. Matta, J. Orr, H. Schachter, A. H. L. Koenderman, and D. van den Eijnden, *Eur. J. Bichem.* **157**, 463 (1986); (25) I. Brockhausen, J. Orr, and H. Schachter, *Can. J. Biochem. Cell Biol.* **62**, 1081 (1984); (26) D. H. van den Eijnden, A. H. L. Koenderman, and W. E. C. M. Schiphorst, *J. Biol. Chem.* **263**, 12461 (1988); (27) A. D. Yates and W. M. Watkins, *Carbohydr. Res.* **120**, 251 (1983); (28) F. Piller, J.-P. Cartron, A. Maranduba, A. Veyrières, Y. Leroy, and B. Fournet, *J. Biol. Chem.* **259**, 13385 (1984); (29) J.-P. Prieels, D. Monnom, M. Dolmans, T. A. Beyer, and R. L. Hill, *J. Biol. Chem.* **256**, 10456 (1981); (30) S. Eppenberger-Castori, H. Lotscher, and J. Finne, *Glycoconjuate J.* **6**, 101 (1989); (31) P. H. Johnson and W. M. Watkins, *Biochem. Soc. Trans.* **10**, 445 (1982); (32) T. A. Beyer, J. E. Sadler, and R. L. Hill, *J. Biol. Chem.* **255**, 5364 (1980); (33) G. D. Longmore and H. Schachter, *Carbohydr. Res.* **100**, 365 (1982); (34) J. Weinstein, U. de Souza-e-Silva, and J. C. Paulson, *J. Biol. Chem.* **257**, 13835 (1982); (35) U. Sticher, H. J. Gross, and R. Brossmer, *Biochem. J.* **253**, 577 (1988); (36) J. C. Paulson, W. E. Beranek, and R. L. Hill, *J. Biol. Chem.* **252**, 2356 (1977); (37) J. E. Sadler, J. I. Rearick, J. C. Paulson, and R. L. Hill, *J. Biol. Chem.* **254**, 4434 (1979); (38) D. M. Carlson, E. J. McGuire, G. W. Jourdian, and S. Roseman, *J. Biol. Chem.* **248**, 5763 (1973); (39) M. L. E. Bergh, G. J. M. Houghwinkel, and D. H. van den Eijnden, *J. Biol. Chem.* **258**, 7430 (1983); (40) H. Higashi, M. Basu, and S. Basu, *J. Biol. Chem.* **260**, 824 (1985).

in acceptor and donor specificities can also be exploited in synthetic applications to give analogs of natural compounds. In the following examples the addition of Gal, GalNAc, Fuc, NeuAc, and GlcNAc is performed by using commercially available and partially purified enzyme preparations. The reaction products can be used to determine the substrate specificity of other glycosyltransferases and glycosidases, for activation and coupling to proteins,[14] to generate carbohydrate-specific antibodies,[15] and for the preparation of matrices for affinity chromatography.[16]

Synthesis of Galβ1 → 4GlcNAcβ-OR (LacNAc-OR)

Principle

Galactosyltransferase catalyzes the following reaction:

$$\text{GlcNAc}\beta\text{-OR} + \text{UDPGal} \xrightarrow{\text{galactosyltransferase}} \text{Gal}\beta1 \rightarrow 4\text{GlcNAc}\beta\text{-OR} + \text{UDP}$$

where R is $(CH_2)_8COOMe$ or methylumbelliferyl.

Reagents

One 1-unit vial of lyophilized bovine milk galactosyltransferase (Sigma, St. Louis, MO)
GlcNAcβ-OR*: 5 mg
UDPGal (sodium salt; Sigma): 15 mg
Calf intestine alkaline phosphatase (molecular biology grade; Boehringer Mannheim, Indianapolis, IN): 4 units
Sodium cacodylate buffer (100 mM), pH 7.5, with 5 mM MnCl$_2$
*R = $(CH_2)_8COOMe$,[17] methylumbelliferyl (Sigma)

Procedure

Incubations are carried out in a total volume of 3 ml in small plastic scintillation vials or plastic test tubes at 37° for 2.5 hr. The time of reaction is calculated as five times that required for conversion based on a unit transferring 1 μmol of galactose from UDPGal to the acceptor per minute. An excess of donor (1.5 equivalents) is added to ensure complete conver-

[14] C. P. Stowell and Y. C. Lee, *Adv. Carbohydr. Chem. Biochem.* **37,** 225 (1980).
[15] R. U. Lemieux, *Chem. Soc. Rev.* **7,** 423 (1978).
[16] J. H. Pazur, *Adv. Carbohydr. Chem. Biochem.* **39,** 404 (1981).
[17] R. U. Lemieux, D. R. Bundle, and D. A. Baker, *J. Am. Chem. Soc.* **97,** 4076 (1975).

sion and to allow for any metal ion-catalyzed decomposition of UDPGal.[18] Reaction progress can also be monitored by thin-layer chromatography (TLC) on glass-backed analytical plates of silica gel 60-F_{254}, 0.25-mm thickness (E. Merck AG, Darmstadt, Germany) with detection by spraying with 5% H_2SO_4 in ethanol followed by slow heating on a hot plate at 200°. Freshly prepared CH_2Cl_2–methanol–H_2O (60:35:6, v/v) is used as the solvent system. The product Galβ1 → 4GlcNAcβ-O(CH_2)$_8$COOMe has an R_f of 0.52 whereas the starting material GlcNAcβ-O(CH_2)$_8$COOMe has an R_f of 0.78. The alkaline phosphatase is added[19] to convert UDP, which is a potent inhibitor of galactosyltransferase, to uridine, which is a considerably weaker inhibitor. The hydrophobic disaccharide products are isolated by adsorption onto reversed-phase C_{18} SepPak cartridges (Waters, Milford, MA) and washing five times with 5 ml of H_2O to remove charged donor, buffer components, and ions. Product is then eluted twice with 5 ml of methanol and concentrated and dried by evaporation under reduced pressure. The residue is dissolved in water (5 ml), passed through a Millex-GV filter (Millipore, Bedford, MA), and lyophilized to give the final product in 80–90% yield. Partial nuclear magnetic resonance (NMR) spectra of the starting monosaccharide and the product disaccharide are shown in Fig. 1. It can be seen in the upper trace of Fig. 1 that a new doublet ($J = 8$ Hz) near 4.5 ppm attributed to H-1 of the βGal residue confirms the addition of this residue to GlcNAc. This is a general procedure for the synthesis and isolation of any products with hydrophobic aglycones, for example, R = 4-methylumbelliferyl. For other aglycones that are not sufficiently hydrophobic for complete adsorption onto reversed-phase cartridges (R = p-nitrophenyl, methyl), an alternate isolation method such as high-performance liquid chromatography (HPLC),[7,8] or chromatography on Dowex or BioGel resins,[9,20] must be used. The use of substrates with hydrophobic aglycones to facilitate product isolation in the assay of glycosyltransferases was pioneered by the Sweeley group.[21] The 8-methoxycarbonyloctyl- or 4-methylumbelliferyl-LacNAc products obtained in this synthesis can be used to assay α(1 → 3)-fucosyltransferases, α(1 → 2)-fucosyltransferases, α(2 → 3)-sialyltransferases, and α(2 → 6) sialyltransferases.[21,22] Galβ1 → 4GlcNAcβ-O(CH_2)$_8$CO can be coupled to

[18] H. A. Nunez and R. Barker, *Biochemistry* **15**, 3843 (1976).

[19] C. Unverzagt, H. Kunz, and J. C. Paulson, *J. Am. Chem. Soc.* **112**, 9308 (1990).

[20] K. Yamashita, T. Mizuochi, and A. Kobata, *in* "Methods in Enzymology" (V. Ginsburg, ed.), Vol. 83, p. 105. Academic Press, New York, 1982.

[21] L. J. Melkerson-Watson, K. Kanemitsu, and C. C. Sweeley, *Glycoconjugate J.* **4**, 7 (1987).

[22] M. M. Palcic, L. D. Heerze, M. Pierce, and O. Hindsgaul, *Glycoconjugate J.* **5**, 49 (1988).

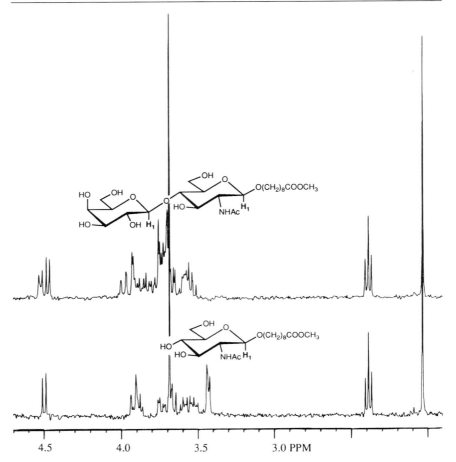

FIG. 1. Partial 360-MHz ^{1}H NMR spectra (in D_2O) of GlcNAcβ-O(CH$_2$)$_8$COOMe (lower trace) and the enzymatically generated product Galβ1 → 4GlcNAcβ-O(CH$_2$)$_8$COOMe (upper trace). GlcNAcβ-OR has the signal of H-1, a doublet with J = 8 Hz, at 4.5 ppm. In the product, a second doublet (J = 8 Hz) near 4.5 ppm, the signal for H-1 of the βGal residue, confirms the addition of that sugar residue.

bovine serum albumin via its acyl azide[23] and serves as a substrate for an enzyme-linked immunosorbent assay for $\alpha(1 \rightarrow 3)$-fucosyltransferases, in which product is detected with a product-specific antibody.[24]

[23] B. M. Pinto and D. R. Bundle, *Carbohydr. Res.* **124**, 313 (1983).
[24] T. Tachikawa, S. Akamatsu, S. Shin, T. Asao, and S. Yazawa, *Glycoconjugate J.* **8**, 264 (1991).

Synthesis of GalNAcβ1 → 4GlcNAcβ-OR

Principle

This synthesis takes advantage of the slow transfer (0.2%) of GalNAc from UDPGalNAc catalyzed by bovine galactosyltransferase[25]:

$$\text{GlcNAc}\beta\text{-OR} + \text{UDPGalNAc} \xrightarrow{\text{galactosyltransferase}} \text{GalNAc}\beta1 \rightarrow 4\text{GlcNAc}\beta\text{-OR} + \text{UDP}$$

where R is $(CH_2)_8COOMe$.

Reagents

Two 1-unit vials of lyophilized bovine milk galactosyltransferase (Sigma)
GlcNAcβ-OR*: 1.2 mg
UDPGalNAc (sodium salt; Sigma): 9 mg
Sodium cacodylate buffer (100 mM), pH 7.5, with 5 mM MnCl$_2$
*R = $(CH_2)_8COOMe$,[17] methylumbelliferyl (Sigma)

Procedure

Incubations are carried out in a total volume of 1.4 ml for 48 hr at 37°. The UDPGalNAc is added in two portions, 4.5 mg at the start of the reaction and 4.5 mg after 24 hr of incubation. A large excess of donor is used to compensate for the metal ion catalyzed decomposition of UDPGal-NAc. Reaction progress can also be monitored by TLC on silica plates, using freshly prepared CH$_2$Cl$_2$–methanol–H$_2$O (60 : 35 : 6, v/v) as the solvent system as described for LacNAc synthesis above. Oligosaccharides can be detected by spraying with 5% H$_2$SO$_4$ in ethanol followed by slow heating on a hot plate at 200°. The product GalNAcβ1 → 4GlcNAcβ-O(CH$_2$)$_8$COOMe has an R_f of 0.57 whereas the starting material GlcNAcβ-O(CH$_2$)$_8$COOMe has an R_f of 0.78. The reaction product is isolated on reversed-phase C$_{18}$ cartridges as described for LacNAc-OR above, with yields of 80–90%. GalNAcβ1 → 4GlcNAcβ-OR can be used as a substrate for a novel β(1 → 3)-galactosyltransferase.[26] Alternate donors that can be used are UDPGLc[27] and UDPGlcNH$_2$ followed by N-acetylation, to give Glcβ1 → 4GlcNAcβ-OR and GlcNAcβ1 → 4GlcNAcβ-OR,[25] respectively.

[25] O. Hindsgaul and M. M. Palcic, *Glycobiology* **1**, 205 (1990).
[26] H. Mulder, H. Schachter, M. De Jong-Brink, J. G. M. Van der Ven, J. P. Kamerling, and J. F. G. Vliegenthart, *Eur. J. Biochem.* **201**, 459 (1991).
[27] L. J. Berliner and R. D. Robinson, *Biochemistry* **21**, 6340 (1982).

Synthesis of NeuAcα2 → 3Galβ1 → 4[Fucα1 → 3]GlcNAcβ-OR
(Sialyl Lewis^x)

Principle

This synthesis employs two succcessive enzyme steps. Sialylation is carried out prior to fucosylation because Galβ1 → 4[Fucα1 → 3]GlcNAcβ-OR is not a substrate for the sialyltransferase.

Galβ1 → 4GlcNAcβ-OR $\xrightarrow{\text{sialyltransferase}}$

$\qquad\qquad\qquad\qquad$ NeuAcα2 → 3Galβ1 → 4GlcNAcβ-OR

NeuAcα2 → 3Galβ1 → 4GlcNAcβ-OR $\xrightarrow{\text{fucosyltransferase}}$

$\qquad\qquad\qquad\qquad$ NeuAcα2 → 3Galβ1 → 4[Fucα1 → 3]GlcNAcβ-OR

Reagents

Galβ1 → 4GlcNAcβ-O(CH$_2$)$_8$COOMe^{17} or Galβ1 → 4GlcNAcβ-OCETE
(Sigma): 3 mg
CMPNeuAc (sodium salt; Sigma): 7.5 mg
GDP-β-Fuc (Sigma or Oxford Glycosystems): 2 mg
α(2 → 3)-Sialyltransferase partially purified from rat liver: 10 milliunits
α(1 → 3/4)-Fucosyltransferase partially purified from human milk: 5 milliunits
Alkaline phosphatase (molecular biology grade; Boehringer Mannheim): 8 units
Sodium cacodylate buffer (25 mM), pH 7.2, with 0.1% Triton X-100 and bovine serum albumin (BSA) (1 mg/ml)
Sodium cacodylate buffer (25 mM) pH 6.5, with 5 mM MnCl$_2$ and BSA (1 mg/ml)

Preparation of α(1 → 3/4)-Fucosyltransferase^{28,29}

A partially purified preparation of fucosyltransferase for synthetic use can be readily obtained from 250 ml of human milk. All steps are carried out at 4° and plasticware is used whenever possible. Human milk is defatted by centrifugation at 7500 *g* for 30 min at 4°, followed by removal of the solid lipid pellet on the surface. The defatted milk is mixed with 2 g/liter of dry SP-Sephadex C-50 (Pharmacia, Piscataway, NJ), and stirred gently at 4° overnight. The resin is washed in a funnel on a filter flask with 3 vol of water and then poured into a small 2.5-cm diameter column. The column is washed with 0.1 M NaCl, then enzyme is eluted with a salt gradient

---

[28] J.-P. Prieels, D. Monnom, M. Dolmans, T. A. Beyer, and R. L. Hill, *J. Biol. Chem.* **256**, 10456 (1981).

[29] S. Eppenberger-Castori, H. Lotscher, and J. Finne, *Glycoconjugate J.* **6**, 101 (1989).

from 0.1 to 0.6 M NaCl. The volume of the gradient is five column volumes, based on the column size after the 0.1 M wash. Enzyme that typically elutes at 0.25–0.4 M NaCl is concentrated by ultrafiltration, and dialyzed into 50 mM sodium cacodylate buffer, pH 7.2, containing 25% (v/v) glycerol and 0.05% (w/v) NaN_3 (buffer A). Dialyzed enzyme is loaded on a 2-ml GDPhexanolamine agarose column[30] (4 μmol/ml), the column is washed with equilibration buffer, then enzyme is eluted with buffer A containing 0.8 M NaCl and 2.5 mM GMP (Sigma). This eluent (20 ml) is brought to 0.14 mg of bovine serum albumin/ml to stabilize the fucosyltransferase, desalted by dialysis against 25 mM sodium cacodylate buffer (pH 6.5) containing 50% glycerol, and concentrated to 3 ml by ultrafiltration. About 20 milliunits of enzyme can be obtained from 250 ml of milk. This enzyme is a mixture of $\alpha(1 \rightarrow 3)$- and $\alpha(1 \rightarrow 3/4)$-fucosyltransferase activities, but is devoid of $\alpha(1 \rightarrow 2)$-fucosyltransferase activity. The $\alpha(1 \rightarrow 2)$-fucosyltransferase is unstable; it is only partially adsorbed onto the SP-Sephadex and any bound enzyme elutes in the 0.1 M NaCl wash. The $\alpha(1 \rightarrow 3/4)$-fucosyltransferase enzyme eluted from the SP-Sephadex column is also suitable for use in preparative synthesis. It contains a small amount of hydrolyzing activity that cleaves the methyl ester of the methoxycarbonyloctyl aglycone on prolonged incubations. About 15% hydrolysis occurs after 48 hr. The fucosyltransferase can readily be assayed by incubating enzyme (20 microunits) with 0.4 mM acceptor [LacNAcβ-OCETE or LacNAcβ-O(CH$_2$)$_8$COOMe], 50 $\mu$$M$ GDPfucose, 75,000 dpm GDP[1-^{3}H]fucose (New England Nuclear, Boston, MA), 0.5 mM ATP, 0.2% (w/v) bovine serum albumin, 20 mM MnCl$_2$, and 20 mM N-2-hydroxylethylpiperazine-N'-2-ethanesulonic acid (HEPES) buffer (pH 7.0), in a total volume of 40 μl in microfuge tubes, for 30 min at 37°. Reaction mixtures are diluted with water, transferred onto C$_{18}$ reversed-phase cartridges (Waters), and washed with water (20–30 ml or until background counts are achieved) to remove unreacted GDPfucose. Radio-labeled products are eluted twice with 4 ml of methanol and counted in 10 ml of ACS liquid scintillation cocktail. One unit of enzyme is defined as the amount of enzyme transferring 1 μmol of fucose per minute.

Preparation of $\alpha(2 \rightarrow 3)$-Sialyltransferase[31–33]

Enzyme is extracted following the procedure of Weinstein *et al.*,[31] with modifications in the volumes of extraction buffer used. Frozen rat livers (500 g obtained from Pel-Freez Biologicals, Rogers, AR) are thawed

[30] T. A. Beyer, J. E. Sadler, and R. L. Hill, *J. Biol. Chem.* **255**, 5364 (1980).
[31] J. Weinstein, U. de Souza-e-Silva, and J. C. Paulson, *J. Biol. Chem.* **257**, 13835 (1982).
[32] U. Sticher, H. J. Gross, and R. Brossmer, *Biochem. J.* **253**, 577 (1988).
[33] M. A. Mazid and M. A. Kashem, PCT Int. Appl. WO 90/12090 (1990).

at 4° overnight, rinsed with cold water, and homogenized with five 4-sec bursts in a blender. Two extractions with 900 ml of 10 mM sodium cacodylate buffer (pH 6.0) and 20 mM MnCl$_2$ per extraction are carried out by homogenizing (four 10-sec pulses, with 30-sec pauses) and centrifuging at 14,000 g for 1 hr at 4°. The pellet is homogenized with 700 ml of 25 mM sodium cacodylate buffer (pH 6.0) containing 10 mM MnCl$_2$ and 75 mM NaCl (two 10-sec bursts), then 20% (w/v) Triton CF-54 is added to give a final concentration of 1.4%. The Triton extract is stirred for 60 min and centrifuged at 14,000 g for 1 hr at 4°. The supernatant is poured through several layers of cheesecloth, and 0.5 M ethylenediaminetetraacetic acid (EDTA) (pH 7.7) is added slowly while the solution is kept at pH 6.0 with the dropwise addition of 0.5 M NaOH, until addition of EDTA causes no further drop in pH. Triton extraction and centrifugation is repeated as above, and the EDTA-treated extracts are combined.

Chromatography of the combined Triton extracts on Cibacron Blue F3GA-Sepharose 6B is carried out as described by Sticher et al.[32] After loading of the extracts onto a 4.6 × 28 cm column, the column is washed with 10 mM sodium cacodylate (pH 6.5), 25% (v/v) glycerol, 150 mM NaCl, and 0.1% (w/v) Triton CF-54 until no further protein elutes (1.5–2 liters of wash buffer). Sialyltransferase is eluted with wash buffer containing 2 M NaCl. The salt eluate, which contains both $\alpha(2 \rightarrow 3)$-sialyltransferase and $\alpha(2 \rightarrow 6)$-sialyltransferase activity, is concentrated by ultrafiltration to about 30 ml and then dialyzed against 10 mM sodium cacodylate buffer (pH 6.5) containing 150 mM NaCl, 0.1% Triton CF-54, and 25% glycerol. An equal volume of dialysis buffer containing 5 mM CDP is added to dialyzed enzyme before loading onto a 15-ml Galβ1 $\rightarrow$ 3GlcNAcβ(CH$_2$)$_8$ CONH(CH$_2$)$_2$NH-Sepharose CL-6B[33] (1.2 μmol/ml) column preequilibrated with dialysis buffer and 2.5 mM CDP. The column is washed with loading buffer until no further protein elutes. $\alpha(2 \rightarrow 6)$-Sialyltransferase does not adsorb to this resin and is removed in the loading and washing steps. $\alpha(2 \rightarrow 3)$-Sialyltransferase is eluted with 50 mM sodium cacodylate buffer (pH 6.5), containing 0.2 M lactose, 0.5 M NaCl, and 0.1% Triton CF-54. Fractions (2.5 ml) are collected with 25 μl of a 10-mg/ml solution of bovine serum albumin in each fraction to stabilize the sialyltransferase. Fractions with activity are pooled, concentrated by ultrafiltration, and dialyzed against 50 mM sodium cacodylate (pH 6.5) with 50% glycerol, 0.3 M NaCl, and 0.1% Triton CF-54 for storage at $-20°$. About 30 milliunits of enzyme can be obtained from 500 g of rat liver. The enzyme can be assayed with 100 μM Galβ1 $\rightarrow$ 3GlcNAcβ(CH$_2$)$_8$COOCH$_3$ or 100 μM Galβ1 $\rightarrow$ 4GlcNAcβ(CH$_2$)$_8$COOCH$_3$ or 100 μM Galβ1 $\rightarrow$ 4GlcNAcβ-OCETE acceptor, 150 μM CMPNeuAc, and 40,000 dpm CMP[9-^{3}H]NeuAc (New England Nuclear), in a total volume of 30 μl of 25 mM

sodium cacodylate buffer (pH 6.0), with 30 mM MnCl$_2$, 0.5% Triton CF-54, and bovine serum albumin (1 mg/ml). Incubations are carried out at 37° for 30 min, then mixtures are diluted with water and applied to C$_{18}$ reversed-phase cartridges, which are washed with water to remove radiolabeled donor and charged decomposition products. Radiolabeled product is eluted twice with 4 ml of methanol and counted in 10 ml of ACS liquid scintillation cocktail. One unit of enzyme is the amount that transfers 1 μmol of sialic acid per minute.

Synthesis of NeuAcα(2 → 3)Galβ(1 → 4)[Fucα(1 → 3)]GlcNAcβ-OR (Sialyl Lewis^x)

Incubations are carried out in 300 μl of 25 mM sodium cacodylate buffer (ph 7.2) containing 0.5% Triton X-100 and bovine serum albumin (1 mg/ml). Acceptor (3 mg), 2.5 mg of CMPNeuAc, 4 units of alkaline phosphatase, and 10 milliunits of sialyltransferase are incubated at 37° for 36 hr. Every 12 hr, 2.5 mg of CMPNeuAc is added to the incubation mixture to compensate for any hydrolysis of the donor. The reaction can be monitored by TLC on silica gel plates as described for LacNAc synthesis above, using CH$_2$Cl$_2$–methanol–H$_2$O (60 : 35 : 6, v/v), with detection by spraying with 5% H$_2$SO$_4$ in ethanol and heating. The acceptor Galβ1 → 4GlcNAcβ-O(CH$_2$)$_8$COOMe has an R_f of 0.52 and the product NeuAcα2 → 3Galβ1 → 4GlcNAcβ-O(CH$_2$)$_8$COOMe has an R_f of 0.22. At the completion of the reaction, samples are diluted to 5 ml with H$_2$O and passed onto a C$_{18}$ cartridge. The cartridge is washed four times with 5 ml of H$_2$O, then three times with 5 ml of methanol to elute product. The methanol eluate is concentrated to dryness and a solution of the residue in H$_2$O (2 ml) is passed onto a 1-ml column of DEAE-Sephadex A-25 (Cl$^-$ form) in a Pasteur pipette. The column is first washed five times with 2 ml of water, then sialylated products eluted three times with 1 M NaCl (2 ml each time), desalted by adsorption onto a C$_{18}$ cartridge, which is washed with 15 ml H$_2$O. Product is eluted with 15 ml of methanol, and concentrated and dried by evaporation under reduced pressure. The residue is dissolved in 5 ml of H$_2$O, passed through a Millex-GV filter, and lyophilized to give the final product in 70% yield.

Preparative fucosylation is carried out by incubating 1 mg of NeuAcα2 → 3Galβ1 → 4GlcNAcβ-OR with 1 mg of GDPfucose and 5 milliunits of fucosyltransferase in 1 ml of sodium cacodylate buffer (pH 6.5), containing 5 mM MnCl$_2$ and BSA (1 mg/ml). After 8 hr at 37° an additional 1 mg of GDPfucose is added and incubation continued for an additional 8 hr. The reaction can be monitored by TLC on silica plates, using freshly prepared CH$_2$Cl$_2$–methanol–H$_2$O (60 : 35 : 6, v/v) as the solvent system and detec-

tion by spraying with 5% H_2SO_4 followed by heating. The product Neu-
Acα2 → 3Galβ1 → 4[Fucα1 → 3]GlcNAcβ-O(CH$_2$)$_8$COOMe has an R_f of
0.15 and the acceptor NeuAcα2 → 3Galβ1 → 4GlcNAcβ-O(CH$_2$)$_8$COOMe
has an R_f of 0.22. At the end of the reaction, the product is isolated on
C_{18} reversed-phase cartridges by diluting them with 5 ml of water and
applying to a conditioned cartridge equilibrated with water. Unreacted
donor and charged materials and buffer are removed by washing with 60
ml of water, then product is eluted with 15 ml of methanol and dried by
evaporation under reduced pressure. The residue is dissolved in 5 ml of
H_2O, passed through a Millex-GV filter, and lyophilized to give the final
product in greater than 90% yield. The same preparative fucosylation
procedure can be used with other acceptors, including Galβ1 → 4Glc-
NAcβ-OR to give Galβ1 → 4[Fucα1 → 3]GlcNAcβ-OR (Le^X) or Galβ1
→ 3GlcNAcβ-OR to give Galβ1 → 3[Fucα1 → 4]GlcNAcβ-OR (Le^a). Sialyl
Lewis^x is a ligand of ELAM-1 (endothelial leukocyte adhesion molecule 1,
E-selectin) and PADGEM (platelet activation-dependent granule external
membrane protein, P-selectin), which mediate the adhesion of neutrophils
and monocytes to endothelial cells, as well as the adhesion of certain tumor
cell lines.[34-37] Sialyl Lewis^x and analogs thereof are useful for studying
carbohydrate recognition by these adhesion molecules and they also
represent targets for the development of novel antiinflammatory and antitumor agents.

Synthesis of GlcNAcβ1 → 2Manα1 → 3[Manα1 → 6]Manβ-OR

Principle

This synthesis employs a readily available extract from rabbit liver to
transfer GlcNAc to the trimannoside core structure of asparagine-linked
oligosaccharides:

$$\text{Man}\alpha1 → 3[\text{Man}\alpha1 → 6]\text{Man}\beta\text{-OR} \xrightarrow{\text{N-acetylglucosaminyltransferase I}}$$
$$\text{GlcNAc}\beta1 → 2\text{Man}\alpha(1 → 3)[\text{Man}\alpha1 → 6]\text{Man}\beta\text{-OR}$$

where R = (CH$_2$)$_8$COOMe

Reagents

UDPGlcNAc (sodium salt; Sigma): 500 mg
Manα1 → 3[Manα1 → 6]Manβ-O(CH$_2$)$_8$COOMe[38]: 98 mg

[34] M. L. Phillips, E. Nudelman, F. C. A. Gaeta, M. Perez, A. K. Shingal, S. Hakomori,
   and J. C. Paulson, *Science* **250**, 1130 (1990).
[35] J. B. Lowe, L. M. Stoolman, R. P. Nair, R. D. Larsen, T. L. Behrend, and R. Marks,
   Cell (Cambridge, Mass.) **63**, 475 (1990).
[36] G. Walz, A. Aruffo, W. Kalanus, M. Bevilacqua, and B. Seed, *Science* **250**, 1132 (1990).
[37] M. Larkin, T. J. Ahern, M. S. Stoll, M. Shaffer, D. Sako, J. O'Brien, S.-T. Yuen, A. M.
   Larson, R. A. Childs, K. M. Barone, P. R. Langer-Safer, A. Hasegawa, M. Kiso, G. R.
   Larsen, and T. Feizi, *J. Biol. Chem.* **267**, 13661 (1992).
[38] K. J. Kaur, G. Alton, and O. Hindsgaul, *Carbohydr. Res.* **210**, 145 (1991).

N-Acetylglucosaminyltransferase I, partially purified from rabbit liver[38,39]: 8 milliunits

Purification of N-Acetylglucosaminyltransferase I

Enzyme is isolated by modification of a published preparation.[38,39] Rabbit liver acetone powder (150 g) is stirred with 900 ml of 0.1 M sodium acetate buffer (pH 6.0) containing 0.2 M NaCl and 0.01 M EDTA and then centrifuged for 15 min at 7100 g at 4°. The supernatant is discarded, and the washing and centrifugation repeated with the same acetate buffer and then water. The pellet is homogenized in 100 ml of 10 mM Tris-HCl (pH 7.8) containing 0.4 M KCl, then is brought to 1% (w/v) Triton X-100 with the addition of 5 ml of 20% Triton X-100. After stirring for 30 min the mixture is centrifuged for 20 min at 7100 g at 4° and the supernatant, which contains the N-acetylglucosaminyltransferase I, is set aside. The homogenization and extraction are repeated. The two extracts are combined and dialyzed against 2 × 4 liters of 50 mM sodium cacodylate buffer (pH 6.5), containing 10 mM MnCl$_2$ and 1% Triton X-100 (buffer A). Any precipitated material that forms during dialysis is removed by centrifugation at 7100 g for 30 min at 4°.

The dialyzed extract is applied to a column (2.5 × 12 cm) of UDPhexanolamine-Sepharose[40] (4–6 μmol/ml) equilibrated with buffer A. The column is washed with 55 ml of buffer A, followed by 120 ml of buffer A containing 0.2 M NaCl. The column is next developed with a linear gradient (200 ml) from 0.2 to 3.0 M NaCl in buffer A. Fractions eluting between 2.1 and 3.0 M NaCl are combined and dialyzed against buffer A to give 8 milliunits of N-acetylglucosaminyltransferase I in a volume of 40 ml for use directly in the preparative synthesis. N-acetylglucosaminyltransferase I can be assayed by incubating with 250 μM Manα1 → 3[Manα1 → 6]Manβ-O(CH$_2$)$_8$COOMe, 75 μM UDPGlcNAc, 110,000 dpm UDP[6-^{3}H]GlcNAc, and buffer A in a total volume of 100 μl. After 1 hr at 37°, the mixture is diluted to 5 ml with water and applied to a C$_{18}$ reversed-phase cartridge, which is washed with approximately 50 ml of water or until all radiolabeled donor is removed. Radiolabeled product is eluted twice with 5 ml of methanol and quantitated by liquid scintillation counting in ACS cocktail. One unit of enzyme is defined as the amount that catalyzes the transfer of 1 μmol of GlcNAc per minute.

Synthesis of GlcNAcβ1 → 2Manα1 → 3[Manα1 → 6]Manβ-OR[38]

A mixture of 98 mg of acceptor, 500 mg of UDPGlcNAc, and 40 ml of enzyme extract is incubated at room temperature for 5 days. The reaction progress can be monitored by TLC on silica gel plates, using

[39] C. L. Oppenheimer and R. L. Hill, *J. Biol. Chem.* **256**, 799 (1981).
[40] R. Barker, K. W. Olsen, J. H. Shaper, and R. L. Hill, *J. Biol. Chem.* **247**, 7135 (1972).

CH_2Cl_2–methanol–H_2O (60 : 35 : 6, v/v) as the solvent system, with detection by spraying and charring with 5% H_2SO_4 in ethanol. The R_f is 0.4 for starting material and 0.26 for product. At the end of incubation, the reaction mixture is applied to a C_{18} column (Waters octadecylsilane, 55–105 μm irregular) (2.5 × 60 cm) and eluted with 3.6 liters of methanol–water (35 : 65, v/v) at a flow rate of 2 ml/min. Fractions (7 ml) are collected and monitored by TLC on silica gel as described earlier. Fractions that contain only product are combined and concentrated by evaporation under reduced pressure. Fractions that contain unreacted starting material and product are concentrated and applied to a column (2.5 × 60 cm) of BioGel P-2, which is eluted with aqueous 10% ethanol. The product that elutes before starting material is combined and concentrated. A yield of 52 mg (41% yield) of product is readily obtained. The product GlcNAcβ1 → 2Manα1 → 3[Manα1 → 6]Manβ-O(CH$_2$)$_8$COOMe can be used to assay N-acetylglucosaminyltransferase II.[38] A facile synthesis of Manα1 → 3[Manα1 → 6]Manβ-O(CH$_2$)$_7$CH$_3$ has been reported.[41]

Acknowledgment

This work was supported by a grant from the Medical Research Council of Canada.

[41] K. J. Kaur and O. Hindsgaul, *Glycoconjugate J.* **8**, 90 (1991).

[19] Glycosidase Inhibitors in Study of Glycoconjugates

By GUR P. KAUSHAL and ALAN D. ELBEIN

Introduction

A number of compounds that inhibit specific steps in N-linked glycoprotein biosynthesis and processing pathway have been identified.[1–3] Because many of these inhibitors are readily taken up by cultured cells, they are potentially valuable tools for studies on glycoconjugate function, targeting, or turnover. Many of these compounds have also been used as potential inhibitors of tumor metastasis and viral replication, including the autoimmune deficiency syndrome (AIDS) virus.[2]

When cultured cells are grown in the presence of a specific inhibitor, they usually synthesize N-linked glycoproteins with one major type of oligosaccharide chain (depending on the site of inhibition). Thus, a variety

[1] A. D. Elbein, *FASEB J.* **5**, 3055 (1991).
[2] A. D. Elbein, *Semin. Cell Biol.* **2**, 309 (1991).
[3] A. D. Elbein, *Annu. Rev. Biochem.* **56**, 497 (1987).

of oligosaccharide structures have been successfully prepared from such glycoproteins. These oligosaccharides are not only useful as substrates for the assay and purification of various processing glycosidases, but they can also be used for various other purposes, including substrate specificity studies or molecular markers. Because most of the inhibitors are alkaloids and contain a nitrogen atom in place of an oxygen atom in the ring structures, this nitrogen can act as a potential site through which these inhibitors can be linked to a solid matrix support. Thus, some of the inhibitors have been successfully employed as affinity ligands for the purification of various processing glycosidases (see below). In addition, inhibitors are useful tools to distinguish glucosidases and α-mannosidases in various subcellular organelles. For example, there are now a number of potent and fairly selective inhibitors that can distinguish some of the processing mannosidases from others. Finally, the structure and specificity of potent inhibitors will provide a greater insight into an understanding of the mechanisms of catalysis of enzymes and also into techniques for designing more specific inhibitors as antitumor, antifungal, or antiviral agents.

The glycoprotein processing pathway and the specific site of action of some of the processing inhibitors are shown in Fig. 1. For more details on this pathway and the various inhibitors, the reader is referred to other detailed reviews on this subject.[1,3]

Preparation of Various Oligosaccharide Substrates with Processing Inhibitors in Cell Culture

There are at least four different glycoprotein processing glycosidases involved in the processing of Glc_3Man_9 $(GlcNAc)_2$ to complex types of structures (Fig. 1). Initial processing occurs by removal of the $\alpha(1\rightarrow2)$-linked glucose by glucosidase I, and two $\alpha(1\rightarrow3)$-linked glucoses by glucosidase II, to produce a $Man_9(GlcNAc)_2$ structure.[4–8] Glucosidases I and II are inhibited by a number of α-glucosidase inhibitors, including castanospermine, deoxynojirimycin, australine, MDL, and 2,5-dihydroxymethyl-3,4-dihydroxypyrrolidine (Fig. 2). All of these inhibitors prevent glycoprotein processing and cause the accumulation of N-linked oligosaccharides with the structure $Glc_3Man_{7–9}(GlcNAc)_2$. Among these inhibitors castanospermine is more effective on glucosidase I than on glucosidase II, whereas MDL is a better inhibitor of glucosidase II than of glucosidase

[4] S. C. Hubbard and P. W. Robbins, *J. Biol. Chem.* **254**, 4568 (1979).
[5] L. S. Grinna and P. W. Robbins, *J. Biol. Chem.* **255**, 2255 (1980).
[6] J. J. Elting, W. W. Chen, and W. J. Lennarz, *J. Biol. Chem.* **255**, 2325 (1980).
[7] J. M. Michael and S. Kornfeld, *Arch. Biochem. Biophys.* **199**, 249 (1980).
[8] D. M. Burns and O. Touster, *J. Biol. Chem.* **257**, 1991 (1982).

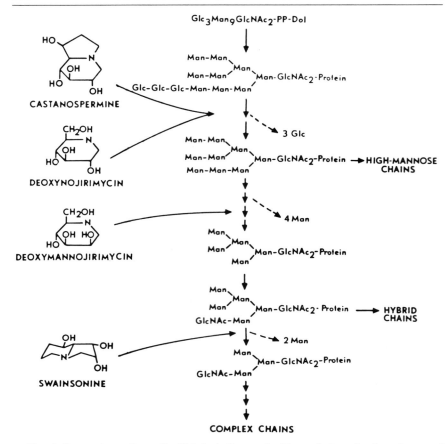

Fig. 1. Processing pathway for N-linked oligosaccharides and sites of action of some of the inhibitors of this pathway.

I.[9] Australine, which is not nearly as potent an inhibitor as some of the others, is specific for glucosidase I and does not inhibit glucosidase II.[10] The inhibition of glucosidase I by castanospermine or other glucosidase inhibitors does not completely prevent the formation of complex types of

[9] G. P. Kaushal, Y. T. Pan, J. E. Tropea, M. Mitchell, P. Liu, and A. D. Elbein, *J. Biol. Chem.* **263,** 17278 (1988).

[10] J. E. Tropea, R. J. Molyneux, G. P. Kaushal, Y. T. Pan, M. Mitchell, and A. D. Elbein, *Biochemistry* **28,** 2027 (1989).

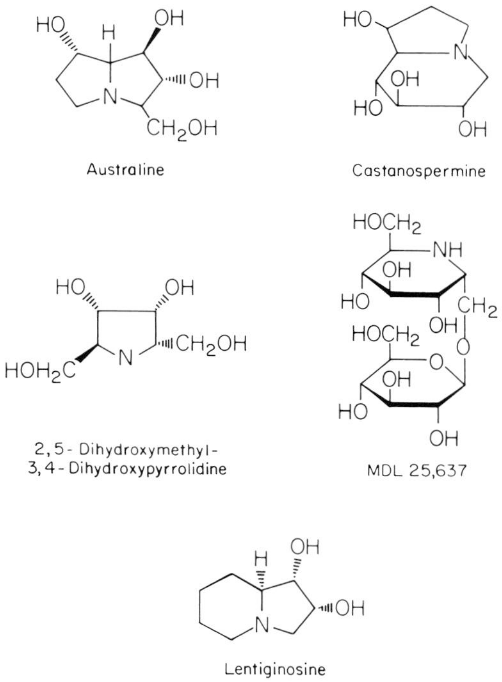

Australine

Castanospermine

2,5-Dihydroxymethyl-
3,4-Dihydroxypyrrolidine

MDL 25,637

Lentiginosine

FIG. 2. Structures of various glycosidase inhibitors.

structures.[11-14] Thus the formation of complex types of structure in the presence of this inhibitor has been shown to be due to the presence of endo-α-mannosidase activity in some mammalian cells that can remove a Glcα1-3Man disaccharide from the Glc$_1$Man$_9$(GlcNAc)$_2$ structure or Glc$_{1-3}$Man saccharides from Glc$_{1-3}$Man$_9$(GlcNAc)$_2$ structures to form Man$_8$(GlcNAC)$_2$ structures.[15,16] As yet, no inhibitor for this enzyme has been identified.

[11] Y. T. Pan, H. Hori, R. Saul, B. A. Sanford, R. J. Molyneux, and A. D. Elbein, *Biochemistry* **22**, 3975 (1983).
[12] R. Datema, P. A. Romero, G. Legler, and R. T. Schwarz, *Proc. Natl. Acad. Sci. U.S.A.* **79**, 6787 (1982).
[13] V. Gross, T. Andus, T. A. Tran-Thi, R. T. Schwarz, K. Decker, and P. C. Heinrich, *J. Biol. Chem.* **258**, 12203 (1983).
[14] P. A. Romero, B. Saunier, and A. Herscovics, *Biochem. J.* **226**, 733 (1985).
[15] W. A. Lubas and R. G. Spiro, *J. Biol. Chem.* **262**, 3775 (1987).
[16] S. E. H. Moore and R. G. Spiro, *J. Biol. Chem.* **265**, 13104 (1990).

After the removal of three glucoses, the high-mannose chain $Man_9(GlcNAc)_2$ is then processed by removal of all of the $\alpha(1{\rightarrow}2)$-linked mannose units by mannosidase I to give rise to a $Man_5(GlcNAc)_2$ structure.[17–19] This enzyme (mannosidase I) is inhibited by deoxymannojirimycin[20,21] and kifunensine.[22] Kifunensine is 50 to 100 times more effective an inhibitor of mannosidase I than is deoxymannojirimycin. In cell culture, kifunensine (1 μg/ml) causes a complete shift in the structure of the N-linked oligosaccharide chains from a complex type of structure to a $Man_9(GlcNAc)_2$ structure. On the other hand, deoxymannojirimycin even at 50 μg/ml does not prevent the formation of all complex-type chains.[22]

The $Man_5(GlcNAc)_2$ structure is the acceptor of a GlcNAc residue donated from UDP-GlcNAc to the 3-linked mannose to produce GlcNAc-$Man_5(GlcNAc)_2$ protein. This reaction is catalyzed by GlcNAc transferase I, which resides in the medial Golgi apparatus. Once this GlcNAc has been added to produce a GlcNAc-$Man_5(GlcNAc)_2$ structure, a Golgi enzyme called mannosidase II removes the two terminal $\alpha(1{\rightarrow}3)$- and $\alpha(1{\rightarrow}6)$-linked mannose units to produce a GlcNAc-$Man_3(GlcNAc)_2$ protein.[23,24] The Golgi mannosidase II is inhibited by swainsonine[25] and by mannostatin A.[26] Mannostatin A is a metabolite produced by the microorganism *Streptoverticillium verticillus*, and is a potent inhibitor of rat epididymal α-mannosidase.[27] In cell culture, both swainsonine[28] and mannostatin A[26] prevent the formation of complex chains and cause the accumulation of hybrid types of N-linked oligosaccharides. The product of mannosidase II, GlcNAc-$Man_3(GlcNAc)_2$, is the precursor of all complex types of oligosaccharides, and is elongated by a variety of glycosyltransferases that add

---

[17] D. R. P. Tulsiani and O. Touster, *J. Biol. Chem.* **263,** 5408 (1988).

[18] W. T. Forsee, C. F. Palmer, and J. S. Schutzbach, *J. Biol. Chem.* **264,** 3869 (1989).

[19] J. Schweden and E. Bause, *Biochem. J.* **264,** 347 (1989).

[20] U. Fuhrmann, E. Bause, G. Legler, and H. Ploegh, *Nature (London)* **307,** 755 (1984).

[21] A. D. Elbein, G. Legler, A. Tlusty, W. McDowell, and R. T. Schwarz, *Arch. Biochem. Biophys.* **235,** 579 (1984).

[22] A. D. Elbein, J. E. Tropea, M. Mitchell, and G. P. Kaushal, *J. Biol. Chem.* **265,** 15599 (1990).

[23] H. Schachter, S. Narasimhan, P. Gleeson, and G. Vella, *Can. J. Biochem. Cell Biol.* **61,** 1049 (1983).

[24] D. R. P. Tulsiani, S. C. Hubbard, P. W. Robbins, and O. Touster, *J. Biol. Chem.* **257,** 3660 (1982).

[25] D. R. P. Tulsiani, T. M. Harris, and O. Touster, *J. Biol. Chem.* **257,** 7936 (1982).

[26] J. E. Tropea, G. P. Kaushal, I. Pastuszak, M. Mitchell, T. Aoyagi, R. J. Molyneux, and A. D. Elbein, *Biochemistry* **29,** 10062 (1990).

[27] T. Aoyagi, T. Yamamoto, K. Kojiri, H. Morishima, M. Nogai, M. Hamada, T. Takeuchi, and H. Umegawa, *J. Antibiot.* **42,** 883 (1989).

[28] A. D. Elbein, P. R. Dorling, K. Vosbeck, and M. Horisberger, *J. Biol. Chem.* **257,** 1573 (1982).

GlcNAc, galactose, and sialic acid to give rise to a great variety of complex types of structures.[29]

By using these specific inhibitors in cell culture, as described above, a large variety of oligosaccharide intermediates can be prepared. The details of procedures for preparing these intermediates are given below.

Preparation of Radiolabeled $Glc_3Man_{7-9}GlcNAc$ and $Glc_3Man_9GlcNAc$ Intermediates

Because castanospermine is a more effective inhibitor than other glucosidase inhibitors, it is a good choice for preparing a radiolabeled Glc_3-$Man_{7-9}GlcNAc$ intermediate to use as an enzyme substrate. Deoxymannojirimycin, a mannosidase I inhibitor [inhibiting the removal of mannoses from the $\alpha(1\rightarrow6)$ branch], is also added to the cell culture along with castanospermine to prevent removal of any mannose residues.

Confluent monolayers of Madin–Darby canine kidney (MDCK) cells grown in modified Eagle's medium with 10% (v/v) calf serum are infected with the NWS strain of influenza virus as described.[11,30] One hour after infection, castanospermine, at a final concentration of 100 μg/ml, and/or deoxymannojirimycin at a final concentration of 25 μg/ml, are added to the infected cell cultures. After preincubation for 2 hr with the inhibitors, either [2-^{3}H]mannose or [6-^{3}H]galactose (both at 25 μCi/ml) is added to label the viral glycoproteins. Infected cells are then allowed to incubate for about 48 hr in order to produce mature virus. The suspension containing the virus and lysed cells is removed and those cells still adhering to the plastic are removed by scraping with a plastic paddle in the presence of phosphate-buffered saline (PBS). The viral and cell suspensions are pooled and centrifuged at 10,000 g for 30 min at 4° to remove cell debris and whole cells. The supernatant liquid is centrifuged overnight at 100,000 g at 4° to pellet the virus. Both the virus pellet and the cell pellet are then digested with pronase to generate labeled glycopeptides having as small a peptide portion as possible. The cell pellets and viral pellets are suspended in pronase digestion solution, containing 20 mM Tris, (pH 7.5), 1 mM CaCl$_2$, and pronase (5 mg/ml). The mixture is incubated for 24 hr at 37° under a toluene atmosphere. At the end of this incubation more pronase solution is added and the mixture is allowed to incubate for an additional 12–16 hr. The pronase digests are precipitated in 10% trichloroacetic acid (TCA) and the glycopeptides are recovered in the TCA-soluble supernatant liquids by centrifugation. After removing TCA by extraction with ethyl ether, the glycopeptides are fractionated on a BioGel P-4 column

---

[29] R. Kornfeld and S. Kornfeld, *Annu. Rev. Biochem.* **54**, 631 (1985).

(200–400 mesh), equilibrated in 0.5% (v/v) acetic acid. The entire glyco-peptide peak is pooled, concentrated, and digested with endo-β-N-acetyl-glucosaminidase H (Endo H) (10 to 20 milliunits/ml) in 50 mM citrate buffer, pH 5.5, for 16 hr. When the Glc$_3$Man$_9$(GlcNAc)$_2$ intermediate is needed, the labeled glycopeptides are treated with peptide N-glycosidase F in 50 mM phosphate buffer, pH 8.0, for 12–16 hr. This enzyme releases the oligosaccharide by cleaving the GlcNAc→Asn bond, and gives an oligosaccharide with two GlcNAc residues. The radiolabeled oligosaccha-rides released by either Endo H or by peptide N-glycosidase F are chro-matographed on a long calibrated column of BioGel P-4 (1.5 × 150 cm, 200–400 mesh), equilibrated in 0.5% acetic acid.[30] The column is usually calibrated with a variety of standard oligosaccharides, including Glc$_3$Man$_9$-GlcNAc, Man$_9$GlcNAc, Man$_5$GlcNAc, and mannose.

The labeled oligosaccharides and glycopeptides can also be chromato-graphed on a small column of concanavalin A (ConA)–Sepharose (2-ml gel in a Pasteur pipette) to compare their elution profiles to those of control oligosaccharides. Concanavalin A columns are equilibrated in 20 mM Tris buffer, pH 7.2, containing 200 mM NaCl, 1 mM CaCl$_2$, 1 mM MgCl$_2$, and 1 mM MnCl$_2$. The radiolabeled oligosaccharides are applied to the column in equilibration buffer and the column is washed. The bound oligosaccha-rides are eluted, first by washing the column with 10 mM α-methylgluco-side in the equilibration buffer to remove biantennary and hybrid struc-tures, and then with 250 mM α-methylmannoside to elute the high-mannose oligosaccharides.[30,31]

Preparation of [2-³H]Mannose-Labeled Man$_9$GlcNAc, Man$_5$GlcNAc, and GlcNAc-Man$_5$GlcNAc Intermediates

The oligosaccharide, [³H]Man$_9$GlcNAc is prepared from influenza vi-rus-infected MDCK cells. The details of the procedure are similar to those already described for the preparation of Glc$_3$Man$_9$GlcNAc, except that deoxymannojirimycin (50 μg/ml)[31] or kifunensine (2–5 μg/ml)[22] is added to the infected cells before the incubation with [³H]mannose. After the incubation, pronase digests of virus and cell pellets are prepared as pre-viously described, and then treated with Endo H or peptide N-glycosidase F. The oligosaccharides released by these enzymes are purified on the calibrated columns of BioGel P-4 (200–400 mesh), as described earlier.

[30] G. P. Kaushal, I. Pastuszak, K. Hatanaka, and A. D. Elbein, *J. Biol. Chem.* **265**, 16271 (1990).
[31] T. Szumilo, G. P. Kaushal, H. Hori, and A. D. Elbein, *Plant Physiol.* **81**, 383 (1986).

The radioactive peak corresponding to [³H]Man₉GlcNAc or [³H]Man₉ (GlcNAc)₂ is collected and used for the preparation of other substrates.[32] To prepare the Man₅GlcNAc, the [³H]mannose-labeled Man₉GlcNAc is incubated overnight with partially purified mannosidase I from mung bean microsomes in 50 mM morpholineethanesulfonic acid (MES) buffer (pH 6.0) in the presence of swainsonine (5 μg/ml) to inhibit any arylmannosidases or mannosidase II that might be present in the enzyme preparation. The resulting [³H]Man₅GlcNAc that is produced in the incubation mixture is purified by gel filtration on a 1.5 × 150 cm calibrated column of BioGel P-4 (200–400 mesh). The [³H]mannose-labeled Man₅GlcNAc is then incubated overnight with GlcNAc-transferase I (a solubilized enzyme prepared from mung bean microsomes) in 50 mM MES buffer (pH 6.5) containing 1 mM UDP-GlcNAc and 5 mM MnCl₂ in the presence of swainsonine.[32] Swainsonine is again added to inhibit arylmannosidases and mannosidase II that might be present in the solubilized enzyme preparation. The reaction is stopped by heating in a boiling water bath for 5 min. The turbid suspension is centrifuged and the clear supernatant is applied to a 1.5 × 150-cm calibrated column of BioGel P-4 (200–400 mesh). The radioactive peak corresponding to GlcNAc-Man₅GlcNAc is pooled, concentrated, and stored at −20° until use.

Use of Inhibitors as Affinity Ligands for Purification of Glycosidases

Among the various processing inhibitors, deoxynojirimycin, an inhibitor of glucosidase I, and deoxymannojirimycin, and inhibitor of mannosidase I, have been successfully used as affinity ligands to purify glucosidase I[33] and mannosidase I,[34] respectively. For the preparation of these affinity ligands, the hexanoic acid derivative of deoxynojirimycin or deoxymannojirimycin is made, and this derivative is then immobilized by coupling it to aminohexyl (AH)–Sepharose (Pharmacia-LKB, Piscataway, NJ) or to Affi-Gel 102 (Bio-Rad, Richmond, CA). Briefly, deoxynojirimycin or deoxymannojirimycin solution in 60% aqueous acetone is reacted with a three- to fourfold excess of 6-bromohexanoic acid for 40 hr at 55° with constant stirring. The reaction mixture is then evaporated to dryness, and the contents are redissolved in water and centrifuged. The supernatant liquid is applied to a Dowex 1-×4 column, which is then washed with water. The bound material is eluted with 2 M acetic acid. The fractions eluted with acid are pooled and extracted with ether to remove unreacted bromo-

[32] G. P. Kaushal, T. Szumilo, I. Pastuszak, and A. D. Elbein, *Biochemistry* **29**, 2168 (1990).
[33] H. Kettkamp, G. Legler, and E. Bause, *Eur. J. Biochem.* **142**, 85 (1984).
[34] J. Schweden and E. Bause, *Biochem. J.* **264**, 247 (1989).

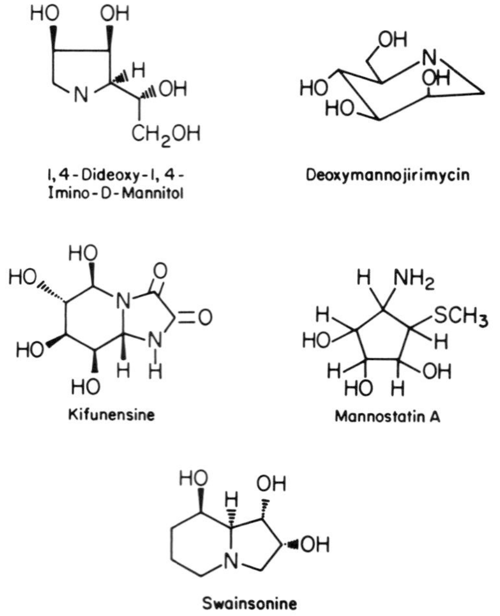

FIG. 3. Structures of various mannosidase inhibitors.

hexanoic acid. The resulting aqueous material is evaporated to dryness. The resulting material containing N-carboxypentyldeoxynojirimycin is dissolved in ethanol and recrystallized from ethanol. N-Carboxypentyl-deoxymannojirimycin is prepared in a similar manner. The purified N-carboxypentyldeoxynojirimycin is then immobilized to a solid matrix by coupling to AH–Sepharose 4B,[33] using the carbodiimide procedure described in the manual of the suppliers (Pharmacia-LKB), or by coupling to Affi-Gel 102 (Bio-Rad).[35] A similar approach can be applied to other inhibitors. For example, such a procedure may be useful for linking kifu-nensine, which is a potent and specific inhibitor of mannosidase I, to a column for the purification of mannosidase I.

Inhibitors as Useful Tools to Distinguish Various Glycosidases

Various mannosidase inhibitors (Fig. 3) have been useful tools to distin-guish individual α-mannosidases. For example, endoplasmic reticulum

[35] K. Shailubhai, M. A. Pratta, and I. K. Vijay, *Biochem. J.* **247,** 555 (1987).
[36] J. Bischoff, L. Liscum, and R. Kornfeld, *J. Biol. Chem.* **261,** 4766 (1986).

(ER)-α-mannosidase[36] is not inhibited by either deoxymannojirimycin or swainsonine, and thus appears to be a different enzyme from mannosidase I and mannosidase II. Swainsonine is an inhibitor of arylmannosidase,[37] lysosomal α-mannosidase,[38] and glycoprotein-processing mannosidase II,[25] but has no activity against $\alpha(1\rightarrow2)$-specific mannosidases such as mannosidase I[31] or ER-α-mannosidase. Moreover, mannostatin A, another potent mannosidase II inhibitor, also inhibits arylmannosidase and lysosomal α-mannosidase.[26] These data imply that mannosidase II shares a common mechanism of catalysis or a similar active site with arylmannosidases, and that this site or mechanism of action is distinct from the mechanism of catalysis of the catalytic intermediate of mannosidase I.

Role of Processing Inhibitors in Glycoprotein Function: Glucosidase I and II Inhibitors

Castanospermine and deoxynojirimycin have been used frequently to study the targeting and function of a various N-linked glycoproteins in cell culture. When either of these inhibitors are administered to the cell culture medium, they prevent glycoprotein processing at the first step in the pathway and cause the formation of N-linked glycoproteins having $Glc_3Man_{7-9}(GlcNAc)_2$ structures.[11,39] Initial studies in the presence of deoxynojirimycin showed that the removal of glucose residues was required for the proper maturation and translocation of specific secretory proteins from the rough endoplasmic reticulum (RER) to the Golgi apparatus.[40] However, in some cases deoxynojirimycin had no effect on transport and surface expression of glycoproteins, as shown with the following systems: vesicular stomatitis virus (VSV) G protein, influenza viral hemagglutinin, human class 1 histocompatibility antigens,[41,42] and the incorrectly processed *erbB* oncogene glycoprotein.[43] The effect of N-methyldeoxynojirimycin and castanospermine on the intracellular migration of glycoprotein E_2 of mouse hepatitis virus was examined.[44] In the presence of these

[37] I. Pastuszak, G. P. Kaushal, K. A. Wall, Y. T. Pan, A. Strum, and A. D. Elbein, *Glycobiology* **1**, 71 (1990).

[38] P. R. Dorling, C. R. Huxtable, and S. M. Colegate, *Biochem. J.* **191**, 649 (1989).

[39] B. Saunier, R. D., Jr., Kilker, J. S. Tkacz, A. Quaroni, and A. Herscovics, *J. Biol. Chem.* **257**, 14155 (1982).

[40] H. F. Lodish and N. Kong, *J. Cell Biol.* **98**, 1720 (1984).

[41] S. Schlesinger, C. Malfer, and M. J. Schlesinger, *J. Biol. Chem.* **259**, 7597 (1984).

[42] B. Burke, K. Matlin, E. Bause, G. Legler, N. Peyrieras, and H. Ploegh, *EMBO J.* **3**, 551 (1984).

[43] J. A. Schmidt, H. Beug, and M. J. Hayman, *EMBO J.* **4**, 105 (1985).

[44] R. Repp, T. Tamura, C. B. Boschek, H. Wege, R. T. Schwarz, and H. Nieman, *J. Biol. Chem.* **260**, 15873 (1985).

processing inhibitors, glycoprotein E_2 was synthesized in normal amounts but it accumulated intracellularly and its transport to the cell surface was greatly delayed.[44] The effect of castanospermine was also examined in 1M-9 lymphocytes to determine whether alteration of carbohydrate structure affects function of the insulin receptor. On treatment of 1M-9 cells with castanospermine a significant decrease in cell surface insulin receptor was observed.[45] Similar observations were made with regard to interleukin 2 (IL-2) receptor.[46] Castanospermine prevented normal processing of the N-linked oligosaccharides of the low-density lipoprotein (LDL) receptor and this alteration resulted in a decrease in the number of receptor molecules at the cell surface.[47] Castanospermine was also shown to inhibit the normal processing of the v-*fms* transforming glycoprotein. In the presence of castanospermine, the v-*fms* transformed cells accumulate altered forms of the glycoprotein that do not reach the cell surface and, thus, the treated cells revert to the normal phenotype.[48] In addition, castanospermine was effective in slowing the growth of v-*fms* transformed cells in the nude mouse, suggesting that this drug may offer an effective therapy against certain tumors.[49]

Both castanospermine and deoxynojirimycin are potent inhibitors of the replication of AIDS virus in tissue culture cells.[50–52] Thus the inhibition of glucosidase I by castanospermine and deoxynojirimycin prevents syncytium formation and inhibits the formation of infectious virus. The decrease in syncytium formation is attributed to the inhibition of processing of the envelope precursor glycoprotein (gp160), with resultant decreased cell surface expression of the mature envelope glycoprotein (gp120). These studies have provided a great impetus in designing more potent and more specific inhibitors that can be useful as chemotherapeutic agents. Thus, various fatty acyl analogs of castanospermine and deoxynojirimycin have been prepared chemically and have been found to be even more potent

[45] R. F. Arakaki, J. A. Hedo, E. Collier, and P. Gorden, *J. Biol. Chem.* **262**, 11886 (1987).

[46] K. A. Wall, J. D. Pierce, and A. D. Elbein, *Proc. Natl. Acad. Sci. U.S.A.* **85**, 5644 (1988).

[47] E. H. Edwards, E. A. Sprague, J. L. Kelley, J. J. Kerbacher, C. J. Schwartz, and A. D. Elbein, *Biochemistry* **28**, 7679 (1989).

[48] A. Hadwiger, H. Niemann, A. Kaebish, H. Bauer, and T. Tomra, *EMBO J.* **5**, 689 (1988).

[49] G. K. Ostrander, N. K. Scribner, and L. R. Rohrschneider, *Cancer Res.* **48**, 1091 (1988).

[50] R. A. Gruters, J. J. Neefjes, M. Tersmette, R. E. Y. de Goede, H. G. Huisman, F. Miedema, and H. L. Ploegh, *Nature* (*London*) **330**, 74 (1987).

[51] B. D. Walker, M. Kowalski, W. C. Goh, K. Kozarsky, M. Kreiger, C. Rosen, L. Rohrschneider, W. A. Haseltine, and J. Sodorski, *Proc. Natl. Acad. Sci. U.S.A.* **84**, 8120 (1987).

[52] D. C. Montefiori, W. E., Robinson, Jr., and W. M. Mitchell, *Proc. Natl. Acad. Sci. U.S.A.* **85**, 9248 (1988).

in inhibiting replication of HIV virus in cell culture.[53,54] These compounds are being tested in clinical trials.

Role of Mannosidase I and II Inhibitors in Glycoprotein Function

A number of compounds that inhibit mannosidase I and II are shown in Fig. 3. Deoxymannojirimycin, the 2-epimer (i.e., mannose analog) of deoxynojirimycin, was synthesized chemically and found to be a specific inhibitor of the Golgi mannosidase I.[20,55] However, it does not inhibit Golgi mannosidase II. The inhibitor causes the accumulation of high-mannose oligosaccharides of the $Man_{8-9}(GlcNAc)_2$ structure in hybridoma cells.[20] However, subsequent studies showed that deoxymannojirimycin caused the accumulation of $Man_9(GlcNAc)_2$ on the glycoproteins of Rous sarcoma virus,[56] influenza virus[21,42] and vesicular stomatitis virus.[42] The inhibitor did not interfere with the biosynthesis or integrity of the Rous sarcoma virus.[56]

The effect of the mannosidase I inhibitor on the role of mannosidase processing in cellular and secreted glycoproteins has been examined in cell culture studies. Deoxymannojirimycin had no effect on the secretion of α_1-proteinase inhibitor and α_1-acid glycoprotein in primary cultures of rat hepatocytes,[57] whereas deoxynojirimycin greatly reduced the rate of secretion of α_1-proteinase inhibitor and α_1-antichymotrypsin by human Hep G2 cells.[40] Deoxymannojirimycin also had no effect on the secretion of the antibodies IgD and IgM in hybridoma cells.[20] On the other hand, deoxynojirimycin greatly reduced the secretion of IgD but not IgM in hybridoma cells.[58] However, some of the effects of deoxymannojirimycin appear to be cell specific. Thus, in human Hep G2 cells, the proteolytic processing of intracellularly retained cathepsin D was retarded, and the amount of cathapsin D that was secreted was increased twofold, whereas in fibroblasts neither secretion nor maturation was affected.[59]

[53] P. S. Sunkara, D. L. Taylor, M. S. Kang, T. L. Bowlin, P. S. Liu, A. S. Tyms, and A. Sjoerdsma, *Lancet* **1,** 1206 (1989).

[54] D. L. Taylor, P. S. Sunkara, P. S. Liu, M. S. Kang, T. L. Bowlin, and A. S. Tyms, *AIDS* **5,** 693 (1991).

[55] G. Legler and E. Julich, *Carbohydr. Res.* **128,** 61 (1984).

[56] J. V. Bosch, A. Tlusty, W. McDowell, G. Legler, and R. T. Schwartz, *Virology* **143,** 342 (1985).

[57] V. Gross, K. Steube, T. A. Trans-Thi, W. McDowell, R. T. Schwartz, K. Decker, W. Gerok, and P. C. Heinrich, *Eur. J. Biochem.* **150,** 41 (1985).

[58] N. Peyrieras, E. Bause, G. Legler, R. Vasilov, L. Claeson, P. Peterson, and H. Ploegh, *EMBO J.* **2,** 823 (1983).

[59] A. Nauerth, P. Lemansky, A. Hasilik, K. von Figura, E. Bause, and G. Legler, *Biol. Chem. Hoppe-Seyler* **366,** 1009 (1985).

Another Golgi mannosidase I inhibitor is kifunensine. This inhibitor is an alkaloid produced by the fungus, *Kitasatosporia kifunense,* and resembles the cyclic oxamide derivative of 1-aminodeoxymannojirimycin in structure.[60] It has been demonstrated that this compound is a specific and potent inhibitor of the glycoprotein-processing enzyme, mannosidase I, and at a concentration of 1 μg/ml in cultured mammalian cells it causes a complete shift in the structure of the N-linked oligosaccharides from complex types to Man$_9$(GlcNAc)$_2$ structures.[22,61] On the other hand, even at 50 μg/ml, deoxymannojirimycin did not prevent the formation of all complex chains. Also, kifunensine was found to be 50 to 100 times more effective than deoxymannojirimycin in inhibiting purified mannosidase I.[22] The role of mannosidase processing on the scavenger LDL receptor was studied in aortic endothelial cells and fibroblasts, using kifunensine to inhibit mannose removal. When endothelial cells were grown in the presence of kifunensine (1 μg/ml), there was a 75% inhibition in the ability of these cells to degrade ^{125}I-labeled acetyl-LDL. On the other hand, kifunensine had little or no effect on the binding or degradation of ^{125}I-labeled LDL in fibroblasts.[61]

Swainsonine was the first compound found to inhibit glycoprotein processing.[28] The historical background of this inhibitor has been described previously.[3] Swainsonine was initially isolated from the toxic Australian plant, *Swainsoma caneseans,* and has since been found in various species of locoweed in this country.[62] This compound is a potent inhibitor of the Golgi mannosidase II, but has no effect on mannosidase I.[63] This compound is also a potent inhibitor of aryl-α-mannosidase, suggesting that mannosidase II and aryl-α-mannosidase share a common mechanism of catalysis that is different from the mechanism of catalysis of mannosidase I. In cell culture, swainsonine caused the inhibition of complex types of structures, which resulted in the accumulation of hybrid types of structures.[28] The change in carbohydrate types did not affect the function of a particular protein.[3] However, swainsonine has been shown to inhibit experimental as well as spontaneous metastasis.[64,65] Thus, when

[60] H. Kayakiri, S. Takese, T. Shibata, M. Okamoto, H. Terano, M. Hashimoto, T. Tada, and S. Koda, *J. Org. Chem.* **54,** 4015 (1989).

[61] A. D. Elbein, J. K. Kerbacher, C. J. Schwartz, and E. A. Sprague, *Arch. Biochem. Biophys.* **288,** 177 (1991).

[62] L. F. James, A. D. Elbein, R. J. Molyneux, and C. D. Warren, "Swainsonine and Related Glycosidase Inhibitors," Iowa State Press, Ames, 1989.

[63] D. R. P. Tulsiani, H. P. Broquist, and O. Touster, *Arch. Biochem. Biophys.* **236,** 427 (1985).

[64] M. J. Humphries, K. Matsumoto, S. L. White, R. J. Molyneux, and K. Olden, *Cancer Res.* **46,** 5215 (1988).

[65] J. W. Dennis, *Cancer Res.* **46,** 5131 (1986).

B16–F10 melanoma cells were grown in the presence of swainsonine and injected into mice, they showed reduced organ colonization as compared to the untreated B16–F10 melanoma cells. Also, the growth of human tumor xenograft in athymic nude mice was reduced by swainsonine treatment.[66]

Another mannosidase inhibitor is mannostatin A. The structure of this inhibitor is also shown in Fig. 3. Mannostatin A is a metabolite produced by the microorganism *S. verticillus,* which was reported to be a potent, competitive inhibitor of rat epididymal α-mannosidase.[27] Like swainsonine, mannostatin A is an effective competitive inhibitor of aryl-α-mannosidase, lysosomal α-mannosidase, and Golgi mannosidase II. This inhibitor is without effect on mannosidase I. In influenza virus-infected MDCK cells, mannostatin A blocked the normal formation of complex types of oligosaccharides on the viral glycoproteins and caused the accumulation of hybrid types of oligosaccharides.[26] As yet, this inhibitor has not been tested to study function of a particular glycoprotein in cell culture.

Future Prospects

Glycosidases are widespread enzymes that play vital roles in catabolism and turnover of many cellular components such as glycoprotein, polysaccharides, glycolipids, and proteoglycans. Inhibitors of these enzymes could serve as potentially beneficial compounds for therapeutic purposes. They could be useful as antihyperglycemic agents, inhibitors of tumor metastasis, antiviral agents, and antifungal agents. Although several inhibitors have been identified for the N-linked glycoprotein-processing enzymes, considerably less work has been done to identify inhibitors for the glycosidases involved in the metabolism of other glycoconjugates such as O-linked glycoproteins, glycolipids, polysaccharides, and proteoglycans.

Although many potent and selective inhibitors of glycoprotein processing enzymes are now available, they generally suffer from a lack of specificity. For example, castanospermine, a potent inhibitor of glucosidase I and II, also inhibits sucrase, maltase, β-glucosidase, and lysosomal α-glucosidase. Thus, when given to animals, it may cause many diverse effects. To avoid these effects, more specific and potent inhibitors need to be designed or isolated from natural sources. Such specific inhibitors will not only be useful in determining the role of carbohydrates in glycoprotein function and their mechanism of action but will also serve as useful chemotherapeutic agents.

[66] J. W. Dennis, K. Koch, S. Yousefi, and I. Vander Elst, *Cancer Res.* **50,** 1867 (1990).

[20] Synthesis and Uses of Azido-Substituted Nucleoside Diphosphate Sugar Photoaffinity Analogs

By ANNA RADOMINSKA and RICHARD R. DRAKE

Introduction

This chapter presents an overview of the synthesis and glycobiological uses of two azido-substituted nucleoside diphosphate sugar photoaffinity analogs, 5-azido-UDPglucose ($5N_3UDPGlc$) and 5-azido-UDPglucuronic acid ($5N_3UDPGlcA$). In previous volumes of this series, the synthesis and uses of 8-azido-ATP and 8-azido-GTP analogs have been reviewed.[1-3] These reviews and a review of the uses of 5-azido-UDPGlc[4] extensively cover general photoaffinity theory and applications. In this chapter, therefore, the enzymatic and chemical synthesis of the nucleoside diphosphate sugar analogs and their use in studying membrane-associated glycosyltransferases will be presented.

In general, membrane-associated glycosyltransferases are difficult to purify and characterize because of their inherent instability after solubilization from their native membrane environment.[5] Advantages in using the nucleoside diphosphate sugar photoaffinity analogs to study glycosyltransferases are that the enzymes do not have to be purified to be characterized and they can be studied in their native membrane environment. In addition to the synthetic protocols, different techniques using the photoaffinity analogs for characterizing membrane-associated glycosyltransferases will be presented. These include their use in identifying the catalytic subunit(s) of glycosyltransferases, monitoring of glycosyltransferase purification, and analyzing the active-site orientation and access of substrates to the glycosyltransferases in their native membranes.

Synthesis of $[^{32}P]5N_3UDPGlc$ and $[^{32}P]5N_3UDPGlcA$

Materials

Enzymes, nucleotides, and chromatography resins are generally purchased from Sigma (St. Louis, MO). All organic reagents are purchased

[1] J. Czarnecki, R. Geahlen, and B. Haley, this series, Vol. 56, p. 642.
[2] R. Potter and B. E. Haley, this series, Vol. 91, p. 613.
[3] B. E. Haley, this series, Vol. 200, p. 477.
[4] R. R. Drake and A. D. Elbein, *Glycobiology* **2**, 279 (1992).
[5] A. D. Elbein and G. P. Kaushal, *Methods Plant Biochem.* **2**, 79 (1990).

METHODS IN ENZYMOLOGY, VOL. 230

from Aldrich (Milwaukee, WI). [32]P is purchased from ICN Radiochemicals (Costa Mesa, CA). Plastic-backed, cellulose thin-layer chromatography (TLC) sheets are from Eastman Kodak (Rochester, NY), and polyethyleneimine (PEI)-cellulose plates are from EM Science. It is recommended that UDPglucose pyrophosphorylase purified from yeast (Sigma) and ammonium bicarbonate (Fisher, Pittsburgh, PA) be used for best results.

Synthesis of $5N_3UMP$

Because the reactions that produce the 5-azido-UDP sugars have thus far been enzymatic, the chemical syntheses of the precursors $5N_3UMP$ and $5N_3UTP$ are the critical steps in these procedures. These reactions can be done in any laboratory equipped with an ultraviolet (UV) spectrophotometer, rotary evaporator, and liquid chromatography systems. $5N_3UMP$ is synthesized as described for $5N_3dUMP$ by Evans and Haley,[6] except for the following modifications. After terminating the nitrosonium tetrafluoroborate nitration reaction of UMP[6] (typically 0.5–1.0 mmol) with the addition of 1 ml of H_2O, 3 ml of 1 N HCl is added to the reaction, which is then evaporated under reduced pressure to a 1-ml final volume. After extraction with diethyl ether, the aqueous solution of $5NO_2UMP$ is then used directly in the zinc reduction procedure. Additional purification of this compound is usually not necessary, and it can be stored at $-20°$. For all synthetic procedures described herein that require water, the purest water (i.e., deionized) should be used.

Reduction of $5NO_2UMP$ to $5NH_2UMP$ and the subsequent isolation of product has been observed to be the most critical step in this synthetic protocol. Prior to reduction, briefly wash 0.5–1 g of granular zinc in 6 N HCl. Remove the acid and add 0.1–0.3 mmol of $5NO_2UMP$ in 5 ml of water. Reduction of $5NO_2UMP$ to $5NH_2UMP$ is monitored by the shifts in the characteristic UV absorbances of both compouds. At pH 1, $5NO_2UMP$ has a λ_{max} at 302 nm and $5NH_2UMP$ has a λ_{max} of 265 nm.[6] On completion (usually 10–15 min), the reaction solution is separated from the zinc by pipette and evaporated under reduced pressure to 1 ml. This is applied to a Dowex 50W H^+ column (1 × 26 cm) (Bio-Rad, Richmond, CA) equilibrated in water. The products are eluted in water in the following order: unreacted $5NO_2UMP$, uncharacterized acidic compound(s) of λ_{max} 264 nm, and the free acid form of $5NH_2UMP$, λ_{max} 290 nm (pH 7). Recovery ratios of these three compounds in the above order of elution are routinely 1:3:6. Only the free acid form of $5NH_2UMP$ is useful for the subsequent azide-exchange reaction. To verify that the

[6] R. K. Evans and B. E. Haley, *Biochemistry* **26**, 269 (1987).

desired product is the free acid of $5NH_2UMP$, acidify a small aliquot, run a UV spectra, and observe any shift in peak absorbance from 290 to 265 nm. The free acid $5NH_2UMP$ fractions are pooled and evaporated under reduced pressure to dryness and then resuspended in 3 ml of 1 N HCl. This solution of $5NH_2UMP$, generally 0.1–0.2 mmol, is then used directly in the diazotization and azide-exchange reactions (performed under a hood) described by Evans and Haley.[6]

To purify $5N_3UMP$, the solution from the exchange reaction is concentrated under reduced pressure to dryness, and resuspended in 150 ml of 10 mM NH_4HCO_3. The pH of this solution is adjusted to above pH 7 by addition of dilute NH_4OH and applied to a DEAE-cellulose column (2 × 10 cm) equilibrated in 10 mM NH_4HCO_3. $5N_3UMP$ can be eluted isocratically with 200 mM NH_4HCO_3 or by gradient elution (10–400 mM NH_4. HCO_3, 400-ml total). Fractions corresponding to $5N_3UMP$, as determined by the characteristic UV absorbance at 288 nm, are pooled in a round-bottom flask and the solution is evaporated to less than 5 ml under reduced pressure. Coevaporation to dryness with 8 ml of triethylamine and repeated additions of 5 ml of methanol are done to remove excess NH_4HCO_3. The sample is resuspended in a final volume of 5 ml of methanol and stored at $-20°$ as the triethylamine salt. The concentration of $5N_3UMP$ is determined by its absorbance at 288 nm, using a molar extinction coefficient of 7600 $M^{-1}cm^{-1}$.[6] To check for photoreactivity, irradiate a small sample of $5N_3UMP$ with UV light, run a UV spectrum, and note any loss of absorbance at 288 nm. Generally, yields of 0.2 mmol of $5N_3UMP$ from 0.5 mmol of UMP can be obtained.

Synthesis of $5N_3UTP$

To synthesize $5N_3UTP$ from $5N_3UMP$ and pyrophosphate, the diphenyl chlorophosphate-coupling procedure of Michelson[7] and a carbonyldiimidazole-coupling procedure of Weckbecker and Keppler[8] can be used. For both reactions, the tributylamine salt of $5N_3UMP$ is prepared by passing 0.1 mmol in 1 ml of water through a 0.5 × 3 cm Dowex 50W H^+ column equilibrated in water. The eluate is collected in a 50-ml round-bottom flask containing 0.3 ml of tributylamine. After removal of water under reduced pressure and repeated drying with 3-ml aliquots of anhydrous methanol, this tributylamine salt of $5N_3UMP$ is resuspended in anhydrous dimethylformamide and used in the coupling procedures. Which coupling procedure to choose depends on the previous experience of the researchers; for us the carbonyldiimidazole method has produced

[7] A. M. Michelson, *Biochim. Biophys. Acta* **91**, 1 (1964).
[8] G. Weckbecker and D. O. R. Keppler, *Biochem. Pharmacol.* **33**, 2291 (1984).

greater yields and is generally easier to perform. The critical aspect of both of these reactions is to use the most anhydrous chemicals available. To purify $5N_3$UTP, the reaction products are diluted in 150 ml of 10 mM NH_4HCO_3, adjusted to greater than pH 7.0, applied to a DEAE–cellulose column (1 $\times$ 15 cm), and eluted by a 400-ml, 10–500 mM NH_4HCO_3 gradient. $5N_3$UTP elutes around 400 mM NH_4HCO_3 and is concentrated and stored as described above for $5N_3$UMP. The purified $5N_3$UTP should have a characteristic UV absorbance of 288 nm[6] and migrate slower than UTP on PEI-cellulose TLC plates developed in 0.4 mM NH_4HCO_3 (R_f values of 0.32 for UTP and 0.25 for $5N_3$UTP).

Synthesis of [β-³²P]5N₃UDPGlc

[β-^{32}P]$5N_3$UDPGlc can be synthesized by the enzymatic coupling of 5-azido-UTP and [^{32}P]Glc-1-P by yeast UDPGlc pyrophosphorylase[9,10]: In a final volume of 0.15 ml and final concentrations as indicated, mix 3 units (approximately 1 mg) of sucrose phosphorylase, 50 mM Tris (pH 8.5), 5 mM $MgCl_2$, 3 μmol of sucrose, 0.5 μmol of sodium phosphate, and 5–10 mCi of ^{32}P$_i$ and allow to react for 15 min. $5N_3$UTP (0.5 μmol) is then dried under nitrogen in a separate tube. To this tube is added the following: the [^{32}P]Glc-1-P reaction mixture, 100 units of UDPglucose pyrophosphorylase, 20 units of inorganic pyrophosphatase, and 5 mM $MgCl_2$ to a final volume of 0.27 ml. The mixture is allowed to stand for 1 hr at room temperature. Both of these enzymes (100 units each) are dissolved in 0.1 ml of 50 mM Tris (pH 8.5) and 10% (v/v) glycerol. The UDPglucose pyrophosphorylase is not stable after 1 day of resuspension. Inorganic pyrophosphatase can be stored at $-20°$ for future use. The production of [^{32}P]Glc-1-P and [β-^{32}P]$5N_3$UDPGlc can be monitored by TLC on cellulose plates developed in methanol–88% formic acid–water (80 : 15 : 5, v/v).[11] [β-^{32}P]$5N_3$UDPGlc is purified as previously described,[10] and then concentrated and stored as described for $5N_3$UMP.

Synthesis of [β-³²P]5N₃UDPGlcA

[β-^{32}P]$5N_3$UDPGlcA can be synthesized by two similar procedures. The original procedure is to mix 0.1–0.2 μmol of [β-^{32}P]$5N_3$UDPGlc with 0.1 mg of UDPGlc dehydrogenase, 2 μmol of NAD, 3 mM $MgCl_2$ and

[9] R. R. Drake, Jr., R. K. Evans, M. J. Wolf, and B. E. Haley, *J. Biol. Chem.* **264,** 11928 (1989).

[10] R. R. Drake, G. P. Kaushal, I. Pastuszak, and A. D. Elbein, *Plant Physiol.* **97,** 396 (1991).

[11] R. S. Bandurski and B. Axelrod, *J. Biol. Chem.* **193,** 405 (1951).

25 mM glycine-NaOH (pH 8.7) in 0.13 ml.[12] UDPGlc dehydrogenase is dissolved in glycine buffer (1 mg/ml) prior to initiation of the reaction. An alternative approach is to combine this reaction with the synthesis of [β-^{32}P]5N$_3$UDPGlc in the following one-reaction protocol. The reactions to prepare [β-^{32}P]5N$_3$UDPGlc are initiated as described earlier, followed immediately by addition of UDPGlc dehydrogenase and NAD (as prepared earlier). After 30 min, 1 unit of alkaline phosphatase is added to the reaction for 1 min to hydrolyze unreacted 5N$_3$UTP. This reaction product is then applied to a DEAE–cellulose column and the products purified as previously described.[12] Both photoprobes are recovered in approximately 30% yields, based on radioincorporation of ^{32}P. An advantage to this procedure is that both photoprobes can be made in one reaction. A potential disadvantage is that the [β-^{32}P]5N$_3$UDPGlc is difficult to separate from NAD during purification. However, thus far the presence of NAD has had no apparent effect on photoaffinity labeling studies using [β-^{32}P]5N$_3$UDPGlc prepared by this method.[13]

General Photoaffinity Labeling Procedures

The following is an outline of the procedures used specifically with [β-^{32}P]5N$_3$UDPGlc and [β-^{32}P]5N$_3$UDPGlcA to photolabel membrane-associated glycosyltransferases. Additional considerations in the use of nucleotide photoaffinity analogs have been presented by Haley.[3] The basic photolabeling reaction consists of incubating the photoprobe with a enzyme preparation in a microcentrifuge tube for 10 sec, followed by irradiation for 90 sec with a hand-held UV lamp (UVP-11; Ultraviolet Products, Inc.). In general, the buffers and reaction conditions used to assay the enzyme of interest can be used for initial photoaffinity experiments. The reactions are terminated by addition of an equal volume of cold 10% (v/v) trichloroacetic acid (TCA) and the enzyme(s) precipitated by centrifugation in a standard microcentrifuge. After careful removal of TCA, precipitated enzyme pellets (sometimes not visible) are resuspended in a protein-solubilizing mixture containing 4 M urea, 20 mM dithiothreitol, 100 mM Tris (pH 8.0), 4% (w/v) sodium dodecyl sulfate (SDS), and bromphenol blue. Frequently, pellets derived from microsomal or detergent-solubilized preparations require sonication in a water bath to achieve resuspension. The resuspended protein samples are separated on an SDS–polyacrylamide gel, followed by conventional staining/destaining, drying, and autoradiog-

---

[12] R. R. Drake, P. Zimniak, B. E. Haley, R. Lester, A. D. Elbein, and A. Radominska, *J. Biol. Chem.* **266**, 23527 (1991).

[13] R. R. Drake and A. Radominska, unpublished observations.

raphy. The relative amount of photolabeling of a particular protein can be determined by densitometric scanning of the autoradiograph.

Identification of Glycosyltransferase Catalytic Subunits

One area in which both $[\beta\text{-}^{32}P]5N_3UDPGlc$ and $[\beta\text{-}^{32}P]5N_3UDPGlcA$ have found great utility has been in the identification and characterization of the catalytic subunits of membrane-associated glycosyltransferases. One of the most effective approaches has been to correlate the effects specific activators or competitive inhibitors of enzyme activity have on the degree of photoincorporation. Examples include the identification of the catalytic subunits of 1,3-β-glucan synthase in red beet membranes[14] and 1,4-β-glucan synthase from *Acetobacter xylinum*[15] by photoincorporation of $[\beta\text{-}^{32}P]5N_3UDPGlc$. In both of these systems, inclusion of known activators (cellobiose for 1,3-β-glucan and cyclic diguanylic acid for (1,4-β-glucan) in the respective photolabeling reactions resulted in enhanced photoincorporation of $[\beta\text{-}^{32}P]5N_3UDPGlc$ into the catalytic subunits.[14,15] The use of effector molecules is a rapid way to identify potential subunits. To aid in the verification of the catalytic subunit, other enzyme properties such as pH optima or metal ion requirements can be manipulated and compared to the degree of photoinsertion. Once a particular subunit is identified, the specificity of this photoincorporation can be demonstrated by saturation of the available sites and inhibition by the natural substrate (i.e., UDPGlc or UDPGlcA).[10,12] In general, optimal conditions for enzyme activity correspond to optimal amounts of photoincorporation. Also, for successfull photoaffinity labeling of membrane-associated glycosyltransferases, the enzyme of interest must be active. Thus, correlation of the degree of enzyme activity with the amount of photoincorporation can also be used to identify potential subunits.

Another related approach for identifying glycosyltransferase catalytic subunits by photoaffinity labeling has been accomplished by comparing the photolabeled proteins in wild-type cells versus those in mutant cells deficient for a particular glycosylating activity. This approach was used with $[\beta\text{-}^{32}P]5N_3UDPGlc$ to identify dolichyl-phosphate β-glucosyltransferase (Glc-P-Dol synthase) in *Saccharomyces cerevisiae*[16] by photolabeling membrane extracts prepared from either a Glc-P-Dol synthase-deficient

---

[14] D. J. Frost, S. M. Read, R. R. Drake, B. E. Haley, and B. P. Wasserman, *J. Biol. Chem.* **265**, 2162 (1990).

[15] F. C. Lin, R. M. Brown, Jr., R. R. Drake, and B. E. Haley, *J. Biol. Chem.* **265**, 4782 (1990).

[16] G. Palamarcyzk, R. R. Drake, B. E. Haley, and W. J. Lennarz, *Proc. Natl. Acad. Sci. U.S.A.* **87**, 2666 (1990).

mutant, *alg5*,[17] or from a mutant with decreased Glc-P-Dol synthase activity, *dpg1*.[18] [β-^{32}P]5N$_3$UDPGlcA has been used to identify multiple subunits of a streptococcal hyaluronate synthase complex by comparison of wild-type strains with strains deficient in hyaluronic acid synthesis.[19] Similar to the use of specific activators or inhibitors, comparison of the photolabeling patterns in wild-type and mutant cells is a rapid way of identifying potential subunits. Again, it is best to correlate other known properties of the wild-type enzyme with the degree of photoincorporation to verify the identity of the enzyme.

Photoaffinity Labeling and Purification of Glycosyltransferases

Mung beans and pig liver have served as model systems for the use of [β-^{32}P]5N$_3$UDPGlc to aid in monitoring the purification of Glc-P-Dol synthase.[4] When used in conjunction with conventional protein purification techniques, photoincorporation of [β-^{32}P]5N$_3$UDPGlc into a progressively enriched 39-kDa protein in mung bean seedlings allowed identification of this protein as Glc-P-Dol synthase.[10] By using a similar approach, a 37-kDa protein has been identified as Glc-P-Dol synthase in pig liver.[4] An example of the combination of purification and photoaffinity techniques is shown in Fig. 1 for pig liver Glc-P-Dol synthase. The Coomassie-stained profiles of microsomal, TritonX-100-solubilized, DE-52, and preparative gel (described below) preparations (Fig. 1, lanes 1–4) photolabeled with [β-^{32}P]5N$_3$UDPGlc are compared to the corresponding autoradiograph (Fig. 1, lanes 5–8). In both the Coomassie and photolabeling profiles, a protein of 37 kDa (Glc-P-Dol synthase) was enriched with purification.

Photoaffinity Labeling and Preparative Gel Electrophoresis

Because photoincorporation of the radiolabeled analog is covalent in nature, this allows the use of purification techniques not normally useful for membrane-associated glycosyltransferases. A new apparatus for preparative gel electrophoresis (from Bio-Rad) that allows simultaneous elution of separated proteins has been used in conjunction with photolabeling and purification of Glc-P-Dol synthase from pig liver. By photolabeling the partially purified enzyme preparation prior to electrophoresis, eluted fractions containing Glc-P-Dol synthase can be isolated by the photoincorporated radioactivity.[4] Figure 1 also illustrates this approach in lanes 4

[17] K. W. Runge, T. C. Huffaker, and P. W. Robbins, *J. Biol. Chem.* **259**, 412 (1984).

[18] L. Ballou, P. Gopal, B. Krummel, H. Tammi, and C. E. Ballou, *Proc. Natl. Acad. Sci. U.S.A.* **83**, 3081 (1986).

[19] I. van de Rijn and R. R. Drake, *J. Biol. Chem.* **267**, 24302–24306.

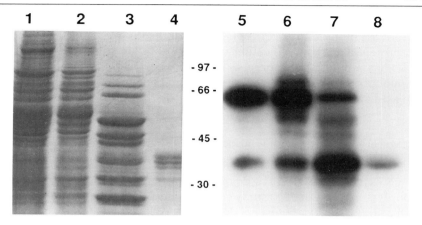

FIG. 1. Photoaffinity labeling of partially purified Glc-P-Dol synthase. Pig liver Glc-P-Dol synthase preparations of increasing purification were photolabeled with 50 μM [β-^{32}P]5N$_3$UDPGlc and separated on a 10% SDS–polyacrylamide gel. The Coomassie-stained profile of this gel is shown in lanes 1–4 and the resulting autoradiograph in lanes 5–8. Lanes 1 and 5, microsomes; lanes 2 and 6, Triton X-100 solubilized; lanes 3 and 7, DE-52; lanes 4 and 8, preparative gel (see details in text).

and 8. A general protocol using this preparative gel unit is as follows. After determining the apparent molecular mass of the protein of interest on SDS–polyacrylamide gels by one of the photoaffinity methods, choose a percentage of acrylamide that gives optimal separation of this protein and neighboring proteins. The length and size of the gel are dependent on the amount of sample to be loaded and can be determined as described in the manufacturer (Bio-Rad) instructions. In our laboratory, up to 100 mg of total protein has been applied to a preparative SDS–polyacrylamide gel.

For the sample in Fig. 1, 30 mg of DE-52-purified pig liver Glc-P-Dol synthase, 2 mg of which had been photolabeled with 50 μM [β-^{32}P]5N$_3$UDPGlc, was separated on a circular 0.75 × 5.5 cm, 14% (w/v) SDS–polyacrylamide gel and eluted at a flow rate of 0.8 ml/min. Prestained protein standards of known molecular mass were included in the sample to monitor the progress of the separation and also to allow discriminant fraction collection of particular protein masses. In this system, the elution buffer is separate and can be different in composition from the electrophoresis buffer. We recommend using an elution buffer lacking SDS to minimize the amount of SDS present in the eluted protein fractions. Representative fractions (approximately 1 ml each) were concentrated to 0.1 ml in Centricon 30 microconcentrator tubes (Amicon, Danvers, MA). The fractions containing the protein of interest can be determined by reseparating a small aliquot of each concentrated sample on a 10% mini-SDS gel. Autoradiography of this gel combined with the known molecular mass of the protein allows identification of the appropriate fractions. This technique has proved useful for isolating highly enriched Glc-P-Dol synthase in amounts sufficient for use in peptide sequencing studies and/or antibody preparation.

Active-Site Topology and Transport Studies

Photoaffinity labeling of intact rat liver microsomal vesicles with $[\beta\text{-}^{32}P]5N_3UDPGlc$ and $[\beta\text{-}^{32}P]5N_3UDPGlcA$ has been combined with procedures commonly used to study glycosyltransferase active-site topologies such as the effects of detergent disruption, protease treatment, and anion transport inhibitors.[20] In these rat liver vesicles, it was observed that both $[\beta\text{-}^{32}P]5N_3UDPGlc$ and $[\beta\text{-}^{32}P]5N_3UDPGlcA$ could be used to photolabel Glc-P-Dol synthase (37 kDa) and the UDPglucuronosyltransferase (UGT) family of enzymes (50–56 kDa).[12,20] A representative photolabeling experiment of intact and detergent-disrupted rat liver vesicles with both photoprobes is shown in Fig. 2. The preparation of intact rat liver microsomes and determination of their mannose 6-phosphatase latency were done as previously described.[20] For preparation of detergent-disrupted vesicles, 0.1 mg of intact vesicles was preincubated in 0.07% (v/v) Triton X-100 for 10 min at room temperature. For photolabeling reactions, 0.1 mg of intact or disrupted vesicles was incubated with 1 40 μM concentration of either photoprobe and 5 mM MgCl$_2$ for 10 sec and UV irradiated for 90 sec. As previously described[20] and shown in Fig. 2, photoincorporation into Glc-P-Dol synthase by both photoprobes was the same in intact or detergent-disrupted vesicles (Fig. 2, lanes 1–6). Photolabeling of rat liver UGTs with $[\beta\text{-}^{32}P]5N_3UDPGlc$ was dependent on detergent disruption of the vesicles (Fig. 2, lanes 2 and 4). Also, photolabeling of a 150-kDa protein only in detergent-disrupted vesicles in the presence of 5 mM calcium was observed (Fig. 2, lane 4). This protein may possibly be the lumenal UDPglucose: glycoprotein glucosyltransferase described by Trombetta and Parodi.[21]

The photolabeling of UGTs with $[\beta\text{-}^{32}P]5N_3UDPGlc$ (Fig. 2, lanes 5–10) was similar in either intact (Fig. 2, lane 5) or disrupted (Fig. 2, lane 6) vesicles. The inclusion of 400 μM UDPGlcA in these reactions competitively inhibited the photolabeling of the UGTs (Fig. 2, lanes 7 and 8).

[20] R. R. Drake, Y. Igari, A. D. Elbein, R. Lester, and A. Radominska, *J. Biol. Chem.* **267**, 11360 (1992).
[21] S. E. Trombetta and A. J. Parodi, *J. Biol. Chem.* **267**, 9236 (1992).

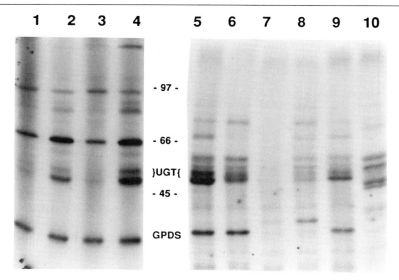

FIG. 2. Photolabeling of intact and detergent-disrupted rat liver vesicles. [β-^{32}P]5N$_3$UDP-Glc (lanes 1–4) and [β-^{32}P]5N$_3$UDPGlcA (lanes 5–10), 40 μM each, were used to photolabel intact (odd-numbered lanes) or 0.07% Triton X-100 disrupted (even-numbered lanes) vesicles (100 μg). Besides 5 mM MgCl$_2$ in each reaction, the following were added as indicated: lanes 3 and 4, 5 mM CaCl$_2$; lanes 7 and 8, 400 μM UDPGlcA; lanes 9 and 10, 100 μg of trypsin (as described in text).

Intact (Fig. 2, lane 9) or disrupted (Fig. 2, lane 10) vesicles photolabeled with [β-^{32}P]5N$_3$UDPGlcA as above were treated with 100 μg of trypsin for 15 min at room temperature prior to acid precipitation. Because the photolabeled UGTs in intact vesicles were not susceptible to trypsin digestion, it has been hypothesized that transport of [β-^{32}P]5N$_3$UDPGlcA into the vesicles was occurring.[20] Subsequent experiments have shown that [β-^{32}P]5N$_3$UDPGlcA photolabeling of the UGTs in intact vesicles was temperature, time, and concentration dependent.[13] These results are consistent with [β-^{32}P]5N$_3$UDPGlcA being transported into the vesicles for access to lumenally oriented UGTs. The use of the photoaffinity analogs to identify and characterize the components of this transport process and studies using cloned and expressed UGTs are currently in progress.

[21] Glycoform Analysis of Glycoproteins

By R. B. PAREKH

Introduction

For a particular glycoprotein, each potentially N-glycosylated asparagine may be fully or partially occupied or unoccupied.[1,2] The structural diversity of O-linked oligosaccharides is comparable to the structural diversity of N-linked oligosaccharides, and the phenomena of site microheterogeneity and variable site occupancy are also common for O-glycosylated serine and threonine residues. Further, reports[3-8] indicate that the attachment of carbohydrate to serine and threonine residues may be even more diverse than hitherto considered. In addition to O-glycosylation through N-acetyl-D-galactosamine[3] and D-mannose,[4] attachment through D-glucose,[5] N-acetyl-D-glucosamine,[6] and L-fucose[7,8] have all been reported on several eukaryotic glycoproteins.

The consequence of microheterogeneity, multiple sites of N- and O-glycosylation along a single polypeptide, and differential occupancy of different sites is that a eukaryotic glycoprotein is not isolated as a single structural entity, but rather as a set of glycosylated variants of a common polypeptide. These glycosylation variants, or simply glycoforms,[9] differ in the number, location, and/or incidence of different oligosaccharide structures. As a minimum, two glycoforms will differ either by carrying oligosaccharides having different structures on at least one glycosylated

[1] A. Kobata, in "Biology of Carbohydrates" (V. Ginsburg and P. W. Robbins, eds.), Vol. 2, p. 87. Wiley, New York, 1984.

[2] G. Pohl, M. Kallström, N. Bergsdorf, P. Wallen, and H. Jornvall, *Biochemistry* **23,** 3701 (1984).

[3] H. Schachter and I. Brockhausen, *Symp. Soc. Exp. Biol.* **43,** 1 (1989).

[4] P. Orlean, H. Schwaiger, U. Appeltauer, A. Haselbeck, and W. Tanner, *Eur. J. Biochem.* **140,** 183 (1984).

[5] S. Hase, H. Nishimura, S. Kawabata, S. Iwanaga, and T. Ikenaka, *J. Biol. Chem.* **265,** 1858 (1990).

[6] G. W. Hart, R. S. Haltiwanger, G. D. Holt, and W. G. Kelly, *Annu. Rev. Biochem.* **58,** 841 (1989).

[7] S. Bjoern, D. C. Foster, L. Thim, F. C. Wiberg, M. Christensen, Y. Komiyama, A. H. Pedersen, and W. Kisiel, *J. Biol. Chem.* **266,**11051 (1991).

[8] H. Nishimura, T. Takao, S. Hase, Y. Shimonishi, and S. Iwanaga, *J. Biol. Chem.* **267,** 17520 (1992).

[9] R. B. Parekh, A. G. D. Tse, R. A. Dwek, A. F. Williams, and T. W. Rademacher, *EMBO J.* **6,** 1233 (1987).

METHODS IN ENZYMOLOGY, VOL. 230

site, or in the relative molar distribution of a common set of oligosaccharide structures associated with at least one glycosylated site. Analysis of the glycosylation of a protein is therefore the analysis of a population of glycoforms.[9,10]

There are a growing number of examples in which the *in vitro* and/or *in vivo* properties of a protein are affected either by changes in occupancy of glycosylated sites and/or by the precise oligosaccharide structures attached to particular sites.[11] Although it is not argued that each and every glycoform is in some sense functionally unique, particular glycoforms of certain proteins can clearly exhibit different biological properties.[12] That is, different glycoforms can exhibit different biological properties. The number and range of examples in which this is so is now sufficient to make it no longer an automatic assumption that glycoforms are functionally equivalent. Furthermore, given the significant contribution of oligosaccharides to the three-dimensional surface of a glycoprotein, and already in a few examples to the conformation of the attached protein, it is postulated that different glycoforms can present different three-dimensional surfaces, and may therefore differ in their ability to engage in a variety of topological interactions.[12]

With current methodologies, a glycoprotein cannot readily be fractionated into its major individual glycoforms by chromatographic, electrophoretic, or mass spectrometric methods. This is due to the number of individual glycoforms constituting most glycoproteins, and the relatively small differences in the physical and chemical properties between many of these. In practice therefore, glycosylation analysis is usually performed on a mixture of glycoforms.

Analytical strategies to perform glycoform analysis are now well defined and involve the following general steps.

1. a determination of the primary amino acid sequence of the protein; this is obviously a prerequisite for subsequent assignment of oligosaccharides to specific glycosylated sites along the polypeptide

2. some characterization of the total pool of oligosaccharides associated with the intact glycoprotein; this is optional, but useful to check the experimental accuracy of site-specific analysis (step 8)

3. identification of those amino acid residues that carry N- or O-linked oligosaccharides

---

[10] R. B. Parekh, R. A. Dwek, J. R. Thomas, T. W. Rademacher, G. Opdenakker, A. J. Wittwer, S. C. Howard, R. Nelson, N. R. Siegel, M. G. Jennings, N. H. Harakas, and J. Feder, *Biochemistry* **28**, 7644 (1989).

[11] T. W. Rademacher, R. B. Parekh, and R. A. Dwek, *Annu. Rev. Biochem.* **57**, 785 (1988).

[12] D. A. Cumming, *Glycobiology* **1**, 115 (1991).

4. a separation of glycosylated sites into glycopeptides carrying just one of each glycosylated site and retaining a sufficient length of peptide to allow an assignment of that glycopeptide along the amino acid sequence of the protein

5. a determination of the degree of occupancy of each glycosylated site

6. analysis of the oligosaccharide structures associated with each glycopeptide

7. a reconstruction of the glycoforms constituting the glycoprotein

8. an experimental check of the accuracy of the analysis

The detailed experimental procedures to achieve steps 1–7 have all been extensively reviewed elsewhere. To avoid repetition, references will be provided for these experimental procedures and only an overview of the practical approaches to glycoform analysis will be provided here.

Determination of Primary Amino Acid Sequence

This is performed with one of various automated apparatus or from a cDNA sequence. In principle, primary amino acid sequence information is not essential to perform site-specific glycosylation analysis. Exhaustive proteolytic digestion can separate glycosylation sites and analysis of individual glycopeptides will therefore give at least some site-specific information. For example, if the number of occupied glycosylation sites is known, and that number of different glycopeptides is obtained in a limited proteolytic digest, site-specific glycosylation analysis can be performed. The value of primary amino acid sequence information residues principally in the following. First, it allows one to ensure that each glycosylation site has been purified away from all others (i.e., that each glycopeptide pool is derived from just one glycosylated site). Second, it allows each glycopeptide to be assigned to a specific glycosylated site along the amino acid sequence, and therefore a measure of the extent to which each potential glycosylation site is actually glycosylated. Although this is essential for full glycoform analysis, available information does not suggest that knowledge of the glycosylation pattern at one site helps in understanding the glycosylation pattern at another. Third, it can facilitate a design of experimental protocols to cleave a glycoprotein into its individual glycosylated sites.

Identification of Glycosylated Sites

There are three main ways in which this is currently done.

1. protein or peptide primary sequence analysis
2. the use of glycosidases and glycopeptidases to leave a marker on

the glycosylation site, in conjunction with protein primary sequence analysis

3. the use of high-performance liquid chromatography/fast atom bombardment-mass spectrometry (HPLC/FAB-MS) to determine the molecular weights of (glyco)peptides before and after treatments with endo(glyco)peptidases.

Primary Sequence Analysis

Occupancy of potentially glycosylated asparagine, serine, and threonine residues can be determined during amino acid sequence analysis. During sequential Edman degradation, as usually performed, the labeled derivatives of glycoamino acids are not recovered to a significant degree unless the attached carbohydrate is only a monosaccharide. Glycosylated amino acid residues can therefore be revealed as "blanks" at the appropriate cycles. Fully occupied asparagine, serine, or threonine residues will be revealed as indistinguishable from background, and hence blank. Note that even if fully occupied, traces of serine or threonine "degradation" products may be detected. Where occupancy of a glycosylation site is partial, a reduced signal will be obtained derived from the unoccupied residue. Except in favorable cases, this method is unreliable for determining accurately the degree of occupancy of partially occupied sites, particularly where the degree of occupancy is low, as is the case, for example, for certain glycoproteins carrying the O-GlcNAc modification. It should be noted that certain forms of glycosylation, such as the O-GlcNAc modification, can reduce the efficiency of the Edman reaction step at which the PTH derivative at the amino terminus become cyclic. The consequent observation is of a dramatic loss of signal, rather than of a blank with little change in the subsequent signals (G. Holt, personal communication, 1993).

Use of Glycosidases and Glycopeptidases to Mark Glycosylation Site

Another approach to determine glycosylation site occupancy is to generate a chemical marker on the protein or peptide at glycosylated sites after treatment with glycosidases.[13] Oligomannose- and hybrid-type structures are sensitive to endoglycosidase H (Endo H), whereas complex-type N-linked oligosaccharides are not.[14] Treatment of glycoproteins or glycopeptides with endoglycosidase H (Endo H; from *Streptomyces plicatus* or *Streptomyces griseus*) leaves behind a GlcNAc residue attached to any asparagine carrying Endo H-susceptible oligosaccharides.[14] Peptide

---

[13] M. W. Spellman, *Anal. Chem.* **62**, 1714 (1990).
[14] A. Tarentino and T. H. Plummer, Jr., this volume [4].

N-glycosidase F (PNGase F; from *Flavobacterium meningosepticum*) is an N^4-(N-acetylglucosaminyl)asparagine aminidase of relatively broad specificity.[15] Its action in releasing susceptible oligosaccharides from accessible sites converts the glycosylated asparagine to an aspartic acid. Sequence analysis of a glycoprotein before treatment with Endo H or PNGase F will indicate blanks at fully occupied N-glycosylation sites, with a reduced asparagine recovery at partially occupied sites. Amino acid analysis after treatment with Endo H will lead to the identification of GlcNAc-Asn at sites that carried Endo H susceptible structures, whereas analysis after treatment with PNGase F will reveal an aspartic acid at previously blank positions. These enzymes can therefore be useful in revealing both occupied N-glycosylation sites and the general features of oligosaccharides at such sites.

Analysis of O-linked *N*-acetyl-D-glucosamine[16] and O-linked fucose[17] modifications involves an essentially similar approach. The relatively low occupancy of residues carrying these modifications poses additional analytical difficulties, which are better solved by first isolating glycopeptides carrying these modifications.

This approach has so far been applied principally to a study of N-glycosylation sites, but in principle is applicable to a study of O-glycosylation sites as the appropriate glycosidases become available.

O-Glycosylation sites can also be identified by amino acid sequence analysis before and after reductive β elimination.[17a] Reductive β elimination releases O-linked oligosaccharides from glycosylated serine and threonine residues, and is associated with the conversion of glycosylated serine residues to alanine, and of glycosylated threonine residues to β-aminobutyric acid. Unglycosylated serine and threonine residues are unaffected by reductive β elimination. Amino acid sequence analysis will therefore show a conversion of glycosylated serine and threonine residues to alanine and β-aminobutyric acid, respectively, after reductive β elimination. It should be noted that reductive β elimination can also cause some peptide bond cleavage.

Use of HPLC/FAB-MS

The ability of FAB-MS to ionize glycopeptides has led to the development of techniques for determining sites of N-linked glycosyla-

[15] A. L. Tarentino and T. H. Plummer, Jr., *in* "Methods in Enzymology" (V. Ginsburg, ed.), Vol. 138, p. 770. Academic Press, Orlando, FL, 1987.

[16] A. J. Reason, I. P. Blench, R. S. Haltiwanger, G. W. Hart, H. R. Morris, M. Panico, and A. Dell, *Glycobiology* **1**(6), 585 (1991).

[17] R. J. Harris, C. K. Leonard, A. W. Guzzetta, and M. W. Spellman, *Biochemistry* **30**, 2311 (1991).

[17a] R. G. Spiro, this series, Vol. 28 [1].

tion following enzymatic cleavage.[18] The glycoprotein is first reduced and alkylated (carboxymethylation or pyridylethylation) and then cleaved with trypsin [tolylsulfonyl phenylalanyl chloromethyl ketone (TPCK)-treated], endoproteinase Lys-C, endoproteinase Asp-N or *Staphylococcus aureus* V8 protease. The molecular weights of the derived peptides are determined by HPLC/FAB-MS. Glycopeptides give weak or nonexistent signals by this technique but the glycosylated peptides can be determined by repeating the analysis after cleavage of the oligosaccharide with PNGase F and comparing the profile with that obtained before sugar release. In addition to the appearance of the previously glycosylated peptides in the profile, their molecular weights are 1 unit higher than expected because of the conversion of asparagine to aspartate by PNGase F. A comparison of the ratio of the asparagine and aspartate forms of the peptide gives an estimate of the extent of site occupancy.

Degree of Occupancy of Each Glycosylated Site

Primary amino acid sequence analysis of the intact glycoprotein, as discussed earlier, can give approximate information on the degree of occupancy of partially glycosylated amino acid residues, but these data are essentially semiquantitative. Where partial site occupancy is indicated, a more accurate method is to perform amino acid sequence analysis on the peptides and glycopeptides obtained after proteolytic or chemical cleavage of the glycoprotein.

During proteolytic or chemical cleavage of a glycoprotein, a partially occupied glycosylation site will be recovered both as part of a peptide and as part of a glycopeptide, which will contain identical or at least partially overlapping amino acid sequences.[9,10] Amino acid sequence analysis of the individual peptides and glycopeptides, separated usually by reversed-phase HPLC, will reveal this. Relative quantitation of the relevant peptide and glycopeptide fractions will then quantify the occupancy of the particular glycosylated residue.

Sodium dodecyl sulfate–polyacrylamide gel electrophoresis (SDS–PAGE) has also been used to quantify the occupancy of N-glycosylation sites. N-Linked oligosaccharides have a molecular weight of 2000, but have a disproportionate effect on the mobility of glycoproteins in SDS–PAGE. Proteins can therefore in some cases be fractionated into distinct bands on the basis of differences in the number of N-linked oligo-

[18] S. A. Carr, J. R. Barr, G. D. Roberts, K. R. Anamula, and P. B. Taylor, *in* "Methods in Enzymology" (J. McCloskey, ed.), Vol. 193, p. 501. Academic Press, San Diego, 1990.

saccharide chains associated with each band.[19,20] The time course of incubation with PNGase F, or partial digestions with PNGase F and analysis of reaction products by SDS–PAGE, can lead to "laddering," in which consecutive protein bands differ by ±1 N-linked oligosaccharide chain. Analysis of lysosomal proteins is a particularly good example of this.[21]

Separation of Glycosylated Sites

Information on the full amino acid sequence of a glycoprotein is essential in designing strategies for proteolytic cleavage of a glycoprotein so as to separate glycosylated sites. Proteolytic cleavage of proteins has been extensively reviewed elsewhere in this series,[21a] and the same procedures apply to glycoproteins. It should be noted that the presence of relatively bulky oligosaccharides can significantly affect the kinetics of proteolysis at individual cleavage sites and extensive optimization of reaction conditions is necessary to ensure full cleavage between glycosylation sites. Chemical cleavage of a glycoprotein is used as a last resort in the absence of a suitable protease. Many conditions for chemical cleavage of glycoproteins are relatively harsh and may modify more labile monosaccharides or their substituents.

Following cleavage of the glycoprotein, the peptide and glycopeptide pool is fractionated by methods dependent on peptide and not carbohydrate. Reversed-phase HPLC is most commonly used. Glycopeptides are often of limited heterogeneity with respect to amino acid sequence and usually of considerable heterogeneity with respect to their attached carbohydrate moieties. Differences in carbohydrate structures do not significantly influence the fractionation of a glycopeptide during reversed-phase HPLC, and the complement of carbohydrate heterogeneity associated with a glycosylated site is often recovered in one of very few glycopeptide pools. Because glycosylated peptides chromatograph on gel-filtration chromatography quite differently from the nonglycosylated form of the same peptide, this approach has also the advantage of allowing the degree of occupancy of a glycosylation site to be estimated with reasonable accuracy. Individual glycopeptides are assigned to the primary structure after amino acid, peptide sequence, or mass analysis. Such analysis will also confirm that glycosylated sites have been separated from one another,

[19] S. Alexander and J. H. Elder, in "Methods in Enzymology" (V. Ginsburg, ed.), Vol. 179, p. 505. Academic Press, San Diego.
[20] J. Arvieux, A. C. Willis, and A. F. Williams, *Mol. Immunol.* **23**, 983 (1986).
[21a] R. C. Judd, this series, Vol. 182, p. 613.

that is, that each glycopeptide carries one occupied amino acid residue only.

Analysis of Oligosaccharide Structures Associated with Each Glycopeptide

Each glycopeptide pool so obtained is usually heterogeneous with respect to attached oligosaccharides and further detailed analysis requires its fractionation on the basis of differences in glycosylation. This can be performed either while the oligosaccharides remain attached to peptide or, more commonly, after release of oligosaccharides from the peptide either by enzymatic or chemical methods (see [4] and [5] in this volume).

Direct fractionation of glycopeptides can be performed by using mass spectrometry with HPLC-MS or capillary electrophoresis (CE)-MS, or through various chromatographic procedures (see [11] and [12] in this volume).

Full glycoform analysis requires not only a determination of the oligosaccharide structures associated with each glycosylated site, but also an accurate determination of the relative molar content of each oligosaccharide in the population of structures at that site. This can be performed accurately only by quantifying each oligosaccharide structure in a sequence-independent manner, which in practice generally involves a titration of the reducing terminus. If such molar quantitation is required, oligosaccharides need to be released from peptide, and the reducing termini labeled by a sequence-independent reaction. Such reactions include reduction with alkaline sodium borotritide[22] or attachment of a fluorophore or chromophore through reductive amination.[23] A further advantage of releasing oligosaccharides from peptide is that this allows the use of various additional chromatographic techniques that have been developed and optimized for carbohydrate fractionation. Methods for the fractionation and structural analysis of oligosaccharides are fully reviewed (see [8] and [9] in this volume).

Glycoform Reconstruction

Because glycoform analysis is performed on a mixture of glycoforms (see earlier), the set of glycoforms constituting a glycoprotein is recon-

---

[22] D. Aminoff, W. D. Gathmann, C. M. McLean, and T. Yadomae, *Anal. Biochem.* **101,** 44 (1980).

[23] S. Hase, T. Ibuki, and T. Ikenaka, *J. Biochem.* (*Tokyo*) **95,** 197 (1984).

structed from a knowledge of the oligosaccharides present at each site.[9,10] As a first approximation, therefore, the nature and percentage relative molar incidence of a particular glycoform is calculated by multiplying the percent molar incidence of structures at each glycosylation site. For example, for a glycoprotein found to carry oligosaccharides A_1–A_n at site A, B_1–B_n at site B, and C_1–C_n at site C, the incidence of the glycoform carrying A_m, B_m, and C_m is the product of the percent molar incidence of A_m at site A, B_m at site B, C_m at site C, and so on for all other glycoforms.

Experimental Check of Accuracy of Glycoform Analysis

For detailed site-specific characterization of protein glycosylation, and subsequent reconstruction of glycoforms, there is no significant alternative to analysis of individual glycopeptides. One potential drawback to this is that specific glycosylation patterns may confer differential susceptibility to proteolysis, leading to glycoform-specific recovery of glycopeptides. It is therefore important with this approach to ensure that the yield of glycopeptides obtained from each site is high. Further, several experimental steps are involved in assigning and separating glycosylated sites, and in analyzing the molar incidence of individual oligosaccharide structures at each site. It is therefore important to ensure that "selective" analysis has not been performed and that the experimental accuracy of the overall analysis is high. This is most easily done by comparing the glycosylation characteristics (such as the content of individual monosaccharides) of the glycoprotein reconstructed from site-specific analysis to corresponding glycosylation characteristics determined experimentally on the intact glycoprotein.[9]

Acknowledgments

I am grateful to David Hawke and Gordon Holt for valuable discussions.

[22] Isolation of Glycosphingolipids

By RONALD L. SCHNAAR

Introduction

Glycosphingolipids are amphipathic glycoconjugates containing a saccharide structure linked to a lipid moiety, ceramide. This simple definition

belies the remarkable complexity and variety of glycosphingolipid structures. Glycosphingolipids are classified primarily on the basis of their
carbohydrate portion, which may vary from a single monosaccharide to
large branched structures composed of >20 monosaccharide units.[1] Techniques for glycosphingolipid purification have largely focused on separations based on saccharide differences, resulting in purification to oligosaccharide homogeneity. There is additional variation in the ceramide, which
is composed of any of several long-chain bases (of which sphingosine is the
most common in mammalian tissues), each of which is further substituted,
through amide linkage, with any of a variety of fatty acids. Although
the functional consequences of variations in the saccharide portion of
glycosphingolipids are well documented,[2,3] the biological significance of
variation in the lipid portion has yet to be thoroughly addressed. Purification of a glycosphingolipid to chemical homogeneity entails resolution
based on both saccharide and lipid variations, although this degree of
resolution may not be required for many experimental goals.

The variety in glycosphingolipid structures is matched by the diversity
in methods reported for their isolation, and there is no one protocol best
suited for all applications. Glycosphingolipid purification involves three
steps: (1) extraction of lipids from a biological source, using organic solvents, (2) bulk separation from major lipid and nonlipid contaminants,
and (3) chromatographic resolution of individual species. Various specific
methods are listed below. These have been culled from research reports
and reviews from leading investigators. The reader is particularly encour-
aged to consult prior reviews in this series,[4-6] as well as the comprehensive
treatise by Hakomori.[7] The choice of a particular protocol depends on
the chemical nature of the glycosphingolipid(s) to be isolated. In this
context it is useful, for pragmatic reasons, to classify glycosphingolipids
into four categories: short neutral glycosphingolipids (having fewer than
five monosaccharide units), long neutral glycosphingolipids (having more

[1] C. L. M. Stults, C. C. Sweeley, and B. A. Macher, in "Methods in Enzymology" (V.
   Ginsburg, ed.), Vol. 179, p. 167. Academic Press, San Diego, 1989.
[2] R. L. Schnaar, Glycobiology 1, 477 (1991).
[3] Y. Igarashi, H. Nojiri, N. Hanai, and S. Hakomori, in "Methods in Enzymology" (V.
   Ginsburg, ed.), Vol. 179, p. 521. Academic Press, San Diego, 1989.
[4] W. J. Esselman, R. A. Laine, and C. C. Sweeley, in "Methods in Enzymology" (V.
   Ginsburg, ed.), Vol. 28, p. 140. Academic Press, New York, 1972.
[5] R. W. Ledeen and R. K. Yu, in "Methods in Enzymology" (V. Ginsburg, ed.), Vol. 83,
   p. 139. Academic Press, Orlando, FL, 1982.
[6] R. Kannagi, K. Watanabe, and S. Hakomori, in "Methods in Enzymology" (V. Ginsburg,
   ed.), Vol. 138, p. 3. Academic Press, Orlando, FL, 1987.
[7] S. Hakomori, in "Sphingolipid Biochemistry" (J. N. Kanfer and S. Hakomori, eds.), p.
   1. Plenum, New York, 1983.

than five monosaccharide units), gangliosides (anionic glycosphingolipids containing sialic acid residues), and sulfated glycosphingolipids. Because many of the methods can be applied to more than one class of glycosphingolipids, they are outlined first with only cursory reference to specific applications. Examples of the use of these methods to purify particular glycosphingolipids or glycosphingolipid classes are described later in the chapter. Although the main focus of this chapter is on the isolation of species with homogeneous carbohydrate structures, separations based on the lipid moiety are addressed at the end of the chapter.

General Considerations

Glycosphingolipid Nomenclature

The complexity of glycosphingolipid saccharide structures has foiled attempts to develop a simple systematic nomenclature. Nevertheless, the IUPAC nomenclature[8] based on characteristic "core" neutral tetrasaccharides, of which other glycosphingolipids are considered derivatives and/ or truncations, has gained increased usage and is preferable to the still frequent use of trivial names for many structures. An exception is the nomenclature of Svennerholm[9] for major ganglioside species, which is widely accepted. A useful listing of glycosphingolipid structures along with their IUPAC and trivial names is published periodically in this series.[1]

The ceramide nomenclature used here designates the long-chain base by the number of hydroxyl groups (e.g., "d" for dihydroxy), the number of carbons and the degree of unsaturation, and the fatty acid amide by the number of carbons and degree of unsaturation. Thus, the most abundant ceramide bases on human gangliosides are d18:1-C18:0 (sphingosine with stearic acid amide) and d20:1-C18:0 (icosasphingosine with stearic acid amide).

Stability

Glycosphingolipids are generally stable, and can be prepared from fresh or frozen tissues or cultured cells. Early studies by Suzuki[10] reported no change in major ganglioside patterns, for instance, when rat brains were kept for 6 hr postmortem at room temperature prior to glycosphingolipid isolation (although this is not generally recommended). In contrast, caution is warranted when using fixed tissues, because time-dependent structural

[8] IUPAC-IUB Commission on Biochemical Nomenclature, *Biochem. J.* **171**, 21 (1978).
[9] L. Svennerholm, *Adv. Exp. Med. Biol.* **125**, 11 (1980).
[10] K. Suzuki, *J. Neurochem.* **12**, 629 (1965).

alterations in glycosphingolipids from tissues in fixatives have been reported. Purified or partially purified glycosphingolipids are generally stable, even for many years, when stored in organic solvent (e.g., chloroform–methanol–water mixtures) at reduced temperature.

Evaporations and Resuspensions

Isolation of glycosphingolipids requires repeated removal of solvents. For small samples this is most conveniently achieved under a stream of nitrogen in a heated block or water bath at ~45°, using a commercial apparatus for multiple small samples [e.g., N-EVAP (Organomation Assoc., Inc., Berlin, MA) or Reacti-Vap (Pierce Chem. Co., Rockford, IL)]. For large numbers of small samples containing volatile salts, which require vacuum for efficient removal (e.g., methanolic ammonium acetate fractions from ion-exchange chromatography), a SpeedVac concentrator (Savant Instruments, Inc., Farmingdale, NY) is recommended.

Larger samples are handled with a rotary evaporator (e.g., a Buchi Rotovapor; Brinkman Instr., Westbury, NY) under reduced pressure at ~45°. The problem of foaming of lipid extracts under vacuum can be lessened by using an inlet feed tube to maintain a low volume of extract in the evaporation flask, or by adding a small amount of isobutyl alcohol or toluene.

Resuspension of the dried lipid residues in small volumes of organic solvents is usually readily accomplished with heating to ~45° and sonication in a bath-type sonicator. However, recoveries should be monitored whenever possible because, for instance, highly sialyated gangliosides that are readily extracted from tissues are selectively lost as films on the bottoms of flasks because of poor solubility of the more purified fractions in organic solvents.[11]

Removal of Salts and/or Small Hydrophilic Contaminants

Extracts containing polar glycosphingolipids are freed of low molecular weight hydrophilic contaminants (salts, sugars, amino acids, etc.) by dialysis or by reversed-phase chromatography. Efficient recovery from dialysis depends on glycosphingolipids existing in micellar form, which is not a problem unless low concentrations of glycosphingolipids are present (glycosphingolipids have low critical micelle concentrations). Nevertheless, reduced molecular weight cutoff tubing (e.g., Spectra/Por molecular weight cutoff 1000 or 3500; Spectrum Medical Industries, Los Angeles, CA) is recommended, and organic solvents should be removed from samples by evaporation prior to dialysis for best results.

[11] H. Miller-Podraza, J.-E. Månsson, and L. Sevennerholm, *Biochim. Biophys. Acta* **1124,** 45 (1992).

An excellent alternative to dialysis is reversed-phase chromatography, which is rapid and simple for removal of polar contaminants even from dilute glycosphingolipid samples, which are concentrated during the process. A convenient method based on the use of Sep-Pak C_{18} cartridges (Millipore Corp., Milford, MA) can be scaled up for larger samples if necessary by using larger amounts of C_{18} sorbent from this or other suppliers.[12,13]

Procedure. A standard Sep-Pak C_{18} cartridge (0.36 g of sorbent) is prewashed by injection of 5 ml each of water, methanol–water (1 : 1, v/v), methanol, chloroform–methanol (1 : 1, v/v), chloroform, chloroform–methanol (1 : 1, v/v), methanol, methanol–water (1 : 1, v/v), and finally chloroform–methanol–water (2 : 43 : 55, v/v). Glycosphingolipid sample in up to 5 ml of chloroform–methanol–water (2 : 43 : 55, v/v) is injected onto the cartridge, and the eluate is collected and reapplied to the column twice, followed by 5 ml of the same solvent. The column is then eluted stepwise with 5 ml each of methanol–water (1 : 1, v/v), methanol–water (4 : 1, v/v), methanol, chloroform–methanol (1 : 1, v/v), and chloroform. Depending on the glycosphingolipid, salt-free samples are eluted in the methanol–water (4 : 1, v/v), the methanol, and/or the chloroform–methanol fraction. Injection of solvents onto the column with a glass syringe is rapid, and the entire procedure can be completed in minutes. Samples may be applied to the column in other solvents, such as the upper phase of a chloroform–methanol–water partition (see below), or methanol–water (1 : 1, v/v). Samples may be applied in water alone to which salt has been added (0.1 M KCl or 0.8 M sodium acetate) to enhance adsorption. The capacity of the Sep-Pak column varies with the extract used, but has been reported as 5 μmol of mixed brain ganglioside sialic acid (equivalent to ~5 mg of ganglioside or extract from 1–2 g of brain tissue) per 0.36-g cartridge.

Monitoring Isolations

Glycosphingolipids and major lipid contaminants are most readily monitored during isolation by thin-layer chromatography (TLC), using methods detailed elsewhere in this volume.[14] Because many of the extraction and isolation methods are sensitive to small variations in solvents and/or sorbents, careful TLC analysis must be performed at each step to confirm

---

[12] M. A. Williams and R. H. McCluer, *J. Neurochem.* **35,** 266 (1980).

[13] R. H. McCluer, M. D. Ullman, and F. B. Jungalwala, *in* "Methods in Enzymology" (S. Fleischer and B. Fleischer, eds.), Vol. 172, p. 538. Academic Press, San Diego, 1989.

[14] R. L. Schnaar and L. K. Needham, this volume [23].

collection of the desired glycosphingolipids and effective removal of contaminants.

Extraction

Glycosphingolipids are extracted along with other lipids from tissues or cells, using organic solvents. Routine distillation of solvents prior to use is recommended. Many variations in extraction procedures have been published, some of the most widely used of which are detailed below. Tissues or cells are generally kept at reduced temperature (0–4°) until organic solvents are added, after which procedures are performed at ambient temperature. Homogenizations are performed for several minutes (unless otherwise indicated) with a Potter–Elvehjem (or similar) homogenizer for small samples, or an appropriate (spark-free) mechanical homogenizer for large samples. Because many of the subsequent partition methods depend on accurate ratios among solvents, care should be taken during extraction to minimize solvent loss by evaporation.

Extraction A: Chloroform–Methanol–Water Method of Folch et al.[10,15]

To the sample (fresh or frozen tissues or cells) add 19 vol (e.g., 19 ml/g tissue or 19 ml/ml packed cells) of chloroform–methanol (2–1, v/v) and homogenize. Separate the solids from the organic extract by centrifugation or filtration, and reextract the solids by adding 0.5 vol (based on the original tissue or cell weight) of water and 9.5 vol of chloroform–methanol (1 : 2, v/v). Remove solids as above and combine the clarified extracts for further processing.

For large tissue samples (kilograms) a dry acetone powder may be prepared[7] prior to extraction as above. When this is done, the first extraction should be chloroform–methanol–water (38 : 19 : 3, v/v).[16]

Efficient extraction of polar and anionic glycosphingolipids is improved by including an additional reextraction (as above) with chloroform–methanol–aqueous 0.4 M sodium acetate (30 : 60 : 8, v/v).

Extraction B: Chloroform–Methanol–Water Method of Svennerholm and Fredman[17]

This procedure, which was optimized for quantitative extraction of gangliosides, uses a chloroform–methanol–aqueous ratio of 4 : 8 : 3 (v/v).

[15] J. Folch, M. Lees, and G. H. Sloane-Stanley, *J. Biol. Chem.* **226,** 497 (1957).
[16] B. L. Slomiany and A. Slomiany, *in* "Cell Surface Glycolipids" (C. C. Sweeley, ed.), p. 149. Am. Chem. Soc., Washington, DC, 1980.
[17] L. Svennerholm and P. Fredman, *Biochim. Biophys. Acta* **617,** 97 (1980).

The order of addition is important to ensure maximal precipitation (removal) of hydrophobic proteins. To the tissue or cell sample add 3 vol ice-cold distilled water and homogenize at ≤4°. Add the homogenate to 10.6 vol of rapidly stirring methanol at ambient temperature. Finally, add 5.3 vol of chloroform to the stirring suspension. The solids are removed by centrifugation or filtration, reextracted with chloroform–methanol–water (4:8:3, v/v), and the extracts combined for further processing.

Other Extractions

Extraction of long neutral glycosphingolipids is improved by using 2-propanol–hexane–water mixtures.[7] Tissues, cells, or acetone powders are extracted with 5–10 vol of 2-propanol–hexane–water (55:20:25 or 55:25:20, v/v) by homogenization, followed by stirring overnight. The solid residue is collected, reextracted, and the clarified extracts combined. The above solvent mixture is two phased; the small upper phase (primarily hexane) is removed prior to use for extraction.[18]

Effective extraction of polar glycosphingolipids has also been reported in tetrahydrofuran–aqueous mixtures.[19] The sample (tissue or cells) is homogenized in 1 vol of 10 mM aqueous potassium phosphate (pH 6.8) for 1 min, then 8 vol of tetrahydrofuran is added and the homogenization continued for an additional 1 min. Solids are collected and the extraction repeated three additional times, using 1 vol of aqueous buffer and 4 vol of tetrahydrofuran. Clarified extracts are combined for further processing.

Lipids may be economically extracted from large amounts of tissue with warm ethanol. The tissue is extracted with 10 vol of 100% ethanol at 60° for 15 min, the solids removed, and reextracted with 90% ethanol under the same conditions.[7] The clarified extracts are combined.

Bulk Separation

The above extracts contain minor amounts of the desired glycosphingolipids compared to contaminating phospholipids, triglycerides, cholesterol, and low molecular weight hydrophilic molecules (sugars, amino acids, small peptides, and salts). Purification of glycosphingolipids requires bulk removal of these contaminants prior to finer chromatographic resolution of species. Partitioning between hydrophilic and hydrophobic organic phases separates the bulk of nonpolar lipids (e.g., phospholipids and cholesterol) from the long neutral lipids and gangliosides that remain

---

[18] E. Nudelman, Biomembrane Institute, Seattle, WA (personal communication, 1992).
[19] G. Tettamanti, F. Bonali, S. Marchesini, and V. Zambotti, *Biochim. Biophys. Acta* **296,** 160 (1973).

in the hydrophilic phase. The short neutral glycosphingolipids and some sulfated glycosphingolipids remain in the hydrophobic phase along with major lipid contaminants. Similarly, anion-exchange chromatography is effective for separation of major lipid contaminants from anionic ganglio-sides and sulfated glycosphingolipids, but not from neutral glycosphingoli-pids, which are eluted in the flow-through fractions along with major neutral and zwitterionic lipid contaminants. Therefore additional bulk separation methods, based on differential hydrolysis of contaminants or adsorption chromatography of underivatized or peracetylated glycosphin-golipids, are required for less polar glycosphingolipids. Various bulk sepa-ration procedures are described below, the applications for which are discussed later in the chapter.

Solvent Partitioning

Partition A: Chloroform–Methanol–Water Method of Folch et al.[10,15] This method follows from extraction A (above), but is also readily applied to lipids extracted by other procedures if the solvents are removed by evaporation and the residue resuspended in about 10 vol (based on original tissue weight) of chloroform–methanol (2 : 1, v/v). If extraction A is used, add chloroform to the combined extracts such that the final chloro-form–methanol ratio is 2 : 1 (v/v). Measure the total volume of this crude extract, and add 0.2-fold that volume of 0.1 M aqueous KCl of NaCl. Mix the resulting solution (e.g., in a separatory funnel), allow the two phases to separate, or centrifuge to accelerate separation. Remove the hydrophilic upper phase (~40% of the total volume) and reextract the hydrophobic lower phase with fresh theoretical upper phase and vice versa. Combine like phases for further processing.

Partition B: Chloroform–Methanol–Water Method of Svennerholm and Fredman.[17] This method follows directly from extraction B, above, but may be applied to samples that have been extracted by other methods, then evaporated to dryness, and resolubilized in chloroform–metha-nol–water (4 : 8 : 3, v/v). Measure the total volume of the combined ex-tracts and add 0.173-fold that volume of water for a final chloroform–meth-anol–water ratio of 4 : 8 : 5.6 (v/v). Mix and allow the two phases to sepa-rate (centrifuge to accelerate this step if feasible). Remove the hydrophilic upper phase, which constitutes ~80% of the volume, and reextract the hydrophobic lower phase with theoretical upper phase made with 10 mM aqueous KCl instead of water. Reextract the upper phase with theoretical lower phase. Combine like phases for further processing.

This method was designed to maximize partioning of gangliosides into the upper phase, and only minor amounts of gangliosides (including some of the G_{M4} and G_{M3}) partition into the lower phase. Partitioning of neutral and sulfated glycosphingolipids in this system has not been fully characterized.

Partitioning with Tetrahydrofuran–Ether–Water. This procedure follows directly from the tetrahydrofuran–aqueous buffer extraction described earlier. Measure the total volume of the combined extracts and add 0.3-fold that volume of ethyl ether. Shake vigorously, separate the phases by centrifugation, collect the hydrophilic lower phase, and reextract the hydrophobic upper phase with 0.1 vol of water. Combine the lower phases. This procedure was designed for isolation of gangliosides, which partition into the hydrophilic phase in a manner comparable to partition B.[20]

Partitioning with Diisopropyl Ether–Butanol–Aqueous Salt.[21] This is a remarkably effective method for separation of gangliosides from phospholipids, sulfatides, and neutral lipids in lipid extracts. Evaporate any of the above lipid extracts to dryness and resuspend by vigorous mixing in 20 vol (based on original tissue weight) of diisopropyl ether–butanol (60:40, v/v). Add 10 vol of 50 mM aqueous sodium chloride and mix vigorously. Separate the phases by centrifugation, remove the hydrophobic upper phase, and reextract the hydrophilic lower phase with 20 vol of the same organic solvent mixture. The resulting lower phase retains nearly all of the gangliosides (along with some long neutral glycosphingolipids) with highly efficient removal of other lipids.

Each of the partitioning methods results in a hydrophilic phase containing polar glycosphingolipids and polar contaminants (sugars, amino acids, etc.) and a hydrophobic phase containing nonpolar glycosphingolipids and lipid contaminants. The hydrophilic phase is evaporated to dryness, and resuspended in water for dialysis or in chloroform–methanol–water (2:43:55, v/v) for reversed-phase chromatography (see above), to remove polar contaminants from the glycosphingolipids. The hydrophobic phase presents more of an experimental challenge, because it must be freed of major contaminating lipids via techniques that will be described later in this section.

Although partitioning of particular glycolipids into hydrophilic and hydrophobic phases can be stated in general terms, considerable variation occurs depending on the amount of other lipids and salts present. Therefore, TLC must be used to follow the desired glycosphingolipid species. Short neutral glycosphingolipids are generally found in the hydrophobic

[20] M. C. Byrne, M. Sbaschnig-Agler, D. A. Aquino, J. R. Sclafani, and R.W. Ledeen, *Anal. Biochem.* **148,** 163 (1985).
[21] S. Ladisch and B. Gillard, *Anal. Biochem.* **146,** 220 (1985).

phase, whereas long neutral glycosphingolipids partition into the hydrophilic phase. Most of the gangliosides are found in the hydrophilic phase, although the smallest ganglioside, G_{M4}, is predominantly found in the hydrophobic phase and G_{M3} splits between the phases, as do some sulfated glycosphingolipids. Salts markedly affect partitioning. For instance, 5–200 mM CaCl$_2$ causes gangliosides, which normally partition into the upper phase of a Folch partitioning (partition A), to appear in the lower phase.[22] Salts also modify sulfatide partitioning.[23]

Bulk Anion-Exchange Chromatography

Anion-exchange chromatography readily separates anionic lipids from neutral and zwitterionic lipids, and is productively used for bulk isolation of anionic glycosphingolipids and/or bulk removal of anionic lipids from neutral glycosphingolipids. Many anion-exchange resins and salt elution protocols are suitable for these purposes; only one will be described here.[5,24]

Convert DEAE–Sephadex A-25 (Pharmacia LKB Biotechnology, Piscataway, NJ) to its acetate form by mixing the dry resin with ≥10 ml/g of solvent B, chloroform–methanol–0.8 M aqueous sodium acetate (30:60:8, v/v). Allow the resin to settle, remove the supernatant, repeat three times with fresh solvent B, and allow the resin to incubate overnight in solvent B. Remove the supernatant and wash the resin three or four times with ≥10 ml/g of original dry resin of solvent A, chloroform–methanol–water (30:60:8, v/v). Make a slurry of the resin in solvent A, load into a suitable glass column, and wash with ≥30 ml/g of original dry resin of solvent A. The minimum amount of resin appropriate for any application depends on the nature of the extract to be applied. When a crude lipid extract is used, Ledeen and Yu[5] report 1 g of dry resin per ~1 g (wet weight) of brain tissue or ~2 g of nonneural tissue. When solvent partitioning is used prior to anion-exchange chromatography, greater relative amounts of extract can be applied (e.g., 1 g of dry resin per ~8.5 g of brain tissue, using Folch upper phase extract.[25] The appearance of gangliosides in the flow-through fraction on TLC analysis is an indication that the column was overloaded.

Adjust the extract to chloroform–methanol–water (30:60:8, v/v) and apply to the resin at a flow rate of ≤1 ml/min, then wash the column with several column volumes of solvent A. Combine the flow-through and

[22] R. Quarles and J. Folch-Pi, *J. Neurochem.* **12,** 543 (1965).
[23] K. Tadano-Aritomi and I. Ishizuka, *J. Lipid Res.* **24,** 1368 (1983).
[24] R. W. Ledeen, R. K. Yu, and L. F. Eng, *J. Neurochem.* **21,** 829 (1973).
[25] M. Iwamori and Y. Nagai, *Biochim. Biophys. Acta* **528,** 257 (1978).

nonsalt wash fractions, which contain neutral and zwitterionic lipids, in-cluding cholesterol, phosphatidylcholine, phosphatidylethanolamine, and neutral glycosphingolipids. Anionic glycosphingolipids are bulk eluted with several column volumes of solvent B.

Many variations in the above technique are tolerated, including variations in the anion-exchange resin (e.g., DEAE–cellulose, Q–Sepharose, etc.), the loading solvent [e.g., chloroform-methanol (1 : 1, v/v), chloro-form–methanol–water (4 : 8 : 3, v/v), etc.], and the organic salt solution used for elution (e.g., ammonium acetate in methanol, etc.). However, solvents higher in chloroform content should be avoided because they cause the resins to float, making the chromatography difficult. Other anion-exchange methods that are used for fine resolution of individual anionic glycosphingolipid species are detailed later in this chapter, and these methods can also be used for bulk isolation.

Mild Alkaline Degradation (Saponification)

The esters of phospholipids and triglycerides are cleaved by treatment with mild alkali under conditions in which the fatty acid amides of glyco-sphingolipids are stable. The resulting degradation products are efficiently separated from glycosphingolipids by silicic acid chromatography (see the next section). Although mild alkaline hydrolysis can be used to remove major lipid contaminants from many glycosphingolipids, when the glyco-sphingolipids to be purified contain alkali labile moieties, such as O-acyl groups, saponification must be avoided. If the target structures are un-known, base should be avoided unless alkali stability of the desired species can be independently confirmed.

Evaporate glycosphingolipid extract or partially purified glycosphin-golipids to dryness and resuspend in 0.1 M sodium hydroxide in methanol. Heat at 37° for 2–3 hr, add 0.1 vol of 1 M aqueous sodium phosphate, pH 7, and 0.1 vol aqueous 1 M hydrochloric acid (addition of the buffer ensures against overacidification). Evaporate to dryness and resuspend in water (for dialysis) or chloroform–methanol–water (2 : 43 : 55, v/v) for reversed-phase removal of salts (as described above). Remove lipid degra-dation products by silicic acid chromatography (see the next section). Alkali hydrolysis can also be performed in chloroform–methanol (1 : 1, v/v),[47] a solvent that may be more suitable for nonpolar short neutral glycosphingolipids.

Bulk Silicic Acid Chromatography

Removal of cholesterol, triglycerides, some phospholipids, and sphin-gomyelin from neutral glycosphingolipids in a crude extract, after solvent partioning, or after saponification (which improves the removal of phospholipids), can be accomplished by bulk silicic acid chromatography. Various similar procedures are useful for this purpose.[4,26]

Evaporate the lipid extract to dryness, resuspend at ~20 mg/ml in chloroform, and load on a column containing ~10 g/g of lipid of silicic acid [e.g., from Mallinckrodt Specialty Chemical Co., St. Louis, MO), sieved to ≥45 μm] that has been previously heated (125°) to activate, cooled, slurried with chloroform, and loaded in an appropriate glass chromatography column. Wash the column with ~10 vol of chloroform and ~10 vol of chloroform–methanol (98:2, v/v), then elute glycosphingolipids with ~10 vol of chloroform–methanol (1:3, v/v). An alternate procedure elutes glycosphingolipids with acetone–methanol (9:1, v/v).

Peracetylation/Bulk Silicic Acid Chromatography/Deacetylation

Thorough separation of glycosphingolipids from other lipid classes can be achieved by peracetylation, adsorption chromatography, and deacetylation.[27] This method can be applied to most glycosphingolipids, although O-acyl groups (or other alkali-sensitive groups) will be lost during deacetylation.

Use redistilled, dried solvents and desalt glycolipids by reversed-phase chromatography for best results. Evaporate the lipid extract to dryness, then in vacuo. Add pyridine–acetic anhydride (3:2, v/v), 0.5 ml for extract from ~1 g of tissue. After 18 hr at ambient temperature, add an excess of toluene and evaporate to dryness. Dissolve the residue in hexane–dichloroethane (1:4, v/v) and load onto a 5-g column of Florisil (e.g., magnesia-silica gel, 600–100 mesh; Mallinckrodt Chemical Specialties) that has been prepared and washed with the same solvent. Wash the column with 3 vol of the same solvent and with 3 vol of dichloroethane (to elute cholesterol), then elute acetylated glycosphingolipids with dichloroethane–acetone (1:1, v/v). Phospholipids remain on the column, and can be eluted with more polar solvents.

An alternate procedure[26] reports acetylation in chloroform–pyridine–acetic anhydride (1:1:1, v/v) to improve solubility. Resolution of acetylated lipids can also be performed by silicic acid chromatography (see above), using chloroform–methanol (98:2, v/v) as the loading solvent and eluting acetylated glycosphingolipids with chloroform–methanol [95:5 (v/v), then 90:10 (v/v)].

[26] K.-A. Karlsson, in "Methods in Enzymology" (V. Ginsburg, ed.), Vol. 138, p. 212. Academic Press, Orlando, FL, 1987.
[27] T. Saito and S. Hakomori, J. Lipid Res. 12, 257 (1971).

Evaporate acetylated glycosphingolipid fractions and deacetylate in a small volume (0.25 ml for extract from 1 g of tissue) of 0.1% sodium methoxide in chloroform–methanol (2 : 1, v/v) for 30 min at ambient temperature. Neutralize by addition of methanol-washed Dowex 50 (H⁺ form) (Bio-Rad, Richmond, CA), then filter. Higher concentrations (up to 2% sodium methoxide) can be used if needed to enhance deacetylation.[28]

Chromatographic Resolution

Preparative resolution of individual glycosphingolipid species on the basis of saccharide structure has been accomplished by anion-exchange chromatography and/or silicic acid chromatography. Purification is monitored by TLC.[14] However, a single TLC band, even in multiple solvent systems, does not guarantee the absence of multiple species, because many closely related glycosphingolipids comigrate and are difficult to separate. Often only compositional analysis, linkage analysis, or mass spectroscopy reveals the presence of multiple species in a sample purified to homogeneity by TLC criteria. Similarly, glycosphingolipids purified to saccharide homogeneity may resolve into multiple bands on TLC because of ceramide differences. Therefore additional analytical procedures must be applied subsequent to chromatographic resolution to ensure homogeneity.

Anion-Exchange Chromatography

Ion-exchange resins can be used in low-pressure or high-pressure liquid chromatography (HPLC) mode as part of a purification scheme for anionic glycosphingolipids. Many resins are suitable for this purpose, and many salt elution protocols have been successfully applied (direct comparisons of some have been reported).[25,29] Salt concentrations necessary for elution of a particular anionic glycosphingolipid will vary with the resin (even when similar resins, e.g., DEAE–Sephadex and DEAE–Sepharose, are used) and with the salt used for elution, necessitating careful analysis of each fractionation by TLC.

Stepwise Elution Using DEAE–Sepharose. Convert DEAE–Sepharose (Fast Flow or CL-6B; Pharmacia LKB Biotechnology) to its acetate form as described for DEAE–Sephadex (see Bulk Separation, earlier). Pack into an appropriate glass chromatography column, then wash out

---

[28] G. K. Ostrander, S. B. Levery, H. L. Eaton, M. E. K. Salyan, S. Hakomori, and E. H. Holmes, *J. Biol. Chem.* **263**, 18716 (1988).

[29] P. Fredman, O. Nilsson, J.-L. Tayot, and L. Svennerholm, *Biochim. Biophys. Acta* **618**, 42 (1980).

any remaining salt with several column volumes of loading solvent, which can be chloroform–methanol–water [e.g., 4 : 8 : 3 (v/v) or 30 : 60 : 8 (v/v)] or methanol, depending on the sample to be fractionated. Apply the lipid sample, for example, the hydrophilic phase from solvent partition, after desalting by reversed-phase chromatography or dialysis (see earlier). If necessary (e.g., after dialysis), evaporate the lipid sample to dryness and redissolve in loading solvent. The capacity of the column will vary depending on the presence of anionic contaminants. For comparison, using the "upper Folch" extract from human brain the capacity was 1 ml of resin per 2 g wet weight of original brain tissue (equivalent to 4 mg of "upper phase" solids or 5 μmol of lipid-linked sialic acid).[25] When loading large amounts of lipid, apply under slow flow conditions and monitor the flow-through fraction to ensure quantitative adsorption of the desired anionic glycosphingolipids.[29] After loading, wash with 1–2 column volumes of loading solvent and 1–2 column volumes of methanol to remove neutral lipids, anionic glycosphingolipids are eluted stepwise with 5–10 column volumes each of increasing concentrations of ammonium acetate in methanol. Although each sample must be analyzed by TLC and/or other analytical methods to identify the elution position of desired glycosphingolipids, in general monosialogangliosides elute at 10 mM ammonium acetate in methanol, monosulfated glycosphingolipids at 20 mM, disialogangliosides at 40 mM, and trisialogangliosides at 100 mM.[30] More highly sialylated gangliosides, multisulfated glycosphingolipids, and glycosphingolipids containing both sulfate and carboxylic acid groups require higher salt concentrations, ranging from 200 mM to as high as 3 M. Alternatively, apply a gradient of ammonium acetate in methanol and analyze fractions by TLC to identify desired fractions for collection.[25] Adjust the desired fractions to chloroform–methanol–aqueous (2 : 43 : 55, v/v) and subject to reversed-phase chromatography to remove salts, or reduce the volume by evaporation, place under vacuum (e.g., in a SpeedVac concentrator) to remove the bulk of the volatile ammonium acetate, then dialyze or subject to reversed-phase chromatography.

Gradient Elution Using Q–Sepharose. Successful large-scale isolation of minor gangliosides has been accomplished by Q–Sepharose chromatography.[31] Convert Q–Sepharose (Pharmacia LKB Biotechnology) to its acetate form as described for DEAE–Sephadex (above), load into an appropriate column, and wash with chloroform–methanol–water

[30] J. L. Magnani, S. L. Spitalnik, and V. Ginsburg, *in* "Methods in Enzymology" (V. Ginsburg, ed.), Vol. 138, p. 195. Academic Press, Orlando, FL, 1987.
[31] Y. Hirabayashi, T. Nakao, M. Matsumoto, K. Obata, and S. Ando, *J. Chromatogr.* **455,** 377 (1988).

(30:60:8, v/v) (solvent A). Load a partially purified glycolipid sample (e.g., Folch upper phase lipids, or total anionic lipids from bulk DEAE–Sephadex) that has been desalted (if necessary) and resuspended at ~10 mg of total lipid per milliliter of solvent A onto a column containing ~1 ml of resin per 10 mg of lipid. Wash the column with 2 column volumes of solvent A, then apply a linear gradient from solvent A to chloroform–methanol–4 M sodium acetate (30:60:8, v/v) with total gradient volume of ≥10 column volumes. Collect fractions and analyze by TLC. Desalt selected fractions by reversed-phase chromatography or by evaporation, resuspension in water, and extensive dialysis against water.

The capacity of Q–Sepharose is high, >20 μmol of ganglioside sialic acid per milliliter of resin. In an interesting related technique, bovine brain gangliosides were purposefully overloaded (>100 mg/ml of Q–Sepharose) under carefully defined conditions, resulting in retention of multisialogangliosides while monosialogangliosides (95% pure G_{M1}) eluted nearly quantitatively in the salt-free run-through fraction.[32]

High-Performance Liquid Chromatography on Amino-Bonded Silica. This technique has the advantages of good resolution and detection of anionic glycosphingolipids by in-line ultraviolet (UV) detection. Although often considered normal-phase chromatography, this weakly basic resin can act as an anion exchanger at acidic pH and so is included in this section. The procedure of Gazzotti *et al.*[33] is detailed, using a LiChrosorb-NH_2 (E. Merck, AG, Darmstadt, Germany) amino-bonded silica column, 250 × 10 mm, as described by Ladisch *et al.*[34] Preequilibrate the column at a flow rate of 6.25 ml/min with acetonitrile–aqueous 5 mM sodium phosphate buffer, pH 5.6 (83:17, v/v) (solvent A). Evaporate salt-free glycosphingolipid sample (e.g., up to 1 μmol of lipid-bound sialic acid) in a microtube, redissolve in 100 μl of water, introduce the solution into the HPLC port, wash the tube with a small volume of additional water, and add to the previous sample in the injection loop. Initiate chromatography using a gradient from solvent A to acetonitrile–20 mM sodium phosphate buffer, pH 5.6 (1:1, v/v) (solvent B), as follows: 7 min at 100% solvent A, a 53-min linear gradient from 0 to 34% solvent B, then a 20-min linear gradient from 34 to 64% solvent B. Collect 0.5-min fractions. Between runs recondition the column for 10 min with solvent B, then for 15 min with solvent A. The procedure can be scaled up or down for larger preparative or smaller analytical runs. Perform in-line analysis for gangliosides with a UV detector at 215 nm. Major brain gangliosides (G_{M3} to G_{Q1b})

---

[32] O. Koul, M. Prada-Maluf, R. H. McCluer, and M. D. Ullman, *J. Lipid Res.* **32**, 1712 (1991).

[33] G. Gazzotti, S. Sonnino, and R. Ghidoni, *J. Chromatogr.* **348**, 371 (1985).

[34] S. Ladisch, C. C. Sweeley, H. Becker, and D. Gage, *J. Biol. Chem.* **264**, 12097 (1989).

elute in order of increasing complexity and are well separated over the course of the chromatographic run, with little or no tailing. Analytical detection limits (using a 250 × 4 mm column running the same gradient at 1 ml/min) are subnanomolar for all gangliosides. Molar absorbance increased with increasing number of sialic acid residues per ganglioside (sulfated glycosphingolipids are also resolved and detected).[35] Selected fractions are evaporated to dryness and desalted by dialysis or reversed-phase chromatography.

Silicic Acid Chromatography

Silicic acid chromatography is an excellent technique for purification of many neutral and anionic glycosphingolipids, underivatized or as their peracetylated derivatives. Although other high-quality porous silica matrixes are suitable,[36] many glycosphingolipid separations have utilized Iatrobeads, beaded porous silica that can be obtained from Iatron Laboratories (Tokyo, Japan).[37] Subsequent to silicic acid chromatography, remove minor contaminants leached from the resins by reversed-phase chromatography.

Low-pressure liquid chromatography is performed with 60-μm-diameter Iatrobeads (6RS-8060), which have been treated at 125° to remove water, cooled, then slurried in initial elution solvent (see below) and loaded into an appropriate chromatographic column. Large, high-capacity columns can be economically prepared with this resin. Columns may be reused, and/or the resin can be regenerated by washing with increasingly polar, then nonpolar, solvents (e.g., water–ethanol), then by oven-drying the resin. Fine resolution of smaller amounts of glycosphingolipids can be accomplished by HPLC on columns packed with 10-μm-diameter Iatrobeads (6RS-8010). Two solvent systems are most useful for glycosphingolipid purification: chloroform–methanol–water and 2-propanol–hexane–water, with the later system giving better resolution for several applications. Glycolipids may be eluted isocratically or with gradients from lower to higher solvent polarity. A few examples serve to illustrate these procedures.

Desalted mixed brain monosialogangliosides (100 mg), purified by stepwise elution from DEAE–Sephadex (see above), were dissolved in

[35] H. Murakami, Z. Lam, B. C. Furie, V. N. Reinhold, T. Asano, and B. Furie, *J. Biol. Chem.* **266,** 15414 (1991).

[36] J. Gottfries, P. Davidsson, J.-E. Månsson, and L. Sevennerholm, *J. Chromatogr.* **490,** 263 (1989).

[37] Address: 11-4 Higashi-Kanda 1-Chome, Chiyoda-ku, Tokyo 101, Japan (phone: 03-3862-1761, fax: 03-3862-1767)

chloroform–methanol–water (65 : 35 : 3, v/v) and loaded onto a 1.2 ×
100 cm column of Iatrobeads 6RS-8060 (50 g) prepared and washed in
the same solvent.[13] Isocratic elution with the same solvent resolved
several of the species. Overlapping fractions were rerun isocratically,
using chloroform–methanol–water (70 : 30 : 3, v/v) to slow elution and
improve separation. In another study,[5] 400 mg of mixed disialogangliosides prepared in a similar manner were dissolved in chloroform–methanol–water (53 : 45 : 2, v/v) and applied to a 1.9 × 115 cm column of
Iatrobeads (6RS-8060) (100 g) in the same solvent. Gradient elution
from the same solvent to chloroform–methanol–water (25 : 73 : 2, v/v)
(1 liter each) was used to resolve some of the disialogangliosides. A
similar gradient was used to resolve sulfated glycosphingolipids from
kidney,[38] whereas less polar gradients of the same solvents (83 : 16 : 0.5
to 55 : 42 : 3, v/v) resolved major short neutral glycosphingolipids from
erythrocytes.[39]

Long neutral glycosphingolipids from erythrocytes (~200 mg) prepared
by solvent partioning followed by reversed-phase chromatography were
dissolved in chloroform–methanol (1 : 1, v/v) and applied to a 1 × 100 cm
column of Iatrobeads (6RS-8060) preequilibrated with 2-propanol–hexane–water (50 : 40 : 5, v/v). Elution with a 400-ml gradient from the same
solvent to 2-propanol–hexane–water (55 : 30 : 15, v/v) resolved several
major species, whereas subsequent HPLC of selected fractions on Iatrobeads (6RS-8010), using a gradient from the same starting solvent to 2-
propanol–hexane–water (55 : 36 : 9, v/v), resolved comigrating species.
Variations of these gradients have been used to purify many novel glycosphingolipids, with more polar glycosphingolipids species eluted with gradients of 2-propanol–hexane–water [from 55 : 40 : 5 (v/v) to 55 : 25 : 20
(v/v), lower phase].[6,7]

Glycosphingolipids that comigrate on HPLC may be resolved as their
peracetylated species. Acetylation (and subsequent deacetylation) of a
fraction of purified glycosphingolipids is performed as described under
Bulk Separations (earlier). The acetylated species in hexane are loaded
onto an Iatrobeads (6RS-8010) column preequilibrated in the same solvent.
Acetylated short neutral glycosphingolipids are eluted with a gradient from
hexane to 2-propanol–hexane (55 : 45, v/v) or 2-propanol–hexane–water
(55 : 45 : 5, v/v), whereas acetylated neutral glycosphingolipids with five
saccharides were separated with a gradient from hexane to 2-propa-
nol–hexane–water (43 : 56 : 1, v/v).[6]

[38] K. Tadano and I. Ishizuka, *J. Biol. Chem.* **257**, 9294 (1982).
[39] S. Ando, M. Isobe, and Y. Nagai, *Biochim. Biophys. Acta* **424**, 98 (1976).

Preparative Thin-Layer Chromatography

Because recovery of resolved glycosphingolipids from TLC plates is often poor and silica-derived contaminants are high compared to HPLC, other methods should be considered prior to initiating preparative TLC. When necessary, the same silica gel 60-based TLC systems used for analytical TLC[14] can be used to resolve species for preparative isolation. A partially purified glycosphingolipid sample is dissolved in a minimum volume and applied in the shortest linear band (up to several centimeters) compatible with good resolution (overloading causes streaking and cross-contamination). Immediately after running, being careful not to let the plate dry completely, the desired area(s) of silica are identified, scraped from the support, and placed in solvent for extraction. Identification of the areas for collection must be rapid (to avoid plate drying, which reduces recovery), and may utilize direct reversible staining or removal of flanking strips on prescored plates for use of more sensitive destructive sprays.[14] Methods for direct reversible staining depend on the amount and nature of the glycosphingolipids. High concentrations of nonpolar glycosphingolipids may be visible as white areas against a translucent background during brief air drying or on spraying with distilled water. Incubation of the partially dried plate in a covered TLC chamber containing solid iodine will result in brown staining of lipids. For higher sensitivity, allow the plate to air dry briefly, spray lightly with a solution of 0.1% (w/v) primuline (e.g., from Aldrich Chem. Co., Milwaukee, WI) in acetone–water (4 : 1, v/v), and identify lipids as light bands against a dark background under long-wave UV light.[40] Thin-layer chromatography scrapings are treated with chloroform–methanol–water (10 : 10 : 1, v/v) or 2-propanol–hexane–water (55 : 25 : 20, v/v) and agitated with a stir bar or in an ultrasonic bath. Recovery of polysialogangliosides is improved by mixing the sorbent first in water, then adding the organic solvents.[5] Extracts must be further purified, for example, by reversed-phase (Sep-Pak) chromatography (see above), to remove contaminants introduced from the silica gel and/or by silicic acid chromatography to remove primuline.[28]

Applications

The particular combination of extraction, bulk separation, and chromatographic resolution techniques to be used will depend on the chemical nature of the target glycosphingolipid(s) and the degree of purification

---

[40] V. P. Skipski, *in* "Methods in Enzymology" (J. Lowenstein, ed.), Vol. 35, p. 396. Academic Press, New York, 1975.

required. For instance, isolation of long and short neutral glycosphingolipids as a group requires a different approach than for isolating a minor polysialylated ganglioside for structural characterization. The purification sequences below are a few examples of combinations of the extraction, separation, and resolution techniques described in earlier sections as applied to particular goals. They represent minor modifications of the referenced published procedures.

Neutral Glycosphingolipids

Large-Scale Purification of Short Neutral Glycosphingolipids.[41] This method was applied for purification of several known neutral glycosphingolipids from blood cell membranes.

1. warm ethanol extraction
2. partition A (use lower phase)
3. acetylation/bulk silicic acid chromatography/deacetylation
4. silicic acid HPLC

Notes: HPLC of 300-μg lower phase lipids was performed on a 500 $\times$ 4 mm columns of Iatrobeads (6RS-8010), applying a 40-min gradient of 2-propanol–hexane–water (55 : 44 : 1 to 55 : 30 : 15, v/v) at 2 ml/min.

Large-Scale Purification of a Long Neutral Glycosphingolipid.[28] This technique is representative of several reports of isolation of novel neutral glycosphingolipids for structure determination.

1. extraction with chloroform–methanol (2 : 1, v/v), then with 2-propanol–hexane–water (55 : 25 : 20, v/v)
2. partition A (use upper phase)
3. bulk DEAE–Sephadex chromatography (use flow-through and salt-free wash fractions)
4. acetylation/bulk silicic acid chromatography/deacetylation
5. silicic acid HPLC
6. preparative TLC.

Notes: HPLC of glycosphingolipid fraction from 500 g of tissue (English sole liver) was performed on a 1 $\times$ 100 cm column of Iatrobeads (6RS-8010), applying a gradient of 2-propanol–hexane–water (55 : 42 : 3 to 55 : 25 : 20, v/v). Preparative TLC utilized high-performance thin-layer chromatography (HPTLC) plates developed in chloroform–methanol–aqueous 0.02% CaCl$_2$ (55 : 40 : 10, v/v); lipids were detected with primuline and sorbent was extracted with 2-propanol–hexane–water (55 : 25 : 20, v/v).

---

[41] K. Watanabe and Y. Arao, *J. Lipid Res.* **22**, 1020 (1981).

Total Neutral Gycosphingolipids (Short and Long).[26] This technique avoids solvent partioning to keep all neutral glycosphingolipids together for subsequent TLC overlay analysis.

1. extraction A
2. saponification
3. bulk silicic acid chromatography
4. bulk anion-exchange chromatography
5. acetylation/bulk silicic acid chromatography/deacetylation
6. bulk anion-exchange rechromatography
7. bulk silicic acid rechromatography

Gangliosides

Large-Scale Purification of Novel Gangliosides

Procedure 1.[42] This procedure is representative of many reports using straightforward and well-established techniques.

1. extraction B
2. partition B
3. saponification
4. step-gradient DEAE–Sephadex chromatography.
5. step-gradient silicic acid chromatography
6. silicic acid HPLC.

Notes: The monosialoganglioside fraction from 2.2 kg of rat spleens was eluted from the DEAE–Sephadex, then applied to a 3.6 × 100 cm Iatrobeads (6RS-8060) column that was eluted stepwise with 2–4 liters each of chloroform–methanol–water (70 : 30 : 4, 60 : 40 : 2, 50 : 50 : 2.5, then 40 : 60 : 3, v/v). High-performance liquid chromatography of a selected fraction was on an Iatrobeads (6RS-8010) column eluted with a gradient of chloroform–methanol–aqueous 7.5 M ammonium hydroxide (60 : 36 : 4 to 30 : 63 : 7, v/v).

Procedure 2.[43] This procedure avoided alkali treatment in order to purify O-acetylated gangliosides.

1. extraction A
2. partition A
3. step-gradient DEAE–Sephadex chromatography
4. silicic acid chromatography

[42] K. Nohara, M. Suzuki, F. Inagaki, H. Ito, and K. Kaya, *J. Biol. Chem.* **265,** 14335 (1990).
[43] S. Ren, J. N. Scarsdale, T. Ariga, Y. Zhang, R. A. Klein, R. Hartmann, Y. Kushi, H. Egge, and R. K. Yu, *J. Biol. Chem.* **267,** 12632 (1992).

Notes: Upper phase partitioned lipids from 1.7 kg of buttermilk, corresponding to >4 g of upper phase lipids, were resolved on a 2 × 100 cm DEAE–Sephadex A-25 column. Desalted disialo- (~700 mg) and trisialoganglioside (~200 mg) fractions were resolved by repeated Iatrobeads column chromatography, first with chloroform–methanol–aqueous 0.5% CaCl₂ gradients, then with chloroform–methanol–water gradients.

Procedure 3.[24,31,44] This procedure was successful in resolving minor species from large-scale extractions.

1. extraction A
2. bulk DEAE–Sephadex chromatography
3. saponification
4. bulk silicic acid chromatography
5. Q–Sepharose chromatography
6. Preparative TLC

Notes: Desalted lipids from step 3 were suspended in chloroform–methanol (1 : 1, v/v), then were adjusted to chloroform–methanol (4 : 1, v/v) for application to silicic acid (step 4). After washing the silica with the same solvent, gangliosides were eluted with chloroform–methanol (1 : 1, v/v). The resulting partially purified ganglioside fraction, containing 5 g of bovine brain lipids, was dissolved in 500 ml of chloroform–methanol–water (30 : 60 : 8, v/v) and applied to a 3 × 75 cm Q–Sepharose column, which was washed with 800 ml of the same solvent and then eluted with a linear gradient (total volume >6 liters) from the same solvent to chloroform–methanol–aqueous 4 *M* sodium acetate (30 : 60 : 8, v/v). Preparative TLC was on HPTLC plates developed in chloroform–methanol–water (5 : 4 : 1, v/v). Lipids were located with iodine vapor, and were eluted from sorbent in methanol–water (1 : 1, v/v).

Small-Scale Isolation of Total Gangliosides for Thin-Layer Chromatography Analysis

Several of the techniques for bulk separation, including partioning and ion exchange, are sufficient for purification of gangliosides from small samples for TLC analysis. One rapid procedure using extraction B and partition B follows.

Procedure.[45] A small tissue sample or cell pellet is adjusted to 0.6 ml with water, transferred to a 3-ml Potter–Elvehjem homogenizer on ice, homogenized (10 strokes), 1.6 ml of methanol is added, and the mixture is agitated vigorously and transferred to a 13 × 100 screw-capped test

---

[44] Y. Hirabayashi, T. Nakao, F. Irie, V. P. Whittaker, K. Kon, and S. Ando, *J. Biol. Chem.* **267**, 12973 (1992).

[45] N. M. Dahms and R. L. Schnaar, *J. Neurosci.* **3**, 806 (1983).

tube with a Teflon-lined cap. Chloroform (0.8 ml) is added and the capped mixture is vigorously agitated at ambient temperature. The tube is centrifuged in a table-top centrifuge (3000 g, 10 min) at ambient temperature, the supernatant transferred to a fresh tube, and the pellet reextracted with 2 ml of chloroform–methanol–water (4 : 8 : 3, v/v). After centrifugation, the second supernatant is added to the first (the remaining pellet may be used for protein determination if desired), the volume of the combined supernatants is measured, and water is added to bring the final chloroform–methanol–water ratio to 4 : 8 : 5.6 (v/v). After vigorous agitation, the two phases are separated by centrifugation (1500 g, 10 min, ambient temperature) and the upper phase is transferred to a fresh tube. The lower phase is reextracted with theoretical upper phase, and the combined upper phases are applied to a prewashed Sep-Pak cartridge, which is subsequently washed with chloroform–methanol–water (2 : 43 : 55, v/v) and methanol–water (1 : 1, v/v) prior to elution of gangliosides in methanol (see General Considerations). The ganglioside-containing fraction is evaporated and resuspended in a small volume of chloroform–methanol–water (4 : 8 : 3, v/v) for subsequent TLC analysis.

Notes. Although most gangliosides are recovered quantitatively in the upper phase of partition B, a portion of the G_{M3} or G_{M4} remains in the lower phase. When these lipids are of interest, techniques that avoid partitioning must be used. One such study, on small tumor samples,[46] used extraction B followed by bulk silicic acid chromatography, followed by peracetylation/bulk silicic acid chromatography, deacetylation, and bulk DEAE ion-exchange chromatography. Total gangliosides were then analyzed by TLC.

Sulfated Glycosphingolipids

Large-Scale Isolation of Novel Sulfated Glycosphingolipids.[47] This procedure starts with partitioning, then uses other bulk separation techniques to combine sulfated glycosphingolipids that split between upper and lower phases.

1. extraction A
2. partition A (use *both* phases)
3. saponification of lower phase from step 2, resuspension in chloroform–methanol–water (86 : 14 : 1, v/v)

[46] J. Gottfries, P. Fredman, J.-E. Månsson, V. P. Collins, H. von Holst, D. D. Armstrong, A. K. Percy, C. J. Wikstrand, D. D. Bigner, and L. Svennerholm, *J. Neurochem.* **55,** 1322 (1990).

[47] N. Iida, T. Toida, Y. Kushi, S. Handa, P. Fredman, L. Svennerholm, and I. Ishizuka, *J. Biol. Chem.* **264,** 5974 (1989).

4. partitioning of solution from step 3 against upper phase from step 2 (use *both* phases)
5. dialysis of upper phase from step 4, removal of solvents
6. combination of residue from step 5 with lower phase from step 4 to constitute total alkali-resistant lipids
7. gradient elution from DEAE–Sephadex
8. silicic acid chromatography
9. silicic acid HPLC

Notes: Lipids from ~600 g of rat kidneys were fractionated on a 2.2 × 33 cm column of DEAE–Sephadex eluted with a 4-liter gradient of chloroform–methanol–aqueous 30 mM ammonium acetate (5 : 10 : 1, v/v) to chloroform–methanol–aqueous 1 M ammonium acetate (5 : 10 : 1, v/v). Desired fractions were applied to a 1.2 × 46 cm Iatrobeads (6RS-8060) column and eluted with a 2-liter gradient of chloroform–methanol–water (190 : 10 : 1 to 70 : 30 : 2, v/v). High-performance liquid chromatography utilized a porous spherical silica column (1 × 30 cm, 5-μm beads) with a 600-ml gradient from chloroform to chloroform–methanol (7 : 3, v/v).

Separations Based on Ceramide Structure

Glycosphingolipids that have been purified to saccharide homogeneity can be resolved, often to ceramide homogeneity, by reversed-phase chromatography on alkylated silica gel HPLC columns. Up to 400 μg of purified monosialoganglioside species has been resolved by isocratic chromatography on a 5-μm particle size C_{18} analytical column (4.6 × 250 mm) eluted with methanol–water at various ratios near 85 : 15 (v/v), depending on the particular ganglioside.[48] Alternatively, isocratic elution with acetonitrile–aqueous sodium phosphate buffer (pH 7) at various ratios near 1 : 1 (v/v) has been used.[49] Eluting gangliosides are detected in line by UV absorbance (195–205 nm). Reverse phase is relatively insensitive to the fine structure of the saccharide, and resolution is dependent primarily on the total number of ceramide methylenes. This results in comigration of gangliosides having d18 : 1-C20 : 0 and d20 : 1-C18 : 0 ceramides. Likewise, gangliosides having d18 : 1-C22 : 0 and d18 : 1-C24–1 ceramides coelute. Nevertheless, reversed-phase chromatography resolves purified human brain G_{M1}, for instance, into at least 15 distinct fractions based on ceramide structure, 9 of which are homogeneous.

[48] H. Kadowaki, J. E. Evans, and R. H. McCluer, *J. Lipid Res.* **25**, 1132 (1984).
[49] P. Palestini, M. Masserini, S. Sonnino, A. Giuliani, and G. Tettamanti, *J. Neurochem.* **54**, 230 (1990).

[23] Thin-Layer Chromatography of Glycosphingolipids

By Ronald L. Schnaar and Leila K. Needham

Introduction

Because of their amphipathic character, glycosphingolipids are well suited for analysis by thin-layer chromatography (TLC), which is useful for monitoring purification, for qualitative and quantitative determination of expression in normal and pathological tissues, for partial structural analysis, and for detecting biological activities, including immunoreactivity and binding activity toward toxins, viruses, bacteria, and eukaryotic cells.

Although this chapter focuses on TLC, complete structural analyses of a glycosphingolipid species require a combination of techniques to determine the composition, sequence, linkage positions, and anomeric configurations of its oligosaccharide moiety, and the fatty acid(s) and long-chain base(s) of its ceramide. Although methods for the detailed determination of glycosphingolipid structure are beyond the scope of this chapter, they are briefly outlined below, with references to direct further reading (also see the comprehensive treatise by Hakomori).[1]

Carbohydrate composition of glycosphingolipids is determined by depolymerization and analysis of the released monosaccharides. Often, purified glycosphingolipids are treated with metanolic HCl followed by N-acetylation of any hexosamines or sialic acids and quantitation of the resulting methylglycosides as their trimethylsilyl derivatives by gas chromatography.[1,2] Alternatively, aqueous acid hydrolysis is performed followed by high-performance liquid chromatography (HPLC) analysis of the released monosaccharides.[3,4] Mass spectroscopy of underivatized glycosphingolipids or of their permethylated derivatives is a powerful tool for carbohydrate sequence determination and ceramide identification.[5,6] Linkage determination is most rigorously performed by methylation analy-

[1] S. Hakomori, *in* "Sphingolipid Biochemistry" (J. N. Kanfer and S. Hakomori, eds.), p. 1. Plenum, New York, 1983.

[2] W. J. Esselman, R. A. Laine, and C. C. Sweeley, *in* "Methods in Enzymology" (V. Ginsburg, ed.), Vol. 28, p. 140. Academic Press, New York, 1972.

[3] K. M. Walton, K. Sandberg, T. B. Rogers, and R. L. Schnaar, *J. Biol. Chem.* **263,** 2055 (1988).

[4] A. E. Manzi, S. Diaz, and A. Varki, *Anal. Biochem.* **188,** 20 (1990).

[5] A. Dell, *Adv. Carbohydr. Chem. Biochem.* **45,** 19 (1987).

[6] B. Domon, J. E. Vath, and C. E. Costello, *Anal. Biochem.* **184,** 151 (1990).

sis,[7] whereas anomeric configurations can be obtained by nuclear magnetic resonance (NMR), given sufficient quantities.[8] Information on carbohydrate sequence, linkage, and anomeric configuration can also be obtained by combining enzymatic hydrolysis, using specific glycohydrolases, with TLC (see below). Release of the entire intact oligosaccharide chains from most glycosphingolipids is readily accomplished using ceramide glycanases (see below),[9,10] and the released oligosaccharides are subject to analysis by methods described elsewhere in this volume.

Treatment of glycosphingolipids with methanolic HCl releases the fatty acid amides in the ceramide as fatty acid methyl esters, which are identified and quantitated by gas chromatography-mass spectroscopy. Similarly, ceramide long-chain bases are released from glycosphingolipids in aqueous/methanolic HCl and identified as N-acetylated, trimethylsilyl derivatives by gas chromatography-mass spectroscopy.[1]

Colorimetric Assays for Glycosphingolipids

Although quantitative TLC can be used to determine the concentration of resolved glycosphingolipid species, it is useful to estimate the total concentration of glycosphingolipid in a sample prior to or in association with TLC, using the methods detailed below. Glycosphingolipid samples must first be freed of major contaminating lipids, proteins, and low molecular weight contaminants by the procedures described in [22] in this volume, which also includes recommendations for sample handling.[11]

Sphingosine

Total sphingolipid concentration is determined by measuring the primary amine on the long-chain base that is liberated from glycosphingolipids by methanolysis. The resulting long-chain base is extracted into ethyl ether and quantitated with fluorescamine.[12] In a series of screw-capped test tubes (13 × 100 mm, with Teflon-lined caps) place aliquots of unknown glycosphingolipid samples or sphingosine standards containing 1–100 nmol of sphingolipid. Evaporate solvents under a stream of nitrogen or *in vacuo*.

[7] S. B. Levery and S. Hakomori, *in* "Methods in Enzymology" (V. Ginsburg, ed.), Vol. 138, p. 13. Academic Press, Orlando, FL, 1987.

[8] R. K.Yu, T. A. W. Koerner, J. N. Scarsdale, and J. H. Prestegard, *Chem. Phys. Lipids* **42**, 27 (1986).

[9] B. Zhou, S.-C. Li, R. A. Laine, R. T. C. Huang, and Y. T. Li, *J. Biol. Chem.* **264**, 12272 (1989).

[10] G. C. Hansson, Y. T. Li, and H. Karlsson, *Biochemistry* **28**, 6672 (1989).

[11] R. L. Schnaar, this volume [22].

[12] M. Naoi, Y. C. Lee, and S. Roseman, *Anal. Biochem.* **58**, 571 (1974).

Add 0.5 ml of methanol–water–concentrated aqueous HCl (82 : 9 : 9, v/v) cap the tubes, mix, and heat at 70° for 18 hr. Allow the tubes to cool, add 0.75 ml of 0.2 M sodium borate buffer (pH 8) and 0.25 ml of 2 M aqueous sodium hydroxide, and mix. Add 1.5 ml of ethyl ether, then add 0.5 ml of fluorescamine (0.15 mg/ml in ethyl ether). Cap the tube and mix vigorously on a vortex mixer. Allow phases to separate, transfer the upper (ether) phase to a glass fluorometer tube or cuvette, and measure the relative fluorescence with excitation set at 385 nm and emission at 480 nm.

Neutral Sugars

Neutral sugar concentration in a glycosphingolipid sample is estimated by a phenol–sulfuric acid assay.[13] In a series of 13 × 100 mm glass test tubes, place unknown glycosphingolipid samples (or standards) containing 20–100 nmol of neutral sugar and evaporate solvents under a stream of nitrogen or *in vacuo*. Add 0.5 ml of water, mix, and sonicate briefly in a bath sonicator to suspend the samples. Handling one sample at a time, add 0.3 ml of 5% aqueous phenol, mix, then rapidly add 1.8 ml of concentrated sulfuric acid and mix vigorously on a vortex mixer (use care, because addition of the concentrated acid causes the sample tube to become hot). Allow the samples to cool and measure the absorbance at 480 nm.

Notes. Convenient addition of phenol and sulfuric acid can be performed with reservoir-type pipettors (e.g., Repipet; Labindustries, Inc., Berkeley, CA). Appropriate standards include previously characterized glycosphingolipids or mixtures of neutral sugars (e.g., glucose and galactose). The assay as described results in tubes that can be placed directly into an adapter of many spectrophotometers (e.g., Spectronic 20, Milton Roy, Rochester, NY), thus avoiding the need to transfer the viscous concentrated acid to a cuvette.

Sialic Acids

The concentration of lipid-bound sialic acid on gangliosides is estimated by a resorcinol–HCl–Cu^{2+} assay[14] after mild periodate treatment.[15] In 13 × 100 mm glass test tubes place ganglioside samples containing 1–12 nmol of sialic acid, evaporate any solvents under a stream of nitrogen or *in vacuo*, and dissolve the residue in 0.3 ml of water. Prepare a series of standard tubes containing 1–12 nmol of sialic acid in 0.3 ml of water. Chill the tubes on ice, then add 60 μl of 40 mM periodic acid and mix. Incubate

[13] J. F. McKelvy and Y. C. Lee, *Arch. Biochem. Biophys.* **132**, 99 (1969).
[14] L. Svennerholm, *Biochim. Biophys. Acta* **24**, 604 (1957).
[15] G. W. Jourdian, L. Dean, and S. Roseman, *J. Biol. Chem.* **246**, 430 (1971).

on ice for 30 min, then add 0.75 ml of ice-cold resorcinol–HCl–Cu^{2+} reagent prepared by mixing 40 ml of 1.5% (w/v) aqueous resorcinol, 0.6 ml of 1% (w/v) aqueous cupric sulfate, and 60 ml of concentrated HCl. Incubate on ice for 5 min, then heat at 100° for 20 min. Allow to cool to ambient temperature, add 1.1 ml of 2-pentanol, mix repeatedly, and centrifuge for 1 min at ~1000 g (ambient temperature) to separate the phases. Remove the upper (organic) phase to a cuvette and measure absorbance at 630 nm.

Notes. A 0.4 *M* stock of periodic acid can be stored frozen, then thawed and diluted to 40 m*M* for use. Resorcinol–HCl–Cu^{2+} reagent should be prepared from stock solutions weekly and stored at 4°; aqueous resorcinol can be stored at 4° for months, but should be replaced if it becomes brown, whereas aqueous cupric sulfate is stable.

Sulfated Glycosphingolipids

Sulfated glycosphingolipids are determined by the Azure A method of Kean.[16] Place a sample containing 1–20 nmol of sulfated glycosphingolipid in a screw-capped tube and evaporate to dryness. Add 2 ml of chloroform–methanol (1 : 1, v/v), 2 ml of 25 m*M* aqueous sulfuric acid, and 0.4 ml of a dye solution consisting of Azure A (Aldrich Chemical Co., Milwaukee, WI), 0.4 mg/ml in 1 m*M* aqueous sulfuric acid. Cap and mix the tubes vigorously for 30 sec, then centrifuge to separate the layers. Measure the absorbance of the lower phase at 645 nm.

Thin-Layer Chromatography

Apparatus

Thin-Layer Chromatography Plates. Commercial precoated silica gel 60 TLC or high-performance TLC (HPTLC) plates (glass support) from E. Merck AG (Darmstadt, Germany) are recommended for most applications. The HPTLC plates (0.2 mm layer, 4 to 7-μm particle size, e.g., Cat. No. 5644; E. Merck AG) offer the best resolution whereas TLC plates (0.25-mm layer, 5 to 20-μm particle size, e.g., Cat. No 5721; E. Merck AG) are convenient for loading larger amounts of sample. The same sorbents coated on aluminum plates are preferred for immunoverlay and some related techniques (see below). Prerunning the plates in the same solvent used for development is recommended, especially for quantitative work. Whether prerun or not, heat the plates in an oven at 125° for >10 min to

---

[16] E. L. Kean, *J. Lipid Res.* **9**, 319 (1968).

remove water (activate), then allow to cool in a sealed dry box. Use at once, or keep in a dry box prior to use.

Chromatography Chambers. Glycolipid TLC frequently requires chloroform–methanol–aqueous mixtures, which form complex vapor–liquid equilibria in the development tank. Therefore tank geometry and conditions during running alter chromatographic migration and resolution. Rectangular tanks, which are available in various sizes (e.g., from Camag Scientific, Muttenz, Switzerland), give the best results. Thin-layer chromatography solvents are added to the chamber to a depth of ~0.5 cm at least 30 min prior to initiating chromatography. Chambers are covered with lids of glass or stainless steel, which may be weighted to improve closure (silicone or petrolatum sealants should be avoided). The chamber is placed away from drafts and sources of heat or cold. Change the solvent daily or after running two or three plates, because most TLC solvent mixtures contain components that are differentially volatile and their composition will change with time and use. Although some investigators place saturation pads (filter paper) on the inside glass of the chamber to help equilibrate the vapor phase, we have not found them to be advantageous.

Solvents

The most widely used TLC developing solvents for glycosphingolipids are mixtures of chloroform, methanol, and water (or aqueous salts), because they form a single phase at a range of hydrophobicities well suited for glycosphingolipid resolution on silica gel TLC plates (Fig. 1). Chloroform–methanol–aqueous mixtures ranging from 70 : 30 : 4 (v/v) (relatively low polarity) to 50 : 40 : 10 (v/v) (relatively high polarity) are chosen depending on the particular glycosphingolipids under study. Chloroform–methanol–aqueous (60 : 35 : 8, v/v) is a good starting point for a wide variety of glycosphingolipids. Smaller, less polar glycosphingolipids will be resolved better in lower polarity solvents, whereas larger, more polar species require more polar solvents.

The use of aqueous salt rather than water is recommended for resolving gangliosides and other anionic glycosphingolipids. Salts alter ganglioside mobility and improve their resolution. Use aqueous KCl (0.25%, w/v) or aqueous $CaCl_2$ (0.02–0.25%, w/v) in place of water for preparation of TLC solvents for gangliosides. Ammonium hydroxide (1–5 M) in the aqueous phase results in changes in the relative mobilities of different gangliosides compared to a neutral aqueous phase, and is particularly useful when multiple solvents are used to determine purity.

Glycosphingolipids are also well resolved in 2-propanol–aqueous mixtures (3 : 1, v/v). Addition of acetonitrile [2-propanol–50 mM aqueous

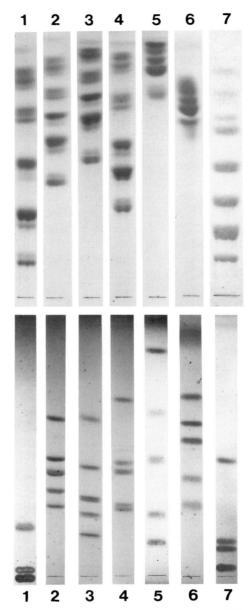

FIG. 1. Thin-layer chromatography of neutral glycosphingolipids and gangliosides in various solvent systems. Mixtures of neutral glycosphingolipids (*top*) and gangliosides (*bottom*) were applied on adjacent 0.5-cm pencil lines on a series of 10 × 10-cm glass prescored

KCl–acetonitrile (67 : 23 : 10, v/v)] has been reported to improve resolution of more polar gangliosides,[17] whereas addition of methyl acetate [2-propanol–6 M aqueous ammonium hydroxide–methyl acetate (15 : 5 : 1, v/v)] improves resolution of neutral glycosphingolipids (Fig. 1).[18]

Thin-layer chromatography of peracetylated glycosphinglipids may separate species that are not resolved as underivatized species. Peracetylated glycosphingolipids are chromatographed in 1,2-dichloroethane–acetone–water (60 : 40 : 0.1, 60 : 40 : 1, or 55 : 45 : 5, v/v) for longer chain acetylated glycosphingolipids,[19,20] in 1,2-dichloroethane–methanol–water (88 : 12 : 0.1, v/v) for shorter chain acetylated glycosphingolipids, or in chloroform–methanol (95 : 5, v/v.)[21]

Technique

Sample Application. For one-dimensional TLC, samples are applied as bands parallel to and 1–2 cm above the bottom of the plate. Mark positions for sample application with a light pencil line or with flanking pencil marks if detection of material at the origin is a concern. Analytical samples are applied as 0.5 to 1-cm bands, whereas preparative samples may be applied in bands up to 18 cm on a 20-cm plate (avoid ~1 cm at

[17] S. Ando, H. Waki, and K. Kon, *J. Chromatogr.* **405,** 125 (1987).

[18] K. Ogawa, Y. Fujiwara, K. Sugamata, and T. Abe, *J. Chromatogr.* **426,** 188 (1988).

[19] M. Gallatin, T. P. St. John, M. Siegelman, R. Reichert, E. C. Butcher, and I. L. Weissman, *Cell* (Cambridge, Mass.) **44,** 673 (1986).

[20] R. Kannagi, K. Watanabe, and S. Hakomori, *in* "Methods in Enzymology" (V. Ginsburg, ed.), Vol. 138, p. 3. Academic Press, Orlando, FL, 1987.

[21] M. E. Breimer, G. C. Hansson, K.-A. Karlsson, and H. Leffler, *J. Biol. Chem.* **257,** 557 (1982).

HPTLC plates. The neutral glycosphingolipid mix contained ~1 nmol/μl each of glucosylceramide, galactosylceramide, lactosylceramide, globotriaosylceramide, globotetraosylceramide, and gangliotetraosylceramide. The ganglioside mix contained 100 pmol/μl of lipid-bound sialic acid each of G_{M3}, G_{M1}, G_{D1a}, G_{D1b}, and G_{T1b}. One microliter of the neutral glycosphingolipid mix and 2 μl of the ganglioside mix were applied per plate. Each plate was developed in one of the solvent systems indicated below. After development, the plate was dried *in vacuo* and broken along the score to separate the neutral glycosphingolipid lane from the ganglioside lane. Neutral glycosphingolipids were detected with the orcinol–sulfuric acid stain and gangliosides with the resorcinol stain, as described in text. The developing solvents were as follows: 1, chloroform–methanol–aqueous 0.25% KCl (70 : 30 : 4, v/v); 2, chloroform–methanol–water (60 : 35 : 8, v/v); 3, chloroform–methanol–aqueous 0.25% KCl (60 : 35 : 8, v/v); 4, chloroform–methanol–2.5 N aqueous ammonium hydroxide (60 : 35 : 8, v/v); 5, chloroform–methanol–aqueous 0.25% KCl (55 : 45 : 10, v/v); 6, 2-propanol–aqueous 0.25% KCl (3 : 1, v/v); and 7, 2-propanol–aqueous 6 M ammonium hydroxide–methyl acetate (15 : 5 : 1, v/v).

each side of the plate, due to edge effects during development). Application of sample (usually in organic/aqueous mixtures) is accomplished with a microliter syringe (701N; Hamilton Co., Reno, NV) with its beveled tip lightly burnished to avoid wounding the silica gel layer. Using a TLC spotting guide (e.g., Analtech, Newark, DE) to help position the syringe tip, evenly apply ~1 μl of the sample per centimeter in a thin line along the premarked band. If >1 μl/cm is to be applied, allow the initial solvent to evaporate and then repeat, using ~1 μl/cm in each application. A forced air dryer can be used to assist in solvent evaporation for larger samples, although heat should be avoided.

Development. Soon after the applied samples are dry, place the plate in a chromatography chamber that has been preincubated with running solvent (see above). The chamber should not be disturbed or opened during the run. Normally, a single run to within 1 cm of the top of the TLC plate is sufficient for the desired resolution, although drying and redevelopment in different solvents improves resolution in some systems.[22,23] The use of short-bed continuous development can also enhance resolution.[24] After running, the plate is removed and solvents are allowed to evaporate in a fume hood. The plate should be thoroughly dry prior to application of analytical detection reagents, because the presence of running solvents will often result in background staining. Drying may be accelerated by placing the plate under vacuum, by mild heating, or by use of a forced air dryer. In contrast, do not allow preparative TLC plates to dry thoroughly or sample recovery will suffer.[1,11]

Reagent Application and Staining. Place the dried, developed plate upright in a spraying box in a fume hood and apply reagents, using a fine mist sprayer (e.g., Cat. No 422500-0125; Kontes, Vineland, NJ). Even application should be accomplished, especially for quantitative work, by initiating the spray off to the side and away from the plate surface, then moving the spray evenly across the plate and beyond it on the other side, then returning in a zigzag pattern until the plate surface is covered. The density of application will depend on the particular reagent, as detailed in the next section. Some reagents require covering the sprayed plate with a clean glass plate of equal size prior to heating. When cleaning cover plates, rinse thoroughly and allow to air dry, avoiding the use of paper towels, which will leave cellulose lint that will char during heating and mottle the chromatogram.

[22] J. K. Yao and G. M. Rastetter, *Anal. Biochem.* **150**, 111 (1985).

[23] S. Harth, H. Dreyfus, P. F. Urban, and P. Mandel, *Anal. Biochem.* **86**, 543 (1978).

[24] W. W. Young, Jr. and C. A. Borgman, *in* "Methods in Enzymology" (V. Ginsburg, ed.), Vol. 138, p. 125. Academic Press, Orlando, FL, 1987.

Chemical Detection of Glycosphingolipids and Lipid Contaminants on
  Thin-Layer Chromatography Plates

Gangliosides

Resorcinol–HCl–Cu^{2+} reagents based on that of Svennerholm[14] remain
among the most selective and sensitive tools for ganglioside detection on
TLC plates. Resorcinol–HCl–Cu^{2+} spray reagent is prepared by mixing
20 ml of 1.5% aqueous resorcinol, 0.3 ml of 1% (w/v) aqueous cupric
sulfate pentahydrate, 30 ml of concentrated aqueous HCl, and 50 ml of
water. This is equivalent to diluting the colorimetric assay reagent with
an equal volume of water (see earlier for storage of reagents). Spray the
dried chromatograph moderately with reagent (the silica gel should not
appear wet or darkened), cover with a clean glass plate, clip the plate in
place with binder clips, and place in an oven at 125° for 20 min. Gangliosides (or other sialic acid-containing substances) appear blue-violet against
a white background, whereas neutral and sulfated glycosphingolipids appear brown. Yellow-brown discoloration of the background may indicate
incomplete evaporation of running solvents prior to spraying. The color
is stable for many days when the plate is kept covered (remove the clamps
and tape the glass cover plate to the chromatogram for storage).

Quantitative detection by scanning densitometry, using a light source
adjusted near 600 nm (with appropriate filters), detects as little as 25 pmol
of lipid-bound sialic acid, and is linear up to 5 nmol.[25] For routine visual
analyses, apply ~200 pmol of lipid-bound sialic acid per ganglioside species. The color yield is approximately proportional to the number of sialic
acids, such that a disialoganglioside, generates about twice as much color
as an equimolar amount of monosialoganglioside, and so on. However,
molar color yield varies between different ganglioside structures and
matched standards should be used when feasible.

Sulfated Glycosphingolipids

Sulfatides, along with other sulfated lipids (e.g., cholesterol sulfate),
are detected with the cationic dye, Azure A.[26] Dissolve 2 g of Azure A
(Aldrich Chem. Co.) per 100 ml of 1 mM aqueous sulfuric acid. Spray
the dried chromatograph heavily with reagent until it appears uniformly
dark, then immediately immerse the plate in 40 mM aqueous sulfuric
acid–methanol (3 : 1, v/v) (for 5 × 10 cm plates, this is conveniently done

[25] N. M. Dahms and R. L. Schnaar, *J. Neurosci.* **3**, 806 (1983).
[26] N. Iida, T. Toida, Y. Kushi, S. Handa, P. Fredman, L. Svennerholm, and I. Ishizuka,
J. Biol. Chem. **264**, 5974 (1989).

in a 15-cm-diameter disposable plastic petri dish). Destain with gentle agitation, changing the washing solution as it discolors until the background appears pale (~1 hr). Sulfated glycosphingolipids appear dark blue against a pale blue background. The color will fade, but can be maintained by wrapping the plate in plastic film and storing at $-20°$. As little as 125 pmol of lipid-bound sulfate is detected and color intensity increases proportionally up to 2 nmol. Azure A interacts selectively with sulfated lipids rather than with those with sialic acids or phosphates,[16] although the selectivity is not absolute and the latter will generate color at higher concentrations.

Neutral or Anionic Glycosphingolipids

Because nearly all glycosphingolipids contain at least one neutral sugar, they can be detected with a number of general carbohydrate stains,[27,28] the most common of which are based on the orcinol–sulfuric acid test of Svennerholm.[29] Dissolve 0.5 g of orcinol in 100 ml of 3 M aqueous sulfuric acid. Slowly add to 100 ml of ethanol. The reagent is stable for several days in the dark at 4°. Spray the dried plate moderately (not visibly moist) and heat at 125° for 5 min. Glycosphingolipids appear purple on an off-white background. Color yield is proportional to the number of neutral sugar residues in the glycosphingolipid structure, with 500 pmol of total neutral sugar per band readily detectable.

Another sensitive assay suitable for quantitative TLC utilizes N-(1-naphthyl)ethylenediamine (NED; J. T. Baker, Phillipsburg, NJ).[30] Prepare a solution of 13 mM NED in methanol-concentrated sulfuric acid (97 : 3, v/v). Spray the dried plate heavily with reagent (~1 ml of reagent for each 10 cm^2 of plate area). Heat at 125° for 10 min. Glycosphingolipids appear pink on a white background. The color fades relatively rapidly, but fading can be slowed by wrapping the plate in plastic and storing at $-20°$. Although sensitivity for different sugars varies, quantitation of less than 100 pmol of saccharide was reported.

General Stains

General stains are useful for detection of glycosphingolipids and contaminants during purification. The simplest of these is ethanol–water–con-

[27] V. P. Skipski, *in* "Methods in Enzymology" (J. Lowenstein, ed.), Vol. 35, p. 396. Academic Press, New York, 1975.
[28] S. K. Kundu, *in* "Methods in Enzymology" (J. Lowenstein, ed.), Vol. 72, p. 185. Academic Press, New York, 1981.
[29] L. Svennerholm, *J. Neurochem.* **1**, 42 (1956).
[30] M. Bounias, *Anal. Biochem.* **106**, 291 (1980).

centrated sulfuric acid (45 : 45 : 10, v/v). The dried plate is sprayed moderately (not visibly moist), then heated at 125° for 20 min. Many organic compounds appear brown against a white background. Higher concentrations of acid are unnecessary, and inclusion of ethanol assists in even spraying of the plate.

Cupric sulfate in aqueous phosphoric acid is a more sensitive char reagent, which can detect 250 pmol of G_{M1} or 1 nmol of sulfatide, and which is suitable for quantitative TLC.[22] In distilled water, prepare an aqueous spray reagent containing 15.6 g of cupric sulfate pentahydrate and 9.4 ml of aqueous concentrated phosphoric acid (85%, w/v) per 100 ml. Spray the dried plate heavily, until saturated with the reagent. Allow the plate, which appears blue, to drain for 5 min, then heat at 150° for 10 min. A broad variety of lipids, including cholesterol, triglycerides, phospholipids, and sphingolipids, appears brown against a white background. The color fades, but can be maintained if the plate is wrapped in plastic and stored at −20°.

Phospholipid Stain

Phospholipids are detected as contaminants of glycosphingolipids by using molybdenum blue reagent, which is available premixed (e.g., Alltech Assoc., Inc., Deerfield, IL) or is prepared as described previously.[31] Spray the dried plate lightly with reagent, then lightly with ethanol–water–concentrated sulfuric acid (45 : 45 : 10, v/v). Phospholipids appear blue on a white background. The color will fade, but can be preserved by wrapping in plastic and storing at −20°. Sensitivity is ~1 nmol of phospholipid.

Reversible General Stains

Reversible stains are useful for preparative TLC.[11] To improve recovery, the plate should not be allowed to dry thoroughly at any time after running. The simplest but least sensitive reversible staining method is to spray the TLC plate heavily with distilled water. The silica becomes translucent while lipids remain white. Iodine vapor detects many lipids, also with low sensitivity. Scatter solid iodine crystals along one side of a double-trough TLC chamber (or place in a beaker in any closed chamber) and cover. After the chamber atmosphere is visibly saturated with iodine vapor, place the TLC plate in the chamber for several minutes, until lipids appear yellow-brown against a tan background.

[31] V. P. Skipski and M. Barclay in "Methods in Enzymology" (J. Lowenstein, ed.), Vol. 14, p. 530. Academic Press, New York, 1969.

The most sensitive reversible stain for lipids is primuline, which can detect 50 ng of glycosphingolipid.[27] After allowing the developed plate to dry briefly, spray lightly with a solution of 0.1% (w/v) primuline (e.g., from Aldrich Chem. Co.) in acetone–water (4 : 1, v/v). Lipids appear as light bands against a dark background under long-wave length ultraviolet (UV) light. After recovery of lipids by extraction with organic solvent, the primuline can be removed by silicic acid chromatography.[32]

Chemical and Enzymatic Modification of Glycosphingolipids

In conjunction with other data (especially compositional analysis), selective chemical or enzymatic modification of glycosphingolipids can generate valuable structural information. Depending on the specificity of the modifications (especially when using enzymes), TLC analysis of the lipid products can reveal the number, sequence, position, and/or anomeric linkage of the removed residues. Comparison of TLC migration of the products with known standards, and/or identification of the partially degraded structures by immunoverlay techniques (see below), help identify the structures.

Chemical Modification

Various mild acid treatments remove sialic acids, fucose residues, and/or sulfates, but leave neutral core structures unaltered.

Partial or complete hydrolysis of sialic acids from gangliosides aids in their identification. Partial hydrolysis is especially useful in analyzing polysialogangliosides.[3,33] Evaporate sample containing ~2 nmol of sialic acid to dryness in a small sealable vial. Dissolve the resulting residue in 100 μl of 10 mM formic acid. Heat at 80° for 30 min. Cool, add 10 μl of 100 mM ammonium hydroxide to neutralize, and evaporate to dryness. Redissolve the residue in 10 μl of chloroform–methanol–water (4 : 8 : 3, v/v) and spot 5 μl for TLC analysis. Complete sialic acid hydrolysis is performed similarly, using 1 M formic acid at 100° for 45 min.

Sulfate groups are removed by treating the dried sample with 50 mM anhydrous methanolic HCl,[34] which is conveniently prepared by adding 0.36 ml of acetyl chloride to 100 ml of methanol. After 3.5 hr at ambient

---

[32] G. K. Ostrander, S. B. Levery, H. L. Eaton, M. E. K. Salyan, S. Hakomori, and E. H. Holmes, *J. Biol. Chem.* **263,** 18716 (1988).

[33] Y. Hirabayashi, T. Nakao, F. Irie, V. P. Whittaker, K. Kon, and S. Ando, *J. Biol. Chem.* **267,** 12973 (1992).

[34] D. K. H. Chou, A. A. Ilyas, J. E. Evans, C. Costello, R. H. Quarles, and F. B. Jungalwala, *J. Biol. Chem.* **261,** 11717 (1986).

temperature, evaporate the solvents and redissolve the residue for TLC analysis. Some sulfated glycosphingolipids also contain glucuronic acid residues, which will be converted to their corresponding carboxylic acid methyl esters by this treatment.

Fucose residues are quantitatively removed by treatment with 0.1 N aqueous trichloroacetic acid at 100° for 2 hr.[35] After incubation, add water and solvents to bring to ~2 ml of chloroform–methanol–water (2 : 43 : 55, v/v). Pass the reaction over a prewashed Sep-Pak C_{18} column, wash with the same solvent and methanol–water (1 : 1, v/v), and elute glycosphingolipids in methanol or chloroform–methanol (1 : 1, v/v) as described.[11] Evaporate to dryness and redissolve in a small volume for TLC analysis.

Although most glycosphingolipids are stable to mild base treatment, those with O-acetyl groups on sialic acids (or elsewhere) or other base-labile structures (e.g., sialic acid lactones) can be characterized by alkaline hydrolysis followed by TLC.[36] Evaporate the sample to dryness and redissolve in a small volume of 0.2 N NaOH in methanol. After 2 hr at 37° add a 1/10 vol of 2 M acetic acid in methanol to neutralize, then isolate the product with a Sep-Pak C_{18} column as described earlier.

Alkaline sensitivity in conjunction with two-dimensional TLC is used to identify glycosphingolipids with alkali-labile groups.[37] Spot a glycosphingolipid sample at one corner of a square TLC plate and develop in any appropriate solvent system (one that does not use ammonium hydroxide). After drying, place the plate in a sealed TLC chamber saturated with ammonia vapor for 5 hr (do not put the TLC plate in direct contact with the ammonium hydroxide solution). Withdraw the plate from the tank and remove any ammonia under a stream of warm air. Redevelop the TLC in the same solvent as that used originally, but at right angles to the original direction. Glycosphingolipids that are base stable form a diagonal line on the plate, whereas those that are base labile migrate off the diagonal.

Enzymatic Modification

Enzymes provide the most specific reagents for TLC structural analysis. Certain exo- and endoglycohydrolases specific for the saccharide structure, linkage position, and anomeric configuration are available.[38] In

[35] K. Stellner, K. Watanabe, and S. Hakomori, *Biochemistry* **12**, 656 (1973).

[36] S. Ren, J. N. Scarsdale, T. Ariga, Y. Zhang, R. A. Klein, R. Hartmann, Y. Kushi, H. Egge, and R. K. Yu, *J. Biol. Chem.* **267**, 12632 (1992).

[37] S. Sonnino, R. Ghidoni, V. Chigorno, M. Masserini, and G. Tettamanti, *Anal. Biochem.* **128**, 104 (1983).

[38] G. S. Jacob and P. Scudder this volume [17].

conjunction with near-quantitative recovery of hydrolyzed glycolipid by reversed-phase chromatography, sequential or combined enzyme treatments followed by TLC provide one of the most sensitive techniques for glycosphingolipid saccharide analysis. Effective hydrolysis of glycosphingolipids by glycohydrolases often requires the presence of detergent, typically sodium cholate or sodium taurodeoxycholate. A few glycohydrolases commonly used for glycosphingolipid analysis are listed in Table I, along with sources and reaction conditions. A general procedure,[32] which may require modification for some enzymes, follows. To a sample of purified glycosphingolipid (e.g., 5–100 nmol) in chloroform–methanol–(water) add approximately an equal weight (e.g., 10–100 μg) of sodium taurodexoycholate (Sigma Chem. Co., St. Louis, MO) in chloroform–methanol (1 : 1, v/v). Dry the sample–detergent mixture under nitrogen or *in vacuo* in a conical bottom microtube (e.g., Reactivial; Pierce Chem. Co., Rockford, IL), add the appropriate buffer (~2 μl/nmol glycolipid), warm to 37°, and sonicate to disperse the sample. Add the enzyme, typically 1–10 milliunits of enzyme per nanomole glycosphingolipid, in a small volume and mix thoroughly. After incubation at 37° with stirring, typically for ~16 hr, add water and solvents to bring to ~2 ml [final chloroform–methanol–water ratio, 2 : 43 : 55 (v/v)]. Load on a prewashed Sep-Pak C_{18} column, wash with the same solvent and with methanol–water (1 : 1, v/v), then elute the glycolipids with methanol and chloroform–methanol (1 : 1, v/v) as described.[11]

Enzyme Release and Isolation of Glycosphingolipid Oligosaccharides

Facile release of the entire intact oligosaccharide from most neutral and acidic glycosphinglipids is achieved with ceramide glycanase.[9,10] The released oligosaccharides are amenable to structural analysis by methods such as those described elsewhere in this volume. Mix glycosphingolipid sample in solvent with an equal weight of sodium cholate in chloroform–methanol (1 : 1, v/v) and evaporate to dryness. Dissolve the residue in ~1 ml/mg glycosphingolipid of 0.1 M sodium acetate buffer, pH 5.0, add 20 milliunits/ml of ceramide glycanase (V-Labs, Covington, LA), and incubate for 24 hr at 37°. Add solvents to obtain an 8 : 4 : 3 (v/v) ratio of chloroform–methanol–water and mix well. Separate the two phases, the lower containing the released ceramide and the upper the oligosaccharide, and evaporate to dryness. Resuspend the upper phase in ~1 ml/mg original glycosphingolipid of methanol–water (9 : 1, v/v) and apply to a solvent-washed/water-washed C_{18} Sep-Pak column. Elute the oligosaccharides in 12 ml of water and evaporate. For large samples, a higher concentration of glycosphingolipid, less enzyme per milligram glycosphingolipid, and

TABLE I

GLYCOHYDROLASES USED FOR GLYCOSPHINGOLIPID MODIFICATION

Enzyme	Biological source	Commercial source	Buffer	pH	Detergent	Ref.[a]
α-L-Fucosidase	Bovine kidney	Sigma (St. Louis, MO)	Sodium citrate (0.2 M)	4.5	Taurodeoxycholate	1
α-Galactosidase	Coffee bean	Sigma (St. Louis, MO)	Sodium acetate (50 mM)	4.9	Taurodeoxycholate	2
β-galactosidase	Jack bean	Sigma (St. Louis, MO)	Sodium citrate (50 mM)	4.0	Taurodeoxycholate	1
β-N-Acetyl-hexosaminidase	Jack bean	Sigma (St. Louis, MO)	Sodium citrate (0.2 M)	4.5	Taurodeoxycholate	i
β-Glucuronidase	Limpet	Sigma (St. Louis, MO)	Sodium acetate (10 mM)	5.0	Taurodeoxycholate	3
Neuraminidase[b]	Vibrio cholera	Calbiochem (La Jolla, CA)	Tris–maleate (0.1 M) 10 mM CaCl$_2$	6.5	None	4,5
Neuraminidase[c]	Newcastle disease virus	Genzyme	Sodium acetate (10 mM)	5.0	Cholate	6
Neuraminidase[d]	Arthrobacter ureafaciens	Boeringer Mannheim (Indianapolis, IN)	Sodium acetate (50 mM)	5.0	Cholate	5
Endo-β-galactosidase[e]	Escherichia freundii	V-Labs (Covington, LA)	Sodium acetate (0.2 M)	5.8	Taurodeoxycholate	7
Ceramide glycanase[f]	Leech	V-Labs	Sodium acetate (0.1 M)	5.0	Cholate	8,9

[a] References cited represent examples of the use of each glycohydrolase for glycosphingolipid TLC analysis. References do not necessarily represent the initial purification or application of the enzyme to glycosphingolipids. Key: (1) G. K. Ostrander, S. B. Levery, H. L. Eaton, M. E. K. Salyan, S. Hakomori, and E. H. Holmes, J. Biol. Chem. 263, 18716 (1988); (2) K. Nohara, M. Suzuki, F. Inagaki, H. Ito, and K. Kaya, J. Biol. Chem. 265, 14335 (1990); (3) D. K. H. Chou, A. A. Ilyas, J. E. Evans, C. Costello, R. H. Quarles, F. B. Jungalwala, J. Biol. Chem. 261, 11717 (1986); (4) N. M. Dahms and R. L. Schnaar, J. Neurosci. 3, 806 (1983); (5) M. Saito, K. Sugano, and Y. Nagai, J. Biol. Chem. 254, 7845 (1979); (6) Y. Suzuki, T. Morioka, and M. Matsumoto, Biochim. Biophys. Acta 619, 632 (1980); (7) M. N. Fukuda, K. Watanabe, and S. Hakomori, J. Biol. Chem. 253, 6814 (1978); (8) B. Zhou, S.-C. Li, R. A. Laine, R. T. C. Huang, and Y. T. Li, J. Biol. Chem. 264, 12272 (1989); (9) G. C. Hansson, Y. T. Li, and H. Karlsson, Biochemistry 28, 6672 (1989).

[b] Although Vibrio cholera neuraminidase has broad specificity for linkage position, it does not cleave the sialic acid from G$_{MI}$ or G$_{M2}$.

[c] Newcastle disease virus neuraminidase has marked preference for α(2 → 3) over α(2 → 6) linkage, although its activity against G$_{MI}$ is negligible.

[d] Arthrobacter ureafaciens neuraminidase readily cleaves the sialic acid from G$_{MI}$ under the conditions described.

[e] Escherichia freundii endo-β-galactosidase cleaves the common structure: R-GlcNAcβ1-3Galβ1-4Glc/GlcNAc.

[f] Leech ceramide glycanase cleaves a broad range of neutral and acidic glycosphingolipids. However, it requires a glucose attached to the ceramide, and will not cleave monoglucosylceramide, certain rare diglycosylceramides, or the "A-4" fucosylated tetraglycosylceramide.

longer times can be used. The reaction can be monitored by TLC in *n*-butanol–acetic acid–water (3:3:2, v/v), in which many oligosaccharides migrate.[39]

Thin-Layer Chromatography Overlay Binding Techniques

The demonstration of selective binding of radiolabeled cholera toxin to gangliosides resolved on a TLC plate by Magnani *et al.*[40] led to the development of related TLC overlay methods for direct binding of antibodies,[41] viruses, bacteria,[42] and eukaryotic cells.[43] These techniques have been useful for identifying glycosphingolipid structures involved in pathological and physiological processes.[44,45] Thin-layer chromatography overlay with structurally specific carbohydrate-directed antibodies[46] can assist in structural determination of minor glycosphingolipid species. When combined with chemical and/or enzymatic modification (see above) immune overlay can be a powerful adjunct to chemical characterization.

Technique

Various TLC overlay techniques share the following steps: Apply sample and develop the TLC as described above, dry thoroughly, dip the plate in solvent-dissolved polymer solution, dry thoroughly, spray or dip in aqueous buffer, overlay with primary binding agent (antibody, cells, etc.), incubate, wash with buffer, if necessary overlay with secondary detecting agents (e.g., secondary labeled antibody), wash, dry for autoradiography on X-ray film or treat for secondary color development (e.g., for peroxidase-based detection). A companion plate is run under identical conditions and subjected to chemical detection.

Choose the proper TLC plate to avoid flaking of the sorbent during multiple aqueous treatments. High-performance TLC sorbent layers

[39] S. Hakomori, E. Nudelman, S. B. Levery, and R. Kannagi, *J. Biol. Chem.* **259,** 4672 (1984).

[40] J. L. Magnani, D. F. Smith, and V. Ginsburg, *Anal. Biochem.* **109,** 399 (1980).

[41] J. L. Magnani, S. L. Spitalnik, and V. Ginsburg, *in* "Methods in Enzymology" (V. Ginsburg, ed.), Vol. 138, p. 195. Academic Press, Orlando, FL, 1987.

[42] K.-A. Karlsson and N. Stromberg, *in* "Methods in Enzymology" (V. Ginsburg, ed.), Vol. 138, p. 220. Academic Press, Orlando, FL, 1987.

[43] P. Swank-Hill, L. K. Needham, and R. L. Schnaar, *Anal. Biochem.* **163,** 27 (1987).

[44] R. H. Quarles, *in* "Methods in Enzymology" (V. Ginsburg, ed.), Vol. 179, p. 291. Academic Press, San Diego, 1989.

[45] M. Tiemeyer, S. J. Swiedler, M. Ishihara, M. Moreland, H. Schweingruber, P. Hirtzer, and B. K. Brandley, *Proc. Natl. Acad. Sci. U.S.A.* **88,** 1138 (1991).

[46] J. L. Magnani, *in* "Methods in Enzymology" (V. Ginsburg, ed.), Vol. 138, p. 484. Academic Press, Orlando, FL, 1987.

bound to aluminum supports (Cat. No. 5547; E. Merck AG) can be cut to size and resist flaking. When using glass supports, silica gel layers impregnated with fluorescent indicator (silica gel 60 F-254, Cat. No. 5635; E. Merck AG) adhere better and are recommended (although the fluorescent indicator is not useful for detection of glycosphingolipids per se, it normally does not interefere with chromatography or staining). Keeping the various aqueous solutions isotonic by adding noninterfering salts (e.g., NaCl) as required also helps resist flaking.

After the plate is developed with any appropriate solvent mixture, dry thoroughly by placing *in vacuo* or in an oven at 50° for 1 hr. Coating the plate with plastic [poly(isobutyl methacrylate) (PIBM), Cat. No. 18, 154-4; Aldrich Chem. Co.] reduces nonspecific binding and enhances the sensitivity of specific binding. Prepare a 10% (w/v) stock solution of PIBM in chlorofrom, which is stable on storage. Dilute into rapidly mixing hexanes for use. The optimal PIBM concentration will vary depending on the particular binding assay; using too little PIBM results in unacceptably high background binding, whereas too much suppresses specific binding. Unless following an established protocol, start with 0.01% PIBM in hexane (1 : 1000 dilution of the above stock), then adjust according to initial results. To ensure even coating, dip the plate rapidly and completely in hexane first to wet and dislodge any air bubbles, then withdraw and transfer to PIBM in hexane for 30 sec. Dipping of plates with dimensions up to 5 × 10 cm is conveniently performed in small cylindrical TLC chambers. Withdraw the plates, and allow them to drain upright until thoroughly dry.

Wet the plate surface by spraying heavily with buffer or placing the plate sorbent side up in a shallow dish of buffer until visibly wet. It may be useful to incubate the plate in buffer containing bovine serum albumin (BSA; up to 50 mg/ml) or polyvinylpyrrolidone (up to 2%, w/v) for 30 min to decrease nonspecific binding. The plate may be washed with fresh buffer prior to addition of binding agent. Subsequent incubation time, secondary detection methods, and washing procedures are dependent on the particular application, examples of which are given below.

Specific Examples

Radioimmunoassay.[41] After chromatography on aluminum-backed plates the plate is dried and dipped in 0.1% PIBM in hexane. After drying, the plate is sprayed with buffer [50 mM Tris-HCl, 150 mM NaCl (pH 7.4), BSA (10 mg/ml)] and then soaked in the same buffer for 15 min. The plate is removed, placed sorbent side up on a pedestal, then overlaid with 60 μl/cm² of monoclonal antibody (2–5 μg/ml) in the same buffer. After 1 hr at ambient temperature, the plate is washed by dipping (four times consecutively, 1 min each) in phosphate-buffered saline, removed, placed sorbent side up, and overlaid with 60 μl/cm^2 of second antibody ($\sim$0.01 μg/ml, $\sim$40 μC/μg of ^{125}I-labeled goat anti-mouse IgG or IgM). After 1 hr at ambient temperature, the plate is washed as before, dried, and placed sorbent side down on X-ray film for 16 hr for detection. Plates may be wrapped in plastic prior to placing on film to avoid sorbent sticking to the film (this is not a problem if the plate is thoroughly dry).

Peroxidase-Based Immunoassay. After chromatography on glass-backed plates (with fluorescence impregnated sorbent) the plate is dried and dipped in hexane, then in 0.01% PIBM in hexane. The plate is sprayed until wet with Ca^{2+}/Mg^{2+}-free phosphate-buffered saline, then placed sorbent side up and overlaid with 60 μl/cm^2 of the same buffer containing monoclonal antibody (1 μg/ml) and BSA (2 mg/ml). After 16 hr at 4° in a humid chamber (a covered 150-mm petri dish), the plates are washed by immersion in the same buffer for 5 min, then by two sequential incubations with fresh buffer for 30 min each. Subsequent treatments are at ambient temperature. The plates are overlaid with 1.5% normal goat serum in the same buffer to block nonspecific binding of the second antibody. After 1 hr the plates are washed with buffer for 5 min, then for 30 min with fresh buffer, then overlaid with either a 1 : 200 dilution of biotinylated goat anti-mouse IgM (Vectastain ABC kit; Vector Laboratories, Burlingame, CA) or with a 1 : 2000 dilution of peroxidase-conjugated goat anti-mouse IgM (Cappel Laboratories, West Chester, PA) in the same buffer. After 1 hr the plates are washed as above, then (in the case of the biotin-conjugated second antibody only) the plate is further incubated with a 1 : 100 dilution (each) of avidin and biotinylated horseradish peroxidase that had been preincubated 30 min prior to addition to the plate. The plate is then washed for 1 hr as above. For both protocols the plate is then immersed in a freshly made solution containing 0.01% (v/v) H$_2$O$_2$ and diaminobenzidine tetrahydrochloride (0.5 mg/ml) in the same buffer. When bands become clearly visible (brown against a light background) the plate is washed in fresh buffer and allowed to air dry.

Thin-Layer Chromatography Overlay with Enzyme Modification. A powerful combination of specific enzymatic modification and TLC overlay involves the use of enzyme overlay to modify resolved glycosphingolipids, which are then detected with specific antibodies. In one example,[47] gangliosides are resolved by HPTLC, the plate dried and dipped in 0.4% PIBM in hexane, then dried and overlaid with *Arthrobacter ureafaciens* neuraminidase (40 mU/ml) in 0.1 M sodium acetate buffer, pH 4.8. After 2 hr, the plate is washed with phosphate-buffered saline, and incubated with

[47] M. Saito, N. Kasai, and R. K. Yu, *Anal. Biochem.* **148**, 54 (1985).

anti-gangliotetraosylceramide rabbit antiserum diluted in 0.3% (w/v) gela-
tin in PBS. After 1 hr the plate is washed and incubated with staphylococcal
[125]I-labeled protein A for 1 hr in the same buffer. Finally the plate is
washed and exposed to X-ray film. Any glycosphingolipid that can be
converted to the gangliotetraosyl ceramide on enzyme treatment is detected. In a further advance on this approach,[48] glycosphingolipids resolved by TLC are exposed to partially purified sialyltransferases in the
presence of a radiolabeled activated sugar donor, CMP[^{14}C]NeuAc. Those
species that are capable of acting as transferase acceptors are selectively radiolabeled.

Limitations of Thin-Layer Chromatography-Derived Data

The combination of TLC resolution, specific chemical and enzymatic
modifications, and direct overlay binding can rapidly generate information
on glycosphingolipid structure and function without large investments in
instrumentation. However, the following precautions must be considered
in analyzing glycosphingolipids by TLC. Purification to a single band,
even in several solvent systems, does not ensure homogeneity. Certain
glycosphingolipids, even some having different monosaccharide components, are difficult to resolve by TLC.[1] On the other hand, glycosphingolipids having the identical saccharide structure often appear as multiple
bands because of variations in the ceramide moiety.[39] Enzyme modification data must be also be interpreted with care, because the structure of
certain glycosphingolipids may mask an otherwise susceptible linkage
from hydrolytic attack. Similar subtleties limit structural assignments
made by immunochemistry. Therefore, even though TLC-based techniques are powerful tools for glycosphingolipid analysis, they must be
linked with other analytical methods, such as those listed at the beginning
of this chapter and elsewhere in this volume, for unambiguous structural
determination. Merging of TLC with other technologies, such as the direct
analysis of glycosphingolipids on TLC plates by mass spectrometry,[49]
promises to enhance its power as an analytical tool.

[48] X. Gu, T. Gu, and R. K. Yu, *Anal. Biochem.* **185,** 151 (1990).
[49] Y. Kushi and S. Handa, *J. Biochem.* (*Tokyo*) **98,** 265 (1985).

[24] Isolation and Characterization of Proteoglycans

By Vincent C. Hascall, Anthony Calabro, Ronald J. Midura, and Masaki Yanagishita

I. Introduction

Complex carbohydrate structures, particularly the glycosaminoglycans, endow proteoglycans with unique properties. In keeping with the glycobiology theme of this volume, we focus attention in this chapter on methods that emphasize the contribution of these complex carbohydrates to the isolation, structure, metabolism, and function of the proteoglycans. The rapid expansion of molecular biological techniques and their use to define the primary sequences of the large number of diverse core proteins are key elements in the final characterization of the manifold proteoglycan gene families. Although the impact of molecular biology on proteoglycan research is vital, it is also beyond the scope of this chapter. Some references to this work are noted in the descriptions of selected proteoglycan types below.[1]

II. Proteoglycan Structure

A. Glycosaminoglycans

Proteoglycans are complex macromolecules that contain a core protein with one or more covalently attached glycosaminoglycan chains.[1a–3] Glycosaminoglycans consist of linear polymers with repeating disaccharides of defined structures that contain one hexosamine and either a carboxylate or a sulfate ester, or usually both. There are four general classes of glycosaminoglycans: (1) hyaluronic acid (HA), (2) chondroitin sulfate (CS) and dermatan sulfate (DS), (3) keratan sulfate (KS), and (4) heparan sulfate (HS) and heparin.

The structures of the characteristic disaccharides for each class are shown in Fig. 1. Hyaluronic acid is the simplest, with alternating gluc-

---

[1] The references cited in this chapter are often review articles that can be consulted to find original citations.

[1a] T. N. Wight, D. K. Heinegård, and V. C. Hascall, *in* "Cell Biology of Extracellular Matrix" (E. Hay, ed.), 2nd Ed., p. 45. New York, 1991.

[2] J. T. Gallagher, *Curr. Opin. Cell Biol.* **1**, 1201 (1989).

[3] T. E. Hardingham and A. J. Fosang, *FASEB J.* **6**, 861 (1992).

uronic acid (GlcUA) and *N*-acetylglucosamine (GlcNAc) residues linked by the indicated glycosidic bonds. Hyaluronic acid does not appear to be synthesized covalently bound to a protein and hence is not classified as a proteoglycan.[1a,4,5] It is synthesized at sites near the plasma membrane, with elongation occurring at the reducing end by a mechanism that differs from all the other glycosaminoglycans. The reducing end of the HA molecule appears to be bound to a uridine diphosphate (UDP) group that is displaced when the next UDPsugar (either UDPGlcNAc or UDPGlcUA) is added [Fig. 1, reaction (1)]. The growing polymer appears to be extruded through the plasma membrane of the cell and can reach impressive sizes (M_r 10 million, ~25,000 disaccharides). Hyaluronic acid does not contain sulfate esters.

Chondroitin sulfate (CS) has the same backbone structure as HA except that the GlcNAc residues are replaced by *N*-acetylgalactosamine (GalNAc). Chondroitin sulfate is synthesized on characteristic linkage oligosaccharides that are assembled on core proteins (see Section II,B), and elongation occurs at the nonreducing end by transfer of the appropriate sugar from a UDPsugar donor as illustrated for KS synthesis [Fig. 1, reaction (2)]. Sulfate esters are added to the growing chains, usually at the C-4 or C-6 positions of the GalNAc. Many cells have an enzyme that can epimerize the carboxyl group on the C-5 carbon of a proportion of the D-GlcUA residues to produce L-iduronic acid (IdoUA) [Fig. 1, reaction (3)]. When this occurs the chain is referred to as DS. The C-2 position on IdoA can also be sulfated. Chondroitin sulfate/dermatan sulfate chains seldom exceed a molecular weight of ~300,000.

Keratan sulfate, like HA and CS/DS, contains alternating $\beta(1 \rightarrow 3)$ and $\beta(1 \rightarrow 4)$ bonds, but with GlcNAc in the hexuronic acid position and galactose in the hexosamine position (Fig. 1). Keratan sulfate is synthesized on two distinctly different linkage oligosaccharides on core proteins (see Section II,B, below). Sulfate esters are generally added to the C-6 position of one or both of the sugar units. Keratan sulfate chains seldom exceed 50,000 in molecular weight.

Heparan sulfate (HS) and heparin have a repeating disaccharide structure distinctly different from the other glycosaminoglycans. The linkage of GlcNAc to GlcUA is $\alpha(1 \rightarrow 4)$ rather than $\beta(1 \rightarrow 4)$ and that between GlcUA and GlcNAc is $\beta(1 \rightarrow 4)$ rather than $\beta(1 \rightarrow 3)$ (Fig. 1). The initial postelongation modification of HS/heparin is the coordinated removal of the *N*-acetyl group and concurrent addition of an *N*-sulfoester to a propor-

[4] T. C. Laurent and R. F. Fraser, *FASEB J.* **6,** 2397 (1992).
[5] P. Prehm, *Ciba Found. Symp.* **143,** 21 (1989).

A. Hyaluronic Acid

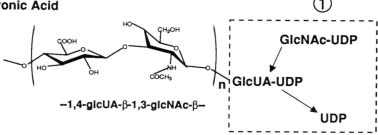

GlcNAc-UDP

GlcUA-UDP

UDP

①

−1,4-glcUA-β-1,3-glcNAc-β−

B. Chondroitin/*Dermatan* Sulfate

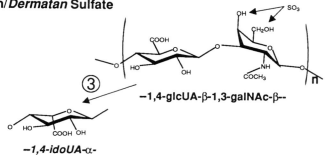

OH ← SO₃

③

−1,4-glcUA-β-1,3-galNAc-β--

−1,4-idoUA-α-

C. Keratan Sulfate

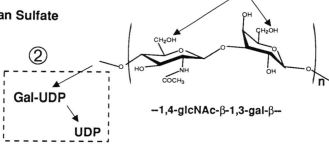

SO₃

②

Gal-UDP

UDP

−1,4-glcNAc-β-1,3-gal-β--

D. Heparan Sulfate/Heparin

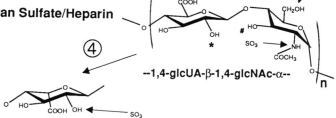

SO₃

④

--1,4-glcUA-β-1,4-glcNAc-α--

−1,4-idoUA-α-

SO₃

tion of the GlcNAc residues (generally less than 50% for HS and 70% or more for heparin). This occurs in block regions such that many adjacent disaccharides will contain N-sulfoesters with intervening block regions that contain only unmodified N-acetyl residues. Subsequent epimerization of GlcUA to IdoUA [Fig. 1, reaction (4)], C-2 O-sulfation of IdoUA, and C-6 O-sulfation of glucosamine residues occur in the block regions that contain the N-sulfoesters. Sulfation occurs infrequently on the C-3 position of GlcNAc (Fig. 1, #) and the C-2 position of GlcUA (Fig. 1, *), but both these substitutions have been shown to have important biological consequences, namely, the ability to bind to anti-thrombin III[6] and involvement in regulation of mitotic activity,[7] respectively. Heparan sulfate/heparin chains seldom exceed a molecular weight of 100,000.

B. Linkage Structures

Each CS/DS and HS/heparin chain is attached to a core protein through a characteristic linkage tetrasaccharide: GlcUAβ1-3Galβ1-3Galβ1-4Xylβ1-O-serine, which is attached by a glycoside bond to the hydroxyl of a serine in an appropriate sequence (usually serine-glycine) that can be recognized by the xylosyltransferase that adds the initial xylose residue.[8] Synthesis then involves specific transferases that sequentially add the other three sugars. The xylose frequently carries a C-2 O-phosphoester.[9] When the core protein is translocated to the appropriate site in the Golgi (HS/heparin) or possibly in the trans-Golgi network (CS/DS),[10] the appropriate backbone disaccharides are elongated on the linkage oligosaccharides.

[6] U. Lindahl, in "Heparin; Chemical and Biological Properties, Clinical Applications" (D. Lane and U. Lindahl, eds.), p. 159. CRC Press, Boca Raton, FL, 1989.

[7] N. S. Fedarko, M. Ishihara, and H. E. Conrad, J. Cell. Physiol. 139, 287 (1989).

[8] V. C. Hascall, D. K. Heinegård, and T. N. Wight, in "Cell Biology of Extracellular Matrix" (E. Hay, ed.), 2nd ed., p. 149. Plenum, New York, 1991.

[9] T. R. Oegema, E. L. Kraft, G. W. Jourdian, and T. R. Van Valen, J. Biol. Chem. 259, 720 (1984).

[10] R. G. Spiro, H. H. Freeze, D. Sampath, and J. A. Garcia, J. Cell Biol. 115, 1463 (1991).

FIG. 1. Backbone structures of glycosaminoglycans. The repeating initial disaccharide structures for the four classes of glycosaminoglycan are shown in parentheses. The mechanisms for elongating HA [reaction (1)] and the other glycosaminoglycans [reaction (2)] are shown in the boxes. Reactions (3) and (4) illustrate the C-5 epimerization reactions that convert some GlcUA residues to IdoUA residues in DS and HS/heparin, respectively. The italicized dermatan and IdoUA [reaction (3)] indicate that DS is defined by the presence of IdoUA residues. Patterns of sulfation are discussed in text. *, #, sites of infrequent sulfation.

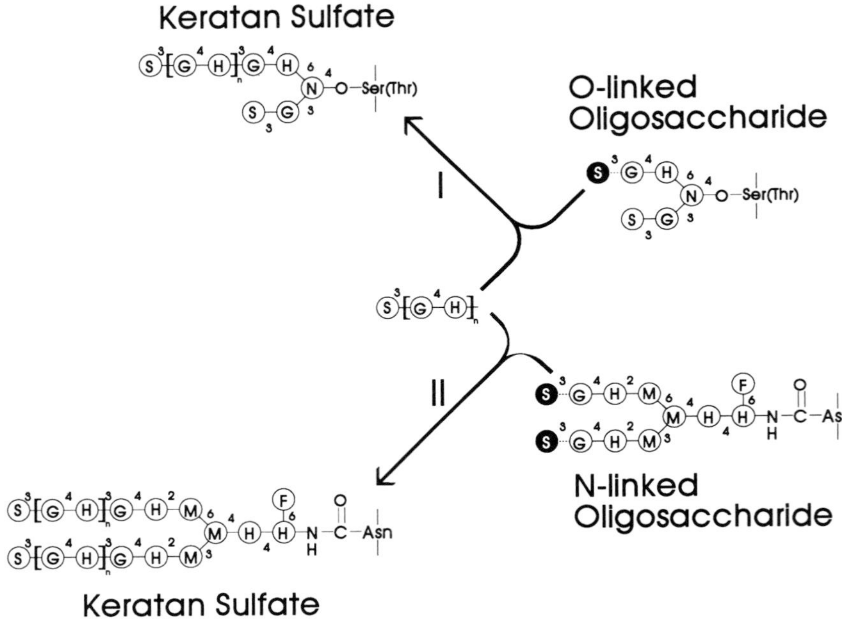

Keratan Sulfate

O-linked Oligosaccharide

N-linked Oligosaccharide

Keratan Sulfate

FIG. 2. Linkage structures for KS. The O-linked hexasaccharide shown is found in proteoglycans such as aggrecan and versican. In hyaline cartilages, it is usually elongated on aggrecan molecules by adding the backbone structure of KS instead of the indicated terminal sialic acid (pathway I). Most KS chains are also capped with a sialic acid residue at their nonreducing terminus. The biantennary complex N-linked oligosaccharide is commonly found on many glycoproteins. In cartilage and cornea, it is often elongated on fibromodulin and lumican, respectively, by adding the backbone structure of KS instead of one or both of the indicated terminal sialic acids (pathway II). The specific sugars are fucose (F), galactose (G), GlcNAc (H), mannose (M), GalNAc (N), and sialic acid (S); the numbers indicate the glycosidic linkage positions.

Keratan sulfate is attached by two distinctly different linkage oligosaccharides whose structures are related to O-linked and N-linked oligosaccharides characteristic of mucins and glycoproteins, respectively (Fig. 2). The O-linked, branched hexasaccharide shown is commonly found on large proteoglycans such as aggrecan and versican and is assembled by sequential glycosyltransferase reactions in a Golgi compartment.[11] When the indicated sialic acid (Fig. 2) is not added, KS chain elongation can occur if the appropriate glycosyltransferases and sulfotransferases are

[11] L. S. Lohmander, V. C. Hascall, M. Yanagishita, K. E. Kuettner, and J. H. Kimura, *Arch. Biochem. Biophys.* **250**, 211 (1986).

present, and if the core protein is targeted to the appropriate Golgi or trans-Golgi network compartment.

The biantennary N-linked oligosaccharide shown in Fig. 2 is commonly found on many glycoproteins and is synthesized by the well-known pathway involving transfer of a mannose-rich oligosaccharide precursor onto an appropriate asparagine in the tripeptide sequence asparagine-xxx-serine (threonine) in a nascent protein in the rough endoplasmic reticulum.[8,12] Subsequent processing steps in the rough endoplasmic reticulum and early Golgi compartments involve trimming to remove outer glucose and mannose residues, and then adding GlcNAc, galactose, and sialic acid residues in later Golgi compartments to yield typical complex N-linked structures. When one or both of the indicated sialic acids (Fig. 2) are not added, the simple biantennary structure shown can be elongated with KS chains on proteoglycans such as fibromodulin and lumican (see Section II,C).

C. Core Proteins

Many distinctly different, genetically unrelated proteins serve as core proteins for proteoglycans.[1,13] The combination of a distinct core protein with the appropriate glycosaminoglycan chains gives an individual proteoglycan its unique structure and biological functions. Core proteins whose primary sequences are currently known range in molecular weight from ~20,000 (serglycin)[14] to more than 200,000 (aggrecan,[15] versican,[16] and perlecan[17]) (Fig. 3). Some contain one or a few glycosaminoglycan chains (decorin and biglycan) whereas others contain many chains (aggrecan and versican). Some contain transmembrane domains that anchor the proteoglycan to the cell surface (the syndecan family[18-20]). Some contain chains from more than one class of glycosaminoglycan (syndecan and aggrecan).

[12] R. Kornfeld and S. Kornfeld, *Annu. Rev. Biochem.* **54**, 631 (1985).

[13] J. R. Hassell, J. H. Kimura, and V. C. Hascall, *Annu. Rev. Biochem.* **55**, 539 (1986).

[14] M. A. Bourdon, *in* "Extracellular Matrix Genes" (L. Sandell and C. Boyd, eds.), p. 157. Academic Press, San Diego, 1990.

[15] K. Doege, C. Rhodes, M. Sasaki, J. R. Hassell, and Y. Yamada, *in* "Extracellular Matrix Genes" (L. Sandell and C. Boyd, eds.), p. 137. Academic Press, San Diego, 1990.

[16] D. R. Zimmermann and E. Ruoslahti, *EMBO J.* **8**, 2975 (1989).

[17] D. M. Noonan, A. Fulle, P. Valente, S. Cai, E. Horigan, M. Sasaki, Y. Yamada, and J. R. Hassell, *J. Biol. Chem.* **266**, 22939 (1991).

[18] M. Bernfield, R. Kekenyesi, M. Kato, M. J. Hinkes, J. Spring, R. L. Gallo, and E. J. Lose *Annu. Rev. Cell Biol.* **8**, 365 (1992).

[19] P. Marynen, J. Zhang, J.J. Cassiman, H. Van den Berghe, and G. David, *J. Biol. Chem.* **264**, 7017 (1989).

[20] T. Kojima, N. W. Shworak, and R. D. Rosenberg, *J. Biol. Chem.* **267**, 4870 (1992).

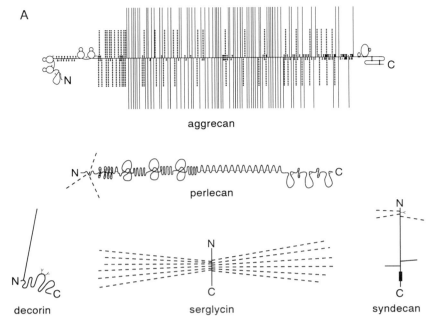

FIG. 3. Model structures for representative proteoglycans. (A) Model structures for aggrecan, perlecan, syndecan, decorin, and serglycin. The protein cores and glycosaminoglycan side chains are shown at approximately the same scale for comparison of each type of proteoglycan. Solid lines on core proteins represent CS/DS chains; dashed lines, HS/heparin chains; and dotted lines, KS chains. (B) Diagrams comparing the homologous structures of the small interstitial proteoglycans: decorin, biglycan, and fibromodulin. (Modified in part from Wight et al.,[1] with permission).

The biological functions of these proteoglycans are manifold, as the following examples indicate. Serglycin, found in storage granules of mast cells, contains an embedded region of serine-glycine repeats that concentrate a large number of long heparin (or oversulfated CS)[21] chains along a minimal length of core protein. This provides a compact molecule with exceptionally high negative charge density. The polyanionic chains can interact with cationic histamines and proteins such as mast cell proteases, carboxypeptidases and glucosaminidases, thereby concentrating them in storage granules and regulating their release when mast cells degranulate during host defense reactions.[22] The small interstitial proteoglycans de-

[21] C. S.-E. Stevens and R. L. Stevens, *Prog. Clin. Biol. Res.* **297,** 131 (1989).
[22] E. Enerback, *in* "Heparin; Chemical and Biological Properties, Clinical Applications" (D. Lane and U. Lindahl, eds.), p. 97. CRC Press, Boca Raton, FL, 1989.

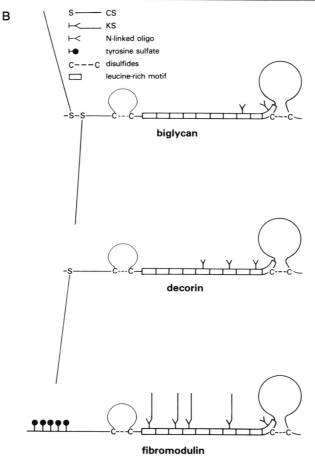

FIG. 3. (*continued*)

corin, fibromodulin (Fig. 3), and lumican, the closely related keratan sulfate proteoglycan in cornea,[23] interact at characteristic, periodic sites on collagen fibrils in most connective tissues[1a,24] and appear to be essential for collagen matrix organization. Aggrecan, which normally contains more than 100 CS and KS chains, functions as the macromolecular shock absorber in cartilages by resisting compressive load and giving resilience to

[23] T. C. Blochberger, J. P. Vergnes, J. Hempel, and J. R. Hassell, *J. Biol. Chem.* **267,** 347 (1992).
[24] J. E. Scott, *Biochem. J.* **252,** 313 (1988).

these tissues.[1a,3,25] Perlecan and related HS proteoglycans are structural components of basement membranes throughout the body and provide their surfaces with an array of negative charges that can act as an anti-thrombogenic barrier, as in the vascular lumen,[26] or as a selective filter, as in the kidney glomerular basement membrane.[27] Cell surface proteoglycans such as the syndecan family provide transmembrane bridges between extracellular interactions (cell–cell, cell–matrix) and intracellular (cytoskeletal) organization.[18,28] They have roles in growth factor binding and function; for example, basic fibroblast growth factor binds to HS, and this interaction is necessary for effective binding to its receptor.[29] Additionally, herpes simplex viruses and some isolates of human immunodeficiency type 2 (HIV-2) viruses bind selectively to cell surface HS proteoglycans, increasing their effectiveness in infectivity.[30,31]

III. Physical Properties of Proteoglycans

The polyanionic characteristics of the glycosaminoglycan chains endow proteoglycans with distinct properties. Thus, proteoglycans have (1) low isoelectric points, (2) large hydrodynamic domains resulting from the relatively extended glycosaminoglycan chains and, often, from the multiplicity of chains, and (3) high buoyant densities if the ratio of glycosaminoglycan mass to protein mass is relatively high. Some, such as the cell surface proteoglycans, are also hydrophobic as a result of transmembrane hydrophobic regions in their core proteins. Each of these properties can be used to advantage in isolating and characterizing different proteoglycans.

Note: Because of their exceptional diversity, no single set of procedures will be suitable for every proteoglycan. Thus, the techniques outlined below illustrate general principles that can be modified and used in different combinations for characterizing a particular proteoglycan.

[25] V. C. Hascall, *ISI Atlas Sci.: Biochem.* **1,** 189 (1988).

[26] S. E. Juul, T. N. Wight, and V. C. Hascall, *in* "The Lung: Scientific Foundations" (R. Crystal and J. West, eds.), p. 1413. Raven Press, New York, 1991.

[27] M. G. Farquhar, *in* "Cell Biology of Extracellular Matrix" (E. Hay, ed.), 2nd ed., p. 365. Plenum, New York, 1991.

[28] A. Woods, J. R. Couchman, and M. Höök, *J. Biol. Chem.* **260,** 10872 (1985).

[29] A. Yayon, M. Klagsbrun, J. D. Esko, P. Leder, and D. M. Ornitz, *Cell (Cambridge, Mass.)* **64,** 841 (1991).

[30] M.-T. Shieh, D. WuDunn, R. I. Montgomery, J. D. Esko, and P. G. Spear, *J. Cell Biol.* **116,** 1273 (1992).

[31] M. Patel, M. Yanagishita, G. Rodriguez, D. C. Bou-Habib, T. Oravecz, V. C. Hascall, and M. A. Norcross, *AIDS Res. Hum. Retrovir.* **9,** 167 (1993).

IV. Procedures

A. Metabolic Labeling

Cell culture and tissue explant systems are frequently used as models for investigating proteoglycans. In these cases, radiolabeling methods provide the most convenient ways to follow proteoglycans through purification steps and to monitor recoveries at each step. [^{35}S]Sulfate is usually used to label proteoglycans selectively, because more than 90% of the incorporated activity with this precursor will usually be in proteoglycans.[32] Frequently, [^{35}S]sulfate is used with a carbohydrate precursor, such as [^{3}H]glucosamine, which will label glycosaminoglycans as well as oligosaccharides on the core protein and other glycoproteins in the system. [^{3}H]Glucosamine is also an excellent precursor for HA synthesis.[33] [2-^{3}H]Mannose is sometimes used as a specific precursor for N-linked oligosaccharides on the core protein, if they are present.[34] [1-^{3}H]Glucose has been used to advantage because this precursor is metabolically converted to almost all the sugar precursors used in complex carbohydrate synthetic pathways. For example, this is the only precursor that is useful for introducing label into the xylose residues of the CS/DS and HS/heparin linkage oligosaccharides because UDPxylose, the biosynthetic precursor, is derived from UDPglucose.[11] After the proteoglycans are purified, these sugar precursors can provide details about the composition, structure, and types of complex carbohydrates on the core proteins.

[^{35}S]Sulfate can also be used with amino acid precursors such as [^{3}H]serine or [^{3}H]leucine. This labels the core proteins as well as other proteins and glycoproteins in the system.[32] Amino acid precursors are useful for evaluating the effectiveness of purification steps, and to define the number of different core proteins in the total proteoglycan population. Further, serines substituted with glycosaminoglycan chains or O-linked oligosaccharides undergo β elimination to form dehydroalanine when treated with alkali. If the treatment is done in the presence of an effective reducing agent, such as borohydride, these previously substituted serines can be quantitatively converted to alanine. Thus, the conversion of [^{3}H]serine to [^{3}H]alanine under these conditions provides a direct measure of the proportion of substituted serines in a given proteoglycan.[11]

---

[32] M. Yanagishita, R. J. Midura, and V. C. Hascall, in "Methods in Enzymology" (V. Ginsburg, ed.), Vol. 138, Part E, p. 279. Academic Press, Orlando, FL, 1987.

[33] M. Yanagishita, A. Salustri, and V. C. Hascall, in "Methods in Enzymology" (V. Ginsburg, ed.), Vol. 179, Part F, p. 435. Academic Press, San Diego, 1989.

[34] M. Yanagishita, Arch. Biochem. Biophys. **251**, 287 (1987).

A typical protocol involves labeling cell or explant cultures in medium appropriate for the culture system. Final radioisotope concentrations between 50 and 250 μCi/ml are convenient, although in short pulse–chase experiments higher amounts are often used.[35] The sulfate concentration in the medium should generally be at least 100 μM to ensure that the glycosaminoglycans are not undersulfated.[36] The specific activity of radiosulfate in the medium should be determined by measuring both the radioactivity and the sulfate concentration in aliquots of the labeling medium at the end of the labeling period. In most cases this provides a direct estimate of the specific activity of sulfate in the metabolic sulfate donor, phosphoadenosine phosphosulfate (PAPS),[33] and provides an estimate of the mass of glycosaminoglycan synthesized during the labeling period (see Section IV,F).

Total sulfate incorporation is usually linear for 24–48 hr after a short equilibration time, usually only a few minutes, between the medium and intracellular precursor pools; this should be determined in preliminary experiments. Labeled macromolecules will be present in both the medium and cell–matrix compartments, and these can be analyzed separately. The other metabolic precursors will generally be diluted by intracellular metabolic pathways, and hence the specific activities of the biosynthetic precursors, such as UDP-N-acetylhexosamines, are generally not easily measured directly. An indirect method useful for sugar precursors is outlined in Section IV,F below. Sugar precursors will also form covalent bonds with amino groups in proteins by nonenzymatic glycation,[37] a particular problem if serum is used in the labeling medium.[34] Control cultures without cells or tissue can be used to estimate the contribution of this process to the macromolecular radioactivity.

B. Extraction/Solubilization

The cell layer or tissue usually contains several different proteoglycans. Some can be integrated in the extracellular matrix by noncovalent interactions with other matrix molecules such as collagen. Others can be associated with the cell surface by hydrophobic interaction with the plasma membrane through either polypeptide intercalation or phosphatidylinositol anchors,[38,39] or by ionic binding with other cell surface molecules. Others

---

[35] M. Yanagishita, and V. C. Hascall, *J. Biol. Chem.* **260,** 5445 (1985).

[36] K. Ito, K. Kimata, M. Sobue, and S. Suzuki, *J. Biol. Chem.* **257,** 917 (1982).

[37] R. E. Bernstein, *Adv. Clin. Chem.* **26,** 1 (1987).

[38] M. Ishihara, N. S. Fedarko, and H. E. Conrad, *J. Biol. Chem.* **262,** 4708 (1987).

[39] M. Yanagishita, *J. Biol. Chem.* **267,** 9499 (1992).

can be sequestered in intracellular compartments, such as in storage or secretory granules,[1,21] or in prelysosomal or endosome compartments.[40]

Usually, the first experimental goal is to solubilize the maximum amount of proteoglycans in the cell–matrix compartments independent of where they reside. The most effective solvents contain both chaotropic reagents, which denature proteins and dissociate most noncovalent interactions, and detergents to dissociate hydrophobic interactions. The most widely used chaotropic agent is 4 M guanidine hydrochloride, which is often used with compatible detergents such as 4% (w/v) 3-[(3)-cholamido-propyl)-dimethyl-ammonio]-1-propanesulfonate (CHAPS) or 2% (w/v) Triton X-100.[32] The zwitterionic detergent CHAPS can be removed more easily in subsequent steps, for example, by dialysis, than can Triton X-100, which forms larger, nondialyzable micelles. If sequential extractions are used, solvents with detergents should be used first. Extraction of cell layers directly with 4 M guanidine hydrochloride in the absence of detergents facilitates artifactual hydrophobic interactions between macromolecules, which are difficult to dissociate with detergents in later steps.[32]

Because proteoglycans are easily degraded by proteases, inhibitors against different classes of proteases are usually added to the extraction solvent,[41] and the extractions are usually done at 4°. The protease inhibitors are particularly important in steps in which the solvent is changed to conditions that favor renaturation before the proteoglycans are fully purified because proteases in the extracts can recover activity. For this reason, initial proteoglycan purification steps are usually done in chaotropic solvents to minimize potential degradation.

For tissue explants, extraction efficiency can be improved if the tissue is frozen and thin slices (less than 1 mm) prepared on a tissue slicer. Otherwise the tissue should be minced finely. For more extensive discussion of extraction methods see Hascall and Kimura.[41]

The following detailed procedure provides an example. An effective solvent is 4 M guanidine hydrochloride, 4% (w/v) CHAPS, 0.1 M sodium acetate (pH 5.8), with a protease inhibitor cocktail of 10 mM 6-aminohexanoic acid (for cathepsin D-like activity), 5 mM benzamidine hydrochloride (for trypsin-like activity), 1 mM phenylmethylsulfonyl fluoride (for serine-dependent proteases), 10 mM disodium ethylenediamine-tetraacetate (EDTA) (for metalloproteases), and 10 mM N-ethylmaleimide (for thiol-dependent proteases). The protease inhibitors are added just prior to use from concentrated stock solutions of 6-aminohexanoic acid

[40] Y. Takeuchi, M. Yanagishita, and V. C. Hascall, *J. Biol. Chem.* **267**, 14677 (1992).
[41] V. C. Hascall and J. H. Kimura, *in* "Methods in Enzymology" (L. W. Cunningham and D. W. Frederiksen, eds.), Vol. 82, p. 769. Academic Press, New York, 1982.

and disodium EDTA and from dried powders for the other inhibitors. The sparingly soluble phenylmethylsulfonyl fluoride (PMSF) and *N*-ethylmaleimide can be dissolved at 100-fold concentration in methanol.

The cell or tissue cultures are extracted at 4° overnight with gentle stirring, using 1–2 ml/35-mm dish for cell cultures or at least 10 vol/g wet weight of tissue. The extract is removed and clarified by low-speed centrifugation if necessary. The residue can be reextracted with another 1/2 vol of extractant for 3–4 hr, which can then be combined with the first. For most cell cultures, the residue after extraction is negligible. However, any proteoglycans remaining on the culture dish can be solubilized by papain digestion and the released glycosaminoglycan peptides quantitated by molecular sieve chromatography. For explants, the residue should be solubilized by papain digestion to quantitate the proportion of residual, nonextracted proteoglycans in the same way. Eighty to 100% of the proteoglycans in the cell matrix compartment is generally solubilized by this extraction procedure.

C. Solvent Exchange and Quantitation of Macromolecular Radioactivity

Both the medium and the cell–matrix extract will contain labeled macromolecules and unincorporated precursors. A molecular sieve step is used to remove unincorporated precursor, to quantitate radioactivity in macromolecules, and to exchange the extraction solvent into one that can be used with anion exchangers in later steps. Sephadex G-50 (fine) (Pharmacia, Piscataway, NJ) columns are prepared in disposable serological pipettes with bed volumes from ~1 to 24 ml, using different size pipettes, depending on the volume to be applied. Baseline separation is achieved between the excluded peak (labeled macromolecules) and the totally included volume (unincorporated precursors), when the sample volume applied is 25% or less of the bed volume. A plastic rack that holds 10 columns spaced appropriately for collecting in rows over standard scintillation vial boxes permits 10 samples to be processed at a time. The columns are often equilibrated with a chaotropic, but nonionic, solvent such as 8 *M* urea or 10 *M* formamide, which will subsquently be suitable for anion-exchange chromatography.

The following procedure provides an example in which 8-ml bed volumes are used in 10-ml pipettes. The tops of 10-ml plastic serological pipettes are scored below the constriction with a triangular file and snapped off. Small glass wool plugs are tamped into the points. The pipettes are placed in a plastic holder for 10 pipettes, which is placed over a collecting tray. A 50% slurry of Sephadex G-50 (fine) previously equilibrated in

0.02% (w/v) NaN_3, is poured into each column until the settled volume of the gel is somewhat above the 2-ml mark (bed volume, 8 ml). Each column is equilibrated with one column volume (8–10 ml) of elution buffer: 10 M formamide, 0.05 M sodium acetate, 0.30 M NaCl, 0.5% (w/v) CHAPS, pH 6, and protease inhibitors (optional). Formamide has better stability than urea, which decomposes with time. Enough gel is removed to make the columns even at the mark. The columns will usually not crack at the top of the gel when drained to the surface, but the tips should be plugged with rubber sleeves until the samples are applied. Two milliliters of sample can be applied to each column. If less than 2 ml is applied, an aliquot of elution buffer is used to wash the sample into the column such that the sample volume plus the aliquot equals 2 ml. The rack is then positioned over vials to collect the excluded volume fractions, which elute between 2.3 and 4.5 ml. Thus the macromolecular fraction is recovered by applying a total of 3 ml of solvent (eluting between 2.0 and 5.0 ml after beginning the sample application), using the first 0.1–0.2 ml to wash in the sample if necessary. The total included peak does not begin before 6 ml; therefore the 3 ml collected will be free of unincorporated radioactivity. The columns, which now contain the large majority of radioactivity used in the labeling step, are then discarded in radioactive waste, effectively minimizing potential contamination of the laboratory. The volume of each excluded fraction can be estimated by weighing the collecting vials before and after sample collection and correcting for the density of the solvent, in this case 1.05 g/ml. An aliquot is counted for radioactivity to determine total recovery of incorporated activity.

D. Anion-Exchange Purification of Proteoglycans

Proteoglycans bind much more tightly to anion-exchange matrices than do proteins and glycoproteins because of the polyanionic glycosaminoglycan chains. Thus anion exchangers are effective for concentrating proteoglycans and for separating them from other macromolecules. Nonspecific adsorption of contaminating macromolecules and reduced recovery of bound proteoglycans are two potential problems. They can be minimized by including detergent and salt in the elution solvents and by using sufficient, but not excessive, amounts of the resin.

The following example describes a batch procedure that is particularly useful when the proteoglycans are in relatively large volumes. Q-Sepharose (Pharmacia) is equilibrated and stored indefinitely in the 10 M formamide solvent described in the previous section. The NaCl concentration can be adjusted up or down as experience is gained with the proteoglycans in the particular system under study. The NaCl content should be as high as possible while still permitting the proteoglycans to bind, thereby minimizing nonspecific adsorption. Portions of the resin are added to samples recovered from the excluded volumes of Sephadex G-50 columns. Typically, 1 ml of resin (settled volume) is sufficient for ~3 mg of proteoglycan. After gently mixing a few times over 60 min, the suspension is centrifuged at low speed. An aliquot of the supernatant is counted for ^{35}S activity to estimate the proportion that has bound to the resin (typically 90–95%). If the amount bound is lower than expected, another portion of resin can be added and the process repeated. After centrifugation, the supernatant is removed and the resin suspended and washed in 10–20 vol of the same solvent for 10 min. The centrifugation and wash steps are repeated twice. Bound proteoglycans are then extracted with a solvent that contains a sufficient salt concentration to reverse their binding, for example a 4 M guanidine hydrochloride solvent, or a 10 M formamide solvent that contains 1 M NaCl. The resin is suspended in 3–5 vol of extraction solvent and mixed intermittently over 30 min. The suspension is then filtered through a small column prepared by putting a glass wool plug in a plastic serological pipette. The retained resin is washed with an additional 2 vol of the extracting solvent, and the filtrate and wash are collected in a preweighed container. The volume of the filtrate is determined, and an aliquot is counted to estimate recovery (typically 85–95%). This procedure is effective for concentrating proteoglycans into smaller volumes.

If a sample contains nucleic acids (e.g., cell extract), they copurify with proteoglycan fractions in anion-exchange chromatography. The presence of nucleic acid may not be apparent when specific metabolic precursors (such as [^{35}S]sulfate and labeled sugar precursors) are used. If necessary, further purification steps (e.g., gel filtration) or enzyme digestion (e.g., DNase) may be required to remove nucleic acids.

E. Separation of Proteoglycans

Different proteoglycans can differ in their average charge densities, average buoyant densities, average hydrodynamic sizes, and the hydrophobicities of their core proteins. Each of these properties can be exploited to identify and separate different proteoglycan classes. The following procedures illustrate examples for each of the properties mentioned earlier.

1. Charge. The batch ion-exchange process described in the previous section normally achieves a reasonably high level of purification of proteoglycans from other types of macromolecules. However, a second ion-exchange step with a continuous salt gradient often achieves better purity and can separate different classes of proteoglycans. In this case, small anion-exchange columns with bound proteoglycans are prepared. A continuous salt gradient is then applied to the column with an appropriate gradient mixer and eluant fractions collected. If the extraction solvent used in the batch step described in the previous section was 10 M formamide with 1 M NaCl, the sample can be diluted with a 10 M formamide solvent without salt to give a final solution with 0.30 M NaCl. Under these conditions the proteoglycans will once again bind to the anion exchanger.

Proteoglycans with different charge densities can resolve, at least partially, into separate peaks, as often occurs when both HS proteoglycans and CS/DS proteoglycans are present.[32] In such systems, anion exchange should be used early in the experimental strategy so that the different proteoglycans can then be further characterized separately.

Many different types of ion exchangers are now available in high-performance liquid chromatography (HPLC) and membrane cartridge forms. Each of these can offer advantages in analysis time, capacity, recoveries, and separation properties. All of these properties can be assessed by procedures such as those described in this chapter.

The following procedure provides an example. A 10-ml plastic pipette is scored and broken to give the bottom third. An aliquot of a Q-Sepharose slurry equilibrated with 10 M formamide, 0.3 M NaCl, 0.5% (w/v) CHAPS, 0.05 M sodium acetate (pH 6) is poured onto a glass wool plug in the bottom of the pipette to form a support layer of $\sim$100 μl. Another aliquot that will give a 1- to 2-ml final packed volume is added to the sample, which has been equilibrated in the same solvent, to bind the proteoglycans as described in the batch protocol above. When the gel settles after the binding step, most of the supernatant can be passed rapidly through the support layer in the column to collect the unbound fraction. The residual gel slurry with bound proteoglycans is added to the column and packed onto the support layer. The column is washed with three to four bed volumes of the same solvent to remove remaining unbound or weakly bound molecules. The large cross-sectional area on top of the column gives better flow rates and improves proteoglycan recoveries. The bottom of the column is attached to a fraction collector via a peristaltic pump, and the top is attached to an appropriate gradient maker. The gradient is prepared from 40 ml of the starting buffer and 40 ml of the same solution, except with 1.2 M NaCl. The gradient is developed over 2–3 hr at a flow rate of 15–20 ml/hr, to give a total of about 80 fractions. Aliquots are counted for radioactivity, with usual recoveries of 85–90%. Figure 4 shows a typical analysis.

2. Buoyant Density. Glycosaminoglycans have much higher buoyant densities in CsCl equilibrium density gradients than do proteins because their carbohydrate backbones have higher intrinsic buoyant densities than

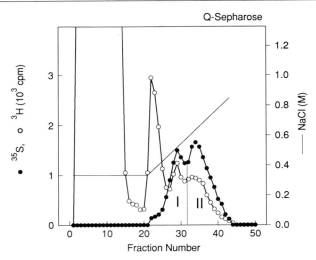

FIG. 4. Resolution of proteoglycan classes by anion-exchange chromatography. Cultures of rat ovarian granulosa cells were labeled with [^{35}S]sulfate and [^{3}H]serine. The cell layer was extracted, eluted on Sephadex G-50 in a 10 M formamide solvent, and then analyzed on a Q-Sepharose anion-exchange column, using an NaCl gradient as described in Section IV,E,1. The ^{35}S activity monitors the elution profiles of the proteoglycan fractions. Two peaks are partially resolved; peak I contains HS proteoglycans and peak II contains DS proteoglycans. Note that less than 1% of the total incorporated serine is eluted with the proteoglycan peaks and that a second anion-exchange step would be required for further purification from other labeled proteins and glycoproteins. See Yanagishita et al.[32] for further details.

proteins, and their numerous anionic groups are associated primarily with dense cesium cations in such gradients. Thus proteoglycans with higher glycosaminoglycan-to-protein ratios have higher buoyant densities. In some cases, differences between two populations are sufficient to resolve them at least partially, in the gradients. Gradients formed in the presence of chaotropic solvents are referred to as *dissociative* gradients whereas those developed in the absence of chaotropic solvents are referred to as *associative* gradients.[41] These gradients were initially developed for studying aggrecan and were essential for defining the process of aggrecan aggregation. Dissociative gradients are often established in the presence of 4 M guanidine hydrochloride, but 8 M urea or 10 M formamide can be used, and the same general principles apply. The presence of 0.5% (w/v) CHAPS improves recoveries, particularly when hydrophobic proteoglycans such as cell surface HS proteoglycans are being analyzed. See Hascall and Kimura[41] for more detailed discussion of this methodology.

The following procedure provides an example of a dissociative gradient. CsCl (0.6 g/g solution) is added to the sample in 4 M guanidine hydrochloride, 0.5% (w/v) CHAPS, 50 mM sodium acetate (pH 6.0) to give an initial density of 1.50 g/ml. Better resolution of proteoglycans in a particular system may be achieved with a lower initial density, and the amount of the CsCl can be adjusted accordingly. Addition of 0.15 g of CsCl per gram of solution increases the buoyant density by approximately 0.1 g/ml under these solvent conditions. Initial densities should be measured by weighing equal volumes of the solution and of H_2O and calculating the ratio. A solvent blank at the same final density can be used to fill the centrifuge tubes to the levels specified by the manufacturer and to prepare matched density blanks for counterbalance if necessary. Samples are centrifuged for 48–60 hr at 10° in an appropriate rotor at a safe speed for the density of the solution [e.g., Beckman (Fullerton, CA) SW 50.1, 35,000 rpm]. *Caution:* If a CsCl pellet is present in the bottom of the gradient, subsequent analyses should be done with either a lower initial buoyant density or a lower rotor speed. Each final gradient is partitioned into 6–12 fractions, typically by gently inserting a long, blunt-ended needle, attached to tubing in a peristaltic pump, into the bottom of the tube and pumping fractions into a fraction collector. Aliquots of the fractions are weighed to determine densities and others are counted to determine the distribution of radiolabel. High-density proteoglycans (such as aggrecan or versican) will be in the bottom fractions whereas low-density proteoglycans (such as decorin or fibromodulin) will be in the upper fractions. Fractions, as deemed appropriate from the distribution of radiolabel, are pooled and dialyzed or ultrafiltered against an appropriate solvent to remove the CsCl. The samples can then be stored at −20° or used directly for further analysis.

3. Size. Molecular sieve chromatography is widely used to separate and characterize intact and selectively degraded proteoglycans. The choice of support matrix, porosity, and elution solvents depends on the properties of the particular proteoglycans being studied. Matrices designed for high pressure and flow rates, such as Superose 6 (Pharmacia), have distinct advantages in terms of speed and tolerance for chaotropic solvents. For proteoglycans with large hydrodynamic volumes, more porous matrices, such as Sephacryl S-500 or S-1000 (Pharmacia), are used.

The following procedure shows an application for separating proteoglycan populations that resolve into more than one peak. A Superose 6 column (1 × 30 cm) with a column guard is equilibrated with solvent: 10 M formamide, 0.30 M NaCl, 0.5% (w/v) CHAPS, 0.05 M sodium acetate (pH 6) that has been prefiltered (0.2-μm pore size). For analytical purposes only, a solvent with 4 M guanidine hydrochloride and 0.5% (w/v)

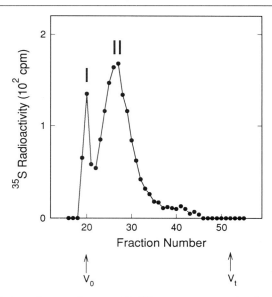

FIG. 5. Resolution of proteoglycans with different average molecular size distributions by molecular sieve chromatography. Dermatan sulfate proteoglycans were purified from the medium of rat ovarian granulosa cell cultures by Q-Sepharose anion-exchange chromatography, using procedures similar to those shown in Fig. 4 (peak II). An aliquot was eluted on Superose 6 with a 4 M guanidine hydrochloride solvent (0.4-ml fractions, 0.4 ml/min). The excluded peak I has characteristics similar to versican and the included peak II has characteristics similar to decorin.

Triton X-100 can be used effectively. The sample (no more than 5 mg) is centrifuged (10,000 g for 5 min) or ultrafiltered (0.2-μm pore size) to clarify. Typically 100–200 μl (up to 500 μl) is injected into the column and 0.4-ml fractions collected at a rate of 0.4 ml/min. Aliquots are counted for radioactivity. The proteoglycans are then concentrated from pooled peak fractions and recovered by the batch anion-exchange method described above. Columns should be used with column guards, and maintained, cleaned, and cared for according to manufacturer recommendations. An example of an analysis is shown in Fig. 5.

4. Hydrophobicity. In chaotropic solvents, core proteins of some proteoglycans can be separated on the basis of differences in their hydrophobic properties. For example, decorin and the KS proteoglycan from cornea elute from octyl-Sepharose at different detergent concentrations during a detergent gradient,[32,42] and different classes of cell surface HS proteoglycans can be resolved by similar procedures.[40]

[42] R. J. Midura and V. C. Hascall, *J. Biol. Chem.* **264,** 1423 (1989).

An example of such an analysis follows. Solvent [10 M formamide, 0.3 M NaCl, 50 mM sodium acetate (pH 6)] is treated with activated charcoal to remove impurities that interfere with the chromatographic steps. Solvents with 4 M guanidine hydrochloride and 8 M urea can be used and are treated in the same way. Octyl-Sepharose CL-4B (Pharmacia) is washed several times with this solvent, and a column with a bed volume of 5 ml is prepared in a plastic pipette. A thin layer of Sephadex G-50 (fine) in the same solvent is applied on top of the column to provide surface stability. Different batches of octyl-Sepharose can have high-affinity binding sites that bind proteoglycans irreversibly. If necessary, these sites can be blocked by preeluting the column with serum albumin (50–100 μg/ml bed volume) dissolved at 1 mg/ml in the same solvent. The sample is diluted with solvent such that the final CHAPS concentration is less than 0.02% (w/v). The final proteoglycan concentrations should be no more than 2 μg/ml of the column bed volume. After sample application, the gradient is developed with 50 ml of solvent and 50 ml of the same solvent with 1.5% (w/v) CHAPS, at 25 ml/hr, with 2-ml fractions collected. An aliquot of concentrated detergent is placed in each fraction tube before the gradient is developed to prevent adsorption of eluted materials, particularly at the beginning of the gradient when the eluting detergent concentration is low. Aliquots are counted for radioactivity. Concentrations of CHAPS in the fractions can be determined with a carbazole assay.[42] Detergents such as Triton X-100 or other nonpolar reagents can be used when CHAPS does not effectively elute or separate proteoglycans. The columns can be regenerated by washing with 10 column volumes of 95% ethanol and then 1-butanol before equilibration with 20 column volumes of the initial solvent. An example of an analysis is shown in Fig. 6.[43]

F. Quantitation of Glycosaminoglycan Synthesis

Environmental sulfate is generally the predominant source for synthesis of sulfoesters on glycosaminoglycans. Even when the concentration of sulfate is limiting such that the glycosaminoglycans synthesized are undersulfated, metabolic sources of sulfate, cysteine, and methionine still contribute only a few percent to the sulfate pool.[33] Additionally, the equilibration of radiosulfate in the medium with the intracellular metabolic precursor, PAPS, is rapid. For these reasons, the specific activity of the radiosulfate in the medium is practically identical to that in the PAPS pool, and it can be used to estimate the mass of glycosaminoglycan synthesized during a labeling period as described later. Sugar precursors in the me-

[43] Y. Takeuchi, M. Yanagishita, and V. C. Hascall, *Arch. Biochem. Biophys.* **298**, 371 (1992).

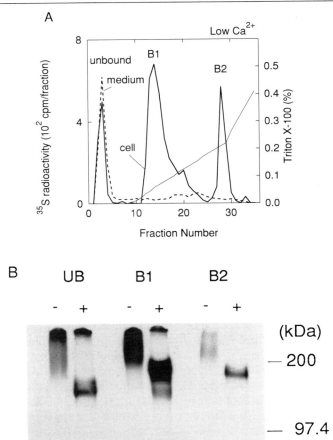

FIG. 6. Resolution of different HS proteoglycan species by hydrophobic chromatography. (A) HS proteoglycans were purified from the medium (dashed line) and cell layer (solid line) of labeled cultures from a rat parathyroid cell line by Q-Sepharose anion-exchange chromatography, using procedures similar to those shown in Fig. 4 (peak I). (B) Aliquots of each sample were eluted on octyl-Sepharose (A) by the procedure described in Section IV,E,4. Portions of the peaks [unbound (UB) and bound (B1 and B2)] from the cell layer

dium, such as [³H]glucosamine, are diluted by intracellular metabolic sources, often by several 100-fold, depending on the metabolic status of the cell and other carbohydrate sources, such as glucose, in the medium. Further, the equilibration of the medium precursor with the metabolic precursor, in this case UDP-N-acetylhexosamines, takes much longer than for sulfate.[11,44] Thus it is not possible to relate the specific activity of the precursor in the medium directly to that of the metabolic precursor. This is a particularly difficult problem for measuring HA because this glycosaminoglycan does not contain sulfoesters.

Most cells that synthesize HA also synthesize CS and/or DS proteoglycans. Thus, radiosulfate in combination with an ³H-labeled sugar precursor ([³H]glucosamine) can be used to double label the CS/DS. After labeling, the samples are digested with a chondroitinase enzyme, and the characteristic disaccharide products separated. Monosulfated disaccharides, which contain 1 mol of sulfoester (labeled from the radiosulfate) and 1 mol of hexosamine (labeled from the [³H]glucosamine precursor), are used to determine the mass of disaccharide synthesized during the labeling period (from the ³⁵S activity) and the dilution of the hexosamine precursor (from the ³H/³⁵S ratio) (Fig. 7).

The following protocol illustrates the use of the double-radiolabel technique to quantitate HA and CS/DS synthesized in culture systems. Medium with [³⁵S]sulfate and [³H]glucosamine is prepared with radioactivities in the range described above or determined from preliminary experiments. It is useful to have the concentrations of the radiolabeled precursors in a range such that the ratio of ³H to ³⁵S in the final monosulfated disaccharide will be between 0.5 and 20, to ensure accuracy in determining the spillover corrections. The specific activity of the precursors can be measured by determining the radioactivity and mass of each precursor in the medium. Sulfate can be determined by an ion chromatography method[40] and glucosamine by standard amino acid/hexosamine procedures. A labeling time of 10–20 hr is usually appropriate. A kinetic experiment can be done to determine that the incorporation of the radiosulfate reaches linearity within a short time (usually minutes) and remains linear until the end of

[44] G. Eriksson, B. Sarnstrand, and A. Malmström, *Arch. Biochem. Biophys.* **235**, 692 (1984).

profile were subjected to SDS–PAGE before (−) and after (+) treatment with heparitinase. Core proteins in the bound fractions elute primarily at higher apparent molecular weights (200,000 and 70,000) than do those in the unbound fraction (180,000 and 64,000). These and other results suggest that the HS proteoglycans in the unbound fraction are derived from those in the bound fractions by removal of hydrophobic portions of the core proteins. (Modified from Takeuchi *et al.*,[43] with permission.)

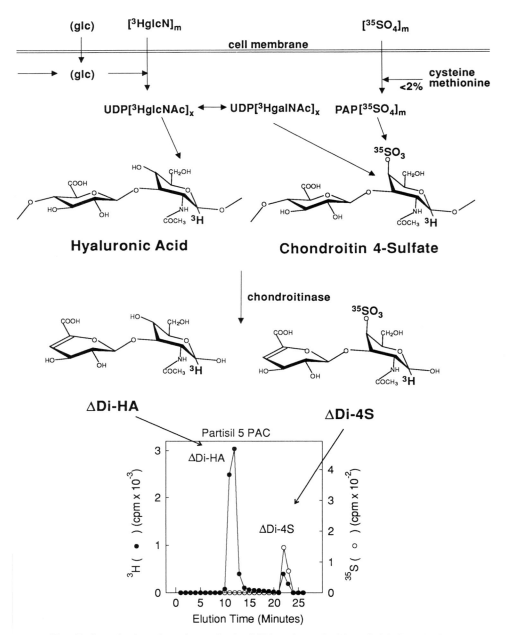

FIG. 7. Quantitation of newly synthesized HA, using a double-radiolabel method. The diagram shows the route of entry of [^{35}S]sulfate into CS and of [^{3}H]glucosamine into CS and HA. The specific activity of the [^{3}H]glucosamine is diluted during its activation to the

the labeling time. The ^{3}H incorporation will reach linearity more slowly (typically 30–90 min) and may not remain linear throughout the labeling time period. After labeling, medium and cell layer compartments are separated. For glycosaminoglycan analyses only, both fractions are treated with protease, typically with papain digestion followed by enzyme inactivation. This ensures better recovery of HA in subsequent steps. Unincorporated precursors are removed by Sephadex G-50 chromatography as described above (Section IV,C), using columns equilibrated with a buffer suitable for chondroitinase digestion: 0.1 M sodium acetate (pH 7.8), for example. The excluded fractions are counted for total ^{35}S recoveries. Other aliquots are digested with chondroitin ABC eliminase (0.01–0.1 unit/ml, 37°, 1 hr). The incubation conditions must be sufficient to digest HA and CS/DS completely. An aliquot of each digest is eluted on a Sephadex G-50 column to determine the proportion of the total incorporated ^{35}S activity susceptible to the enzyme and eluting near the total volume of the column. The digests are ultrafiltered, using low molecular weight cutoff membrane (such as Centricon 3k; Amicon, Danvers, MA), to recover the digestion products in the filtrate. Disaccharides can be separated by various HPLC procedures, such as that shown in Fig. 7, which resolves the ΔdiHA disaccharide and the Δdi4S monosulfated disaccharide. The mass of the chondroitin 4-sulfate (C-4-S) (or C-6-S) derivative synthesized during the labeling period is determined by calculating the total ^{35}S in the original samples that would be in the unsaturated 4-S disaccharide (Δdi4S) and dividing by the specific activity of the ^{35}SO$_4$ in the medium. The mass of HA synthesized is then determined by multiplying the ratio of ^{3}H counts in ΔdiHA versus Δdi4S by the mass of C-4-S (or C-6-S) disaccharide synthesized.[33] See figure 7 for structures.

G. Enzymatic Digestion of Glycosaminoglycans

A number of commercially available enzymes can selectively degrade one or more classes of glycosaminoglycan. Company catalogues provide useful information concerning specificities, purities, digestion conditions, and pertinent references for these enzymes.

Several eliminase enzymes are commonly used to digest HA, CS, and/or DS. They catalyze the cleavage of the N-acetylhexosamine bond

---

UDPsugar precursors for glycosaminoglycan synthesis, yielding an unknown specific activity (indicated by the subscript x). The characteristic disaccharide products from the chondroitin ABC eliminase digestion are indicated, and an example of their resolution on Partisil 5-PAC is shown in the elution profile.

through a β-elimination reaction that leaves a 4,5-unsaturated hexuronic acid on the nonreducing ends, as shown in Fig. 7. Chondroitin ABC eliminase (*Proteus vulgaris*) digests all three of these glycosaminoglycans, yielding primarily a mixture of Δdisaccharides characteristic of the HA and proteoglycans in the reaction mixture. The disaccharides can be separated to determine the average composition of the chains. Chondroitin AC eliminase I (*Flavobacterium heparinum*) or chondroitin AC eliminase II (*Arthrobacter aurescens*) digest HA and CS, but do not cleave linkages to IdoUA in DS chains. The AC-I enzyme digests linkages to GlcUA in DS chains to a greater extent than does the AC-II enzyme. Differential analyses of chondroitin ABC and AC eliminase digests are frequently used to estimate the proportion of idoUA linkages in the chains. Chondroitin B eliminase (*F. heparinum*), conversely, digests linkages to IdoUA, but not to GlcUA. Examples of the use of chondroitinase digestions are shown in Figs. 7 and 8. The eliminase, hyaluronidase (*Streptococcus dysgalactiae*), digests only HA and is often used diagnostically to determine the proportion of HA in a glycosaminoglycan mixture.

A different group of eliminases has specificities restricted to HS/heparin. Heparitinase (*F. heparinum*) catalyzes the eliminative cleavage of GlcNAc linkages with or without a C-6 sulfate ester and adjacent to a nonsulfated hexuronic acid. Thus it effectively digests block regions in HS that lack *N*-sulfoesters. Designation of the different heparitinases may differ between different companies. Heparinase (*F. heparinum*) requires a C-2-sulfated IdoUA next to an N-sulfated GlcN linkage. Thus this enzyme is relatively selective for heparin, digesting different HS species poorly if at all. An example of the use of heparitinase for revealing the characteristics of the core proteins in an HS proteoglycan mixture is shown in Fig. 6.

The keratanase enzymes are hydrolases that specifically digest KS. Keratanase (*Pseudomonas* spp.) is an endo-β-galactosidase that requires a C-6-sulfated GlcNAc next to an unsulfated galactose in the linkage that is cleaved. Keratanase II (*Bacillus* spp.) is an endo-β-glucosaminidase that requires that the GlcNAc be C-6 sulfated. The galactose at the cleavage site can be C-6 sulfated or not. Endo-β-galactosidase (*Escherichia freundii*) cleaves bonds in which neither the GlcNAc nor the galactose is sulfated. An example of the use of these enzymes to reveal properties of the core protein of the corneal KS proteoglycan is shown in Fig. 8.

A wide variety of solvents and digestion conditions is used for these enzymes, depending on pH optima, substrate specificities, mass of glycosaminoglycan, protease contaminants, and so on, and references to specific applications can be consulted. If the digestion products are the main objective, then digestion times and enzyme concentrations should be cho-

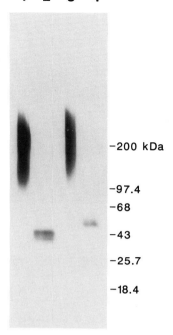

FIG. 8. The use of selective enzyme digestion to characterize the core protein of a DS-proteoglycan and a KS-proteoglycan. The DS-proteoglycan (primarily decorin) and KS-proteoglycan (lumican) were labeled in chick cornea explants, purified by anion-exchange chromatography, and separated from each other by octyl-Sepharose chromatography, using procedures similar to those used in Figs. 4 and 6. The purified DS-proteoglycans (lanes 1 and 2) and KS-proteoglycans (lanes 3 and 4) were subjected to SDS–PAGE either before (lanes 1 and 3) or after treatment with chondroitin ABC eliminase (lane 2) or keratanase (lane 4). The enzymes are specific for the respective proteoglycans and reveal the apparent molecular weights for the core proteins (~43,000 for the DS proteoglycan, ~51,000 for the KS proteoglycan). See Ref. 42 for further details. (Modified from Midura and Hascall,[42] with permission.)

sen to ensure complete digestion, and centrifugal ultrafiltration and washing should be used to separate products from proteins in the enzyme preparations (often primarily serum albumin used as a stabilizer) and from residual core proteins and other undegraded macromolecules in the original sample, which could interfere with the subsequent analytical procedures. The digestion products can then be analyzed further by a variety of separation techniques. If studies of the core proteins are a major objective, then the amount of enzyme used should be minimal and protease

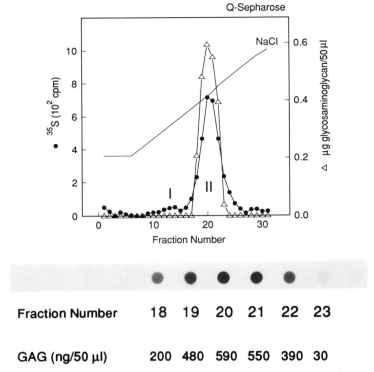

Fraction Number	18	19	20	21	22	23
GAG (ng/50 µl)	200	480	590	550	390	30

FIG. 9. The use of a dye precipitation method to determine chemical properties of a CS proteoglycan. Proteoglycans were labeled with [^{35}S]sulfate in cultures of a rat osteoblastic sarcoma cell line (UMR-106). Proteoglycans in the medium were purified by Sephadex G-50 and Q-Sepharose anion-exchange chromatography. The Q-Sepharose profile of radioactivity shows two peaks (peak I, HS proteoglycans; peak II, CS proteoglycans). Safranin O dye precipitation (Section IV,H) revealed significant amounts of glycosaminoglycan in the fractions coeluting with the DS proteoglycans (peak value of ~600 ng) and no detectable glycosaminoglycan content in the fractions corresponding to the HS proteoglycans, which were present below the detection limit of the assay (~10 ng).

inhibitors should be used to prevent core protein degradation.[45] Preliminary experiments comparing a series of enzyme dilutions for a fixed incubation time, using SDS–PAGE and fluorography, can be of advantage.

H. Quantitation of Chemical Amounts of Proteoglycans

Tissue explants or large numbers of cells often contain sufficient amounts of proteoglycans to quantitate them with chemical analyses on

[45] Y. Oike, K. Kimata, T. Shinomura, K. Nakazawa, and S. Suzuki, *Biochem. J.* **191,** 193 (1980).

microscales. This can allow the properties of radiolabeled species and the resident proteoglycans to be compared. Procedures that involve saturation binding of cationic dyes to the anionic groups on the glycosaminoglycans precipitate the proteoglycans. If the precipitated dye is a chromophore that can be quantitated, the amount of the dye precipitated can be measured, for example, by reflectance absorption of the precipitate or by resolubilization of the complex and absorbance spectroscopy at an appropriate wavelength. Among procedures described, the ones using Safranin O appear to have the best sensitivity and give low backgrounds.[46]

   The following protocol can detect ~5 ng of sulfated glycosaminoglycan and is linear up to ~2 μg. A polyvinylidene difluoride membrane (Immobilon) is wetted in 100% methanol and washed several times in water (5 min/wash). It is then placed on moistened filter paper cut to the same size, and both are positioned in an appropriate Hydri-dot manifold. Vacuum is applied to dry the wells. Aliquots (50 μl) such as those in fractions from an ion-exchange analysis (Fig. 9) and glycosaminoglycan standards (10 ng–2 μg) in 50 μl of the same solvent are placed on the sides of individual wells in the manifold, where they remain by surface tension without sealing the well to the vacuum. An aliquot (400 μl) of Safranin O reagent [0.02% (w/v) Safranin O, 50 mM sodium acetate (pH 4.8)] is injected into each well with a multiwell repeating pipette. The reagent essentially instantaneously mixes and dilutes the sample, precipitates the glycosaminoglycans and proteoglycans in the well, and seals the well to the vacuum. The solvent is drained from the wells, leaving a uniform magenta dot of variable intensity, depending on the concentration of glycosaminoglycan and/or proteoglycan in each fraction. The membrane is removed from the manifold and washed twice (5 min each) in 100% methanol to remove background staining. After air drying, the dots are quantitated by a densitometer. This procedure will give quantitative precipitation of dye–glycosaminoglycan/proteoglycan even if the original 50-μl aliquots contain up to 4 M guanidine hydrochloride or other salts.

---

[46] M. Lammi and M. Tammi, *Anal. Biochem.* **168**, 352 (1988).

[25] Structural Analysis of Glycosylphosphatidylinositol Anchors

By ANANT K. MENON

Introduction

Glycosylated phosphoinositides constitute a diverse family of molecules found in prokaryotes and eukaryotes. This chapter is concerned with methods for identifying and characterizing members of this family defined by the sequence EtN-PO_4-6Manα1-2Manα1-6Manα1-4GlcNα1-6-*myo*-inositol-1-PO_4–lipid. These structures, variously referred to as glycosylphosphatidylinositols (GPIs), glycoinositol phospholipids (GIPLs), or phosphatidylinositol glycans (PI-Gs), were first discovered covalently linked to eukaryotic cell surface glycoproteins and recognized to be an important alternative mechanism for anchoring proteins to cell membranes.[1] Over 100 GPI-anchored proteins have been described to date and a number of comprehensive reviews on the subject have appeared.[2–6] Related glycoconjugates that are not involved in protein anchoring but which contain the limited structural motif Manα1-4GlcNα1-6-*myo*-inositol-1-PO_4–lipid will not be explicitly considered here, although some of the methodology described will clearly be relevant to the study of these structures as well.

Glycosylphosphatidylinositols: Structure and Biosynthesis

The addition of a GPI anchor to protein occurs soon after the polypeptide is translocated across the membrane of the endoplasmic reticulum (ER). The reaction proceeds by cleavage of a carboxyl-terminal signal sequence[3,7] and attachment of a GPI precursor (via ethanolamine) to the newly exposed α-carboxyl group of the polypeptide. The GPI precursor itself is synthesized in the ER by stepwise addition of components (donated by UDPGlcNAc, dolichol-PO_4-mannose, and phosphatidylethanolamine)

[1] M. G. Low, M. A. J. Ferguson, A. H. Futerman, and I. Silman, *Trends Biochem. Sci.* **11**, 212 (1986).
[2] M. A. J. Ferguson and M. A. J. Williams, *Annu. Rev. Biochem.* **57**, 285 (1988).
[3] G. A. M. Cross, *Annu. Rev. Cell Biol.* **6**, 1 (1990).
[4] M. A. J. Ferguson, *Curr. Opin. Struct. Biol.* **1**, 522 (1991).
[5] M. G. Low, *Biochim. Biophys. Acta* **988**, 427 (1989).
[6] J. R. Thomas, R. A. Dwek, and T. W. Rademacher, *Biochemistry* **29**, 5413 (1990).
[7] S. Udenfriend, R. Micanovic, and K. Kodukula, *Cell Biol. Int. Rep.* **15**, 739 (1991).

to phosphatidylinositol (PI). Reviews may be consulted for further details of anchor biosynthesis.[3,8,9]

A limited library of GPI structures is currently available (Fig. 1).[10–25] All protein-linked GPIs contain the conserved sequence discussed above, in which the terminal ethanolamine residue is amide linked to the C-terminal α-carboxyl group of the mature protein. The core GPI structure may be elaborated in a variety of ways and the GPI anchor of a given protein may exist as a spectrum of glycoforms (Fig. 1). Substituents on the core glycan (including phosphoethanolamine, mannose, galactose, GalNAc, and sialic acid) and the precise nature and composition of the lipid moiety (diacylglycerol, monoacylglycerol, alkylacylglycerol, ceramide) vary with protein and cell type. An important variation from the point of view of structural analysis is the acylation of the inositol ring found in some GPI structures. Esterification of one of the inositol hydroxyls with

[8] T. L. Doering, W. J. Masterson, G. W. Hart, and P. T. Englund, *J. Biol. Chem.* **265,** 611 (1990).

[9] A. K. Menon, *Cell Biol. Int. Rep.* **15,** 1007 (1991).

[10] S. W. Homans, M. A. J. Ferguson, R. A. Dwek, T. W. Rademacher, R. Anand, and A. F. Williams, *Nature (London)* **333,** 269 (1988).

[11] M. L. S. Güther, M. L. Cardoso de Almeida, N. Yoshida, M. A. J. Ferguson, *J. Biol. Chem.* **267,** 6820 (1992).

[12] N. Stahl, M. A. Baldwin, R. Hecker, K.-M. Pan, A. L. Burlingame, and S. B. Prusiner, *Biochemistry* **31,** 5043 (1992).

[13] M. A. Deeg, D. R. Humphrey, S. H. Yang, T. R. Ferguson, V. N. Reinhold, and T. L. Rosenberry, *J. Biol. Chem.* **267,** 18573 (1992).

[14] T. Kamitani, A. K. Menon, Y. Hallaq, C. D. Warren, E. T. H. Yeh, *J. Biol. Chem.* **267,** 24611 (1992).

[14a] P. Schneider, J. E. Ralton, M. J. McConville, and M. A. J. Ferguson, *Anal. Biochem.* **210,** (in press).

[15] S. Hirose, G. M. Prince, D. Sevlever, L. Ravi, T. L. Rosenberry, E. Ueda, and M. E. Medof, *J. Biol. Chem.* **267,** 16968 (1992).

[16] M. A. J. Ferguson, S. W. Homans, R. A. Dwek, and T. W. Rademacher, *Science* **239,** 753 (1988).

[17] M. C. Field, A. K. Menon, and G. A. M. Cross, *EMBO J.* **10,** 2731 (1991).

[18] M. A. J. Ferguson, *Biochem. J.* **284,** 297 (1992).

[19] W. L. Roberts, S. Santikarn, V. N. Reinhold, and T. L. Rosenberry, *J. Biol. Chem.* **263,** 18776 (1988).

[20] S. Mayor, A. K. Menon, and G. A. M. Cross, *J. Biol. Chem.* **265,** 6174 (1990).

[21] M. C. Field, A. K. Menon and G. A. M. Cross, *J. Biol. Chem.* **266,** 8392 (1991).

[22] W. L. Roberts, J. J. Myher, A. Kuksis, M. G. Low, and T. L. Rosenberry, *J. Biol. Chem.* **263,** 18766 (1988).

[23] P. Schneider, M. A. J. Ferguson, M. J. McConville, A. Mehlert, S. W. Homans, and C. Bordier, *J. Biol. Chem.* **265,** 16955 (1990).

[24] W. L. Roberts, J. J. Myher, A. Kuksis, and T. L. Rosenberry, *Biochem. Biophys. Res. Commun.* **150,** 271 (1988).

[25] A. Conzelmann, A. Puoti, R. L. Lester, and C. Desponds, *EMBO J.* **11,** 457 (1992).

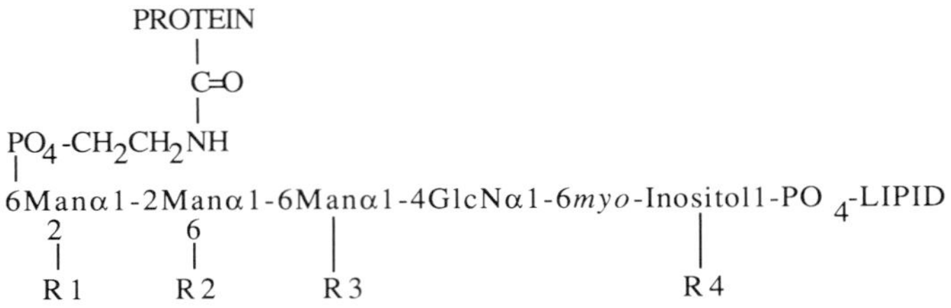

PROTEIN
|
C=O
|
PO$_4$-CH$_2$CH$_2$NH
|
6Manα1-2Manα1-6Manα1-4GlcNα1-6myo-Inositol1-PO$_4$-LIPID
2   6 | |
| | R 3   R 4
R 1   R 2

R1: α-mannose
R2: EtN-P
R3: EtN-P, β-GalNAc, Gal-GalNAc, SA-Gal-GalNAc, α-Gal$_{2-4}$
R4: hydroxyester-linked palmitic acid
LIPID: dimyristoylglycerol, 1-stearoylglycerol, 1-alkyl-2-acyl-
       glycerol, ceramide

FIG. 1. Glycosylphosphatidylinositol structures: evolutionarily conserved GPI sequence. The carboxyl-terminal amino acid of the mature protein is amide linked to the ethanolamine residue. The sequence is modified by various substituents (R1–R4), as observed in the anchors analyzed thus far. Because the anchors exist as a spectrum of glycoforms, all substituents may not necessarily be present in each of the molecular species of the anchor. The R1 substituent, α-mannose, has been found in rat brain Thy-1,[10] *Trypanosoma cruzi* 1G7 antigen,[11] and in some GPI glycoforms of the scrapie prion protein.[12] The R2 substituent, phosphoethanolamine, has been found in the human erythrocyte acetylcholinesterase anchor,[13] and in some mammalian non-protein-linked GPIs.[14,15] R3 substituents are varied. Phosphoethanolamine is found in rat brain Thy-1,[10] human erythrocyte acetylcholinesterase,[13] all GPI glycoforms of the scrapie prion protein,[12] and in some mammalian non-protein-linked GPIs.[14,15] β-GalNAc is found in rat brain Thy-1,[10] and in some GPI glycoforms of the scrapie prion protein.[12] Gal-GalNAc and SA-Gal-GalNAc are found in some GPI glycoforms of the scrapie prion protein.[12] α-Gal is found in the GPI anchors of the variant surface glycoproteins of bloodstream form *Trypanosoma brucei* (except variant 118).[16] R4 is a hydroxyester-linked palmitic acid found in some GPI anchors (e.g., human erythrocyte acetylcholinesterase, *T. brucei* procyclic acidic repetitive protein),[13,17–19] and in mammalian and trypanosomal nonprotein-linked GPIs.[14,15,20,21] The lipid moiety can be diacylglycerol,[16] monoacylglycerol,[17] alkylacylglycerol,[22–24] and ceramide.[25] EtN-P, Phosphoethanolamine; GalNAc, *N*-acetylgalactosamine; Gal, galactose; GlcN, glucosamine; SA, sialic acid.

palmitic acid[17,18,20–22] renders the GPI structure resistant to cleavage by phosphatidylinositol-specific phospholipases C (PI-PLCs) and GPI-specific phospholipase C (GPI-PLC), enzymes often used to demonstrate the presence of a GPI anchor. Palmitoylation of the inositol ring occurs early in the assembly of the GPI precursor[26,27]; biosynthetic information on the addition of the other glycan substituents is not currently available.

[26] L. C. Costello and P. Orlean, *J. Biol. Chem.* **267**, 8599 (1992).
[27] M. Urakaze, T. Kamitani, R. DeGasperi, E. Sugiyama, H.-M. Chang, C. D. Warren, and E. T. H. Yeh, *J. Biol. Chem.* **267**, 6459 (1992).

General Strategy for Identifying and Characterizing Glycosylphosphatidylinositol-Anchored Proteins

A variety of techniques can be used to determine whether a protein is GPI anchored.[28] In many cases GPI-anchored proteins can be selectively released into the culture medium on treatment of intact cells with bacterial PI-specific phospholipases C. If specific antibodies are available, loss of a particular protein from the cell surface can be monitored by fluorescence-activated cell sorter (FACS) analysis.[29,30]

Another approach commonly used to identify GPI-anchored proteins involves detergent partitioning with Triton X-114.[31,32] If cell extracts are subjected to Triton X-114 phase separation, GPI-anchored proteins (along with most membrane proteins) will partition into the detergent phase. Treatment with PI- or GPI-specific phospholipases C will remove the hydrophobic component of the GPI anchor (*note:* inositol-acylated GPI anchors are not susceptible to PI-PLC) and the resulting hydrophilic proteins will partition into the detergent-poor aqueous phase if the sample is once again subjected to detergent phase separation. The detergent and aqueous phases can be separately analyzed by sodium dodecyl sulfate-polyacrylamide gel electrophoresis (SDS-PAGE) to assess the cleavage and to identify the susceptible polypeptides. These experiments are typically performed with cells metabolically labeled with radioactive amino acids (e.g., [^{35}S]methionine).[33] Alternatively, surface-radioiodinated[34] or surface-biotinylated[35] cells may be used for the analysis. If the protein of interest is an enzyme (e.g., alkaline phosphatase), or if antibodies to the protein are available, then analysis of a labeled sample is not essential as enzymatic activity or immunoreactivity can be followed instead.

Other techniques for identifying GPI-anchored proteins involve hydrophobic interaction chromatography,[28,36] nondenaturing polyacrylamide gel electrophoresis,[28,37] in conjunction with phospholipase treatment, and de-

[28] N. M. Hooper, *in* "Lipid Modification of Proteins: A Practical Approach" (N. M. Hooper and A. J. Turner, eds.) p. 89. IRL Press, Oxford, 1992.

[29] M. G. Low and P. W. Kincade, *Nature (London)* **318**, 62 (1985).

[30] M. G. Low and P. W. Kincade, *Nature (London)* **318**, 687 (1985).

[31] C. Bordier, *J. Biol. Chem.* **256**, 1604 (1981).

[32] J. G. Pryde, *Trends Biochem. Sci.* **11**, 160 (1986).

[33] A. Conzelmann, H. Riezman, C. Desponds, and C. Bron, *EMBO J.* **7**, 2233 (1988).

[34] A. Conzelmann, A. Spiazzi, R. Hyman, and C. Bron, *EMBO J.* **5**, 3291 (1986).

[35] M. P. Lisanti, M. Sargiacomo, L. Graeve, A. R. Saltiel, E. Rodriguez-Boulan, *Proc. Natl. Acad. Sci. U.S.A.* **85**, 9557 (1988).

[36] Y. Takesue, K. Yokota, Y. Nishi, R. Taguchi, and H. Ikezawa, *FEBS Lett.* **201**, 5 (1986).

[37] J.-P. Toutant, W. L. Roberts, N. R. Murray, and T. L. Rosenberry, *Eur. J. Biochem.* **180**, 503 (1989).

tection of the cross-reacting determinant (CRD) by SDS-PAGE and Western blotting after PI-PLC treatment.[23,28,38]

Metabolic labeling with [³H]ethanolamine is particularly useful in identifying GPI structures because metabolic conversion of the precursor is limited and few molecules other than GPIs are labeled. With the exception of a single hydrophilic protein [elongation factor 1α (EF-1α)] that can be labeled with [³H]ethanolamine in animal cells,[39-41] [³H]ethanolamine incorporation into the cellular protein fraction is confined to GPI-anchored proteins. Inhibition of [³H]ethanolamine incorporation into protein by D-mannosamine[42] may be used as an additional criterion of labeling specificity. In addition, because [³H]ethanolamine provides a specific tag for the carboxy terminus of a GPI-anchored protein, GPI-linked carboxy-terminal peptides can be readily identified following proteolysis of a purified [³H]ethanolamine-labeled GPI-anchored protein and reversed-phase high-performance liquid chromatography (HPLC) analysis of the proteolytic products.[43] Metabolic labeling with [³H]ethanolamine is also useful in identifying nonprotein-linked GPIs. Although radiolabeled phosphatidylethanolamine accounts for a large proportion of the radioactivity incorporated into the lipid fraction, it can be separated from polar lipids (e.g., ethanolamine-containing GPIs) by a simple differential extraction protocol and by thin-layer chromatography (TLC) (see below).

³H-Labeled fatty acid samples are essential for complete structural analysis. However, ³H labeling of fatty acids is not specific for GPIs and GPI-anchored proteins and is therefore not the best choice for initial screening of radiolabeled mixtures. In addition, identification of non-protein-linked GPIs via ³H labeling of fatty acids is problematic because GPI species are relatively minor membrane components and total lipid extracts contain mostly radiolabeled phospholipids. However, separation of GPIs from phospholipids can be achieved by differential extraction protocols and by TLC in order to purify samples for structural analysis.

Identification of GPIs and GPI-anchored proteins via metabolic labeling with [³H]mannose may be complicated unless labeling is performed

[38] S. E. Zamze, M. A. J. Ferguson, R. Collins, R. A. Dwek, and T.W. Rademacher, *Eur. J. Biochem.* **176**, 527 (1988).

[39] E. J. Tisdale and A. M. Tartakoff, *J. Biol. Chem.* **263**, 8244 (1988).

[40] T. L. Rosenberry, J. A. Krall, T. E. Dever, R. Haas, D. Louvard, and W. C. Merrick, *J. Biol. Chem.* **264**, 7096 (1989).

[41] S. W. Whiteheart, P. Shenbagamurthi, L. Chen, R. J. Cotter, and G. W. Hart, *J. Biol. Chem.* **264**, 14334 (1989).

[42] M. P. Lisanti, M. C. Field, I. W. Caras, A. K. Menon, and E. Rodriguez-Boulan, *EMBO J.* **10**, 1969 (1991).

[43] P. Moran, H. Raab, W. J. Kohr, and I. W. Caras, *J. Biol. Chem.* **266**, 1250 (1991).

in the presence of tunicamycin (roughly 1 μg/ml) to eliminate protein labeling due to N-linked glycosylation, as well as labeling of dolichol-linked oligosaccharide species (dolichol-PP-GlcNAc$_2$Man$_{1-9}$, and dolichol-PP-GlcNAc$_2$Man$_9$Glc$_{1-3}$; dolichol-PO$_4$-mannose will still be labeled but can be easily identified by TLC and by diagnostic cleavage reactions; O-mannosylation of proteins in yeast is also not inhibited by tunicamycin). Screening of [^{3}H]glucosamine-labeled protein and lipid samples is additionally complicated because labeled O-linked sugars on proteins and some radiolabeled glycosphingolipids will be generated even if labeling is performed in the presence of tunicamycin.

Protocols for metabolic labeling with radiolabeled monosaccharides, inositol, ethanolamine, and fatty acids are described in detail elsewhere[44-48] and will not be elaborated here. The main points to note are that (1) uptake of [^{3}H]mannose and [^{3}H]glucosamine is inhibited by glucose and therefore labeling must be performed in low-glucose (e.g., 0.1 mg/ml) medium; (2) inositol-free medium must be used for [^{3}H]inositol labeling; and (3) in all cases, only dialyzed calf serum should be used in the labeling medium. Typically, ~10^7 cells are incubated with ^{3}H-labeled precursor (specific activity, 20–50 Ci/mmol; concentration in the labeling medium, 5–100 μCi/ml) for 1–2 hr or overnight. For African trypanosomes, incorporation of radioactivity (per 10^7 cells) into ethanolamine-containing GPI precursors (i.e., non-protein-linked GPIs) during a 1- to 2-hr incubation ranges from ~10^5 cpm for [^{3}H]myristic acid labeling to ~10^4 cpm for [^{3}H]glucosamine labeling.[20] In murine thymocytes a spectrum of non-protein-linked GPIs can be labeled with [^{3}H]mannose, yielding ~10^6 cpm (per 10^7 cells) in short-term labeling.[14] Numerous examples of GPI labeling may be found in the primary literature.[20,49-52]

Once these preliminary analyses are performed and there is a clear indication of a GPI anchor, material can be prepared for detailed structural

[44] T. L. Doering, J. Raper, L. U. Buxbaum, G. W. Hart, and P. T. Englund, *Methods* **1**, 288 (1990).

[45] M. C. Field and A. K. Menon, in "Lipid Modification of Proteins: A Practical Approach" (N. M. Hooper and A. J. Turner, eds.), p. 155. IRL Press, Oxford, 1992.

[46] T. L. Rosenberry, J.-P. Toutant, R. Haas, and W. L. Roberts, *Methods Cell Biol.* **32**, 231 (1989).

[47] A. Varki, *FASEB J.* **5**, 226 (1991).

[48] P. D. Yurchenco, C. Ceccarini, and P. H. Atkinson, in "Methods in Enzymology" (V. Ginsburg, ed.), Vol. 50, p. 175. Academic Press, New York, 1978.

[49] A. Conzelmann, C. Fankhauser, and C. Desponds, *EMBO J.* **9**, 653 (1990).

[50] S. H. Fatemi, R. Haas, N. Jentoft, T. L. Rosenberry, and A. M. Tartakoff, *J. Biol. Chem.* **262**, 4728 (1987).

[51] R. K. Margolis, B. Goossen, and R. U. Margolis, *Biochemistry* **27**, 3454 (1988).

[52] N. Takami, S. Ogata, K. Oda, Y. Misumi, and Y. Ikehara, *J. Biol. Chem.* **263**, 3016 (1988).

characterization of the GPI moiety. This chapter describes methods for characterizing radiolabeled GPI structures isolated from cells metabolically labeled with GPI components such as ^{3}H-labeled fatty acids, [^{3}H]mannose, or [^{3}H]glucosamine. The analytical approach involves selective chemical or enzymatic fragmentation of radiochemically pure material, followed by analysis of the released glycan and lipid fragments. The analysis of chemically pure GPIs by modern physical methods will not be explicitly considered here (see [12], [9], and [8] in this volume for descriptions of high-pH anion-exchange chromatography, nuclear magnetic resonance, and mass spectrometry techniques, respectively).

Isolation of Glycosylphosphatidylinositol Structures from Metabolically Labeled Cells

Glycosylphosphatidylinositol-Anchored Proteins

In the absence of specific reagents and conventional protein purification strategies, protein samples of adequate purity may be obtained by SDS-PAGE of metabolically labeled cell lysates, followed by electroelution of the polypeptides of interest and extensive washing to remove radioactive impurities. Removal of radioactive impurities is particularly important in the case of material labeled with ^{3}H-labeled fatty acids, [^{3}H]inositol, or [^{3}H]ethanolamine, because contamination of the protein fraction with radiolabeled phospholipids is inevitable even if the sample has been isolated via SDS-PAGE.

Pellet the radiolabeled cells by centrifugation and solubilize the cell pellet by boiling in 0.375 mM Tris-HCl (pH 6.8) containing 4% SDS (w/v). Protease inhibitors may be included if necessary. Resolve the lysate by SDS-PAGE.[53] Stain the gel with Coomassie Brilliant Blue. With a razor blade excise pieces of the gel containing the polypeptides of interest and recover the polypeptides by electroelution of the gel pieces as described.[54] The electroeluted protein is recovered in 0.02% SDS, 10 mM NH$_4$HCO$_3$. The sample should then be dried in a SpeedVac evaporator (Savant Instruments, Hicksville, NY) and the SDS and radiolabeled lipid impurities removed by acetone precipitation and ethanol washes. Incubate the dried sample with 10 vol of ice-cold acetone at $-20°$ for 3 hr, and centrifuge the precipitate at 13,000 g for 15 min at 4°. Wash the precipitate with ice-

---

[53] D. E. Garfin, in "Methods in Enzymology" (M. Deutscher, ed.), Vol. 182, p. 425. Academic Press, San Diego, 1990.

[54] M. W. Hunkapiller, E. Lujan, F. Ostrander, and L. E. Hood, in "Methods in Enzymology" **91**, (C. Hirs and S. Timasheff, eds.), Vol. 91, p. 227. Academic Press, New York, 1983.

cold acetone and then with 90% ethanol to ensure removal of radiolabeled lipids. Aliquots of the acetone and ethanol washes should be taken for liquid scintillation counting to monitor the removal of radiolabeled lipids, and the washes should be repeated if necessary. Carrier protein (e.g., bovine serum albumin, or cytochrome c, 10 μg) may be added to the sample to facilitate precipitation and minimize loss of material.

Lipids

Free GPIs may be recovered from metabolically labeled cells by conventional lipid extraction procedures. A differential extraction protocol using chloroform–methanol [CM (2:1, v/v)] and chloroform–methanol–water [CMW (10:10:3, v/v)] is particularly useful in fractionating the lipids into two broad categories on the basis of polarity. Individual species in each extract can then be isolated by TLC, anion-exchange chromatography, and/or hydrophobic interaction chromatography.

Add 1.5 ml of ice-cold CM to pelletted cells in a glass tube (sufficient CM must be used to minimize the effect of water in the cell pellet and to maintain a single phase; 1.5 ml of CM is appropriate for up to 70 μl of aqueous sample). Mix by stirring with a long glass Pasteur pipette; do not vortex or shake. Centrifuge the sample (1500 g, 10 min) to pellet the debris, remove the CM supernatant carefully (using a glass Pasteur pipette), and repeat the extraction several times (less CM may be used in the second and subsequent extracts). Take aliquots of each extract for liquid scintillation counting and stop the CM extraction when the radioactivity in the final extract is <0.1% of that in the first extract. Pool the CM extracts and leave on ice. Continue the extraction process by stirring the partly delipidated cell debris with 1 ml of ice-cold CMW. Centrifuge the sample to pellet the debris and reextract with CMW. Pool the CMW extracts. Remove water-soluble contaminants in both the CM and CMW extracts by separate washing procedures. For the CM extract, add 0.2 vol of 4 mM MgCl$_2$, vortex to mix, and separate the phases by centrifugation. Remove the upper phase and reextract the lipid-containing chloroform-rich lower phase with mock upper phase prepared by mixing fresh CM with 4 mM MgCl$_2$ in a separate tube. The CMW extract should be processed by drying in a SpeedVac and resuspending the residue by vortexing in 500 μl each of n-butanol and water and centrifuging to separate the phases. Lipids will be quantitatively recovered in the upper, butanol-rich phase.

Lipids in the CM extract or in the CMW-derived butanol extract can be resolved by TLC [silica 60; solvent system A (see next section) is generally useful] and purified from the TLC plate by scraping and extraction.[45] Although this method is convenient, yields of >60% cannot be

expected. Other lipid purification techniques involving liquid chromatography are discussed in detail elsewhere.[44,45,55]

Analysis of Glycosylphosphatidylinositol Hydrophobic Domain

Information on the structure of the GPI hydrophobic domain may be
obtained via chemical and enzymatic cleavage analysis of appropriately
radiolabeled GPI material. Biosynthetic labeling with radioactive fatty
acids is the most convenient way of generating labeled material for this
purpose, although in some cases exogenous labeling may be necessary,
using radioiodinated photoactivatable probes such as 3-(trifluoromethyl)-
3-(m-[125I]iodophenyl)diazirine) ([125I]TID) that react preferentially with
hydrophobic structures.[56–58]

Lipid fragments may be generated from a GPI-anchored protein in a
variety of ways (Fig. 2). The discussion below focuses on GPI-anchored
proteins containing a glycerophospholipid moiety, particularly 1-alkyl-2-
acylglycerol, as in the mammalian GPIs characterized to date. The anchors
may or may not contain acylinositol. Ceramide-based GPIs (found in
budding yeast)[25] will not be explicitly considered although it will be seen
that many of the procedures described here are applicable to these structures as well.

The generation and analysis of GPI lipid fragments involves the following general experimental protocol. The radiolabeled GPI sample is resuspended in a detergent-containing aqueous buffer, reagent (chemical or
enzyme) is added, and the sample is incubated for a period of time at
ambient temperature or at 37°. The released lipid fragments are extracted
by adding an organic solvent (hexane, toluene, or n-butanol as described),
vortexing to mix, and separating the aqueous and organic phases by centrifugation. Typically, an aliquot of each phase is taken for liquid scintillation
counting to assess cleavage efficiency (a negative control with an identical
sample but no added cleavage reagent is essential to obtain reliable results). The remainder of the organic phase is taken for TLC analysis (a
list of TLC solvent systems is given under Thin-Layer Chromatography:
Materials and Solvent Systems, below). If radioactivity on the TLC plate
is to be detected by scanning (using, e.g., a Berthold LB2842 automatic
TLC linear analyzer, Wallac Inc., Gaithersburg, MD), the entire analysis

[55] K. Aizetmüller, in "Practice of High Performance Liquid Chromatography" (H. Engel-
hardt, ed.), p. 287. Springer-Verlag, New York, 1986.
[56] J. Brunner and G. Semenza, Biochemistry 20, 7174 (1981).
[57] W. L. Roberts and T. L. Rosenberry, Biochemistry 25, 3091 (1986).
[58] S. Steiger, U. Brodbeck, B. Reber, and J. Brunner, FEBS Lett. 168, 231 (1984).

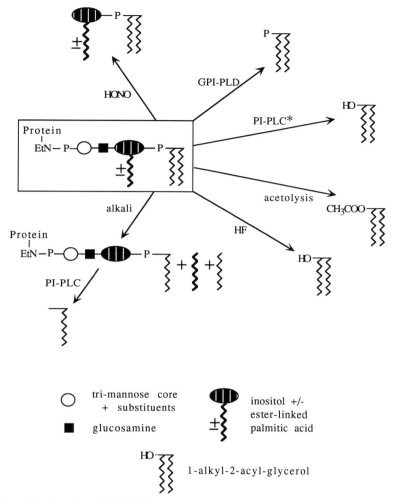

FIG. 2. Production of GPI lipid fragments: cleavages for a GPI-anchored protein contain-ing inositol ± hydroxyester-linked palmitic acid, and 1-alkyl-2-acyl-glycerol; only lipid prod-ucts are shown. Phospholipid fragments may be generated by nitrous acid (HONO) treatment, or by cleavage with GPI-specific phospholipase D (GPI-PLD). Glycerides may be generated by enzymatic treatment, using bacterial PI-specific phospholipases C (PI-PLC*; the asterisk indicates that PI-PLC will cleave only GPI structures that are not inositol acylated), or by chemical treatment via acetolysis or aqueous HF dephosphorylation. Alkali treatment removes the ester-linked fatty acids (including the inositol-linked fatty acid if present), and PI-PLC treatment of the resulting *lyso*-GPI anchor generates 1-alkylglycerol. Glyceride fragments and free fatty acids are extracted into hexane, toluene, or chloroform–methanol (2 : 1, v/v). Phospholipid fragments are extracted into water-saturated *n*-butanol. If the sample is a non-protein-linked GPI rather than a GPI-anchored protein, then any uncleaved starting material will also be extracted into *n*-butanol (but not into hexane or toluene) and can be distinguished from the cleavage product by TLC.

may be performed with as little as 1000 cpm of radiochemically pure starting material. Somewhat more radioactivity is needed to obtain rapid results via autoradiography.

Cleavage reactions to generate GPI lipid fragments are illustrated schematically in Fig. 2. Treatment of GPI-anchored proteins with PI- or GPI-specific phospholipases C results in the release of a glyceride fragment. However, because the phospholipases C will not cleave GPIs containing acylinositol, it may be necessary to subject the anchor to acetolysis or chemical dephosphorylation in order to generate glycerides for analysis. Alternatively, phospholipase C cleavage may be attempted after treating the anchor with mild base to deacylate the inositol residue. Phospholipid fragments may be generated from a GPI anchor by phospholipase D cleavage, or by treatment with nitrous acid. The GPI-specific phospholipase D found in mammalian serum cleaves GPIs irrespective of inositol acylation to give phosphatidic acid. Nitrous acid treatment of GPIs results in deamination of the glucosamine residue and cleavage of the glucosamine-inositol glycosidic bond to release an inositol-containing phospholipid fragment. The various treatments may be attempted in any sequence, although if the amount of sample is limited it is probably most informative to begin with the phospholipases C and D.

Phosphatidylinositol-Specific Phospholipases C Treatment

Reagents

Bacillus thuringiensis PI-PLC (Cat. No. GPI-02; Oxford Glycosystems Inc., Rosedale, NY) or *Bacillus cereus* PI-PLC (Cat. No. 11403-069; Boehringer Mannheim, Indianapolis, IN)

Buffer: 0.1 M Tris-HCl (pH 7.4), 0.1% (w/v) sodium deoxycholate

Procedure. Dissolve the GPI-anchored protein or free GPI in 100 μl of buffer, add PI-PLC (to a final concentration ranging from 0.1 to 1 unit/ml), vortex briefly to mix, and incubate at 37° for 1–4 hr. Terminate the reaction by adding 5 μl of glacial acetic acid and extract the reaction products twice with 100 μl of toluene or hexane. Take an aliquot of the combined organic phases for liquid scintillation counting. Dry the remainder of the organic extract in a SpeedVac evaporator (Savant Instruments) and resuspend the residue in 15 μl of toluene or hexane for TLC analysis [silica G thin-layer plate, using solvent system D (see later section)].

Note: No systematic information is available on the effect of different detergents on the action of the *Bacillus* PI-PLCs. Triton X-100 at concentrations greater than 0.1% (w/v) appears to inhibit the *B. thuringiensis* enzyme. In any case if other detergents are to be used in the assay, it is best to keep the detergent concentration as low as possible without compromising the solubility of the substrate.[59]

Alkaline Hydrolysis to Remove Ester-Linked Fatty Acids

   Reagents

Methanol–30% ammonia, 1 : 1 (v/v), freshly prepared
Buffer: 0.1 *M* Tris-HCl (pH 7.4), 0.1% (w/v) sodium deoxycholate
   Procedure. Resuspend the GPI sample in 200 μl of the methanol–ammonia mixture and incubate the sample at 37° for 2 hr (the sample may also be left at room temperature and incubated longer). At the end of the incubation, evaporate the reaction mixture in a SpeedVac evaporator and resuspend the residue in 200 μl of buffer. Extract released fatty acids (some proportion of these will be in the form of methyl esters) by vortexing with 200 μl of toluene or hexane, and centrifuging to separate the aqueous and organic phases. Repeat the extraction if necessary. As illustrated in Fig. 2, the deacylated GPI sample remaining in the aqueous phase may be further treated with PI-PLC to give a *lyso*-alkyl glycerol.

Acetolysis

   Reagents

Acetic acid–acetic anhydride, 3 : 2 (v/v)
Acetic acid–pyridine, 1 : 1 (v/v)
   Procedure. Dry the GPI sample in a ReactiVial (Pierce Chemical Co., Rockford, IL), add 200 μl of the acetic acid–acetic anhydride mixture, and incubate at 105° for 3 hr. Dry the sample in a SpeedVac evaporator and resuspend the residue in chloroform–methanol (2 : 1, v/v) for TLC analysis [silica gel G plates, using solvent system D (see below)]. Glyceride acetate standards for TLC may be prepared by acetylating the corresponding glycerides, using the acetic anhydride–pyridine mixture, for 16 hr at room temperature.
   Note: Some isomerization of the glyceride acetates may be observed.[20]

Chemical Dephosphorylation with Aqueous Hydrofluoric Acid

   The procedure is described in the section, Aqueous Hydrofluoric Acid Dephosphorylation (below). Dry the radioactive fatty acid-labeled GPI sample in a plastic microcentrifuge tube, using a SpeedVac, cool the tube by placing it in an ice–water mixture, and add 50 μl of ice-cold aqueous

---

[59] M. G. Low, *in* "Lipid Modification of Proteins: A Practical Approach" (N. M. Hooper and A. J. Turner, eds), p. 117. IRL Press, Oxford, 1992.

HF. After incubating the sample in the ice–water mixture for 60 hr, neutralize the reaction mixture with saturated LiOH, adjust the pH as described, and extract the released glyceride fragment by vortexing the sample with a small volume of toluene or hexane. Glycerides recovered in the organic phase can be analyzed as described above.

Glycosylphosphatidylinositol-Specific Phospholipase D Treatment

Reagents

Rabbit, rat, or bovine serum (containing roughly 10^3 units of GPI-PLD/ml, where 1 unit is defined as 1% degradation of GPI substrate per minute).[60] Store the serum in aliquots at $-20°$. Once an aliquot has been thawed for use, it should be stored in a refrigerator and may be used for a period of up to 2 weeks

Buffer: 0.1 M Tris-HCl (pH 7.4), 2.5 mM CaCl$_2$, containing 0.1% (w/v) sodium deoxycholate or Nonidet P-40 (NP-40)

Procedure. Dissolve the protein precipitate or free GPI in 100 μl of buffer, add 2 μl of serum, vortex briefly to mix, and incubate at 37° for 1–4 hr. Terminate the reaction by adding 5 μl of glacial acetic acid and extract the reaction products twice with 200 μl of water-saturated *n*-butanol. Take an aliquot of the combined butanol phases for liquid scintillation counting. Dry the remainder of the butanol extract in a SpeedVac evaporator (Savant Instruments) and resuspend the residue in 15 μl of water-saturated *n*-butanol for TLC analysis [silica 60 with solvent system A, B, or C (see below); run phosphatidic acid and other phospholipid standards alongside].

Note: GPI-PLD activity can be completely inhibited by preincubating the serum for 30 min on ice with buffer containing 0.5 mM 1,10-*o*-phenanthroline (Cat. No. P9375; Sigma, St. Louis, MO). Ethylenediaminetetraacetic acid (EDTA) (1 mM) in the assay buffer also inhibits the enzyme.[61–63]

Nitrous Acid Deamination

Reagents

Solutions: 0.2 M sodium acetate [CH$_3$COONa·3H$_2$O (27.22 g/liter), or anhydrous CH$_3$COONa (16.4 g/liter)], 0.2 M acetic acid (11.5 ml of glacial acetic acid per liter)

[60] M. G. Low and K.-S. Huang, *Biochem. J.* **279**, 483 (1991).
[61] M. A. Davitz, J. Hom, and S. Schenkman, *J. Biol. Chem.* **264**, 13760 (1989).
[62] K.-S. Huang, S. Li, W.-J. C. Fung, J. D. Hulmes, L. Reik, Y.-C. E. Pan, and M. G. Low, *J. Biol. Chem.* **265**, 17738 (1990).
[63] M. G. Low and A. R. S. Prasad, *Proc. Natl. Acad. Sci. U.S.A.* **85**, 980 (1988).

Buffer: 0.2 *M* sodium acetate, pH 3.7 (prepared by mixing 10 ml of 0.2
   M sodium acetate with 90 ml of 0.2 *M* acetic acid)
Nonidet P-40 [10% (w/v) solution in water]
Sodium nitrite [Cat. No. S2252 (Sigma); Cat. No. 10256 (BDH Chemicals, Poole, England)]
   Procedure. Dissolve the dried GPI sample in 100 μl of buffer plus 2
μl of 10% Nonidet P-40. Prepare (fresh) a 400 m*M* solution of sodium
nitrite in buffer, and add 100 μl to the GPI solution. Check that the pH
of the sample is ~4 (the pH of the mixture is critical as the reaction will
not occur outside a narrow range of pH centered at 4). Leave the sample
at room temperature. After a few hours (e.g., 4 hr), add 50 μl of a freshly
prepared 1 *M* solution of sodium nitrite in buffer and let the reaction
continue at room temperature for an additional 4–8 hr. The sample should
gradually acquire a pale yellow color. Also, if the reaction is being performed in a plastic microcentrifuge tube it will be noticed that the tube
becomes slightly discolored. Terminate the reaction by extracting twice
with 200 μl of water-saturated *n*-butanol. Pool the two extracts and take
an aliquot for liquid scintillation counting and another for TLC analysis
[on silica 60 plates, using solvent system B or A (see later section)].
   Note: Detergents other than NP-40 may be used to solubilize the
sample.

Analysis of Hydroxyester-Linked Fatty Acids

Reagents

Methanol–30% ammonia, 1 : 1 (v/v), prepared fresh
HCl (10 m*M*)
Diethyl ether
NaCl (5 *M*) in 0.5 *M* acetic acid
BF₃–methanol (12%, w/w) (Cat. No. 3-3020; Supelco, Bellefonte, PA)
Phosphomolybdic acid spray reagent [10% phosphomolybdic acid in
   ethanol; Cat. No. P1518 (Sigma)]
   Procedure. Release hydroxyester-linked fatty acids by alkaline hy-
drolysis of the GPI sample as follows. Incubate the dried sample with 100
μl of methanol–ammonia at 37° for 2 hr. Dry the sample in a SpeedVac,
add 100 μl of 50% methanol, and dry again to remove residual ammonia.
Partition the products between 250 μl of 10 m*M* HCl and 250 μl of diethyl
ether. Transfer the ether phase to a clean ReactiVial (Pierce Chemical
Co.) and reextract the aqueous phase twice with 250 μl of ether. The
released fatty acids will be quantitatively recovered in the ether phase.
Pool the ether phases and dry under a stream of nitrogen.

Esterify the fatty acids by heating the dried residue with 200 μl of BF$_3$–methanol at 100° for 2 min. Add 200 μl of 5 M NaCl in 0.5 M acetic acid and extract the fatty acid methyl esters (FAMES) into toluene. An aliquot of the toluene phase should be taken for TLC analysis [silica 60, solvent system E (see below)] to assess the completeness of the esterification reaction (FAMES R_f = 0.97; free fatty acids R_f = 0.45).

To analyze the FAMES, chromatograph the remainder of the toluene phase, along with standards, on RP-18 F$_{254}$S thin-layer plates (20 × 20 cm; EM Science), using solvent system F (see below). Detect nonradioactive FAME standards by spraying the plates with the phosphomolybdic acid spray reagent and heating at 120° for a few minutes until the best spot formation is attained. Methyl esters of unsaturated fatty acids will give dark blue spots whereas those of saturated fatty acids will show up as relatively unstained spots in the blue background of the plate. Radioactive samples can be detected by scanning or autoradiography, although it should be noted that the scanner detector response is somewhat less for the RP plates than for the silica 60 plates used in other analyses.

Thin-Layer Chromatography: Materials and Solvent Systems

Thin-layer chromatography plates may be obtained from a variety of sources including EM Science (Merck, Rahway, NJ), J. T. Baker Chemical Co. (Phillipsburg, NJ), and Phase Separations, Inc. (Norwalk, CT). Lipid standards for TLC may be obtained from Avanti Polar-Lipids (Pelham, AL), Genzyme Corporation (Cambridge, MA), Sigma Chemical Company, Serdary Research Laboratories (London, Ontario, Canada), Indofine Chemical Company, Inc. (Somerville, NJ), and Larodan Fine Chemicals (U.S. distributor: Accurate Chemical & Scientific Corp., Westbury, NY). Lipid standards may be detected by exposing the plate to iodine vapor in a closed tank, or by using a variety of spray reagents. The solvent systems listed below are given as volume ratios.

Solvent A: Chloroform–methanol–water, 4 : 4 : 1 or 10 : 10 : 3 (v/v)
Solvent B: Chloroform–methanol–acetic acid–water, 25 : 15 : 4 : 2 (v/v)
Solvent C: Chloroform–methanol–90% formic acid, 50 : 30 : 7 (v/v)
Solvent D: Petroleum ether–diethyl ether–acetic acid, 80 : 20 : 1 (v/v)
Solvent E: Chloroform–methanol–30% ammonia, 65 : 25 : 5 (v/v)
Solvent F: Acetonitrile–acetic acid, 1 : 1 (v/v)

Production and Analysis of Glycosylphosphatidylinositol
  Glycan Fragments

Figure 3 illustrates routes for the production of GPI glycan fragments. The analyses require [^{3}H]mannose- or [^{3}H]glucosamine-labeled material.

For non-protein-linked GPIs, glycans are most simply prepared by nitrous acid deamination of the sample, or by treatment with (G)PI-specific phospholipases C or D (Fig. 3A). However, the most general and perhaps most useful way to produce a GPI glycan fragment involves cleaving the phosphodiester bonds linking the glycan to the lipid moiety and to the one or more ethanolamine residues (Fig. 3B). This is achieved by chemical dephosphorylation, using ice-cold aqueous hydrofluoric acid. The glycan fragment is then converted to a neutral species by N-acetylating or deaminating the glucosamine residue (Fig. 3B). The resulting neutral glycan product is then analyzed by a variety of chromatographic procedures, including BioGel (Bio-Rad, Richmond, CA) P-4 gel-filtration chromatography, high-pH anion-exchange chromatography, and TLC, in conjunction with methylation linkage analysis, and exoglycosidase digestion.

Exogenous radiolabeling by reductive radiomethylation of the GPI ethanolamine and glucosamine residues may be used as an alternative to metabolic labeling and has proved useful in a number of instances. In addition, hydrolysis of the radiomethylated glycan in trifluoroacetic acid produces a characteristic GPI glycan fragment [GlcN(CH$_3$)$_2$-inositolphosphate] that can be analyzed by high-pH anion-exchange chromatography. These methods will not be discussed here and the reader is referred to a number of publications for further details.[13,64,65]

Aqueous Hydrofluoric Acid Dephosphorylation

Reagents

Methanol–30% ammonia, 1 : 1 (v/v)
Aqueous HF (50%) (Cat. No. HX0621; EM Science): in a −20° freezer
LiOH (saturated solution)
NaHCO$_3$ (saturated solution)
Trichloroacetic acid (100%): 500 g of acid/227 ml of water
Dowex AG 3-X4 (OH$^-$) ion-exchange resin (Cat. No. 140-5341; Bio-Rad)

Procedure. Dry the [^{3}H]mannose- or [^{3}H]glucosamine-labeled GPI sample in a plastic microcentrifuge tube, using a SpeedVac, and treat the residue with 50 μl of methanol–ammonia for 2 hr at 37° to remove hydroxyester-linked fatty acids. Dry the sample in a SpeedVac, add 100 μl of 50% methanol, and dry again to remove residual ammonia. Place the tube in an ice–water mixture in a Dewar flask. Add 50 μl of ice-cold 50% aqueous HF, taking care to wipe the outside of the pipette tip and to deliver the entire amount of liquid into the tube. Accuracy at this stage

[64] W. L. Roberts and T. L. Rosenberry, *Biochemistry* **25**, 3091 (1986).
[65] R. Haas, P. T. Brandt, J. Knight, and T. L. Rosenberry, *Biochemistry* **25**, 3098 (1986).

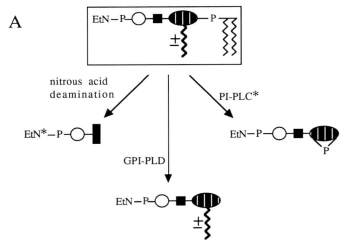

FIG. 3. Production of GPI glycan fragments. (A) Glycans may be generated from non-protein-linked GPIs by nitrous acid deamination, PI-PLC treatment (the asterisk indicates that PI-PLC will cleave only GPI structures that are not inositol acylated), or GPI-PLD treatment. If the sample is inositol acylated, then the lipophilic glycan generated by GPI-PLD treatment can be deacylated by alkali treatment to give a water-soluble product. If the original sample contains acyl inositol and a glyceride moiety with only ester-linked fatty acids, then the GPI-PLD-generated lipophilic glycan should be separated from the starting material by TLC [silica 60, Solvent A; the lipophilic glycan stays close to the origin whereas the intact GPI moves up the plate] or by hydrophic interaction chromatography[45] before deacylation. (B) A general procedure for generating neutral glycan fragments from GPI anchors or non-protein-linked GPIs involves deacylation of the structure to improve its solubility, dephosphorylation with ice-cold aqueous HF (note: phosphoethanolamine substituents will be removed by this procedure) and conversion to a neutral species via N-acetylation or nitrous acid deamination/sodium borohydride reduction. EtN*, deaminated ethanolamine.[20]

is important, as it simplifies the subsequent neutralization procedure. Along with the GPI sample, prepare a set of 5–10 "blank" tubes, each containing 50 μl of aqueous HF, and place these in the ice–water mixture. Leave the flask in a refrigerated cold room for 48–60 hr.

At the end of the incubation, neutralize the reaction mixture with a saturated solution of LiOH. To assess accurately the amount of LiOH required for the neutralization, it is best to perform a series of trial neutralizations with the "blank" tubes. Add 270 μl of LiOH to one of the blanks, vortex to mix, spin in a microcentrifuge to pellet the LiF precipitate, and check the pH of the supernatant by spotting aliquots onto pH paper (pH 0–6 range; EM Science). Using the other blanks, increase or decrease the amount of LiOH used until the final pH is in the range 3 < pH < 7.

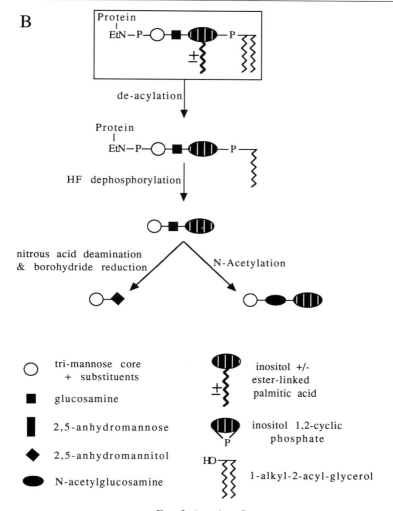

FIG. 3. (*continued*)

Estimate the amount of LiOH that will slightly underneutralize the HF to give a final pH in the range 3–5. Freeze this volume of saturated LiOH solution in a fresh microcentrifuge tube placed on dry ice. Transfer the HF-treated GPI sample onto the frozen LiOH and let the sample thaw on ice (this minimizes heating of the reaction mixture during neutralization and concomitant degradation of the GPI glycan). Vortex the sample to mix, spin in a microcentrifuge, and return the supernatant and a 100-μl water wash of the LiF precipitate to the original reaction tube. Check the pH of the sample by spotting a small (<5-μl) aliquot onto pH paper; the pH can be adjusted if necessary, using a saturated solution of NaHCO$_3$ (if the pH exceeds 7 it can be lowered with glacial acetic acid).

If the starting material was a GPI-anchored protein, remove contaminating polypeptide in the dephosphorylated sample by precipitation with trichloroacetic acid (TCA) as follows (for lipid samples, i.e., non-protein-linked GPIs, proceed directly to the ion-exchange step below): Cool the sample on ice, add ice-cold 100% TCA to a final concentration of 5%, and maintain the sample on ice for 3 hr to precipitate protein (10 μg of bovine serum albumin may be included in the sample to facilitate precipitation). Centrifuge to remove the protein precipitate. Pass the supernatant over a 1-ml column of Dowex AG 3-X4 (OH$^-$) ion-exchange resin [the column is most easily prepared with a disposable plastic column, e.g., the Poly-Prep chromatography column, Cat. No. 731-1550 (Bio-Rad)] and elute three times with 1 ml of water each time. Pool the eluate and washings and adjust the pH to 4 with glacial acetic acid. Freeze-dry the sample (or evaporate to dryness in a SpeedVac), resuspend the residue in water, and store at $-20°$ until required for analysis.

Note: If the incubation with HF is performed for less than 60 hr it is possible to generate partially dephosphorylated species. For example, in dephosphorylation experiments with GPI anchor precursors in African trypanosomes, it was found that the phosphodiester-linked terminal ethanolamine residue could be quantitatively cleaved from the GPI glycan in <24 hr, whereas the cleavage of the phosphodiester bond linking inositol to glycerol took 60 hr to go to completeness.[66] In another application, an incubation time of 8 hr was used to generate a spectrum of dephosphorylated species from mammalian GPIs containing three phosphoethanolamine groups.[14]

N-Acetylation

  Reagents

  Acetic anhydride
  NaHCO$_3$ (saturated solution)
  Chelex 100 (Na$^+$) (Cat. No. 142-2832; Bio-Rad) and Dowex AG 50W-X12 (H$^+$) (Cat. No. 142-1651; Bio-Rad) ion-exchange resins
  Procedure. Dry the sample in a microcentrifuge tube, resuspend the residue in 50 μl of saturated NaHCO$_3$ solution, and cool on an ice–water

[66] S. Mayor, A. K. Menon, G. A. M. Cross, M. A. J. Ferguson, R. A. Dwek, and T. W. Rademacher, *J. Biol. Chem.* **265,** 6164 (1990).

mixture. Add acetic anhydride at 10-min intervals (three additions, 2 μl each) while maintaining the sample on ice. Leave the sample at ambient temperature for 30 min after the last addition of acetic anhydride, then desalt (at ambient temperature) by passage through a tandem ion-exchange column consisting of 0.2 ml of Chelex 100 (Na$^+$) layered over 0.4 ml of Dowex AG 50W-X12 (H$^+$). Wash the column four times with 600 μl of water. Pool the eluate and washings and dry in a SpeedVac. Flash evaporate the residue three times with 50 μl of toluene to remove residual acetic acid, then redissolve the sample in water. Filter the sample through a 0.45-μm pore size filter [e.g., Acrodisc syringe filters, (Cat. No. 4452 or 4473; Gelman Sciences, Ann Arbor, MI)] and store at $-20°$ until required for analysis.

Nitrous Acid Deamination and Sodium Borohydride Reduction

Reagents

Deamination buffer: 0.2 M sodium acetate, pH 3.7 (prepared by mixing 10 ml of 0.2 M sodium acetate with 90 ml of 0.2 M acetic acid)
Sodium nitrite [Cat. No. S2252 (Sigma); Cat. No. 10256 (BDH Chemicals)]
Boric acid (0.4 M)
NaOH (1 M)
Sodium borohydride (2 M) in 0.3 M NaOH (prepared fresh)
Chelex 100 (Na$^+$) (Cat. No. 142-2832; Bio-Rad), Dowex AG 50W-X12 (H$^+$) (Cat. No. 142-1651; Bio-Rad), Dowex AG 3-X4 (OH$^-$) (Cat. No. 140-5341; Bio-Rad), QAE–Sephadex A-25 (OH$^-$) (Cat. No. 17-0190-01; Pharmacia LKB, Piscataway, NJ) ion-exchange resins

Procedure. Dry the dephosphorylated GPI glycan and resuspend in 200 μl of deamination buffer. Add 200 μl of a 0.5 M solution of sodium nitrite (prepared fresh in deamination buffer) and leave the reaction of ambient temperature for 4–8 hr. Terminate the reaction by adding 300 μl of 0.4 M boric acid, followed by sufficient 1 M NaOH to raise the pH to 10 (use control samples with buffer only to assess the amount of NaOH needed). Add 100 μl of the sodium borohydride solution and allow the reduction to proceed for 12 hr at ambient temperature. Stop the reaction by adding 20 μl of glacial acetic acid (bubbling will be observed briefly). Desalt the sample on a 1-ml column of Dowex AG 50W-X12 (H$^+$). Wash the column three times with 1 ml of water, pool the eluate and washings, and dry in a SpeedVac. Flash evaporate the dried residue with 5% acetic acid in methanol (twice, 200 μl each time) (to remove boric acid) and then with toluene (three times, 20 μl each time) (to remove residual acetic acid). Redissolve the sample in 100 μl of water, filter through a 0.45-μm pore size filter [e.g., Acrodisc syringe filters, (Cat. No. 4452 or 4473; Gelman Sciences)], and store at −20° until required for analysis.

Note: Deamination of the glucosamine residue gives 2,5-anhydromannose; subsequent reduction converts this to 2,5-anhydromannitol (AHM) (Fig. 3B). The pH of both the deamination and reduction reactions is critical. The pH of the sample in deamination buffer should be adjusted to just under 4 if necessary before adding sodium nitrite. Likewise, it is important to have the sample at pH 10–11 before adding borohydride.

Thin-Layer Chromatography

Some neutral as well as charged GPI glycan fragments may be resolved by TLC on silica 60 plates developed three times in chloroform–methanol–water (10:10:3, v/v) with air drying between each development. This system has been used to resolve Man_1AHM, Man_2AHM, and Man_3AHM, as well as negatively charged glycans prepared by deaminating mammalian GPIs containing one to three phosphoethanolamine substituents (Fig. 3A).[14] Other TLC systems for GPI glycan separation have also been described.[14a]

BioGel P-4 Gel-Filtration Chromatography

Analysis of neutral glycans by BioGel P-4 gel filtration gives a direct measurement of size (hydrodynamic volume) and, in conjunction with exoglycosidase digestion experiments, provides a powerful means of structure determination. Detailed methodology is given in [11] in this volume and elsewhere in this series.[67] GPI-neutral glycan fragments prepared by dephosphorylation, deamination, and reduction elute as follows (size expressed in glucose units): AHM, 1.7; $Man\alpha1$-4AHM, 2.3; $Man\alpha1$-6$Man\alpha1$-4AHM, 3.2; $Man\alpha1$-2$Man\alpha1$-6$Man\alpha1$-4AHM, 4.2; $Man\alpha1$-2$Man\alpha1$-2$Man\alpha1$-6$Man\alpha1$-4AHM, 5.2.[11,16,66,68]

Dionex Anion-Exchange High-Performance Liquid Chromatography

Equipment and Buffers

Dionex Bio-LC system, HPIC AS6 column plus HPIC AG6 guard column, pulsed amperometric detector (PAD), anion micromembrane suppressor (AMMS) (Dionex Corp., Sunnyvale, CA)

[67] K. Yamashita, T. Mizuochi, and A. Kobata, *in* "Methods in Enzymology" (V. Ginsburg, ed.), Vol. 83, p. 105. Academic Press, New York, 1982.
[68] M. A. J. Ferguson, *in* "Lipid Modification of Proteins: A Practical Approach" (N. M. Hooper and A. J. Turner, eds.), p. 191. IRL Press, Oxford, 1992.

NaOH, 50% solution (w/w) (Cat. No. SX0597-3; EM Science): Prepare a 0.1 M solution by diluting 20.94 ml to 4 liters in water

Sulfuric acid (ULTREX ultrapure reagent, Cat. No. 4802-05; J. T. Baker, Phillipsburg, NJ): Prepare a 0.1 M solution by diluting 22 ml to 4 liters in water

Sodium acetate, 0.5 M in 0.1 M NaOH

Glucose Oligomer Standards. Glucose oligomers are prepared by partial acid hydrolysis of dextran.[67,68] Dissolve 200 mg of dextran (BDH grade C, Cat. No. 38016; Gallard-Schlesinger Industries, Inc., Carle Place, NY) in 1 ml of 0.1 M HCl and incubate at 100° for 4 hr. Pass the sample through 0.5 ml of Dowex AG 3-X4 (OH⁻) resin (Cat. No. 140-5341; Bio-Rad) and elute with 3 ml of water. The sample contains a series of glucose oligomers ranging from Glc_1 to approximately Glc_{30} at 50 mg/ml. Glucose oligomers may also be purchased from Oxford GlycoSystems (Cat. No. HS-01500).

Gradient Program. Begin with 100% solvent A (0.1 M NaOH), 0% solvent B (0.5 M CH_3COONa, 0.1 M NaOH) up to 3 min after sample injection, followed by a linear change to 55% solvent A, 45% solvent B at 33 min, then 0% solvent A, 100% solvent B at 38 min. The flow rate is 1 ml/min.

Chromatography. Analyze the desalted radioactive glycan samples with the Dionex Bio-LC system, an HPIC AS6 column, and the gradient elution program described above. Glucose oligomers should be injected along with the sample and detected with the pulsed amperometric detector supplied with the Dionex instrument. Collect fractions (typically 0.4 min/fraction) directly into 7-ml scintillation vials. Unusually high backgrounds are observed with some scintillation cocktails if the high pH of the column eluant is not reduced; this may be achieved by using an on-line anion micromembrane suppressor (Dionex) with 0.1 M sulfuric acid as the counterflow regenerant, or by manually adding acetic acid to each fraction. Recovery of radioactivity is typically 35–50%.

Typical Retention Times. The following retention times (in minutes)[45,66,69] give an indication of the type of separation achieved on the Dionex system, using the conditions described above.

Glucose oligomers: Glc_1 (6.0), Glc_2 (10.4), Glc_3 (14.8)

GPI glycan fragments: GlcNAcα1-6inositol (2.6), AHM (5.0), Manα1-4AHM (6.0), Manα1-6Manα1-4AHM (10.5), Manα1-2Manα1-6Manα1-4AHM (13.5)

Other: GlcNAc-ol (2.6), GlcNAcβ1-4GlcNAc-ol (3.5), GlcNAc (6.0)

[69] A. K. Menon, R. T. Schwarz, S. Mayor, and G. A. M. Cross, *J. Biol. Chem.* **265,** 9033 (1990).

Note: Other examples of anion-exchange chromatography with Dionex CarboPac columns may be found in the literature.[11–13,22,68]

Methylation Linkage Analysis of [³H]Mannose-Labeled Glycans

Reagents

Iodomethane (methyl iodide) (Cat. No. 28956-6; Aldrich Chemical Co., Milwaukee, WI)
Trifluoroacetic acid (Cat. No. 28902; Pierce)
Dimethyl sulfoxide (DMSO)
Sodium thiosulfate, 100 mg/ml
Aqueous methanol, 50%

Procedure. The procedure is derived from the work of Ciucanu and Kerek.[70] Dry the [³H]mannose-labeled neutral glycan sample in a 2-ml ReactiVial (Pierce Chemical Co.). Add 50 μl of DMSO and stir continuously with a magnetic stirrer (Pierce) designed to fit the conical interior of the vial. Grind NaOH pellets with a mortar and pestle and use the fine powder to make a 120-mg/ml slurry in DMSO. Add 50 μl of the slurry to the sample and continue stirring for 20 min. Then add three 10-μl additions iodomethane at 10-min intervals and terminate the reaction 10 min after the last addition by adding 250 μl of chloroform and 1 ml of sodium thiosulfate solution. The permethylated glycans will be recovered in the lower chloroform phase. Wash the lower phase five times with water (use 1 ml water, vortex to mix, and centrifuge to separate the phases each time), then take an aliquot for liquid scintillation counting to assess recovery of material (typically 50–65%).

Evaporate the chloroform in a SpeedVac and hydrolyze the dried residue in 200 μl of 2 N trifluoroacetic acid for 4 hr at 120°. After hydrolysis, dry the sample in a SpeedVac and flash evaporate residual acid three times with 100 μl of methanol. Resuspend the residue in 50% aqueous methanol and analyze by TLC on a silica 60 thin-layer plate, using benzene–acetone–water–30% ammonia (50 : 200 : 3 : 1.5, v/v).[71]

Mixtures of partially O-methylated methylmannosides may be prepared as described by Hull and Turco[72] and hydrolyzed as described above to generate standards for TLC. Alternatively, completely characterized [³H]mannose-labeled glycans from N-linked glycans, dolichol-linked oligosaccharides, or GPI anchors may be analyzed in parallel with the sample of interest to provide a spectrum of standards.

[70] I. Ciucanu and F. Kerek, *Carbohydr. Res.* **131,** 209 (1984).
[71] E. Li, I. Tabas, and S. Kornfeld, *J. Biol. Chem.* **253,** 7762 (1978).
[72] S. R. Hull and S. J. Turco, *Anal. Biochem.* **151,** 554 (1985).

Exoglycosidase Digestion

Enzymes and Other Materials

Jack bean α-mannosidase (Cat. No. 107 379; Boehringer Mannheim, Indianapolis, IN)

Aspergillus saitoi α-mannosidase (Cat. No. X-5009; Oxford Glyco-Systems)

Sodium acetate buffer (0.2 *M*, pH 5.0)

Dowex AG 50W-X12 (H$^+$) (Cat. No. 142-1651; Bio-Rad)

Procedure. Exoglycosidase digestions should be performed on [^{3}H]mannose- or [^{3}H]glucosamine-labeled GPI glycans. The glycans should be analyzed by TLC, anion-exchange HPLC, or BioGel P-4 gel filtration before and after treatment. Conditions for the two α-mannosidases useful in the analysis of GPI glycans are described below. Other enzymes that may be useful include coffee bean α-galactosidase [Cat. No. G8507 (Sigma), Cat. No. 105023 (Boehringer Mannheim), Cat. No. X-5007 (Oxford GlycoSystems)] and jack bean β-*N*-acetylhexosaminidase [Cat. No. A2264 (Sigma), Cat.No. X-5003 (Oxford GlycoSystems)]. Reaction conditions for these and other glycosidases may be found in the catalog *Tools for Glycobiology* available from Oxford GlycoSystems.

Jack bean α-mannosidase acts on terminal, unsubstituted α-D-mannose (pyranose) residues. Although its fine specificity for different mannose α linkages can be exploited for structural analysis, overnight (>12-hr) incubations result in the sequential release of virtually all α-mannose residues. The enzyme is supplied as a suspension in ammonium sulfate and should be dialyzed against acetate buffer prior to use, and adjusted to a concentration of about 30 units/ml. Resuspend the dried [^{3}H]mannose- or [^{3}H]glucosamine-labeled GPI glycan sample in 20 μl of the enzyme solution and incubate for 12–16 hr at 37°. Terminate the digestion by heating the reaction to 100° for 2 min. Desalt the mixture by passage through a 0.2-ml column of Dowex AG 50W-X12 (H$^+$), wash the column three times with 0.2 ml of water, pool the eluate and washings, and evaporate to dryness in a SpeedVac. Flash evaporate the residue with toluene (twice, 50 μl each time) to remove residual acetic acid. Resuspend in water and store at −20° until required for analysis.

Aspergillus saitoi α-mannosidase is specific for terminal, unsubstituted mannose α(1 → 2) linkages.[73] Resuspend the GPI glycan sample in 20 μl of acetate buffer containing the α-mannosidase and incubate for 2 hr at ambient temperature. Process the digest as described above.

---

[73] A. Kobata and J. Amano, *in* "Methods in Enzymology" (V. Ginsburg, ed.), Vol. 138, p. 779. Academic Press, Orlando, FL, 1987.

Anion-Exchange Chromatography of Negatively Charged Glycans

Glycosylphosphatidylinositol glycans prepared by nitrous acid treatment of non-protein-linked GPIs (Fig. 3A) can be analyzed by conventional anion-exchange chromatography to determine their net negative charge.[14] The analysis is conveniently performed with a Mono Q column (HR 5/5, bed size 5 × 50 mm) and a gradient fast protein liquid chromatography (FPLC) system (Pharmacia, Piscataway, NJ).

Buffers

Buffer A: Filtered deionized distilled water
Buffer B: 0.5 M ammonium acetate [take 30 ml of glacial acetic acid, add 900 ml of water, adjust the pH with 30% ammonia (roughly 37 ml will be needed), then make up the volume to 1 liter]. Filter the buffer through a 0.45-μm pore size polytetrafluorethylene membrane. Do not degas

Gradient Elution Program. Begin with 100% solvent A up to 5 min after sample injection, then a linear change to 75% solvent A, 25% solvent B at 30 min, followed by changes to 50% solvent A, 50% solvent B at 40 min, and 0% solvent A, 100% solvent B at 45 min. The flow rate is 1 ml/min.

Chromatography. Collect 1-ml fractions during the run and measure radioactivity by liquid scintillation counting. Appropriate radioactive standards include [^{3}H]mannose, inositol-PO$_4$-glycerol [prepared by deacylating [^{3}H]inositol-labeled phosphatidylinositol (Cat. No. TRK. 713; Amersham Corp., Arlington Heights, IL)], and UDP[^{3}H]galactose. [^{3}H]Mannose and other neutral molecules will elute in the void volume (fractions 1–2), inositol-PO$_4$-glycerol elutes in fractions 12–13, and UDP[^{3}H]galactose elutes in fractions 34–35. GPI glycans prepared by deaminating non-protein-linked GPI containing one, two, or three phosphoethanolamine substituents elute in fractions 12, 20–21, and 37–38, respectively.[14]

Acknowledgment

This work was supported by NIH Grant AI28858 and by an award from the Irma T. Hirschl Trust. I thank George A. M. Cross, Michael A. J. Ferguson, Mark C. Field, Satyajit Mayor, and Ralph T. Schwarz for their contributions to the work presented here, and Jolanta Vidugiriene for comments on the manuscript.

[26] Detection of O-Linked *N*-Acetylglucosamine (*O*-GlcNAc) on Cytoplasmic and Nuclear Proteins

By ELIZABETH P. ROQUEMORE, TEH-YING CHOU, and GERALD W. HART

Introduction

We have described an abundant form of intracellular protein glycosylation (termed *O*-GlcNAc) consisting of *N*-acetylglucosamine moieties O-glycosidically linked to serine and threonine residues.[1,2] *O*-GlcNAc occurs on a myriad of both nucleoplasmic and cytoplasmic proteins. However, only a few of these glycoproteins have been identified (Table I).[3–33] *O*-GlcNAc is present in all eukaryotes from yeast to humans, but has not yet been found in prokaryotes. *O*-GlcNAc-bearing proteins include

[1] G. W. Hart, R. S. Haltiwanger, G. D. Holt, and W. G. Kelly, *Annu. Rev. Biochem.* **58,** 841 (1989).

[2] R. S. Haltiwanger, W. G. Kelly, E. P. Roquemore, M. A. Blomberg, L.-Y. D. Dong, L. Kreppel, T.-Y. Chou, and G. W. Hart, *Biochem. Soc. Trans.* **20,** 264 (1992).

[3] G. W. Holt, R. S. Haltiwanger, C.-R. Torres, and G. W. Hart, *J. Biol. Chem.* **262,** 14847 (1987).

[4] K. Nyame, R. D. Cummings, and R. T. Damian, *J. Biol. Chem.* **262,** 7990 (1987).

[5] T. Luthi, R. S. Haltiwanger, P. Greengard, and M. Bahler, *J. Neurochem.* **56,** 1493 (1991).

[6] W. G. Kelly and G. W. Hart, unpublished.

[7] C.-F. Chou and M. B. Omary, *J. Biol. Chem.* **268,** 4465 (1993).

[8] I. A. King and E. F. Hounsell, *J. Biol. Chem.* **264,** 14022 (1989).

[9] C.-F. Chou, A. J. Smith, and M. B. Omary, *J. Biol. Chem.* **267,** 3901 (1992).

[10] M. L. Privalsky, *J. Virol.* **64,** 463 (1990).

[11] L.-Y. D. Dong, Z.-S. Xu, M. R. Chevrier, R. J. Cotter, D. W. Cleveland, and G. W. Hart, *J. Biol. Chem.* **268,** in press (1993).

[12] T.-Y. Chou, C. V. Dang, J. Eiden, and G. W. Hart, in preparation.

[13] J. Hagmann, M. Grob, and M. M. Burger, *J. Biol. Chem.* **267,** 14424 (1992).

[14] K. G. Mullis, R. S. Haltiwanger, G. W. Hart, R. B. Marchase, and J. A. Engler, *J. Virol.* **64,** 5317 (1990).

[15] B. Datta, M. K. Ray, D. Chakrabarti, D. E. Wylie, and N. K. Gupta, *J. Biol. Chem.* **264,** 20620 (1989).

[16] D. M. Benko, R. S. Haltiwanger, G. W. Hart, and W. Gibson, *Proc. Natl. Acad. Sci. U.S.A.* **85,** 2573 (1988).

[17] W. Meikrantz, D. M. Smith, M. M. Sladicka, and R. A. Schlegel, *J. Cell Sci.* **98,** 303 (1991).

[18] M. Whitford and P. Faulkner, *J. Virol.* **66,** 3324 (1992).

[19] C. M. Starr and J. A. Hanover, *J. Biol. Chem.* **265,** 6868 (1990).

[20] G. D. Holt, C. M. Snow, A. Senior, R. S. Haltiwanger, L. Gerace, and G. W. Hart, *J. Cell Biol.* **104,** 1157 (1987).

[21] L. I. Davis and G. Blobel, *Proc. Natl. Acad. Sci. U.S.A.* **84,** 7552 (1987).

TABLE I
KNOWN O-GlcNAc PROTEINS

Protein	Ref.	Protein	Ref.
Erythrocyte band 4.1	3	Many schistosome proteins	4
Synapsin I	5	Many trypanosome proteins	6
Cytokeratins 13, 8, 18	7–9	v-*erb*-A oncogene protein	10
Neurofilaments	11	c-*myc* oncogene protein	12
Talin	13	Adenovirus fiber	14
67-kDa red blood cell elongation factor kinase	15	Cytomegalovirus tegument protein	16
65-kDa nuclear tyrosine phosphatase	17	Baculovirus tegument protein	18
Nuclear pore complex proteins	19–24	NS26 protein of rotavirus	25
RNA polymerase II catalytic subunit	26	92-kDa SER protein	27
Many RNA polymerase II transcription factors	28–31	Many chromatin proteins in *Drosophila*	32
Lens α-crystallins (small heat shock proteins)	33		

transcriptional regulatory proteins, RNA polymerase II,[26] and cytoskeletal, oncogene, heat shock, viral, and nuclear pore proteins.

Even though these glycoproteins display a diverse range of functions and intracellular localizations, they share two common features: All of them are also phosphoproteins, and all form reversible multimeric complexes depending on their phosphorylation states and/or the stage of the cell cycle. These shared features suggest that O-GlcNAc may be involved in modulating the formation of a number of specific multimeric structures.

[22] J. A. Hanover, C. K. Cohen, M. C. Willingham, and M. K. Park, *J. Biol. Chem.* **262,** 9887 (1987).

[23] M. K. Park, M. D'Onofrio, M. C. Willingham, and J. A. Hanover, *Proc. Natl. Acad. Sci. U.S.A.* **84,** 6462 (1987).

[24] M. Schindler, M. Hogan, R. Miller, and D. DeGaetano, *J. Biol. Chem.* **262,** 1254 (1987).

[25] S. A. Gonzalez and O. R. Burrone, *Virology* **182,** 8 (1992).

[26] W. G. Kelly, M. E. Dahmus, and G. W. Hart, *J. Biol. Chem.* **268,** 10416 (1993).

[27] C. Abeijon and C. B. Hirschberg, *Proc. Natl. Acad. Sci. U.S.A.* **85,** 1010 (1988).

[28] S. P. Jackson and R. Tjian, *Cell (Cambridge, Mass.)* **55,** 125 (1988).

[29] S. P. Jackson and R. Tjian, *Proc. Natl. Acad. Sci. U.S.A.* **86,** 1781 (1989).

[30] S. Lichtsteiner and U. Schibler, *Cell (Cambridge, Mass.)* **57,** 1179 (1989).

[31] A. J. Reason, H. R. Morris, M. Panico, R. Marais, R. H. Treisman, R. S. Haltiwanger, G. W. Hart, W. G. Kelly, and A. Dell, *J. Biol. Chem.* **267,** 16911 (1992).

[32] W. G. Kelly and G. W. Hart, *Cell (Cambridge, Mass.)* **57,** 243 (1989).

[33] E. P. Roquemore, A. Dell, H. R. Morris, M. Panico, A. J. Reason, L.-A. Savoy, G. J. Wistow, J. S. Zigler, Jr., B. J. Earles, and G. W. Hart, *J. Biol. Chem.* **267,** 555 (1992).

- **Nuclear Pore 62 kDa** (one of several sites) - ...MAGGPADT**S**DPL...

- **Erythrocyte 65 kDa Cytosolic Protein** - ...D**S**PVSQPSLVGSK....

- **Human Erythrocyte Band 4.1** - ...AQTIT**S**ETPSSTT...

- **Bovine Lens α-A-Crystallin** - ...^{158}DIPV**S**REEK^{166}...

- **Serum Response Transcription Factor** -
  ...^{269}VTNLPGTT**S**TIQTAPSTSTTMQVSSGPSFPITNYLAPV**S**A**S**VSPSAV
  SSANGTVLKSTGSGPVSSGGLMQLPTSFTLMPGGAVAQQVPVQAIQVHQA
  PQQASPSRDSSTDLTQT<u>**SSS**</u>GTVTLPATIMTSSVPTT^{402}...

- **Rat Spinal Cord Neurofilament (NF-L)** -
  ...^{18}YVE**T**PRVHI**S**SVR^{30}.......SAYSSYSAPV<u>**SSS**</u>LSVR....

- **Rat Spinal Cord Neurofilament (NF-M)** -
  ...^{44}GSPS**T**VSSSYK^{54}......^{427}QPSV**T**ISSK^{435}....

FIG. 1. Identified sites of *O*-GlcNAc attachment *in vivo*. For references see Table I. Boldface type indicates major sites mapped; underlining of multiple residues indicates major site (SRF) or uncertainty (NFL).

Present data suggest that *O*-GlcNAc is as abundant and as widespread as protein phosphorylation. In addition, several studies indicate that this saccharide is highly dynamic. The levels of *O*-GlcNAc modification of several proteins are responsive to cellular metabolism[34,35] or stage of the cell cycle. Thus, our working hypothesis is that *O*-GlcNAc is a regulatory modification that conceptually has more in common with phosphate than it has with other types of protein glycosylation.

Several studies have identified sites of *O*-GlcNAc attachment on proteins (Fig. 1). On the basis of these data, synthetic peptides have been prepared and used to purify an *O*-GlcNAc : polypeptide *N*-acetylglucosaminyltransferase to homogeneity.[36] Identified sites and peptide substrate specificity studies with this purified enzyme indicate that the sites of *O*-GlcNAc attachment are similar to those recognized by a number of the "growth factor" or "proline-specific" kinases, suggesting that in some instances phosphorylation and *O*-GlcNAc addition may be reciprocal. Indeed, the glycosylation and phosphorylation of the C-terminal do-

[34] K. P. Kearse and G. W. Hart, *Proc. Natl. Acad. Sci. U.S.A.* **88,** 1701 (1991).
[35] C.-F. Chou, A. J. Smith, and M. B. Omary, *J. Biol. Chem.* **267,** 3901 (1992).
[36] R. S. Haltiwanger, M. A. Blomberg, and G. W. Hart, *J. Biol. Chem.* **267,** 9005 (1992).

main of RNA polymerase II do appear to be reciprocal.[26] However, in studies of mitotic arrested epithelial cells, the glycosylation and phosphorylation of cytokeratins increase in parallel at G_2/M.[7] It seems likely that analogous to phosphorylation the specific function(s) of O-GlcNAc will be protein or even glycosylation site specific. A neutral cytosolic N-acetylglucosaminidase (GlcNAcase) with selectivity for O-GlcNAc-peptides has also been purified.[37] The rapid dynamics of O-GlcNAc residues has led to the postulation that O-GlcNAc transferases/GlcNAcases may be analogous to the kinases/phosphatases of phosphorylation systems.

Here we describe methods for the detection and initial analysis of O-GlcNAc on proteins. All of these procedures use commercially available reagents and do not require any expertise or equipment beyond that already present in virtually all biochemistry or molecular biology laboratories. Although the galactosyltransferase/UDP[^{3}H]galactose probe method is highly specific (provided product analysis is performed) it requires amounts of protein in the low microgram range. We also describe a coupled transcription/translation/lectin chromatography method for the detection of O-GlcNAc on low-abundance proteins (e.g., oncogene proteins or transcription factors) for which a cDNA is available. This method relies on the observation that the reticulocyte lysates commonly used for *in vitro* translation already contain sufficient sugar nucleotide and O-GlcNAc transferase(s) to glycosylate the translation products efficiently.[19] Finally, because O-GlcNAc is virtually the only GlcNAc-containing molecule in the cytoplasmic or nuclear cellular compartments, metabolic radiolabeling with [^{3}H]glucosamine, in conjunction with subcellular fractionation, is also a useful method for identifying O-GlcNAc-bearing proteins.

Detection by Galactosyltransferase Labeling

Galactosyltransferase is a specific and sensitive probe frequently used in the detection of O-GlcNAc on cytoplasmic and nuclear proteins.[38] The enzyme is used to catalyze the following reaction:

$$\text{UDP[}^3\text{H]Gal} + \text{GlcNAc-R} \xrightarrow{\text{galactosyltransferase}}$$

(sugar donor)   (acceptor)

$$\text{[}^3\text{H]Gal}\beta\text{1,4-GlcNAc-R} + \text{UDP}$$

(sugar acceptor)

[37] L.-Y. D. Dong and G. W. Hart, *Glycoconjugate J.* **8,** 211 (1911).

[38] S. W. Whiteheart, A. Passaniti, J. S. Reichner, G. D. Holt, R. S. Haltiwanger, and G. W. Hart, *in* "Methods in Enzymology" (V. Ginsburg, ed.), Vol. 179, p. 82. Academic Press, San Diego, 1989.

The radiolabeled products are then analyzed to determine saccharide linkage (O- or N-linkage) and structure. In some cases, this method is sensitive enough to detect *O*-GlcNAc in the subpicomole range.

Methods for Labeling

The use of glycosyltransferases as probes has been described previously[2,38] and is also discussed briefly in [3] and [18] of this volume. A procedure designed specifically for the detection of O-linked GlcNAc, based on the method of Holt *et al.*,[3] is described below. A kit for the detection of O-GlcNAc is also commercially available (Oxford Glycosystems, Oxford, UK).

Materials

Bovine milk GlcNAc β-1,4-galactosyltransferase (lactose synthase) (Sigma, St. Louis, MO): Prior to use, the enzyme is autogalactosylated by incubating 25 units for 30 min at 37° in 1 ml of 50 mM Tris-HCl (pH 7.3), 5 mM MgCl$_2$, 1 mM 2-mercaptoethanol, 1% (v/v) aprotinin, 0.4 mM UDPgalactose. The enzyme is then concentrated by precipitation with 85% saturated ammonium sulfate. The enzyme is stable at 20–30 U/ml in 25 mM N-2-hydroxyethylpiperazine-N'-2-ethanesulfonic acid (HEPES)-NaOH (pH 7.3), 5 mM MnCl$_2$, 50% glycerol at −20° for at least 1 year

Labeling buffer (10×): 100 mM HEPES-NaOH (pH 7.3), 100 mM galactose, 50 mM MnCl$_2$

Protease inhibitor stock solutions (1000×):
  PIC 1: Leupeptin (1 mg/ml), antipain (2 mg/ml), benzamidine (10 mg/ml), dissolved in aprotinin (10,000 units/ml)
  PIC 2: Chemostatin (1 mg/ml), papstatin (1 mg/ml), dissolved in dimethyl sulfoxide (DMSO)
  Phenylmethylsulfonyl fluoride (PMSF): 0.1 M, dissolved in 95% ethanol

Uridine diphospho[6-³H]galactose (5–20 Ci/mmol; Amersham, Arlington Heights, IL); UDP[1-³H]Gal and UDP[4,5-³H]Gal are available from New England Nuclear (Boston, MA): the radiolabel is lyophilized to dryness, and resuspended in 25 mM 5'-adenosine monophosphate (5'-AMP)

Stop solution (10×): 0.1 M ethylenediaminetetraacetic acid (EDTA), 10% (w/v) sodium dodecyl sulfate (SDS)

Gel-filtration chromatography: A 1 × 30 cm column of Sephadex G-50 (coarse) is equilibrated in 50 mM ammonium formate, 0.1% SDS, 0.02% (w/v) sodium azide

Sample Preparation. The protein sample should be fully soluble in a solution that has an ionic strength of less than 0.2 M and is free of EDTA, which will chelate manganese required by the enzyme. Nonionic detergents such as Triton X-100 or Nonidet P-40 may be added up to levels of 2% to increase solubility. If SDS is used to solubilize the protein, a sevenfold excess (v/v) of nonionic detergent must be added to protect the galactosyltransferase from inactivation by SDS. Protease inhibitors (PIC 1, PIC 2,[3] and PMSF) should also be included.

Controls. Ovalbumin (fraction V; Sigma), which contains N-linked terminal GlcNAc moieties, is a suitable positive control for the reaction. A negative control, containing no sample, should be included.

Procedure

1. To 50 μl of solubilized sample or control, add the galactosyltransferase (determined empirically for each sample preparation; typically 25–100 mU) diluted in 10 μl of 10× labeling buffer. Adjust the volume to 90 μl with deionized water.

2. Initiate the reaction by adding 10 μl of 25 mM 5'-AMP containing 2–3 μCi of UDP[³H]Gal to bring the final concentration to 2.5 mM 5'-AMP. If acceptor proteins are to be analyzed quantitatively (see Procedural Comments, below), radiolabel should be isotopically diluted to approximately five times the K_m for UDPgalactose ($K_m = 6.0 \times 10^{-5} M$).

3. Incubate the sample under the desired conditions (typically 1 hr at 37°; see Procedural Comments, below).

4. Terminate the reaction by adding 11 μl of 10× stop solution. Boil for 3 min.

5. To separate unincorporated radiolabel from the ³H-galactosylated species, apply the sample to the Sephadex G-50 desalting column, and collect 1-ml fractions. Locate and pool radiolabeled macromolecules by counting aliquots of each fraction.

6. Lyophilize the pooled material to dryness, resuspend in a small volume of deionized water, aliquot as desired for product analysis (see below), and precipitate with 8 vol of acetone at −20° for 6 hr. Collect the precipitate by centrifugation at 1500 g for 20 min at 4°.

Procedural Comments. If product analysis is to be performed (see below), then enzyme concentration, labeling time, and incubation temperature of the reaction should be adjusted for maximal incorporation of radiolabel. Reaction volumes may also be varied from 25 to 250 μl, as necessary. If, instead, acceptor proteins in the sample are to be examined or compared quantitatively, the conditions of the experiment must be optimized to ensure that acceptor sites are saturated. Saturation conditions are established empirically by determining dependence of label incorporation on time, enzyme concentration, and sugar-nucleotide concentration. In addition, it must be demonstrated that acceptor concentration is within the linear range of the assay.

Product Analysis

The O-linkage of GlcNAc to protein is demonstrated by its characteristic resistance to the enzyme PNGase F (EC 3.5.1.52), which specifically releases N-linked oligosaccharides, concomitant with its sensitivity to base-catalyzed β elimination. O-Linked carbohydrates released by β elimination are then examined by one or more conventional techniques to confirm the saccharide structure.

Materials

PNGase F reagents:
    Buffer A: 1% SDS, 1% 2-mercaptoethanol
    Buffer B: 150 mM sodium phosphate (pH 8.6), 15 mM EDTA, 5% (v/v) Nonidet P-40 (NP-40)
Gel electrophoresis apparatus
β-Elimination reagents:
    β-Elimination buffer (made fresh just before use): 1 M NaBH$_4$, 0.1 M NaOH
    Neutralization solution: 4 M ice-cold acetic acid
Gel-filtration chromatography: A 1 × 30 cm column of Sephadex G-50 (coarse) equilibrated in 50 mM ammonium formate, 0.1% (w/v) SDS, 0.02% (w/v) sodium azide

Peptide N-Glycosidase F (PNGase F) Treatment

1. Resuspend the acetone-precipitated sample in 200 μl of PNGase buffer A, boil for 5 min, and cool to room temperature. Divide the sample into two 100-μl portions, one of which will serve as the no-enzyme negative control. ^{3}H-Galactosylated ovalbumin (prepared previously as a positive control for the galactosyltransferase reaction) may be used as a positive control for the PNGase F reaction.

2. Add 200 μl of PNGase buffer B to each 100-μl sample and vortex.

3. Initiate the reaction by adding 100–500 mU (as defined by Boehringer Mannheim, Indianapolis, IN) of PNGase F to each sample. Add an equal volume of deionized water to the negative control. Incubate the samples at room temperature with constant mixing. After 24 hr, add another 100 mU of PNGase F, and continue the incubation for an additional 24 hr.

4. Stop the reaction by adding 300 μl of distilled water and 60 μl of 20% SDS.

5. Apply the samples to the G-50 desalting column to separate the released oligosaccharides from the resistant material.

6. Pool PNGase F-resistant material in the void volume of the column, lyophilize, and precipitate the macromolecules with 8 vol of acetone as described before.

7. Resolve the resistant products on sodium dodecyl sulfate-polyacrylamide gel electrophoresis (SDS–PAGE), and visualize by autofluorography after impregnation with En³Hance (New England Nuclear).

β Elimination and Saccharide Analysis

1. Resuspend acetone-precipitated samples in 500 μl of freshly made β-elimination buffer, and incubate at 37° for 18–48 hr. Check the pH of the reaction mixture after several hours of incubation; if it is less than 13, increase the alkalinity by addition of more β-elimination buffer.

2. After the incubation is complete, transfer the sample to a 15-ml plastic tube, set the sample on ice, and cool. If necessary, adjust the pH to greater than 13 with β-elimination buffer.

3. Over the course of 1 hr, stop the reaction by intermittent addition of 4 M ice-cold acetic acid, dropwise with vortexing, until the pH is between 6 and 7.

4. Resolve the reaction products by Sephadex G-50 gel filtration. Recover released radiolabeled structures from the included volume of the column, and lyophilize to dryness.

5. Analyze the β-eliminated material to determine the saccharide structure. If the starting material contained O-linked GlcNAc, then the expected β-elimination product is the disaccharide [³H]Gal-β-1,4GlcNAcitol. A variety of conventional techniques can be used to identify the radiolabeled β-eliminated products. These include gel-filtration chromatography,[39] high-voltage paper electrophoresis,[40] and Dionex (Sunnyvale, CA) high-performance liquid chromatography (HPLC).[33,39,40]

Procedural Comments. Because polypeptides are destroyed under alkaline conditions, proteins generally cannot be analyzed by SDS–PAGE after β elimination.

Glycosylation Site Analysis

To determine sites of *O*-GlcNAc addition, ³H-galactosylated peptides are generated by enzymatic or chemical proteolysis. In the example below, trypsin is used to catalyze the proteolysis. Other methods of generating peptides include treatment with cyanogen bromide, chymotrypsin, and

[39] C.-R. Torres and G. W. Hart, *J. Biol. Chem.* **259**, 3308 (1984).
[40] G. D. Holt and G. W. Hart, *J. Biol. Chem.* **261**, 8049 (1986).

prolidase. Peptides are then purified by repeated rounds of HPLC, and analyzed by a variety of techniques, including gas-phase sequencing, manual Edman sequencing, and mass spectrometry.

Sample Preparation

Typically, the sample is labeled with [³H]galactose as described above, and N-linked oligosaccharides are removed by treatment with PNGase F, if necessary. The protein may then be reduced and alkylated[41] to maximize accessibility of enzymatic or chemical cleavage sites. Prior to proteolysis, the sample is desalted by Sephadex G-50 gel filtration, followed by lyophilization and acetone precipitation, as described earlier.

Generation of ³H-Galactosylated Tryptic Peptides

1. Prepare a stock solution of tolylsulfonyl phenylalanyl chloromethyl ketone (TPCK)-treated bovine pancreas trypsin (Sigma) in 1 m*M* HCl, such that the amount of trypsin in 10 μl is 1–2% (w/w) of the protein to be digested. This solution may be stored at 4° for up to 24 hr.

2. Suspend the protein in 480 μl of 0.1 *M* ammonium bicarbonate. If the sample cannot be fully solubilized in the ammonium bicarbonate, it will usually become soluble as the digestion proceeds.

3. Initiate the reaction by adding 10 μl of trypsin stock solution. Add a small stir bar to the reaction, and stir magnetically for 8 hr at room temperature. Add another 10 μl of trypsin stock solution, and continue the reaction for 16 hr. A blank sample containing trypsin only should also be incubated in parallel as a control.

4. Terminate the digestion by lyophilizing to dryness.

High-Performance Liquid Chromatography Separation of Peptides

Radiolabeled peptides are purified by reversed-phase HPLC, using a series of different gradients and solvent systems. The use of reversed-phase HPLC for the localization of sites on widely different glycoforms is facilitated by the fact that saccharide structure has comparatively little influence on the elution times of most glycopeptides (either N- or O-linked).[26,41–44] Figure 2 illustrates the final steps from a multistep purification and sequencing of a tryptic *O*-GlcNAc-bearing peptide of bovine lens α-crystallin.

[41] S. J. Swiedler, J. H. Freed, A. L. Tarentino, T. H. Plummer, Jr., and G. W. Hart, *J. Biol. Chem.* **260,** 4046 (1985).

[42] S. J. Swiedler, G. W. Hart, A. L. Tarentino, T. H. Plummer, Jr., and J. H. Freed, *J. Biol. Chem.* **258,** 11515 (1983).

[43] L. D. Powell, K. Smith, and G. W. Hart, *J. Immunol.* **139,** 1206 (1987).

[44] N. M. Dahms and G. W. Hart, *J. Biol. Chem.* **261,** 13186 (1986).

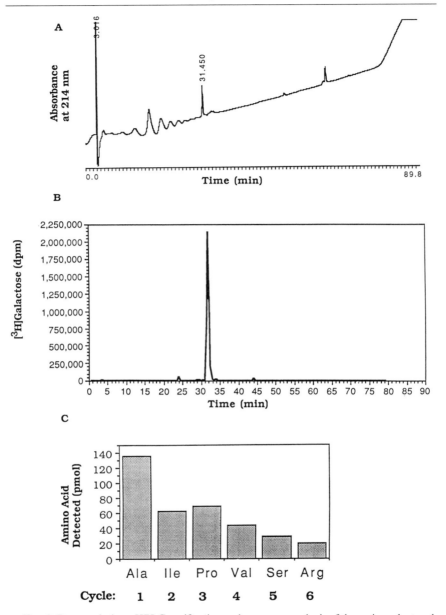

FIG. 2. Reversed-phase HPLC purification and sequence analysis of the major galactosyl-transferase-radiolabeled bovine lens α-crystallin tryptic glycopeptide. The major [³H]galactose-radiolabeled α-crystallin tryptic peptide was first partially purified by reversed-phase HPLC (RP-HPLC), using a linear gradient of acetonitrile in 0.1% trifluoroacetic acid, and

Analysis of Purified Peptides

Gase-Phase Sequencing. Frequently, automated Edman degradation with a gas-phase sequenator is the first step used to identify a labeled peptide. Mass amounts of at least 20 pmol are usually necessary for definitive results. If the sequenced peptide is found to contain more than one possible glycosylation site, a second enzyme may be chosen to redigest the peptide into fragments containing only one possible site of glycosylation. This technique requires mass amounts in the picomolar or high nanomolar range. Alternatively, manual Edman degradation may be used to identify sites of glycosylation on a peptide of known sequence found to contain multiple serine and/or threonine residues.

Manual Edman Degradation. This method has been used to determine sites of *O*-GlcNAc on RNA polymerase II,[26] and has been used to localize sites of phosphorylation.[45] The technique is most successful for short peptides containing at least 5000 cpm of tritium. A convenient kit is commercially available (Millipore, Bedford, MA). As described by Kelly *et al.*,[26] each cycle of manual Edman degradation is analyzed for tritium content by scintillation counting. Cycles in which tritium is detected are correlated with gas-phase sequencing data to determine which sites are glycosylated.

Mass Spectrometry. Gas chromatography mass spectrometry (GC-MS; see [7] in this volume) and fast atom bombardment-mass spectrometry (FAB-MS; see [8] in this volume) may also be used to identify *O*-GlcNAc-bearing peptides. Laser-desorption mass spectrometry has been adapted for the detection of low mass amounts of glycosylated peptides.

Special Considerations

Acceptor Accessibility. The success of galactosyltransferase labeling for the detection of *O*-GlcNAc is dependent in part on the accessibility of *O*-GlcNAc residues to the enzymatic probe. To maximize exposure of *O*-GlcNAc residues to the enzyme, samples may be boiled in 0.1% SDS for 5 min. In this case, at least a sevenfold excess of nonionic detergent must be added prior to labeling to protect the enzymatic probe from

[45] S. Sullivan and T. W. Wong, *Anal. Biochem.* **197**, 65 (1991).

---

then rechromatographed on RP-HPLC under different conditions until constant specific activity [radioactivity/ultraviolet (UV) absorbance (A_{210})] was achieved. (A) Ultraviolet profile of the second round of glycopeptide purification by RP-HPLC with a linear gradient of 0–12% acetonitrile in 10 m*M* triethylamine, pH 5.5. (B) Tritium profile of the second round of purification. (C) Gas-phase sequence analysis of the major peak eluting at 33 min.

inactivation by SDS. In some cases, O-GlcNAc is not fully accessible to the galactosyltransferase even after SDS denaturation.[14,26] Alternatively, accessibility of O-GlcNAc residues may be improved by proteolyzing the sample prior to galactosyltransferase labeling.

Hexosaminidase Activity. Many sample preparations contain substantial hexosaminidase activity, which may rapidly remove O-GlcNAc before it can be detected by the galactosyltransferase probe. Samples can be tested for hexosaminidase activity by the following colorimetric assay: Dilute an aliquot of sample to a 100-μl final volume with 0.1 M citrate/phosphate buffer (pH 5.3), 1 mM p-nitrophenyl-GlcNAc. After incubation for 1 hr at 37°, stop the reaction by the addition of 900 μl of 0.5 M sodium carbonate. Compare the absorbance at 400 nm against a control containing no sample.

Detection by Wheat Germ Agglutinin

Wheat germ agglutinin (WGA) is a lectin that recognizes both terminal β-N-acetylglucosamine and sialic acid residues. Two techniques for the use of WGA in identifying proteins bearing O-GlcNAc are described below. (See [6] in this volume for a general discussion of the use of lectins in glycoconjugate analysis.) Results using WGA for the detection of O-GlcNAc are variable. Observations in our laboratory suggest that WGA binds N-linked oligosaccharides and clustered O-GlcNAc residues much more avidly than it binds isolated O-GlcNAc moieties. Because this lectin detects both N- and O-linked GlcNAc, it must be used in conjunction with either PNGase F treatment, or hexosaminidase treatment and product analysis, for definitive identification of O-GlcNAc.

Probing Western Blots with Wheat Germ Agglutinin

Reagents

Transblot apparatus: Any commercially available apparatus may be used; we use Genie (Idea Scientific, Corvallis, OR) or Mini-Transblot (Bio-Rad, Richmond, CA)

Polyvinylidene difluoride (PVDF) membrane (Millipore): Membrane is cut to size, prewetted by soaking for 5 sec in 100% methanol, rinsed for 5 min in 500 ml of deionized water, and equilibrated for 15–20 min with transfer buffer

Transfer buffer: 192 mM glycine, 25 mM Tris, 20% methanol; final pH should be 8.3, without need of adjustment

TBST: 10 mM Tris (pH 7.9), 150 mM NaCl, 0.1% (v/v) Tween-20

HS-TBST: 1 M NaCl in TBST

Blocking buffer: 4% bovine serum albumin (fraction V; Sigma) in TBST

HS-buffer: Blocking buffer made 1.0 M with NaCl

Horseradish peroxidase-coupled wheat germ lectin (WGA-HRP; E-Y Laboratories, San Mateo, CA)

Enhanced chemiluminescence (ECL) reagents (Amersham)

X-Ray film (X-OMAT; Kodak, Rochester, NY)

Sample Preparation. Protein sample is resolved by SDS–PAGE along with molecular weight standards containing ovalbumin glycoprotein (Bethesda Research Laboratories, Gaithersburg, MD) as a positive control for the WGA-HRP. Two identical gels should be prepared: one for incubation with WGA-HRP alone, the other for incubation in the presence of competing GlcNAc monosaccharide. The gels are then equilibrated for 15 to 20 min in transfer buffer.

Procedure

1. Electroblot proteins and molecular weight standards to prewetted, preequilibrated PVDF membrane according to manufacturer instructions. Typical transfer periods range from 30 min to 1 hr at 12 V.

2. Rinse the blots once with blocking buffer, immerse in blocking buffer, and incubate with constant mixing for at least 4 hr at 4°.

3. Rinse the blots thoroughly with TBST (see Procedural Comments, below), and immerse in HS-buffer containing WGA-HRP (1 μg/ml) in the presence or absence of 0.5 M competing GlcNAc monosaccharide. Incubate for at least 2 hr with mixing at room temperature.

4. Wash the blots extensively with HS-TBST, followed by TBST.

5. Visualize the bound WGA-HRP by enhanced chemiluminiscence, using ECL reagents (Amersham) according to the instructions of the manufacturer, followed by exposure to X-ray film.

Procedural Comments. Adequate washing of the blots after blocking and after incubation with WGA-HRP is crucial to the success of the procedure. For best results, try the following procedure: Rinse both sides of each blot thoroughly by applying the wash solution liberally from a spray bottle. After spraying, immerse the blot in fresh water solution and mix for 20 min on a shaker. Rinse the blots once more by spraying before proceeding to the next step.

The use of sodium azide in buffer solutions should be avoided, because it quenches the ECL reaction.

Wheat Germ Agglutinin–Sepharose Chromatography

Proteins of particularly low abundance or short half life may be difficult to obtain in sufficient quantities for O-GlcNAc detection. If the cDNA of such a protein is available, coupled transcription/translation followed by WGA affinity chromatography is a promising alternative. This rapid and sensitive detection technique uses a commercially available rabbit reticulocyte lysate transcription/translation system (Promega, Madison, WI) to generate [^{35}S]Met-labeled protein. A WGA-conjugated Sepharose column is then used to capture glycosylated protein. Figure 3 shows an example of this technique, using the cDNA for rat p62 nuclear pore glycoprotein[46,47] and for the transcription factor c-Fos.

Materials

Coupled *in vitro* transcription/translation system (Promega)
cDNA: cDNA of protein of interest is cloned downstream of either the SP6, T3, or T7 promoter in an appropriate expression vector
Wash buffer: 0.01 M sodium phosphate buffer (pH 7.4), 0.15 M NaCl, 0.2% Nonidet P-40
Gal-elution buffer: 1 M D-(+)-galactose (Sigma) in wash buffer
GlcNAc-elution buffer: 1 M N-acetyl-D-glucosamine (Sigma) in wash buffer
WGA affinity chromatography: A 150-μl WGA–Sepharose (E-Y Laboratories) column is prepared in a Pipetman (Rainin, Emeryville, CA) tip plugged with nylon wool, and is equilibrated in wash buffer
N-Acetyl-β-D-glucosaminidase [from jack bean (Sigma) or beef kidney (Boehringer Mannheim)]

Procedure

1. Prepare plasmid DNA by the alkaline lysis method or by using a commercially available purification column kit (Qiagen, Chatsworth, CA).

2. Begin the transcription/translation reaction: A standard 50-μl reaction consists of 25 μl of rabbit reticulocyte lysate, 2 μl of reaction buffer, 1 μl of SP6, T3, or T7 RNA polymerase, 1 μl of 1 mM methionine-free amino acid mixture, 4 μl of ^{35}S methionine (1000 Ci/mmol) at 10 mCi/ml, 1 μl of ribonuclease inhibitor (40 U/μl), 1 μg of plasmid DNA, and nuclease-free water to final volume. Incubate the reaction mixture for 1–2 hr at 30° (optimal reaction time is determined empirically to minimize proteolysis and maximize product). The luciferase gene included in the kit is used as a positive control. A negative control containing no DNA should also be included.

[46] M. D'Onofrio, C. M. Starr, M. K. Park, G. D. Holt, R. S. Haltiwanger, G. W. Hart, and J. A. Hanover, *Proc. Natl. Acad. Sci. U.S.A.* **85,** 9595 (1988).
[47] C. M. Starr, M. D'Onofrio, M. K. Park, and J. A. Hanover, *J. Cell Biol.* **110,** 1861 (1990).

A

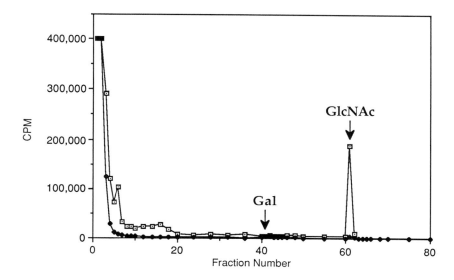

B

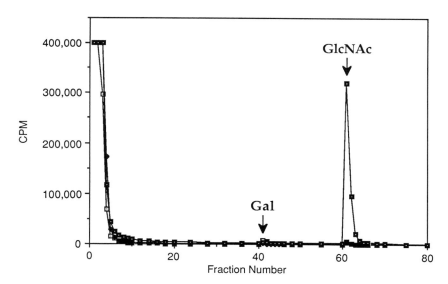

Fig. 3. Wheat germ agglutinin affinity chromatography of glycoproteins from coupled *in vitro* transcription/translation in rabbit reticulocyte lysate. (A) Results of WGA–Sepharose chromatography when c-Fos cDNA (pGEMMmFos) is the template. (□) c-Fos; (◆) c-Fos plus hexosaminidase. (B) Results of WGA–Sepharose chromatography when p62 cDNA (pGEMc62) is the substrate. (□) No DNA; (◆) luciferase; (■) p62. Gal, 1 *M* galactose; GlcNAc, 1 *M* N-acetylglucosamine.

3. Treat 30 μl of the reaction products with N-acetyl-β-D-glucosamini-dase to remove terminal GlcNAc molecules. The reaction contains 30 μl of sample, 30 μl of citrate-phosphate buffer (pH 4.0) containing 1 U of enzyme, 0.01 U of aprotinin (Sigma), 1 μg of leupeptin (Sigma), and 1 μg of α_2-macroglobulin (Boehringer Mannheim). React for 12–24 hr at 37°. Analyze 1–5 μl of the end reaction products by SDS–PAGE to determine if proteolysis has occurred. Apply the remainder to the WGA–Sepharose column, as described in step 4.

4. Apply 30 μl of the transcription/translation reaction products, or 60 μl of N-acetyl-β-D-glucosaminidase reaction products, to the WGA–Sepharose column, and allow to stand for 30 min.

5. Wash the column with 15 ml of wash buffer at a flow rate of 10 ml/hr. Collect 0.5-ml fractions.

6. Load the column with 300 μl of Gal-elution buffer, and allow to stand for 20 min. Elute with 5 ml of Gal-elution buffer, collecting 0.5-ml fractions. (This step serves as a specificity control.)

7. Repeat step 6, using GlcNAc-elution buffer to elute specifically bound material.

8. Transfer 25 μl of each fraction to a scintillation vial, and analyze in a scintillation counter for ^{35}S. Pool the peak fractions of radiolabeled material from steps 5–7, evaporate them to dryness, or precipitate the proteins with trichloroacetic acid, and analyze by SDS–PAGE.

Procedural Comments. Surprisingly, the reticulocyte lysate used in this method appears to contain sufficient polypeptide : O-GlcNAc glyco-syltransferase and UDPGlcNAc for efficient glycosylation.[19] Addition of purified rat liver polypeptide : O-GlcNAc transferase and exogenous UDP-GlcNAc does not generally increase glycosylation efficiency.

Because WGA dimers dissociate below pH 3.5, the pH of all buffers used in WGA affinity chromatography should be neutral.

O-GlcNAc on Cytoplasmic and Nuclear Proteins: Subcellular Fractionation of Metabolically Labeled Proteins

Metabolic labeling of O-GlcNAc-bearing proteins, followed by subcel-lular fractionation into cytoplasmic and nuclear pools, may be a useful means of observing the gross dynamic changes in O-GlcNAc levels and subcellular localization that occur during cell cycling or in response to extracellular stimulation.[34] In the protocol described below, lymphocytes that have been metabolically labeled with [^{3}H]glucosamine are permeabil-ized with digitonin to release cytoplasmic proteins, which can then be separated from nuclear and membrane-bound proteins by centrifugation.

Materials

D-[6-³H]Glucosamine (22 Ci/mmol; Amersham)

T cell or B cell hybridomas: $1-5 \times 10^7$ cells at 5×10^6 cells/ml cultured in appropriate growth medium with 10% fetal calf serum (FCS) at 37° in an atmosphere of 5% CO_2/95% air

Labeling medium: Glucose-free medium containing D-[6-³H]glucosamine (22 Ci/mmol; Amersham) (50 μCi/ml), 5% FCS

Phosphate-buffered saline (PBS)

Permeabilization medium: PBS containing digitonin (100–500 μg/ml) (see Procedural Comments, below)

Trypan blue (GIBCO, Grand Island, NY)

Buffer H: 50 mM HEPES, 500 mM NaCl, 2% (v/v) Triton X-100, protease inhibitors (see Detection by Galactosyltransferase Labeling, above)

Gel-filtration chromatography: A 1×30 cm column of Sephadex G-50 (coarse) equilibrated in 50 mM ammonium formate, 0.1% SDS, 0.02% (w/v) sodium azide

Procedure

Metabolic labeling of cultured cells with [³H]glucosamine is discussed in detail in [2] of this volume. Digitonin permeabilization of lymphocytes and T cell hybridomas has been described by Kearse and Hart.[34]

1. Metabolic labeling: Culture cells for 5–24 hr in labeling medium. Refer to [2] in this volume for technical details.

2. Wash metabolically labeled cells twice by resuspending in 5 ml of PBS and centrifuging for 10 min at 500 g and 4°.

3. Resuspend cells in 1 ml of permeabilization medium, and incubate at room temperature for 10 min, with occasional mixing. Assess the efficiency of permeabilization by the inability of the cells to exclude trypan blue.

4. Pellet nuclear and membranous materials by centrifugation at 200 g for 10 min at 4°.

5. Aspirate the soluble portion, and precipitate the proteins with 8 vol of acetone at $-20°$ for 6 hr.

6. Solubilize the pelleted fraction in 1 ml of buffer H. Apply the sample to the desalting column. Pool radiolabeled proteins in the void volume, and acetone precipitate the proteins as above.

7. Remove N-linked carbohydrates from the samples by PNGase F treatment, as described previously (see Detection by Galactosyltransferase Labeling).

8. Analyze the radiolabeled macromolecules by two-dimensional electrophoresis, using nonequilibrium pH-gradient electrophoresis in the first dimension,[34] followed by SDS–PAGE. Stain the gel with Coomassie Brilliant Blue, incubate with En³Hance, and visualize by autofluorography.

Procedural Comments

Digitonin must be titrated carefully for each situation, depending on both cell number and cell type, to avoid excessive contamination of the cytosolic fraction by extracellular or lumenal N-linked glycoproteins. Also, digitonin is a substrate for galactosyltransferase and in some cases can interfere, if not adequately removed prior to labeling with galactosyltransferase.

Acknowledgments

We thank Dr. J. A. Hanover for providing pGEMc62, and Dr. Y.-L. Fong for providing pGEMmFos. We also thank Drs. R. E. Willoughby, Jr., and W. G. Kelly for helpful suggestions for the WGA affinity chromatography. Original work supported by NIH R01 CA42486 and NIH R01 HD13563 to G.W.H.

[27] Identification of Polysialic Acids in Glycoconjugates

By Jean Ye, Ken Kitajima, Yasuo Inoue, Sadako Inoue, and Frederic A. Troy II

Introduction

This chapter provides readers with a description of the strategies and methods used in the authors' laboratories for detecting, analyzing, and modifying polysialic acids (polySia) and their derivatives. Polysialic acids are a structurally unique group of carbohydrate chains that consist of N-acetylneuraminic acid (Neu5Ac) or N-glycolylneuraminic acid (Neu5Gc) joined internally by α-2,8-, α-2,9-, or alternating α-2,8/α-2,9-ketosidic linkages. 2-Keto-3-deoxy-D-glycero-D-galacto-nononic acid (KDN) is a special deaminoneuraminic acid, and its polymers share many properties in common with polySia. Polysialic acid chains covalently mod-

METHODS IN ENZYMOLOGY, VOL. 230

ify surface glycoconjugates on cells that range in evolutionary diversity from microbe to human, and thus appear to function in a remarkably wide range of important biological processes.[1,2] For example, the polySia capsules of neurotropic *Neisseria meningitidis* group B and *Escherichia coli* K1 are virulence determinants associated with neonatal meningitis.[3] In fish, polysialoglycoproteins (PSGPs) are functionally implicated in fertilization. KDN-rich glycoproteins that contain α-2,8-linked oligo/polyKDN chains, designated KDN-gp, may mediate egg–sperm interaction.[4,5] Polysialic acid is also an oncodevelopmental antigen in human kidney and brain,[6–8] and its surface expression on a number of tumors may enhance the metastatic potential of some human cancers.[9,10] Finally, polysialylation of the α subunit of the sodium channel in adult rat brain suggests that polysialylation of surface glycoconjugates could regulate a number of important, and heretofore unknown, membrane-mediated events.[11]

As a consequence of the wide occurrence of polySia in nature, studies to understand polysialylation have emerged as an exciting area of glycobiology that continues to impact on contemporary studies in microbiology, molecular, cell, and developmental biology, neurosciences, and oncology. Much of the effort has sought to determine the molecular involvement of these novel sugar chains in bacterial pathogenesis, fertilization, cell adhesive interactions, neural development, and malignancy.[2] Three types of prokaryotic-derived reagents that are highly specific and sensitive for either recognizing, synthesizing, or depolymerizing α-2,8-linked polySia

[1] F. A. Troy, *Glycobiology* **2**, 5 (1992).

[2] J. Roth, U. Rutishauser, and F. A. Troy, eds., "Polysialic Acid: From Microbes to Man." Birkhaeuser, Basel, 1993.

[3] L. D. Sarff, G. H. McCracken, M. S. Schiffer, M. P. Glode, J. B. Robbins, I. Ørskov, and F. Ørskov, *Lancet* **1**, 1099 (1975).

[4] S. Inoue, A. Kanamori, K. Kitajima, and Y. Inoue, *Biochem. Biophys. Res. Commun.* **153**, 172 (1988).

[5] Y. Inoue, in "Polysialic Acid: From Microbes to Man" (J. Roth, U. Rutishauser, and F. A. Troy, eds.), p. 171. Birkhaeuser, Basel, 1993.

[6] J. Roth, D. J. Taatjes, D. Bitter-Suermann, and J. Finne, *Proc. Natl. Acad. Sci. U.S.A.* **84**, 1969 (1987).

[7] R. D. McCoy, E. R. Vimr, and F. A. Troy, *J. Biol. Chem.* **260**, 12695 (1985).

[8] S. Hoffman, B. C. Sorkin, P. C. White, R. Brackenbury, R. Mailhammer, U. Rutishauser, B. A. Cunningham, and G. M. Edelman, *J. Biol. Chem.* **257**, 7720 (1982).

[9] M. Lipinski, M. R. Hirsch, H. Deagostini-Bazin, O. Yamada, T. Tursz, and C. Goridis, *Int. J. Cancer* **40**, 81 (1987).

[10] M. C. Glick, B. D. Livingston, G. W. Shaw, J. L. Jacobs, and F. A. Troy, in "Advances in Neuroblastoma Research" (A. E. Evans, A. G. Knudson, R. C. Seeger, and G. J. D'Angio, eds.), p. 267. Wiley-Liss, New York, 1991.

[11] C. Zuber, P. M. Lackie, W. A. Catterall, and J. Roth, *J. Biol. Chem.* **267**, 9965 (1992).

chains have made many of these studies possible.[12] These reagents are (1) endo-*N*-acylneuraminidase (Endo N), the soluble form of a bacteriophage-derived endosialidase produced by infection of *E. coli* with a lytic bacteriophage; (2) the anti-polySia antibody, H.46, a polyclonal IgM prepared by immunizing a horse with *N. meningitidis* group B; and (3) a CMPNeu5Ac : poly-α-2,8-sialosylsialyltransferase (polyST) from *E. coli* K1 that can transfer Sia from CMPSia (or sialylmonophosphorylundecaprenol) to endogenous or exogenous acceptors that contain preexisting oligo- or polySia residues.[13] Thus, the most direct and sensitive assay to identify polySia is to determine if H.46 immunoreactivity is abolished by pretreating the sample with Endo N. This can be done by dot-blot or Western blot analysis. Confirmation of this conclusion should involve structural studies to establish that concomitant with the loss of anti-polySia immunoreactivity is the formation of sialyl oligomers. Further confirmation should be provided by showing that the presumed polysialylated glycoconjugate can function as an exogenous acceptor in the *E. coli* K1 polysialyltransferase reaction.[12]

The major goal of this chapter is to describe developments in the use of these reagents and methods to identify and study polySia in glycoconjugates. We also describe methods for detecting KDN-containing glycoconjugates, and conclude by describing methods to prepare some of the specific reagents that will allow others to identify and study polySia.

As the reagents described here become more readily available, and newer and more sensitive methods are developed, it is anticipated that our knowledge of the occurrence and function of polysialylated glycoconjugates will continue to expand. It is also likely that these studies will reveal an even greater role of polySia in biological processes, and a greater understanding of how altered expression of these molecules may be associated with a number of pathophysiological conditions, involving perhaps all organ systems.

Properties of Specific Reagents Used to Identify Polysialic Acids

Anti-Polysialic Acid Antibodies

Several highly specific and sensitive anti-polySia and anti-KDN antibodies have been developed to identify α-2,8-linked polySia, (KDN)G_{M3},

---

[12] E. R. Vimr, R. D. McCoy, H. F. Vollger, N. C. Wilkison, and F. A. Troy, *Proc. Natl. Acad. Sci. U.S.A.* **81,** 1971 (1984).

[13] F. A. Troy and M. A. McCloskey, *J. Biol. Chem.* **254,** 7377 (1979).

TABLE I
PRIMARY AND SECONDARY ANTIBODIES FOR DETECTING POLYSIALIC ACID AND
KDN GLYCOCONJUGATES

Primary antibody	Immunogen	Primary antibody immunoglobulin	Minimum structure recognized	Matching secondary antibody[a]	Ref.[b]
H.46	Group B meningococci	Polyclonal horse IgM	$\alpha2,8$-OligoSia, DP > 9	Anti-horse IgG (50% cross-reactivity)	1
735	Group B meningococci	Monoclonal mouse IgG_{2a}	$\alpha2,8$-OligoSia	Anti-mouse IgG (γ-chain specific)	2
RM	KDN-gp[c]	Polyclonal rabbit IgG	$(\rightarrow 8KDN\alpha2)_n$, $KDN\alpha2 \rightarrow 3Gal$	Anti-rabbit IgG	3
kdn3G	$(KDN)G_{M3}$[d]	Monoclonal mouse IgG	$KDN\alpha2 \rightarrow 3Gal$	Anti-mouse IgG (γ-chain specific)	4
kdn8kdn	KDN-gp[c]	Monoclonal mouse IgM	$(8KDN\alpha2 \rightarrow)_n$ $(n = 2-6)$, $\gg KDN\alpha2 \rightarrow 3Gal$	Anti-mouse IgM (μ-chain specific)	5

[a] The biotinylated secondary antibodies are used and are available from Vector Laboratory (see text).

[b] Key to references: (1) L. D. Sarff, G. H. McCracken, M. S. Schiffer, M. P. Glode, J. B. Robbins, I. Ørskov, and F. Ørskov, *Lancet* **1**, 1099 (1975); (2) M. Frosch, I. Görgen, G. J. Boulnois, K. N. Timmis, and D. Bitter-Suermann, *Proc. Natl. Acad. Sci. U.S.A.* **82**, 1194 (1985); (3) A. Kanamori, K. Kitajima, Y. Inoue, and S. Inoue, *Glycoconjugate J.* **8**, 222 (1991); (4) Y. Song, K. Kitajima, and Y. Inoue, *Glycobiology* **3**, 31 (1993); and (5) A. Kanamori and Y. Inoue, personal communication (1992).

[c] The oligosaccharide moiety of KDN-gp (where $n = 2-6$) is

$$KDN\alpha2 \rightarrow [\rightarrow 8KDN\alpha2 \rightarrow]_n$$
$$\searrow$$
$$\overset{6}{\underset{3}{}}GalNAc\alpha1 \rightarrow Thr/Ser$$
$$\nearrow$$
$$KDN\alpha2 \rightarrow 3Gal\beta1 \rightarrow 3GalNAc\alpha1$$

[d] $(KDN)G_{M3}$ is $KDN\alpha2 \rightarrow 3Gal\beta1 \rightarrow 4Glc\beta1 \rightarrow$ ceramide.

or oligoKDN expression in prokaryotic and eukaryotic cells (Table I). The anti-polySia antibodies H.46 and monoclonal antibodies (MAb) 735 have been used to identify polySia in a variety of sources ranging from microbes to human, including a number of normal and pathological states.[1,2] The minimum length of α-2,8-polySia for the binding of H.46 and MAb 735 antibodies is a degree of polymerization (DP) of ~8–10 Sia residues.[14,15,15a] The unusual length is required for the antibody to recog-

[14] J. Finne and P. H. Mäkelä, *J. Biol. Chem.* **260**, 1265 (1985).

[15] H. J. Jennings, R. Roy, and F. Michon, *J. Immunol.* **134**, 2651 (1985).

[15a] J. Finne, D. Bitter-Suermann, C. Goridis, and U. Finne, *J. Immunol.* **138**, 4402 (1987).

nize a conformational epitope contained within the internal six sialyl residues.[16] The immunospecificity of H.46 toward fish egg PSGPs containing exclusively poly(Neu5Gc) chains, a mixture of molecules containing poly(Neu5Ac) and poly(Neu5Gc) chains, and a copolymer of poly (Neu5Ac, Neu5Gc) chains was studied.[17,18] H.46 was found to react only with poly(Neu5Ac) and not with poly(Neu5Gc) chains. This specificity was used to localize the poly(Neu5Ac) antigen in trout egg PSGPs within the Golgi-derived secretory vesicles (cortical alveoli) in the unfertilized eggs of lake trout fish.

Anti-KDNα2,3Gal and Anti-α2,8-Linked OligoKDN Antibodies

Three antibodies specific for detecting KDN and oligoKDN structures in glycoconjugates have been developed by Inoue and colleagues (Table I). The first, designated kdn3G, was prepared by fusion of spleen cells from a BALB/c mouse immunized with $(KDN)G_{M3}$ ($KDN\alpha2$-$3Gal\beta1$-$4Glc\beta1$-1ceramide) and mouse myeloma cells P3-X63 Ag8.U1 (P3U1). This yielded a hybrid cell line that produced an IgG MAb that bound to $(KDN)G_{M3}$ but not to $(Neu5Ac)G_{M3}$.[19] The specificity of the antibody was determined by an enzyme-linked immunosorbent assay. kdn3G was found to react most strongly with $(KDN)G_{M3}$ and less strongly with a glycoprotein containing a number of $KDN\alpha2$-$3Gal\beta1$-$3GalNAc\alpha1$-$3[(-8KDN\alpha2)_n$-$6]Gal$-$NAc\alpha1$- chains. These results indicate that kdn3G specifically recognizes the disaccharide structure, $KDN\alpha2,3Gal\beta1$, and can discriminate between $KDN\alpha2$-$3Gal\beta1$- and $Neu5Ac\alpha2$-$3Gal\beta1$-.[19] kdn3G was used to localize $(KDN)G_{M3}$ to the external surface of the entire plasma membrane in rainbow trout sperm, by an indirect immunofluorescence procedure.

Two antibodies that recognize $(-8KDN\alpha2-)_n$ oligosaccharides have been prepared with KDN-gp as immunogen (Table I). The structure (where $n = 2$–6) of the oligo(KDN)-containing glycan moiety of this glycoprotein[5] is

$$KDN\alpha2 \rightarrow [\rightarrow 8KDN\alpha2 \rightarrow]_n$$
$$\searrow$$
$$^6_3GalNAc\alpha1 \rightarrow Thr/Ser$$
$$\nearrow$$
$$KDN\alpha2 \rightarrow 3Gal\beta1 \rightarrow 3GalNAc\alpha1$$

---

[16] F. Michon, J. R. Brisson, and H. J. Jennings, *Biochemistry* **26,** 8399 (1987).

[17] K. Kitajima, S. Inoue, Y. Inoue, and F. A. Troy, *J. Biol. Chem.* **263,** 18269 (1988).

[18] C. Sato, K. Kitajima, S. Inoue, Y. Inoue, and F. A. Troy, *J. Biol. Chem.* **268,** in press (1993).

[19] Y. Song, K. Kitajima, and Y. Inoue, *Glycobiology* **3,** 31 (1993).

The first of these two antibodies is a polyclonal IgG raised in a rabbit, and is designated RM.[20] RM recognizes α-2,8-linked oligoKDN chains but not oligo(Neu5Ac) or oligo(Neu5Gc) residues.[20] The antibody has been used to immunocytochemically localize KDN-gp in the outer layer (second layer from outside) of the vitelline envelope of rainbow trout eggs.[5,20] RM also reacts with (KDN)G_{M3}, and thus can be regarded as a specific reagent for detecting KDNα2,3Gal and α-2,8-linked oligoKDN groups.[20] The second antibody raised in response to KDN-gp is a mouse monoclonal IgM, designated kdn8kdn[21] (Table I). This MAb is highly specific and sensitive for recognizing the (-8KDNα2-)$_n$ epitope on KDN-gp, being able to detect as little as 0.1 ng of KDN-gp (see the following section). kdn8kdn shows some cross-reactivity with (KDN)G_{M3} but only at high concentrations of glycolipid.

The anti-polySia and anti-KDN antibodies described above are currently being used in our laboratories to search for new families of polySia and KDN-containing glycoconjugates in a variety of cells, tissues, and developmentally regulated systems.

There are no antibodies that specifically recognize poly(Neu5Gc) chains, even though these structures occur in rainbow trout PSGPs.[5,22] Further, the anti-polySia antibodies require a minimum DP of 8 to 10 Sia residues for immunodetection. Hence, poly(-8Neu5Acα2-) chains are the only polySia that have been described to date in mammalian systems. This likely results from the limitation of our present methodologies to identify these structures, rather than from the possibility that poly(Neu5Ac) or oligoKDN structures are uniquely restricted to teleost fish. Thus, as newer and more sensitive methods to identify these structures become available, we can anticipate an even wider occurrence in nature of poly(Neu5Gc), oligoKDN, and perhaps even alternating α2,8/α2,9 and α2,9 polySia. Although the continued development of MAbs will broaden our armamentarium of reagents to search for new classes of polySia, it is important to recognize that positive immunoreactivity will have to be confirmed by rigorous chemical, biochemical, and physical studies. To facilitate such studies, micromethods for some of the procedures described here will have to be developed so that nanogram amounts, or less, of oligo/polySia can be detected. Capillary electrophoresis coupled with tandem mass spectrometry (MS), for example, electrospray ionization, is a promising method for development to meet this need.

[20] A. Kanamori, K. Kitajima, Y. Inoue, and S. Inoue, *Glycoconjugate J.* **8**, 222 (1991).

[21] A. Kanamori and Y. Inoue, personal communication (1992).

[22] D. Nadano, M. Iwasaki, S. Endo, K. Kitajima, S. Inoue, and Y. Inoue, *J. Biol. Chem.* **261**, 11550 (1986).

Endo-N-acylneuraminidase (Endo N): Specific Molecular Probe to Detect and to Modify Polysialic Acids Selectively

Endo N is the key enzyme for confirming the presence of polySia. It is specific for hydrolyzing oligo- or polySia residues in sources as distinct as bacteria and human brain.[13] Endo N catalyzes the depolymerization of polySia according to the following reaction:

$$(\text{-8Neu5Ac}\alpha2\text{-})_n - X \rightarrow (\text{-8Neu5Ac}\alpha2\text{-})_{2-4} + \text{Neu5Ac} + X$$

In the polySia capsule of *E. coli* K1, n is ~200 Sia residues and X represents endogenous acceptor molecules.[23-26] The polySia chain in *N*-CAM was shown to contain at least 55 Sia residues, but its actual length is estimated to be closer to 200.[27] X in *N*-CAM is a complex-type *N*-asparaginyl-linked oligosaccharide chain. In fish PSGPs, oligo(Neu5Ac), oligo(Neu5Gc), and oligoKDN chains are attached to other sugar residues of an oligosaccharide chain O-linked to protein.[5] Endo N can hydrolyze the α-2,8 linkage in *E. coli* polySia capsules that contain alternating α-2,8 and α-2,9 ketosidic linkages.[1,28] Further, Endo N can also depolymerize the poly(Neu5Gc) chains of rainbow trout PSGPs, yielding as the major product (Neu5Gc)$_3$.[17] Thus, although H.46 is a probe specific for poly-(Neu5Ac), Endo N can be used as a reagent for identifying both poly(Neu5Ac) and poly(Neu5Gc) chains. In all of our studies we have used the soluble form of the enzyme that is derived from a bacteriophage (K1F) lysate of *E. coli* K1.[12] The K1F phage was isolated from the University of California, Davis sewage after enrichment on *E. coli* K-235 by standard procedures.[29] K1F encodes for the synthesis of both a soluble and phage-bound form of the enzyme. The soluble form has a pH optimum of 7.4. It does not have any known cofactors. It is stable when stored in the freezer as lyophilized powder or in solution. Other phage-bound endosialidases, some with different properties, have been described.[14,30,31]

[23] T. E. Rohr and F. A. Troy, *J. Biol. Chem.* **255**, 2332 (1980).

[24] E. C. Gotschlich, B. A. Fraser, O. Nishimura, J. B. Robbins, and T.-Y. Liu, *J. Biol. Chem.* **256**, 8915 (1981).

[25] M. A. Schmidt and K. Jann, *FEMS Microbiol. Lett.* **14**, 69 (1982).

[26] C. Weisgerber and F. A. Troy, *J. Biol. Chem.* **265**, 1578 (1990).

[27] B. D. Livingston, J. L. Jacobs, M. C. Glick, and F. A. Troy, *J. Biol. Chem.* **263**, 9443 (1988).

[28] P. C. Hallenbeck, E. R. Vimr, F. Yu, B. Bassler, and F. A. Troy, *J. Biol. Chem.* **262**, 3553 (1987).

[29] R. J. Gross, T. Cheasty, and B. Rone, *J. Clin. Microbiol.* **6**, 548 (1977).

[30] B. Kwiatkowski, B. Boscheck, H. Thiele, and S. Stirm, *J. Virol.* **43**, 697 (1982).

[31] S. Tomlinson and P. W. Taylor, *J. Virol.* **55**, 374 (1985).

Method to Assay Endo N Activity

The K1F Endo N has been widely used in both immunodetection procedures and polysialyltransferase reactions to prove the existence of α-2,8-polySia by catalyzing its depolymerization. Oligo- or polySia from *E. coli* K1 is depolymerized by either the soluble or phage-bound Endo N. The reaction is followed by quantitating the loss of radioactivity from [^{14}C]polySia, or by measuring the amount of product ([^{14}C]sialyl oligomers) formed. Reaction products vary with depolymerases derived from different K1 specific phages,[14,28,30,31] and with the length of the different oligo- or polySia used as substrate. Suitable substrates include unlabeled or radiolabeled oligomers with a DP of 5 or higher, and polySia (DP, ~200). [^{14}C]Polysialic acid ([^{14}C]polySia) may be synthesized *in vivo* or *in vitro*.[13,23] Sialyl oligomers, such as colominic acid, can be generated by either mild acid hydrolysis[13] or Endo N depolymerization of polySia.[28] A typical protocol for using Endo N to carry out the depolymerization is in 20 m*M* Tris-HCl, pH 7.4, at 33–37°. Reducing agents or ethylenediaminetetraacetic acid (EDTA) have no apparent effect on Endo N activity. The optimum digestion time and enzyme concentration can vary according to the substrate and should be determined empirically. Generally, samples from mammalian sources require more time (up to 3–6 hr) or more enzyme than those from prokaryotic sources (30 min–1 hr).

Escherichia coli K1 Polysialyltransferase Activity

The *E. coli* K1 polyST is an inner membrane-bound enzyme complex consisting of at least two enzymes, designated NeuE and NeuS. NeuE, a 22 to 26-kDa transmembrane protein, has been implicated in polySia chain initiation.[1] NeuS, a protein with a deduced molecular mass of 47 kDa, has been implicated in chain polymerization.[1,32] Neither enzyme has been solubilized and purified. The most convenient source of the enzyme is inside-out vesicles (IOVs) prepared from *E. coli* K1 or its mutant strains.[1] The functional domain of the polyST is thus localized on the external surface of sealed IOVs. These vesicles can synthesize polySia when supplied with CMP[^{14}C]Neu5Ac by catalyzing sialylation of either endogenous or exogenous acceptors.[1] The IOVs are stable when stored at −20°. Repeated freeze-thawing should be avoided.

One of the important properties of the polyST with practical value is that it can attach Neu5Ac from CMPNeu5Ac to exogenous acceptors containing at least three α-2,8-linked sialyl residues. This property serves two purposes. First, *E. coli* EV11, a mutant strain derived from a hybrid

[32] C. Weisgerber, A. Hansen, and M. Frosch, *Glycobiology* **1**, 357 (1991).

of *E. coli* K12 and K1, is defective in catalyzing the endogenous synthesis of polySia.[33] Inside-out vesicles from EV11 can sialylate exogenous acceptors.[12] Thus, [^{14}C]polySia synthesis in EV11 IOVs, due to the addition of exogenous acceptors, is a sensitive indicator for determining if the acceptor substrate contains preexisting oligo/poly(-8Neu5Acα2-) residues.[12,27] Second, the *E. coli* K1 polyST can also be used as a synthetic reagent for the synthesis of structurally unique polysialylated glycosphingolipids and other glycoconjugates.[34]

Immunochemical Methods for Identification of Poly(-8Neu5Acα2-) and OligoKDN Residues

Background

We have described immunochemical methods for rapidly detecting α-2,8-linked oligo/polySia chains in cell or tissue homogenates that allow this epitope to be readily detected by Western blot or rocket immunoelectrophoresis, using the H.46 antibody.[12,27] Similar procedures using monoclonal antibodies, enzyme-linned immunnosorbent assays (ELISAs), and immunocytochemistry have also been described.[35,36] The presence of polySia is confirmed if pretreating the sample with the diagnostic enzyme, Endo N, abolishes H.46 immunoreactivity, and the products of Endo N depolymerization are structurally characterized.[7,12] The Endo N control is important when using antibodies to detect polySia because an IgM has been reported that cross-reacts with polynucleotides.[37] H.46 and the MAbs that have been described are highly specific for the poly(-8Neu5Acα2-) epitope.[12,14,15] Both H.46 and the MAbs have been used to identify polySia in a variety of pathological conditions, attesting to the simplicity and potential of the method.[1]

Tissue as Positive Control

A convenient source of tissue as a positive control for polySia expression is 8- to 19-day-old embryonic chicken brain or 8-day-old postnatal rat brain. Chicken heads can be stored in a $-80°$ freezer for several years without losing H.46 immunoreactivity of the polysialylated *N*-CAM

[33] E. R. Vimr and F. A. Troy, *J. Bacteriol.* **164,** 854 (1985).

[34] J. W. Cho and F. A. Troy, *Glycoconjugate J.* **8,** 153 (1991).

[35] J. Roth, C. Zuber, P. Wagner, I. Blaha, D. Bitter-Suermann, and P. U. Heitz, *Am. J. Pathol.* **133,** 227 (1988).

[36] D. Bitter-Suermann and J. Roth, *Immunol. Res.* **6,** 225 (1987).

[37] E. A. Kabat, K. N. Nickerson, J. Liao, L. Grossbard, E. F. Osserman, E. Glickman, L. Chess, J. B. Robbins, R. Scheerson, and Y. Yang, *J. Exp. Med.* **164,** 642 (1986).

molecules. The apparent molecular mass as determined by sodium dodecyl sulfate–polyacrylamide gel electrophoresis (SDS–PAGE) may change,
however, due to general acid catalysis of the inherently labile α-2,8 ketosidic linkages in the long-chain polymers. When needed, the forebrain is
removed from the head for preparation of homogenates. *Escherichia coli*
K1 cells expressing the polySia capsule can also be used as a positive
control.

Experimental Procedures

Western Blot. The following protocol describes our modified Western
blot procedure for the immunodetection of oligo/poly(-8Neu5Acα2-) residues, using the anti-polySia antibodies H.46 and MAb 735. Immunochemical methods for detecting KDN-containing glycoconjugates with two anti-
KDN antibodies (kdn8kdn and kdn3G) are also described.

1. Cells or tissue are first homogenized in 4 to 6 vol of ice-cold HENT
buffer [50 mM N-2-hydroxyethylpiperazine-N'-2-ethane sulfonic acid
(HEPES) (pH 7.5), containing 2 mM EDTA, 50 mM NaCl, and 2% Triton
X-100, 1 mM dithiothreitol (DTT), and 1 mM phenylmethylsulfonyl fluoride (PMSF)]. Preferably, additional protease inhibitors should be added,
including aprotinin (500 KIU/ml final concentration), leupeptin (40 μg/
ml), and 1 μM pepstatin A. The tissue homogenates are prepared with a
Polytron (Brinkmann, Westbury, NY) or Dounce (Wheaton, Millville, NJ)
homogenizer. After disruption the homogenates are centrifuged at 14,000
g for 1 min. The supernatant fraction is used for the next step.
2. A typical Endo N incubation mixture is included that consists of
90 μl of homogenate, 10 μl of Endo N (1 U/ml), or 10 μl of Tris-HCl (pH
7.4) for "minus Endo N" control. The reaction is carried out at 33–37°
for about 3 hr.
3. After incubation, 14.3 μl of a modified Laemmli sample buffer[38]
consisting of 40% glycerol, 8% (w/v) SDS, and 20% 2-mercaptoethanol
(GSM) is added and incubated at 37° for 1 hr. The sample is then loaded
on to a 4–20% gradient SDS polyacrylamide gel or a discontinuous SDS
polyacrylamide gel with an 8% (w/v) acrylamide resolving gel. The Western blot is carried out according to Towbin,[39] using a positively charged
nylon membrane (Zeta-Probe; Bio-Rad, Richmond, CA) as the blotting
membrane.
4. Immunostaining is carried out with a commercially available biotinylated secondary antibody (see Table I) and Vectastain Elite ABC kit

---

[38] U. K. Laemmli, *Nature* (*London*) **227,** 680 (1970).
[39] H. Towbin, T. Staehelin, and J. Gordon, *Proc. Natl. Acad. Sci. U.S.A.* **76,** 4350 (1979).

(Vector Laboratory, Burlingame, CA), as described in the manual provided by the manufacturer. A 1 : 500 dilution of H.46 and a 1 : 1000 dilution (protein concentration, approximately 1 μg/ml) of kdn3G and kdn8kdn in TTBS (0.5% Tween-20 in Tris-buffered saline containing 50 mM Tris-HCl, pH 7.5, and 150 mM NaCl) are used. The blots are incubated at room temperature for 1 hr. The MAb 735 is diluted 1 : 1000 in TTBS. The blots are incubated at room temperature for 3 hr.

Dot-Blot Assay. A variation of the Western blot procedure described above is used for the dot-blot determination of polySia. A solution consisting of 2% SDS and 5% 2-mercaptoethanol(2-ME) is added to the samples after Endo N digestion (1 : 3, v/v). The sample is incubated at 37° for 1 hr, then centrifuged at 8000 g for 2 min to remove any particulate material. This sample is then serially diluted at 25° in 0.5% SDS containing 1.25% 2-ME and spotted directly on a Zeta-Probe membrane. Immunodetection is then carried out as described above.

Trifluoroacetic Acid Hydrolysis Control. A specific exo- or endo-KDNase has not been reported. Therefore, to confirm that the immunoreactivity observed with the anti-KDN antibodies is specific for KDN-containing glycoconjugates, it is important to show that immunoreactivity is lost when KDN is released from their polymers by mild acid hydrolysis. KDN-containing glycoconjugates are not hydrolyzed by Endo N[39a] and are relatively resistant to the commercially available sialidases. The procedure we have developed, using trifluoroacetic acid (TFA), can also be used to release oligo/polyNeu5Ac and Neu5Gc residues from their polymers.

1. To 20 μl of homogenate, 6.7 μl of 0.4 M TFA is added (final concentration of TFA is 0.1 M). The hydrolysis is carried out at 80° for 30 min. The sample is then neutralized by adding 3.3 μl of 0.8 M NaOH; 2% SDS/5% 2-ME is added to the sample (1 : 3, v/v) and incubated at 37° for 1 hr. The sample is then centrifuged at 8000 g for 2 min at 25° to remove any particulate material. The supernatant is transferred to a new tube, and used in the dot blot.

2. Alternatively, the sample is denatured first with 2% SDS/5% 2-ME at 37° for 1 hr, centrifuged as above, and the supernatant spotted directly on the Zeta-Probe membrane. The membrane is then fully saturated with 0.1 M TFA by immersing it in 0.1 M TFA at room temperature.

3. The TFA-saturated membrane is then placed between two pieces of glass and incubated in an 80° oven for 1 hr. Immunostaining is then carried out as described above.

[39a] S. Inoue, M. Iwasaki, K. Kitajima, A. Kanamori, S. Kitazume, and Y. Inoue, *in* "Polysialic Acid: From Microbes to Man" (J. Roth, U. Rutishauser, and F. A. Troy, eds.), p. 183. Birkhaueser, Basel, 1993.

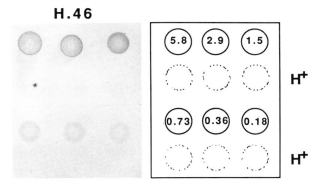

FIG. 1. Immunodetection of polySia by dot-blot analysis. Embryonic chick brain homogenate was dot blotted, using the anti-polySia antibody H.46. The numbers circled on the diagram to the right of the H.46 filter are the amounts of proteins (in micrograms) in the brain homogenate. The H^+ noted on the right-hand side of the dotted circles indicates the controls in which the homogenates containing the polySia antigens were first hydrolyzed with 0.1 M trifluoroacetic acid at 80° for 1 hr before immunoblotting with H.46.

Results

Results from three representative immunodetection experiments are given below.

Dot-Blot Assay for Detecting Poly(-8Neu5Acα2-) Residues. Dot-blot assays were carried out on detergent-solubilized, 19-day-old embryonic chick brains to test the sensitivity of this method for detecting poly (-8Neu5Acα2-) residues. As shown in Fig. 1, H.46 can detect polySia in as little as 0.18 μg of chick brain protein. As demonstrated previously, this immunoreactivity is sensitive to pretreatment with Endo N, thus confirming the presence of poly(-8Neu5Acα2-) chains in chick membrane.[12] The results in Fig. 1 also show that the H.46 immunoreactivity is sensitive to mild acid hydrolysis with 0.1 M TFA.

Dot-Blot Assay for Detecting OligoKDN and KDNα2,3Gal Residues. The dot-blot assay was also used to test the sensitivity of the two KDN-specific antibodies, kdn3G and kdn8kdn. When tested against the purified KDN-containing glycoprotein, KDN-gp, both the kdn3G and kdn8kdn antibodies could detect the KDNα2,3Gal and (-8KDNα2-)$_n$ epitopes in as little as 0.78 and 0.1 ng protein, respectively (Fig. 2). This immunoreactivity is also sensitive to 0.1 M TFA hydrolysis, a result consistent with the expectation that these antibodies recognize the KDN-containing epitopes in this glycoconjugate.

Under the conditions of the immunoblot assay described above, the extent of immunoreactivity is a semiquantitative measure of the amount

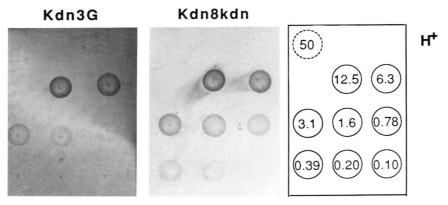

FIG. 2. Immunodetection of oligoKDN and KDNα2,3Galβ1- structures by dot-blot analysis. The KDN-containing antigen applied to the two filters denoted kdn3G and kdn8kdn was KDN-gp. The numbers circled on the right of the two filters are the amounts of KDN-gp (in nanograms). kdn3G and kdn8kdn were used as primary antibodies. The H+ noted on the right-hand side of the dotted circles indicates that the antigens were first hydrolyzed with 0.1 M trifluoroacetic acid at 80° for 1 hr before immunoblotting with kdn3G and kdn8kdn.

of polySia present. This follows because limited quantities of antigen are applied to the Zeta-Probe membrane, resulting in an excess of H.46, kdn3G, and kdn8kdn antibodies.

Western Blot Analysis for Detecting Poly(-8Neu5Acα2-) Chains. The anti-polySia antibody H.46 can detect polySia residues expressed on embryonic chick brain cells by Western blotting (Fig. 3). As shown in lane 1 (Fig. 3), the major band immunoreactivity migrates in SDS-PAGE with an apparent molecular mass of ca. 180–250 kDa, the molecular mass expected for the embryonic (polysialylated) form of N-CAM.[40] The immunoreactivity is abolished by prior treatment of the brain homogenate with Endo N (lane 2, Fig. 3), thus confirming that H.46 recognizes the poly (-8Neu5Acα2-) epitope in N-CAM.[12]

Procedures to Confirm Presence of Polysialic Acid and OligoKDN Residues in Glycoconjugates

Background

Three additional experiments can be carried out to confirm the presence of polySia in glycoconjugates. First, studies should be carried out

[40] J. B. Rothbard, R. Brackenbury, B. A. Cunningham, and G. M. Edelman, *J. Biol. Chem.* **257,** 11064 (1982).

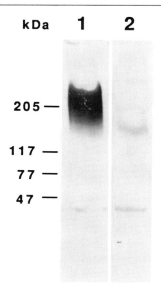

FIG. 3. Immunodetection of polySia in embryonic chick brain by Western blot analysis. Embryonic chick brain homogenate was immunoblotted after transfer with H.46 (lane 1) The sample in lane 2 was pretreated with Endo N before SDS-PAGE (control). The amount of protein in the brain homogenate was about 200 μg. As noted, immunoreactivity was associated with a polydispersed band at about 180–250 kDa, which is the expected molecular mass for embryonic N-CAM.

to establish that concomitant with the loss of anti-polySia immunoreactivity by Endo N is the formation of sialyl oligomers.[12] These oligomers can be fractionated by anion-exchange chromatography and characterized further by exosialidase digestion, and other structural methods.[7,12] Second, the tissue that contains the presumed polysialylated glycoconjugate can be assayed to determine if it has endogenous polyST activity, that is, an activity that can incorporate [14C]Neu5Ac from CMP[14C]Neu5Ac into sialyl polymers that are sensitive to Endo N.[7] Third, the presumed polysialylated glycoconjugate can be tested to determine if it can function as an exogenous acceptor in the *E. coli* K1 polyST reaction.[7,12] Polysialic acid synthesis catalyzed by the *E. coli* K1 polyST can be stimulated by addition of exogenous acceptors containing α-2,8-linked oligo- or polySia to the enzyme complex.[13,41] Because of the specificity and sensitivity of this reaction, the *E. coli* K1 polyST can be used to prove the presence of preexisting α-2,8-linked polySia chains in glycoconjugates.[12]

[41] C. Whitfield, D. A. Adams, and F. A. Troy, *J. Biol. Chem.* **259**, 12769 (1984).

The methods for carrying out these three types of experiments are described below. Greater details have been published in an earlier volume of this series.[42,43] Most of the procedures described were developed with fetal rat and embryonic chick brain, and they work well to identify polySia in tumor cells.[1] Changes may be required, however, to adapt the methods to other sources of biological samples.

Procedure 1: Endo N Treatment of Brain Membranes to Release Sialyl Oligomers

Brains are homogenized in ice-cold phosphate-buffered saline (PBS) supplemented with protease inhibitors, as described earlier. Membrane fractions are isolated by differential centrifugation[44] and resuspended to 20–40 mg of protein per milliliter in 10 mM Tris-HCl, pH 7.6. Endo N is added and the membranes are incubated at 33° for 2 hr. Membranes are then centrifuged at 145,000 g for 2 hr at 4°. The supernatant fraction containing sialyl oligomers is lyophilized, redissolved in water, and applied to a DEAE-Sephadex A-25 column (0.9 × 30 cm). Sialyl oligomers are eluted with 150 ml of a 0–0.4 M NaCl gradient in 10 mM Tris-HCl, pH 7.6. Alternatively, and sialyl oligomers can be resolved by high-performance liquid chromatography (HPLC).[45] The column fractions are subjected to acid hydrolysis in 0.05 M H$_2$SO$_4$ for 4 hr at 80° or 0.1 M TFA at 80° for 1 hr. The amount of Sia present is then determined by the thiobarbituric acid (TBA) procedure.[46] Results obtained from this type of experiments have shown that greater than 80% of the total Sia acid measured by TBA elutes from the DEAE column with a DP of 3–5 Sia residues.[7,12] These results show that the H.46-specific, Endo N-sensitive brain material consists of multimers of Sia. The oligomeric nature of this material is established by showing that the oligomers are sensitive to exosialidase, after reduction of the oligomers with KB^{3}H$_4$.[12]

Procedure 2: Polysialylated Glycoconjugate as Exogenous Acceptor in Escherichia coli K1 PolyST Assay

Escherichia coli K1 polyST is prepared from a K12-K1 wild-type strain (RHM18) or a hybrid mutant strain, EV11.[33] Cells are grown in trypticase soy broth (BBL Microbiology Systems, Cockeysville, MD) at 33° on a

[42] F. A. Troy, P. C. Hallenbeck, R. D. McCoy, and E. R. Vimr, this series, Vol. 138, p. 169.
[43] R. D. McCoy and F. A. Troy, this series, Vol. 138, p. 627.
[44] D. J. Morre, *in* "Molecular Techniques and Approaches in Developmental Biology" (M. J. Chrispeels, eds.), p. 1. Wiley (Interscience), New York, 1973.
[45] P. C. Hallenbeck, F. Yu, and F. A. Troy, *Anal. Biochem.* **161,** 181 (1987).
[46] L. Warren, *J. Biol. Chem.* **234,** 1971 (1959).

rotary shaker and harvested in the late logarithmic phase of growth by centrifugation at 4000 g. After washing in TM buffer [50 mM Tris-HCl (pH 8.0), 25 mM MgCl$_2$],[23] the cells are resuspended (30% cells by volume) in TMD buffer (1 mM DTT in TM) and disrupted in an Aminco French press cell (American Instrument, Silver Spring, MD) at 7000 psi. Disruption of cells in a French press gives rise to a population of sealed vesicles that are composed of about 85–90% inside-out vesicles (IOVs). DNase is added to a final concentration of 0.5 mg/ml and the vesicles are incubated on ice for 10 min. Any unbroken cells are removed by centrifugation at 4000 g. The supernatant and "fluffy pellet" are transferred to a new tube, and the IOVs are recovered as a pellet after ultracentrifugation at 50,000 g for 1 hr at 4° in a Beckman (Fullerton, CA) 60Ti rotor. Incorporation of [^{14}C]Neu5Ac from CMP[^{14}C]Neu5Ac into polySia is carried out in 300-μl incubation mixtures consisting of about 200 μg of membrane protein (as IOVs) and 20 nmol of CMP[^{14}C]Neu5Ac (500 dpm/pmol) in TMD buffer, as described.[23,47]

Presumed exogenous sialyl acceptors can be added to the above incubation mixture. If they contain oligo- or polySia in α-2,8-ketosidic linkages, they will stimulate sialyl polymer synthesis. As a control, colominic acid is added as an exogenous acceptor.[13] Colominic acid consists of a population of sialyl oligomers with a DP ranging from about 3 to 40. Fetal rat brain membrane can also function as an exogenous acceptor, whereas rat liver cannot.[12] For use as exogenous acceptors, brain membranes are prepared as described in procedure 1 (above). The pH optimum for the poly-α2,8-sialyltransferase activity in fetal rat brains is 6.0.[7] The $E.\ coli$ K1 polyST assay is carried out at pH 8.0. Therefore, there is no contribution of polyST activity from fetal rat brains in the $E.\ coli$ K1 polyST assay. In testing other cells or tissues for exogenous acceptor activity, however, a control for endogenous polyST activity should be included.

Procedure 3: Endogenous polyST Activity in PolySia-Expressing Fetal Brains

The polyST assay as originally described for fetal rat brains[7] is carried out as follows. Brains are homogenized in ice-cold MEG buffer [50 mM morpholineethanesulfonic acid (MES), pH 6.0, containing 10% (v/v) glycerol]. Protease inhibitors are supplemented as described above. DNase I is added to a final concentration of 0.5 mg/ml and incubated on ice for 10 min. A standard incubation mixture contains the following: 2 to 5 mg protein (homogenate), 200 nmol of CMP[^{14}C]Neu5Ac (Du Pont Company, New England Nuclear Research Products, Boston, MA). The specific

---

[47] F. A. Troy, I. K. Vijay, M. A. McCloskey, and T. E. Rohr, this series, Vol. 83, p. 540.

activity of CMP[^{14}C]Neu5Ac is adjusted to 1.2×10^4 dpm/nmol by the addition of unlabeled CMPNeu5Ac. MEG buffer is added so that the final volume is 350 μl. The reaction is run at 33° for a 20-hr period. At selected time intervals, 30-μl aliquots are removed and spotted on Whatman (Clifton, NJ) 3MM paper. At the end, the remaining incubation mixture is ultracentrifuged to separate CMP[^{14}C]Neu5Ac from the membrane fraction. The membranes are resuspended in 20 mM Tris, pH 7.4, and divided into two aliquots. Endo N, or an equal volume of 20 mM Tris-HCl, pH 7.4, is added. The Endo N reaction is run for 1–3 hr and aliquots are spotted on Whatman 3MM paper. The chromatography papers are subject to descending chromatography in ethanol–1 M ammonium acetate, pH 7.5 (7 : 3, v/v). Radioactivity remaining at the origin is quantitated by liquid scintillation counting.[7]

Chemical Methods to Identify and Study Polysialic Acid and OligoKDN-Containing Glycoconjugates

Mild Acid Hydrolysis

In the detection of polySia, considerable advantage is taken of differences in the rates of hydrolysis of the different ketosidic linkages to effect relatively selective degradations, thus leading to the isolation of sialyl oligomers whose structures can be determined with certainty.[17] The following conditions for the controlled acid hydrolysis of polySia chains permit liberation of sialyl oligomers with negligible amounts of monomers.

1. Incubation of rainbow trout PSGP for 3 days in sodium acetate buffer (pH 4.8) at 37° results in 60% release of Neu5Gc from the core oligosaccharide chains. Ninety percent of the released Neu5Gc is recovered as oligo(Neu5Gc).[17,48]

2. Milder acidic conditions (pH 5.6, 37°, 24 hr or pH 5.6, 80°, 1–3 hr) increases the recovery of the higher sialyl oligomers, both from PSGP and colominic acid.[48]

3. Mild acid hydrolysis (pH 6.0) is recommended to obtain a series of oligoKDN from KDN-gp.[48] Trifluoroacetic acid is recommended for the complete hydrolysis of polySia and oligo/polyKDN-containing glycoconjugates. This is best carried out by the addition of 200 μl of 0.1 M TFA to a dried sample containing 5–200 μg of Sia or KDN. The sample is heated at 80° for varying periods of time (up to 3 hr). The liberated Sia or KDN is quantitated by a modification[48] of the original TBA method of Aminoff.[49]

[48] S. Kitazume, K. Kitajima, S. Inoue, and Y. Inoue, *Anal. Biochem.* **202,** 25 (1992).
[49] D. Aminoff, *Biochem. J.* **81,** 384 (1961).

Colorimetric Reactions for Identification of KDN in OligoKDN

Both free and bound KDN exhibit color in the TBA reaction without prior hydrolysis (hereafter referred to as the "direct" TBA reaction).[50] A vicinal hydroxyl group is present only in the nonreducing terminal residue of α-2,8-linked polyNeu5Ac or polyNeu5Gc chains. By contrast, a periodate-sensitive vicinal hydroxyl group is present in every internal KDN residue, that is, HO-C_4-C_5-OH, in α-2,8-linked oligoKDN chains. Because 50 μg of KDN monomer gives no color in the resorcinol or orcinol reaction, it can be readily differentiated from Neu5Ac or Neu5Gc.[50] The chromophore formed from KDN has nearly the same absorption maximum at 552 nm in the mixed solvent of water and 2-methoxyethanol as Neu5Ac and Neu5Gc, with a molar extinction coefficient of 9.1×10^{-7} cm²/mol) (SD $\pm$ 0.4) [cf. Neu5Ac ($8.5 \pm 0.2 \times 10^{-7}$ cm²/mol) and Neu5Gc ($4.9 \pm 0.3 \times 10^{-7}$ cm²/mol).[50] No quantitative data are available regarding how O-acetylation of KDN affects the molar extinction coefficients. KDN can be acetylated at C_4 and C_{7-9}.[51] The chromogen involved in the TBA reaction is believed to be 3-formylpyruvate, which reacts with 2 mol of TBA to yield the chromophore with the absorption maximum at 552 nm. The color yield obtained from oligoKDN chains by the direct TBA reaction is 80% or more of the color yield obtained after mild acid hydrolysis of oligoKDN. Because no other colorimetric method is available to estimate the concentration of oligoKDN without prior hydrolysis, estimation by the direct TBA method is sufficient for ordinary purposes.[50]

A problem encountered in studies of the kinetics of nonenzymatic hydrolysis of KDN-glycans, or in the assay of enzymatic hydrolysis of KDN-glycans in searching for specific glycohydrolase(s), is the determination of free KDN (product of reaction) in the presence of ketosidically bound KDN (substrate). To circumvent this problem, quantitative analytical methods involving the use of enzymes that will react specifically only with free KDN, or pyruvate therefrom, are needed. No exo- or endo-KDNases have been reported, and KDN-glycoconjugates are relatively resistant to hydrolysis by the commercially available sialidases.

Mild Methanolysis and Gas-Liquid Chromatography Analysis of Polynonulosonates Containing Neu5Ac, Neu5Gc, and KDN

Use of gas-liquid chromatography (GLC) for analysis of polySia requires conversion of each of the highly polar, ionic residues Neu5Ac, Neu5Gc, and KDN to nonionic derivatives of sufficient volatility. The polySia is first converted to the respective methyl ester methylketosides

---

[50] K. Kitajima, S. Inoue, S. Kitazume, and Y. Inoue, *Anal. Biochem.* **205**, 244 (1992).
[51] M. Iwasaki, S. Inoue, and F. A. Troy, *J. Biol. Chem.* **265**, 2596 (1990).

(α and β forms) by methanolysis, followed by derivatization of all free hydroxyl groups to trimethylsilyl (TMS) ethers by reaction with a mixture of pyridine–trimethylchlorosilane–hexamethyldisilazane. The following procedure is recommended.

1. Methanolic HCl (50 mM) (0.5 ml) is added to dried samples containing 10–200 μg of Sia or KDN-containing glycoconjugates. Five to 10 μg of *myo*-inositol is added as an internal standard.

2. After heating at 80° for 1 hr, the methyl ester methylketosides of Sia and KDN are converted to the TMS derivatives, as described by Yu and Ledeen.[52]

3. The TMS derivatives are analyzed on a glass column (3 mm × 1 m) containing either 1.5% OV-17 or 2% OV-101 on Chromosorb W (John Manville Co., New York, NY) or Gas Chrom Q (Applied Science Co., Tampa, FL) with a temperature gradient from 130 to 230°. The peak of the KDN derivative appears before the peak of the Neu5Ac derivative.[50]

Trifluoroacetic Acid Hydrolysis prior to Mild Methanolysis for
   Quantitative Determination of Sialic Acid and
   KDN in Glycoconjugates

The conditions required to obtain the maximum yield of nonulosonate monomers from the polynonulosonates without cleaving N-acyl groups were determined.[50] The conditions recommended previously by Yu and Ledeen[52] for the differential quantitation of nonreducing terminal Sia residues (0.05 N methanolic HCl, 80°, 1 hr) were found to cleave only 3 and 16% of the ketosidic linkages in poly(Neu5Ac) (colominic acid) and rainbow trout PSGP [oligo/poly(Neu5Gc)], respectively.[50] Extending the methanolysis for up to 4 hr resulted in an increase in these values to 7 and 30%, respectively, yields still too low for quantitative analysis.[50] Therefore we determined that mild acid hydrolysis (0.1 N TFA, 80°, 3 hr) followed by mild methanolysis (0.05 N methanolic HCl, 80°, 1 hr) were optimal conditions for cleaving the interketosidic linkages in polySia.[50] Under these conditions, the mild TFA hydrolysis is effective in cleaving ketosidic linkages while preserving the identity of the N-acyl groups. Methanolysis is necessary to convert the Sia/KDN monomers to their methyl ester methylketosides. This procedure has proved useful for poly(Neu5Ac), poly(Neu5Gc), and fish egg PSGP. Under these conditions of mild hydrolysis and mild methanolysis (MH-MM), 80% of the Sia residues in fish egg PSGP were converted to the monomers. In the differential determination of individual Sia residues in polySia containing both

---

[52] R. K. Yu and R. W. Ledeen, *J. Lipid Res.* **11**, 506 (1970).

Neu5Ac and Neu5Gc, the MH-MM method is clearly superior to the conventional MM method.[50]

In contrast to poly(Neu5Ac) and poly(Neu5Gc), the KDN residues in oligo/polyKDN are liberated in high yield (98%) by mild methanolysis (0.05 N methanolic HCl–methanol, 80° for 3 hr). The corresponding value obtained after a 1-hr methanolysis is only about 60% of that obtained by the MH-MM method.[50] A high yield of KDN monomer (94%) is also attained by using the standard conditions of methanolysis generally used in carbohydrate compositional analysis of glycoproteins (0.5 N methanolic HCl, 65°, 16 hr). These results are expected because of differences in the rate of hydrolysis between N-acylneuraminates and KDN polymers that result from differences in the acid lability of the ketosidic linkages. KDN is also more stable in acid because it lacks an N-acyl group.

Detection of Oligo(Neu5Ac), Oligo(Neu5Gc), and OligoKDN Liberated by
  Partial Acid Hydrolysis of Polysialic Acid

Thin-Layer Chromatography

Polysialic acid chains in glycoconjugates can be determined by analyzing oligoSia formed by controlled acid hydrolysis. These products can be analyzed by thin-layer chromatography (TLC) on silica gel plates in 1-propanol–25% aqueous ammonia–water (6 : 1 : 2.5, v/v), as described above. Oligo(Neu5Ac), oligo(Neu5Gc), and oligoKDN are separated, based on their DP, and on differences in the substituent at C-5. Because detection by H_2SO_4 or resorcinol is relatively insensitive, the method can require as much as 1 mg of polySia-containing glycoconjugates and 2.5 mg of KDN polymers, depending on the extent of sialylation.

High-Performance Liquid Chromatography

Adsorption-partition chromatography with HPLC is effective in separating Sia monomers and oligo(Neu5Ac), oligo(Neu5Gc), and oligoKDN with a DP < 7.[48] Like TLC, HPLC separates the oligomers on the basis of chain length and the nature of the substituent at C-5.[48] Oligomers can be separated on a Jasco HPLC (Tokyo, Japan), using a Shodex RS pak (150 × 6 mm i.d.) polystyrene cation-exchange resin gel (DC-613; Showa, Denko, Tokyo). Samples in 1–10 μl are injected and eluted isocratically with a 1 : 2 to 1 : 3 mixture of 0.02–0.025 M sodium phosphate buffer (pH 7.4) and acetonitrile at 35°. The flow rate is 1.0 ml/min and the column is run under pressure at 11–14 kg/cm². The elution pattern is monitored by ultraviolet (UV) absorption below 220 nm. Ten to 100 ng of Neu5Ac and

Neu5Gc and 100 ng to 1 mg of KDN are detectable. The O-acetyl derivatives are also separated from the parent oligoSia by this procedure.[48] The
HPLC method is superior to TLC in the lower limit of detection, but the
detection lacks specificity.

Anion-Exchange Chromatography

Anion-exchange chromatography can be used to separate the oligoSia
chains both preparatively and on a microanalytical scale.[45,48] Anion-exchange HPLC on Mono-Q HR5/5 of ^{3}H-labeled sugar alcohols derived
from oligoSia may be the most sensitive chemical method for analysis of
oligoSia formed by controlled acid hydrolysis or Endo N depolymerization
of poly(Neu5Ac) and poly(Neu5Gc). The elution of oligoKDN is retarded
compared with that of oligo(Neu5Ac) or oligo(Neu5Gc) of the same DP.[48]

Preparation of Reagents Specific for Identification of Polysialic
Acid in Glycoconjugates

Although several commercial firms have expressed an interest in providing the anti-polySia antibodies and Endo N, these reagents are not
presently available. Procedures for the preparation and assay of *E. coli*
K1F Endo N are described below. Also described are methods to prepare
radiolabeled polySia.

Procedures for Preparing Radiolabeled Oligo- and PolySia

Unlabeled, [3H]- or [U-14C]polySia can be isolated from *E. coli* K1
culture filtrates by a modification of our previous method.[13,47] Cells are
grown in 1 liter of modified M9 medium containing the following components (grams per liter): Na_2HPO_4, 6; KH_2PO_4, 3; NaCl, 0.5; NH_4Cl, 1;
casamino acids, 4; glucose, 4. After autoclaving, 10 ml of 10 mM $CaCl_2$
and 1 ml of 1 M $MgSO_4$ are added. For preparing radiolabeled polymers,
1 mCi of [U-14C]glucose (or 2 mCi of [3H]glucose) in 1.0 ml of 90% ethanol
is added directly to the sterile medium. Cells are grown at 37° on a rotary
shaker. The polySia capsular polymers are removed from the surface of
late log-phase cells by treating the culture in a Polytron model PCU-2
(three to five times, for 40 sec each). The acapsulated cells are removed
by centrifugation (16,000 g for 25 min at 4°) and the polymers precipitated
by cetyltrimethylammonium bromide [CETAB, 0.3% (v/v) final concentra-
tion]. After 24 hr at 4°, the polySia is extracted from the CETAB complex with 1 M $CaCl_2$ and fractionally purified by ethanol precipitation.[53]

[53] W. F. Vann and K. Jann, *Infect. Immun.* **25**, 85 (1979).

[^{14}C]Polysialic acid at this stage of purification is sufficiently pure for use as a substrate in the Endo N assays described below, and may be further purified by published procedures.[54,55] Alternatively, *in vitro* synthesized [^{3}H]- or [^{14}C]polySia, prepared as previously described,[23] may be used. Unlabeled sialyl oligomers are commercially available (colominic acid; Sigma Chemical Co., St. Louis, MO). These preparations contain variable degrees of sialyl oligomers. Sialyl oligomers can also be generated by either mild acid hydrolysis[13,50] or Endo N depolymerization of polySia.[28] Unlabeled colominic acid can be labeled either in the reducing terminus by reduction with KB^{3}H$_4$, as described previously,[13] or in the nonreducing terminus by sequential periodate oxidation and KB^{3}H$_4$ reduction.[23,56]

Purification of Endo N

A purification procedure for purifying Endo N to homogeneity was first published in 1987.[28,42] The procedure was later modified.[57] Interested readers should consult these original papers for details regarding preparation of the *E. coli* K1F phage lysate, and the purification procedures.

Method to Assay Endo N Activity

Qualitative Assay. A qualitative assay to measure Endo N activity that does not require radiolabeled substrate is to incubate the enzyme at various dilutions with colominic acid (50 μg; Sigma Chemical Co.) in 20 mM Tris-HCl, pH 7.4, at 37°. At chosen time intervals (up to 60 min) an aliquot containing up to about 10 μg of substrate (based on Sia) is spotted on silica gel TLC plates (Kiesel gel 60; Merck, Rahway, NJ). The plate is developed, ascending in 1-propanol–25% aqueous ammonia–water (6 : 1 : 2.5, v/v) for 8–10 hr, removed, and air dried. Sialyl oligomers with DPs up to at least 15 residues are readily separated, and if they have been depolymerized by Endo N appear as short oligomers (DP 2–4) and Sia. The sialyl oligomers and their depolymerized products are visualized by spraying the plate with orcinol in sulfuric acid or the resorcinol reagent, and heating at 80° until the spots develop (15–30 min).[48,58]

In principle, this assay can also be used to determine if a presumed polysialylated glycoconjugate contains polySia. Following incubation of

[54] E. C. Gotschlich, T.-Y. Liu, and M. S. Artenstein, *J. Exp. Med.* **129**, 1349 (1969).
[55] R. Schneerson, M. Bradshaw, J. K. Whisnant, R. C. Myerowitz, J. C. Parke, and J. B. Robbins, *J. Immunol.* **108**, 1551 (1972).
[56] L. van Lenten and G. Ashwell, *J. Biol. Chem.* **246**, 1889 (1971).
[57] E. R. Vimr, W. Aaronson, and R. P. Silver, *J. Bacteriol.* **171**, 1106 (1989).
[58] R. Schauer, *in* "Methods in Enzymology" (V. Ginsburg, ed.), Vol. 50, p. 64. Academic Press, Orlando, FL, 1987.

the glycoconjugate with Endo N, the appearance of sialyl oligomers with the expected DP is strong presumptive evidence of polysialylation. When carried out kinetically the loss in appearance of higher sialyl oligomers (DP > 10–20) concomitant with the formation of smaller DPs (2–4) provides additional confirmatory evidence. Sensitivity in detecting sialyl oligomers by the orcinol or resorcinol reagent is a possible limitation of this method. If only small amounts of material are available, this limitation can be overcome, by first reducing the sialyl oligomers with $NaBT_4$ and autoradiography.[27] Alternatively, the free reducing groups could be derivatized by reductive amination with an ultrasensitive fluorophore, for example, CBQCA [3-(4-carboxylbenzoyl)-2-quinolinecarboxaldehyde].[59] The CBQCA derivatives can be detected at attomole levels.[59] Nonradioactive labeling of the reducing sialyl oligomers could also be carried out with PMPMP [1-(p-methoxy)phenyl-3-methyl-5-pyrazolone], as described by Lee and colleagues.[60] The sensitivity of detection by UV with this method is in the subnanomole range.

Quantitative Assay. The reaction is followed by quantitating the loss of radioactivity from [^{14}C]polySia (substrate depletion) or by measuring the amount of product ([^{14}C]sialyl oligomers) formed (product formation). The details of this assay have been published.[42]

1. *Assay of Endo N by substrate depletion:* Endo N activity can be determined by measuring the depolymerization of radiolabeled polySia. Standard incubation mixtures (25 μl) containing 67 μg of [^{14}C]polySia (~92 × 10^3 dpm/mg polymer) in 20 mM Tris-HCl, pH 7.4, are incubated at 37° with varying dilutions of enzyme. After 15 min, 20 μl is spotted on Whatman 3MM paper and chromatographed in ethanol–1 M ammonium acetate, pH 7.5 (7:3, v/v).[47] [^{14}C]Polysialic acid that remains at the origin is quantitated by scintillation counting. Endo N activity is calculated from the highest dilution of enzyme that gives at least a 20% loss in radioactivity from the substrate (radioactivity that remains at the origin). In principle, this method can be used to measure Endo N activity by quantitating oligoSia formation but this requires radiochromatographic scanning or the tedium of cutting and counting multiple strips. Moreover, this method does not yield initial reaction rates because Endo N preferentially depolymerizes higher molecular weight polymers, and the chromatography assay favors mobility of limit digestion products. For this reason, substrate depletion underestimates Endo N activity. In this assay, Endo N activity is expressed as a function of ^{14}C-labeled sialyl oligomers released from

[59] J. P. Liu, O. Shirota, D. Wiesler, and M. Novotny, *Proc. Natl. Acad. Sci. U.S.A.* **88,** 2302 (1991).
[60] K. Kakehi, S. Suzuki, S. Honda, and Y. C. Lee, *Anal. Biochem.* **199,** 256 (1991).

the polymer. One unit is defined as the amount of enzyme required to mobilize 50,000 cpm of sialyl oligomers per minute. This is equivalent to about 1 μmol of sialic acid reducing terminus released per minute.

2. *Assay of Endo N activity by product formation:* Alternatively, Endo N activity can be determined by measuring the number of α-2,8 linkages hydrolyzed. Because each cleavage results in the formation of a new reducing terminus, which in the acyclic form is sensitive to periodate oxidation between C-6 and C-7, Endo N activity can be quantitated by the thiobarbituric acid (TBA) method.[61] The 2-acetamido-4-deoxyhexos-5-uluronic acid resulting from periodate oxidation readily undergoes an aldol cleavage to form β-formyl pyruvate, which condenses with TBA.[62] An advantage of this method is that it can be carried out with unlabeled sialyl polymers. Controls (no enzyme) are essential to determine the background of free reducing ends in the undegraded substrate, which is a more serious problem the smaller the DP of the substrate. This background can be decreased by prior reduction of the sialyl oligomers with borohydride.[23] After reduction, the oligomers no longer react with TBA. A potential disadvantage of this method for following Endo N activity through the purification scheme is interference in formation of the TBA chromophore, particularly by salts, for example, $(NH_4)_2SO_4$.

Standard 1.0-ml incubation mixtures containing 2 to 6 mg of oligo- or polySia in 20 mM Tris-HCl, pH 7.4, are incubated at 37° with appropriately diluted Endo N solutions. The reaction is terminated after 1.0, 2.5, 5.0, 7.5, 10, and 15 min by the addition of ethanol to 50% final concentration. The number of reducing ends is quantitated by the TBA assay, omitting the initial acid hydrolysis step. The reaction is linear for at least 10 min at 0.2–4 mg of polySia/ml and 2 milliunits (mU) of enzyme. For determining initial kinetics, the $OD_{540 \, nm}$ characteristically changes from ~0.02 to >0.3–0.6. Under these conditions, less than 10% of the polySia linkages are cleaved. One unit of Endo N forms 1 μmol of product (as TBA-reactive material) per minute per milligram protein at 37°.

Acknowledgments

We would like to thank the following people for helpful suggestions and collaboration regarding this work: Akiko Kanamori, Yu Song, Shinobu Kitazume, Takaho Terada, Chihiro Sato, Takashi Angata, Mariko Kudo, Kim Deckard Janatpour, and Jin-Won Cho. This research was supported in parts by Research Grant AI-09352-21 from the United States Public Health Service, National Institutes of Health (F.A.T.), and by Grants-in-Aid for

[61] L. Skoza and S. Mohos, *Biochem. J.* **159**, 457 (1976).
[62] R. Kuhn and P. Lutz, *Biochem. Zeitschrift* **338**, 554 (1963).

Developmental Scientific Research (02558014), for General Scientific Research (04453160), and for the International Scientific Research Program: Joint Research (04044055) from the Ministry of Education, Science, and Culture of Japan, and by a fund from the Mitsubishi Foundation (to Y.I.). The expert editorial assistance of Ann Marie Godwin in preparing this manuscript is gratefully acknowledged.

[28] Neoglycolipids: Probes of Oligosaccharide Structure, Antigenicity, and Function

By TEN FEIZI, MARK S. STOLL, CHUN-TING YUEN, WENGANG CHAI, and ALEXANDER M. LAWSON

Introduction

Understanding the roles of oligosaccharides N- and O-glycosidically linked to protein, and of the diverse sequences in their core, backbone, and peripheral domains, has been one of the challenges in modern cell biology.[1] The neoglycolipid technology as described in this chapter was introduced[2] to address the need for a micromethod to evaluate individual N- and O-linked oligosaccharide species released from glycoproteins and proteoglycans as antigens and as ligands for carbohydrate-binding proteins. There is a pressing need for such a technology because procedures commonly used to assess roles of oligosaccharides of glyco-proteins as recognition structures are for the greater part indirect: deglycosylation by chemical or enzymatic means, inhibition or modifica-tion of glycosylation during biosynthesis, site-directed mutagenesis to delete specific glycosylation sites, or inhibition with high concentrations of monosaccharides. However, for precise assignments of specificities, it is necessary to perform quantitative inhibition experiments with a series of structurally defined oligosaccharides rather than monosaccha-rides. This approach has been used with success to define the combining specificities of antibodies to the blood group antigens and related carbohydrate differentiation antigens[1,3,4] (and references therein). This is not always feasible, however, because oligosaccharides derived from glycoproteins are frequently heterogeneous and difficult to purify, and

[1] T. Feizi, *Nature (London)* **314,** 53 (1985).

[2] P. W. Tang, H. C. Gooi, M. Hardy, Y. C. Lee, and T. Feizi, *Biochem. Biophys. Res. Commun.* **132,** 474 (1985).

[3] W. M. Watkins, *in* "Advances in Human Genetics" (H. Harris and K. Hirschhorn, eds.), pp. 1 and 379. Plenum, New York, 1980.

[4] E. A. Kabat, *Am. J. Clin. Pathol.* **78,** 281 (1982).

once removed from protein the amounts required for inhibition are generally substantially greater (up to 10,000 times greater with *O*-linked oligosaccharides derived from mucin-type glycoproteins) than the amounts on the proteins,[5] due predominantly to the loss of the cooperative effects of multivalence. Mono- and oligosaccharides may be rendered multivalent by a variety of conjugation procedures,[6] for example, by conjugating to proteins (neoglycoproteins). Neoglycoproteins have found considerable use in studies of carbohydrate-binding proteins and monoclonal antibodies and are available as commercial products. However, this approach is unsuitable for assigning binding specificities to individual components within mixtures of oligosaccharides.

The special feature of the neoglycolipid technology is that it is applicable not only to mixtures of oligosaccharides released from proteins by established procedures [hydrazinolysis (N-linked chains) or alkaline borohydride degradation (O-linked chains) or by various partial degradation procedures, including endoglycosidase digestions] but is also applicable to any desired oligosaccharide sequence, be it natural or synthetic. By a microconjugation procedure each oligosaccharide (within mixtures or individually) is joined to a lipid molecule. The resulting neoglycolipids[2,7] serve as immobilized probes on silica gel chromatograms, on plastic plates, or as microparticulate probes after incorporation into liposomes, and they are readily amenable to assay procedures established for glycosphingolipids.[8–14] In the form of neoglycolipids, the oligosaccharides are presented at the matrix surface in a clustered state. This is an important requirement for interactions with carbohydrate-binding proteins, including antibodies, which frequently have low affinities and give no detectable binding to monovalent carbohydrate ligands under conventional binding assay conditions. With the oligosaccharides displayed in a clustered state, avidities of binding are increased because of the cooperative effects of multivalence.

The lipid most extensively used for generating neoglycolipids has been

[5] E. Wood, E. F. Hounsell, and T. Feizi, *Carbohydr. Res.* **90**, 269 (1981).

[6] Y. C. Lee, *Ciba Found. Symp.* **145**, 80 (1989).

[7] P. W. Tang and T. Feizi, *Carbohydr. Res.* **161**, 133 (1987).

[8] J. Holmgren, *Infect. Immun.* **8**, 851 (1973).

[9] D. M. Marcus and G. A. Schwarting, *Adv. Immunol.* **23**, 203 (1976).

[10] W. W. Young, Jr., E. M. S. MacDonald, R. C. Nowinski, and S. Hakomori, *J. Exp. Med.* **150**, 1008 (1979).

[10a] E. Wood, J. Lecomte, R. A. Childs, and T. Feizi, *Mol. Immunol.* **16**, 813 (1979).

[11] H. Leffler and C. Svanborg Eden, *FEMS. Microbiol. Lett.* **8**, 127 (1980).

[12] J. L. Magnani, D. F. Smith, and V. Ginsburg, *Anal. Biochem.* **109**, 399 (1980).

[13] G. C. Hansson, K.-A. Karlsson, G. Larson, N. Strömberg, J. Thurin, C. Örvell, and E. Norrby, *FEBS Lett.* **170**, 15 (1984).

[14] L. M. Loomes, K.-I. Uemura, and T. Feizi, *Infect. Immun.* **47**, 15 (1985).

the aminophospholipid L-1,2-dipalmitoyl-*sn*-glycero-3-phosphoethanol-amine

$$CH_2OCO(CH_2)_{14}CH_3$$
$$CHOCO(CH_2)_{14}CH_3$$
$$CH_2OP(O)(OH)OCH_2CH_2NH_2$$

(abbreviated to PPEADP or DPPE; the latter abbreviation is used hereafter). The aldehyde group, present as a low concentration equilibrium form at the reducing end of the oligosaccharide, is joined[2] to the amino group of the lipid by reductive amination. For O-linked oligosaccharides that are released with alkaline borohydride and are in the reduced state (and thus do not contain reactive aldehydes), a mild periodate oxidation procedure was introduced[2] to generate reactive aldehydes prior to conjugation.

In its original form the procedure was applied with success to the following.

1. Individual or mixtures of O-linked oligosaccharides released from human fetal gastrointestinal mucins (meconium), and chemically synthesized oligosaccharides: Specific binding of a monoclonal antibody, anti-IMa, to its known trisaccharide epitope was shown.[2]

2. Free oligosaccharides isolated from milk: Specific binding of the monoclonal antibody directed to the stage-specific embryonic antigen (SSEA-1), and of the plant lectin *Ricinus communis* agglutinin to their respective oligosaccharide determinants was observed.[2]

3. N- and O-linked oligosaccharides released from human milk galacto-syltransferase: Blood group Le^a-active oligosaccharides were identified among both types of chain.[7]

4. Oligosaccharides obtained by endo-β-galactosidase treatment of bovine corneal keratan sulfate: Reciprocal expression of the keratan sulfate-specific antigens and the i antigen was shown on the native and desulfated oligosaccharides.[15]

Development of Neoglycolipid Technology

In the course of detailed analyses of reaction products under different conjugation conditions, it was observed that the water content in the reductive amination reaction mixture has a marked influence on the conju-

---

[15] P. W. Tang, P. Scudder, H. Mehmet, E. F. Hounsell, and T. Feizi, *Eur. J. Biochem.* **160**, 537 (1986).

gation rate of reducing oligosaccharides to the lipid molecule.[16] Because the conjugation step involves water elimination

$$R_1-NH_2 + CHO-R_2 \xrightarrow{\text{NaBH}_3\text{CN}} R_1-NH-CH_2-R_2 + H_2O$$

(where R_1-NH_2 is DPPE, and $CHO-R_2$ is reducing sugar), anhydrous conditions would be expected to enhance the reaction rate. Borch and colleagues[17] made use of this principle by including water-absorbent molecular sieves to accelerate conjugation reactions between highly hindered cyclic ketones, such as norbornanone and ammonia, or dimethylamine. Wiegandt and Ziegler[18] reported that successful conjugation of sialyllactose to stearylamine (50% yield) was achieved by refluxing in methanol for 2 hr. For conjugating dextran oligomers to stearylamine in water-containing reaction mixtures [25% (v/v) water in tetrahydrofuran] Wood and Kabat[19] observed that an incubation time of 3 weeks was required. Similarly, for conjugating oligosaccharide fragments from a bacterial polysaccharide to N-(1-deoxyalditol-1-yl)octadecylamine, in aqueous tetrahydrofuran, van Dam et al.[20] observed that incubation times of 5–14 days were required. Our studies[16] clearly showed that almost complete conjugation of di- to hexasaccharides to DPPE can be achieved in 16 hr at 50° by using an organic solvent, chloroform–methanol (1 : 1, v/v), in the absence of water. A possible additional factor in the enhanced conjugation rate with DPPE using anhydrous organic conditions is that the reductive amination of aliphatic amines is favored.[21] With longer oligosaccharides, however, a lack of solubility in chloroform–methanol means that the inclusion of some water or the use of alternative solvent systems and reaction conditions is required. Thus, a hepta- to decasaccharide mixture of high mannose-type N-linked chains could be successfully conjugated (about 80% yield) in the presence of water (5%, v/v) and by raising the incubation temperature to 60°. Traces of monoacyl products derived from DPPE and from the neoglycolipids are found under both the hydrous and anhydrous conjugation conditions. These by-products, on chromatography, yield R_f values about one-half of those in the parent compound. This is avoided if

[16] M. S. Stoll, T. Mizuochi, R. A. Childs, and T. Feizi, *Biochem. J.* **256,** 661 (1988).
[17] R. F. Borch, M. D. Bernstein, and H. D. Durst, *J. Am. Chem. Soc.* **93,** 2897 (1971).
[18] H. Wiegandt and W. Ziegler, *Hoppe-Seyler's Z. Physiol. Chem.* **355,** 11 (1974).
[19] C. Wood and E. A. Kabat, *J. Exp. Med.* **154,** 432 (1981).
[20] J. E. G. van Dam, A. A. M. Maas, J. P. Kamerling, and J. F. G. Vliegenthart, *Carbohydr. Res.* **187,** 25 (1989).
[21] E. Kallin, H. Lonn, and T. Norberg, *Glycoconjugate J.* **3,** 311 (1986).

an alternative lipid, L-1,2-dihexadecyl-*sn*-glycero-3-phosphoethanolamine (DHPE),[22,23] is used:

$$CH_2O(CH_2)_{15}CH_3$$
$$CHO(CH_2)_{15}CH_3$$
$$CH_2OP(O)(OH)OCH_2CH_2NH_2$$

As a result this lipid is being used increasingly for generating neoglyco-lipids.

In some of the original experiments, oligosaccharides were first re-solved on silica gel thin-layer plates and subjected to the lipid conjugation reaction *in situ*. Although strong antigenic activities were generated by this approach,[2,7] due to the clustered oligosaccharide presentation, the yield of lipid-linked oligosaccharides was only on the order of 5–10% (M. S. Stoll, unpublished observations, 1987). In solution derivatization, followed by chromatography of the reaction products,[16] is preferable be-cause of the high yields of derivative as discussed above. Moreover, by chromatography, the nonconjugated oligosaccharides and lipids are readily separated from the neoglycolipids.[16]

An important advance in the neoglycolipid technology has been with respect to structural analysis of the oligosaccharide moieties by mass spectrometry. Thus the technology originally designed for detecting oligo-saccharide ligands also constitutes a microsequencing strategy, for it was observed[24] that these lipid-linked oligosaccharides have exceptional ion-ization properties in mass spectrometry, and they can be detected with greater sensitivity than natural glycolipids, and far greater sensitivity than free oligosaccharides. The sensitivity of detection by liquid secondary ion mass spectrometry (LSIMS) of oligosaccharides taken through the conjugation and analysis procedure (as described under Procedure, below) is higher than that reported for any other oligosaccharide derivative in terms of information on molecular weight, despite the increase in mass created by the lipid moiety (see Ref. 24 and references cited therein). This may be attributed in part to the favorable surface activity of the neoglycolipids and the charged phosphate group. Thus, by an approach

[22] G. Pohlentz, S. Schlemm, and H. Egge, *Eur. J. Biochem.* **203**, 387 (1992).

[23] C.-T. Yuen, A. M. Lawson, W. Chai, M. Larkin, M. S. Stoll, A. C. Stuart, F. X. Sullivan, T. J. Ahern, and T. Feizi, *Biochemistry* **31**, 9126 (1992).

[23a] P. J. Green, T. Tamatani, T. Watanabe, M. Miyasaka, A. Hasegawa, M. Kiso, C.-T. Yuen, M. S. Stoll, and T. Feizi, *Biochem. Biophys. Res. Commun.* **188**, 244 (1992).

[24] A. M. Lawson, W. Chai, G. C. Cashmore, M. S. Stoll, E. F. Hounsell, and T. Feizi, *Carbohydr. Res.* **200**, 47 (1990).

previously applied to analyses of natural glycolipids,[25] molecular weight as well as monosaccharide sequence information can be obtained by LSIMS of the neoglycolipid components directly on the chromatogram surface. One to 200 pmol of the neoglycolipids on the chromatogram surface and subpicomole amounts on the target probe give informative spectra, using secondary electron multiplier detection.[24] Improvements in sensitivity would be achieved by introduction of a multichannel array detector.[26]

The mild periodate oxidation–conjugation procedure for reduced O-linked oligosaccharides in solution has been optimized so that overoxidation does not occur; as a precaution, excess oxidant is quenched with 2,3-butanediol prior to the conjugation step, and ethanol is used in place of methanol in the reaction mixtures in order to prevent methylation side reactions[27] (as described under Procedure, below). The need for anhydrous conditions is much less marked for reductive amination of the periodate-oxidized oligosaccharide alditols, because unlike the naturally reducing oligosaccharides they are predominantly in the aldehyde form.

The mechanism of the oxidative cleavage by periodate has been analyzed by the thin-layer chromatography (TLC)-LSIMS procedure, and it has been shown[27,28] that under these conditions cleavage occurs specifically at the *threo*-diol C-4–C-5 bond of the core N-acetylgalactosaminitol, giving rise to two cleavage products with distinctive mass spectra (see Procedure, below). Although this cleavage would be a drawback in carbohydrate recognition systems that require the entire oligosaccharides, it does have the advantage of allowing unambiguous assignment of positions of linkage (at C-3 or C-6) of oligosaccharide sequences to the core N-acetylgalactosamine. For example, from an oligosaccharide disubstituted at the core, two lipid-linked oligosaccharides are obtained corresponding to the branches C-3- and C-6-linked to the N-acetylgalactosaminitol, as depicted below:

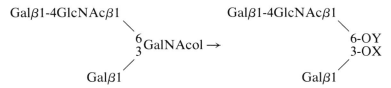

[25] Y. Kushi and S. Handa, *J. Biochem.* (*Tokyo*) **98**, 265 (1985).

[26] L. Poulter, J. P. Earnest, R. M. Stroud, and A. L. Burlingame, *Biomed. Environ. Mass Spectrom.* **16**, 25 (1988).

[27] M. S. Stoll, E. F. Hounsell, A. M. Lawson, W. Chai, and T. Feizi, *Eur. J. Biochem.* **189**, 499 (1990).

[28] W. Chai, M. S. Stoll, G. C. Cashmore, and A. M. Lawson, *Carbohydr. Res.* **239**, 107 (1993).

where OX is —OCH[CH(NHAc)CH$_2$OH]CH$_2$-DPPE and OY is —OCH$_2$CH$_2$-DPPE. Another advantage is that oligosaccharide alditol mixtures that are difficult to separate in the free state by existing chromatographic procedures can often be resolved into clearly separated bands after the controlled oxidation and conjugation to lipid, and they can be identified and sequenced by LSIMS, using submicrogram amounts of carbohydrate material.[27] This technology has been crucial for deriving sequence information on the core and backbone sequences of oligosaccharide mixtures that are available only in limited amounts from mucin-type glycoproteins, and are too complex for characterization by conventional mass spectrometry and nuclear magnetic resonance (NMR) alone.[27,29] Hanisch and Peter-Katalinić have also used this approach for core typing of O-linked oligosaccharides from human amniotic fluid.[30]

The periodate oxidation procedure, as applied to reduced oligosaccharides, results in sialic acid modification to the extent of about 40%; however, 60% remains intact to give sialic acid containing neoglycolipids after conjugation.[28] In the case of NeuAcα2-6GalNAcol, neoglycolipids containing intact sialic acid and degraded sialic acid were readily separated by TLC. A nonreductive method for the release of O-linked oligosaccharides, which obviates the need for periodate oxidation, is under development (see Procedure, below).

A potential drawback of the reductive amination for carbohydrate recognition studies with reducing oligosaccharides is the inevitable modification (ring opening) of the core monosaccharide. An alternative strategy that preserves the core monosaccharide ring structure exists,[31] and requires development to become a simple "one test-tube experiment" comparable to the present neoglycolipid procedure.

Procedure

Oligosaccharide Preparations

For generating neoglycolipids, oligosaccharides may be prepared by many of the established methods: they may be released from proteins by chemical or enzymatic means, isolated from body fluids, or synthesized chemically. For best results in the conjugation reaction, the oligosaccharide samples should be rendered free from noncarbohydrate contaminants. Oligosaccharides that have been conjugated with success have included

---

[29] A. M. Lawson, E. F. Hounsell, M. S. Stoll, J. Feeney, W. Chai, J. R. Rosankiewicz, and T. Feizi, *Carbohydr. Res.* **221,** 191 (1991).

[30] F.-G. Hanisch and J. Peter-Katalinić, *Eur. J. Biochem.* **205,** 527 (1992).

[31] E. Kallin, H. Lönn, T. Norberg, and M. Elofsson, *J. Carbohydr. Chem.* **8,** 597 (1989).

the following: (1) mixtures of O-linked oligosaccharides released by alkaline borohydride degradation from sheep gastric mucins or human milk galactosyltransferase and passed over a Dowex AG 50W (H⁺) column (Bio-Rad, Richmond, CA),[2,5,7] (2) individual oligosaccharides or oligosaccharide fractions released by alkaline borohydride degradation of human meconium glycoproteins and subjected to sequential chromatographies [Dowex AG 50W (H⁺), BioGel P-4 (Bio-Rad), and high-performance liquid chromatography (HPLC),[2,27,29,32,33] (3) mixtures of N-linked oligosaccharides released by hydrazinolysis from bovine RNase B,[16] human and mouse immunoglobulins G (IgG), human transferrin, and hen ovalbumin,[34] and the envelope glycoprotein gp120 of the human immunodeficiency virus,[35-37] and passed over Bio-Rad AG 50W (H⁺) and subjected to paper chromatography; sialooligosaccharides were desialylated with sialidase and passed over Bio-Rad AG 50W (H⁺) and AG 3-4A (OH⁻); (4) tri- and tetraantennary sialooligosaccharides released from proteins by N-glycanase and purified by Dionex (Sunnyvale, CA) chromatography (gifts of R. Townsend, University of San Francisco); (5) lactose, maltotriose, maltopentaose, and maltoheptaose purchased from Sigma (St. Louis, MO), and various oligosaccharides isolated from human milk and urine[16,38,39] or purchased from Biocarb (Lund, Sweden); (6) oligosaccharide oligomers obtained by acid hydrolyses of dextran[40] and chitin[41] and passed over Bio-Rad AG 3-4A (OH⁻) and AG 50W (H⁺); and (7) oligosaccharides synthesized chemically.[2,38]

A convenient way of separating reduced oligosaccharides from reduc-

[32] E. F. Hounsell, A. M. Lawson, J. Feeney, H. C. Gooi, N. J. Pickering, M. S. Stoll, S. C. Lui, and T. Feizi, *Eur. J. Biochem.* **148,** 367 (1985).

[33] E. F. Hounsell, A. M. Lawson, M. S. Stoll, D. P. Kane, G. C. Cashmore, R. A. Carruthers, J. Feeney, and T. Feizi, *Eur. J. Biochem.* **186,** 597 (1989).

[34] T. Mizuochi, R. W. Loveless, A. M. Lawson, W. Chai, P. J. Lachmann, R. A. Childs, S. Thiel, and T. Feizi, *J. Biol. Chem.* **264,** 13834 (1989).

[35] T. Mizuochi, M. W. Spellman, M. Larkin, J. Solomon, L. J. Basa, and T. Feizi, *Biochem. J.* **254,** 599 (1988).

[36] T. Mizuochi, T. J. Matthews, M. Kato, J. Hamako, K. Titani, J. Solomon, and T. Feizi, *J. Biol. Chem.* **265,** 8519 (1990).

[37] M. Larkin, R. A. Childs, T. J. Matthews, S. Thiel, T. Mizuochi, A. M. Lawson, J. S. Savill, C. Haslett, R. Diaz, and T. Feizi, *AIDS* **3,** 793 (1989).

[38] J. C. Solomon, M. S. Stoll, P. Penfold, W. M. Abbott, R. A. Childs, P. Hanfland, and T. Feizi, *Carbohydr. Res.* **213,** 293 (1991).

[39] M. Larkin, T. J. Ahern, M. S. Stoll, M. Shaffer, D. Sako, J. O'Brien, C-T. Yuen, A. M. Lawson, R. A. Childs, K. M. Barone, P. R. Langer-Safer, A. Hasegawa, M. Kiso, G. R. Larsen, and T. Feizi, *J. Biol. Chem.* **267,** 13661 (1992).

[40] M. S. Stoll and E. F. Hounsell, *Biomed. Chromatogr.* **2,** 249 (1988).

[41] R. W. Loveless, T. Feizi, R. A. Childs, T. Mizuochi, M. Stoll, R. G. Oldroyd, and P. J. Lachmann, *Biochem. J.* **258,** 109 (1989).

ing oligosaccharides, amino acids, and glycopeptides is by chromatography on a phenylboronic acid Bond Elut column (Analytichem International, Inc., Harbour City, CA).[40]

To these established methods may be added an alkaline β-elimination procedure (W. Chai, T. Feizi, C.-T. Yuen, and A. M. Lawson, unpublished, 1993), still under development,[23] to release nonreduced O-linked oligosaccharides from mucin-type glycoproteins by using 70% (w/v) ethylamine. "Peeling" of the released oligosaccharides occurs to varying degrees with different glycoproteins, but a substantial proportion of the oligosaccharides are either not peeled (contain the core N-acetylgalactosamine) or have a reducing galactose or N-acetylglucosamine that allows conjugation to lipid without resorting to periodate oxidation (see also Carbohydrate Recognition Studies).

Preparation of Neoglycolipids

Stock Solutions for Conjugation Procedures

L-1,2-Dipalmitoyl-sn-glycero-3-phosphoethanolamine (DPPE, given as phosphatidylethanolamine dipalmitoyl; Sigma (5 mg/ml): Dissolve 50 mg of DPPE in 10 ml of chloroform–methanol (1 : 1, v/v) (see comments under "Choice and Modifications of Procedure" for use of ethanol instead of methanol). Sonicate in a sonic water bath at 60° until dissolved. Store at − 20°. Solution will require resonication before use

L-1,2-Dihexadecyl-sn-glycero-3-phosphoethanolamine   (DHPE)   (5 mg/ml): Dissolve 50 mg of DHPE (Fluka Chemicals, Ltd., Glossop, Derbyshire, England) in 10 ml of chloroform–methanol (1 : 1, v/v) and store as above

DPPE (or DHPE) (1 mg/ml): Dissolve 10 mg of DPPE (or DHPE) in 10 ml of chloroform–ethanol (1 : 1, v/v). Store at − 20°

Imidazole buffer (40 mM): Dissolve 272 mg of imidazole in about 80 ml of water and adjust to pH 6.5 with HCl. Make volume up to 100 ml with water. Store at 4°

2,3-Butanediol reagent: Dissolve 146 mg of 2,3-butanediol in 10 ml of water. Store at 4°

Conjugation Procedures for Reducing Oligosaccharides

The oligosaccharides are dissolved in water, transferred into glass microvials (fitted with Teflon-lined caps), and dried under N_2. The reaction mixtures at the end of the conjugation procedure (see below) may be conveniently stored in these same vials at − 20° for many months.

Procedure 1. To the dry oligosaccharide(s) add 5-mg/ml DPPE (DHPE may be used instead of DPPE in all of the conjugation reactions) solution to give ≥8 : 1 mole ratio of DPPE to oligosaccharide. Seal and sonicate at 60° for 5 min and heat at 60° for about 1 hr. Add freshly prepared sodium cyanoborohydride, 10 mg/ml in methanol, to give a 1:2 mole ratio of sodium cyanoborohydride to DPPE. Heat at 60° for about 16 hr.

Procedure 2. To the dry oligosaccharide(s), add 10 μl of water and 5-mg/ml DPPE solution to give ≥32:1 mole ratio of DPPE to oligosaccharide. Evaporate the mixture to dryness in a stream of N_2 at 60°. Add water followed by chloroform–methanol (1:1, v/v) to give a final concentration of 20 mg of DPPE/ml in chloroform–methanol–water (95:95:10). Sonicate for 5 min and heat at 60° for about 1 hr, add freshly prepared sodium cyanoborohydride as above but to give a 1:8 mole ratio of sodium cyanoborohydride to DPPE, and heat at 60° for about 3 days.

Choice and Modifications of Procedure. For neutral oligosaccharides up to tetrasaccharides, use procedure 1. For acidic oligosaccharides up to tetrasaccharides and neutral oligosaccharides up to octasaccharides, use procedure 1, but add enough water to the dry oligosaccharides to give a concentration of 5% (v/v) in the final reaction volume. For larger neutral or acidic oligosaccharides, use procedure 2. For oligosaccharides of unknown sizes use procedure 2, and to ensure an excess of reagents, calculate ratios on the basis of a molecular weight corresponding to disaccharide. With certain oligosaccharide preparations (e.g., certain preparations of high mannose-type oligosaccharides obtained by hydrazinolysis of RNase B) some methylation of neoglycolipids has been observed. This can be eliminated by substituting ethanol for methanol as the reaction solvent, but at the expense of oligosaccharide solubility. Neoglycolipids derived from oligosaccharides having *N*-acetylglucosamine at the reducing end undergo a variable degree of monodehydration at the lipid-linked *N*-acetylglucosamine, as detected by TLC-LSIMS analysis (see Characterization of Neoglycolipids by Mass Spectrometry, below). The dehydration process may be pushed to near completion by adding a 10-fold excess of chloroform–methanol (1:1, v/v) to the reaction mixture at the end of the conjugation procedure, and heating for a further 3 days at 60° (see also Addendum).

Scale of Conjugation Procedure. For neutral di- to tetrasaccharides up to 1 mg has been successfully conjugated; for acidic oligosaccharides or neutral oligosaccharides larger than tetrasaccharides, up to 200 μg; when the amounts of oligosaccharides are limited, an excess of reagents may be used (e.g., 20-μl total reaction volume) to provide a volume convenient for handling. Vials should be no more than 10 times the volume of the reaction mixture.

Conjugation Procedure for Reduced Oligosaccharides

To the dry reduced oligosaccharide(s) on ice add ≥2 molar excess of freshly prepared sodium periodate as a 1.25-mg/ml solution in imidazole buffer at 0°. Mix thoroughly. After 5 min in the dark add 2,3-butanediol reagent to give a 2 M excess with respect to periodate. Keep on ice in the dark for 40 min, then add enough of the 1-mg/ml solution of DPPE to give a 5 M excess with respect to periodate, and then add freshly prepared sodium cyanoborohydride, 10 mg/ml in ethanol, to give a 25 M excess with respect to periodate, and heat the mixture at 50° for about 16 hr.

Thin-Layer Chromatography and Chemical Detection of Neoglycolipids

Stock Solvents for Thin-Layer Chromatography

Solvent A: Chloroform–methanol–water (50 : 55 : 18, v/v)
Solvent B: Chloroform–methanol–water (105 : 100 : 28, v/v)
Solvent C: Chloroform–methanol–water (55 : 45 : 10, v/v)
Solvent D: Chloroform–methanol–water (60 : 35 : 8, v/v)
Solvent E: Chloroform–methanol–water (130 : 50 : 9, v/v)
Solvent F: Chloroform–methanol–water (25 : 25 : 8, v/v)

Thin-Layer Chromatography

High-performance TLC (HPTLC) plates (E. Merck AG, Darmstadt, Germany) of aluminum-backed 5-μm silica (typically 10 × 10 cm) are used (or glass-backed if resorcinol spray is to be used). Neoglycolipids [either purified and dissolved in solvent F, or as the entire reaction mixture after addition of water to a final concentration of 15% (v/v)] are applied as spots or bands. Chromatography is usually performed at ambient temperatures (15–22°). Solvent is equilibrated in the chromatography tank for at least 1 hr. The choice of development solvent depends on the nature of the oligosaccharide portion of the neoglycolipid. The more polar the neoglycolipid the more polar the solvent should be. Solvents A to E are in decreasing order of polarity. Solvent A is suitable for large acidic oligosaccharide–lipid conjugates (e.g., sialylated multiantennary oligosaccharides); solvent B is suitable for large neutral oligosaccharide–lipid conjugates (e.g., asialobiantennary and multiantennary oligosaccharides of complex type, and high mannose-type oligosaccharides); solvents C and D are suitable for neutral tri- and tetrasaccharides or acidic di- and tetrasaccharides linked to lipid; and solvent E is suitable for monosaccharides and disaccharides linked to lipid, and those derived from periodate-oxidized oligosaccharide alditols (for best results the latter derivatives should be freed from salt after the conjugation reaction (see Purification and Further Treatments of Neoglycolipids, below).

Unreacted lipid has the highest mobility, the neoglycolipids have intermediate mobilities, and the free oligosaccharides have the lowest mobilities. Derivatives of DHPE have slightly higher mobilities than those of the corresponding DPPE derivatives.

Stock Solutions for Chemical Detection of Neoglycolipids

All solutions are stored at 4°.

Primulin reagent: Dissolve 100 mg of primulin (Sigma) in 100 ml of 10% acetone in water (v/v)

Orcinol reagent: Dissolve 900 mg of orcinol (Aldrich, Milwaukee, WI) in 25 ml of water. Add 375 ml of ethanol. Cool on ice. Gradually add 50 ml of 18 M H_2SO_4 with stirring, maintaining the temperature below 10°

Resorcinol reagent: To 5 ml of resorcinol solution [2% (w/v) in water] add 45 ml of 5 M HCl. Add 125 μl of 0.1 M $CuSO_4$ solution. Keep for 4 hr before use. (Stable for up to 1 week)

Ninhydrin reagent: Dissolve 300 mg of ninhydrin in 100 ml of 3% acetic acid in 1-butanol (v/v)

Phosphate-buffered saline (PBS): 20 mM sodium phosphate buffer containing 150 mM NaCl, pH 7.4

Visualization of Lipid (Conjugated and Nonconjugated). Prepare working primulin reagent by diluting stock reagent 1 : 100 with acetone–water (4 : 1, v/v). Spray with working reagent until the surface appears slightly wet. Dry in a stream of warm air. View under long-wavelength ultraviolet light. Lipids fluoresce light blue against a dark blue background.

Visualization of Nonconjugated Lipid. Ninhydrin, which reacts with a primary amino group, can be used to visualize free DPPE or its degradation products still containing a free amino group. In conjunction with primulin staining, conjugated and nonconjugated lipids can be differentiated. Spray lightly with ninhydrin reagent until the plate appears slightly wet. Heat in a vented oven at 105° for about 2 min, until the violet color given by the amino lipid is maximal.

Visualization of Hexose. Spray with orcinol reagent until the plates appear slightly wet. Heat in a vented oven at 105° for about 5 min or until the violet color given by hexose is maximal. For densitometry, the plates should be stored in a box containing silica gel, in the dark for 2 hr.

The orcinol response of the ring-opened core hexose differs with various oligosaccharides. For example, with lactose, and other milk oligosaccharides with a lactose core structure, the ring-opened glucose reacts (surprisingly) as a normal hexose, but with maltose Glcα1-4Glc or isomaltose Glcα1-6Glc and the disaccharide GlcNAcβ1-3Gal, the core hexoses do not react. Thus it is important to use reference compounds whose core structures resemble those of the test compounds.

Staining in sequence with primulin spray, ninhydrin, and orcinol may be performed on the same chromatogram if it is desired to visualize lipid (conjugated and nonconjugated), amino lipid (nonconjugated), and hexose.

Visualization of Sialic Acid. Spray with resorcinol reagent until the plate appears slightly wet. Clamp the plate between glass plates to prevent evaporation of reagent and heat in a vented oven at 140° for about 10 min, or until the purple color given by sialic acid is maximal. Plates may be used for densitometry immediately.

Primulin Staining of Lipid for Densitometry. An immersion procedure that gives a more uniform lipid staining for quantitation of neoglycolipids is the preferred method provided that subsequent chemical visualization of the water-soluble nonconjugated oligosaccharides is not required.

Prepare working reagent by diluting 1 ml of stock primulin reagent with 100 ml of PBS. Slowly lower the plate into PBS, being careful to avoid trapping air bubbles. Leave to soak for 10 min. Transfer the plate without drying to the working reagent. Leave to soak for 20 min in the dark. Transfer to fresh PBS without drying, and soak for 10 min in the dark. Drain and air dry the plate.

Densitometry. A Shimadzu CS-9000 scanner (Shimadzu Corporation, Kyoto, Japan) may be used for densitometry. For lipids stained with primulin it is used in linear reflectance fluorescence mode with excitation at 370 nm and detection through a filter with a cutoff at ~460 nm. For carbohydrates stained with orcinol or resorcinol, densitometry is performed in flying-spot reflectance absorption mode at 550 nm. Test samples are measured against standard neoglycolipids with essentially similar structural features. Because hexose stain is unstable, an average value should be obtained for standards scanned before and after scanning of the test samples.

Purification and Further Treatments of Neoglycolipids

The lipids in current use do not contain an ultraviolet (UV) or fluorescent chromophore for sensitive detection in HPLC. This is an area under development. Described below are microseparation methods in current use for desalting or for group separation or isolation of individual neoglycolipids.

Stock Solutions for Column Chromatography

Solvent G: Chloroform–methanol–water (15 : 70 : 30, v/v)
Solvent H: Chloroform–methanol–water (20 : 70 : 30, v/v)
Solvent I: Chloroform–methanol–water (25 : 70 : 30, v/v)
Solvent J: Chloroform–methanol–water (40 : 70 : 30, v/v)

Solvent K: Chloroform–methanol (30 : 70, v/v)

Boric acid (0.1 M): Dissolve 3.1 g of boric acid in 500 ml of water

Desalting of Reaction Mixture. The presence of salts can interfere with HPTLC, especially with reaction mixtures in which periodate oxidation has been performed. The salts can be removed by adsorption of lipids onto C_{18}–silica, washing, and then eluting.

Wash a 1-ml C_{18}-silica column (Bond Elut, Analytichem International, Inc.) with 2 ml of solvent D and then with 5 ml of solvent G. Dry the reaction mixture, dissolve in the minimum volume of solvent G, and apply to the column. Collect the run-through and wash the column with 1 ml of solvent G. Elute the lipids from the column with solvent E (2 ml) and collect two 1-ml fractions. Test aliquots of each fraction by HPTLC. Salt will be found in the run-through and the wash fractions. Neoglycolipids and free lipids should be found predominantly in the first eluate fraction, with some trace in the second eluate fraction.

Separation of Neoglycolipids from Reaction Mixtures by C_{18}–Silica Adsorption. Proceed as for Desalting of Reaction Mixture (above) to the end of the run-through stage; then, instead of eluting all lipid with solvent E, perform stepwise elution as follows: five 400-μl aliquots of solvent G, ten 400-μl aliquots of solvent H, two 400-μl aliquots of solvent I, two 400-μl aliquots of solvent J, and two 400-μl aliquots of solvent E. This gives a total of 22 fractions (including run-through). An aliquot of each fraction is subjected to TLC. A chromatogram of fractions 1–20, obtained from a reaction mixture of asialooligosaccharides from IgG and stained with primuline followed by orcinol, is shown in Figs. 1A and B, respectively. In general neoglycolipids containing large oligosaccharides elute early and those containing small oligosaccharides elute later. For mono- and disaccharides some overlap with DPPE will occur. For acidic oligosaccharides some overlap with salt may occur unless more polar solvents are used. Variations of the elution protocol can be made to suit the particular neoglycolipids.

Using Phenylboronic Acid–Silica Bonding. Neoglycolipids containing a vicinal diol system in the reduced monosaccharide to which lipid has been attached by reductive amination can be reversibly bonded to phenylboronic acid (PBA), thus providing a means to separate neoglycolipids selectively from the other components in a reaction mixture. Bonding occurs under alkaline conditions and release occurs under mild acid conditions, especially with boric acid.

Wash a 1-ml PBA column (Bond Elut) with 2 ml of chloroform–methanol (1 : 1, v/v) containing 0.1% ammonia [prepared by adding 15 μl of concentrated ammonia solution (specific gravity 0.880) to 5 ml of chloroform–methanol (1 : 1, v/v)]. Render the reaction mixture alkaline by adding

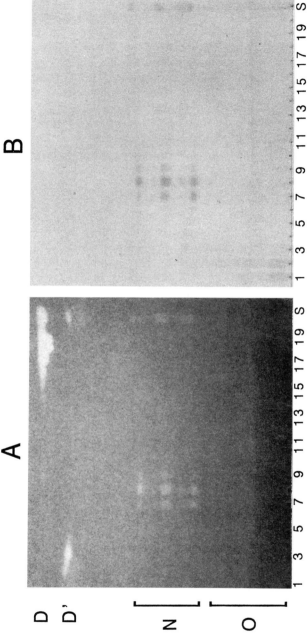

FIG. 1. Separation of neoglycolipids from salts and free lipids, using a C_{18}–silica column. The reaction mixture, after conjugation of 10 μg of asialooligosaccharides derived from human IgG, was chromatographed on a 1-ml C_{18}–silica column (Bond Elut) as described in text, and fractions collected. Aliquots of fractions 1–20 and of the starting material (S) were subjected to TLC in solvent B and stained with primulin (A) followed by orcinol (B). The positions of free DPPE and monoacyl-DPPE (D and D', respectively), the neoglycolipids (N), and the free oligosaccharides (O) are indicated. Chromatography was upward.

1 part of concentrated ammonia solution to 300 parts of reaction mixture. Add water to give a final water content of 15% (v/v). Apply this to the column and allow it to pass through under gravity. Add 1 ml of 0.1% ammonia in chloroform–methanol (1 : 1, v/v). Collect all of the above washings as run-through. Add a further 1 ml of the same solvent followed by 1 ml of chloroform–methanol (1 : 1, v/v). Collect these as wash. Add 2 ml of freshly prepared chloroform–methanol–0.1 M boric acid (30 : 70 : 30, v/v) (prepared by adding 3 ml of 0.1 M boric acid to 10 ml of solvent K) and collect as eluate 1. Add a further 2 ml of the same solvent and collect as eluate 2. Evaporate the four fractions to dryness in a stream of N_2. To each of the eluate fractions add 400 μl of 1% acetic acid in methanol (v/v) and evaporate with N_2 at 60°. To remove boric acid by conversion to methyl borate, add 400 μl of methanol and evaporate to dryness; add methanol and evaporate once further.

Semipreparative High-Performance Thin-Layer Chromatography. For isolating individual neoglycolipids, semipreparative HPTLC can be performed. After development and drying, the positions of bands can be found by spraying with water, whereupon prominent lipid bands can be seen as white against a gray background. Alternatively the band positions can be identified and marked after spraying with primulin; the plate is then rechromatographed in chloroform/methanol (2 : 1, v/v) to remove primulin. After drying again, the required strips of plate can be cut out with a scalpel. The silica gel is scraped from the strips into glass tubes and both the gel and aluminum are extracted together with aliquots of solvent E, centrifuging between each extraction. Pooled extracts are evaporated to dryness in a glass vial. The residue containing neoglycolipid and some silica is reextracted with solvent E and transferred to a second vial. This provides purified neoglycolipid containing only traces of silica, most of the dissolved silica remaining attached to the surface of the first vial.

Enzymatic and Chemical Modifications. As with natural glycolipids, the neoglycolipids may be subjected to partial degradations by glycosidases, on the chromatogram surface or in solution. Specific examples are given here. For removal of peripheral galactose residues $\beta(1 \rightarrow 4)$-linked to *N*-acetylglucosamine,[34] a mixture of neoglycolipids derived from asialooligosaccharides of human IgG was subjected to TLC; the area of the chromatogram containing the neoglycolipids was cut, dipped for 1 min in a 0.05% (w/v) solution of Plexigum P28 (Cornelius Chemical Co., Ltd., Romford, England) in *n*-hexane and immersed for 2 hr in bovine serum albumin (3%, w/v) in phosphate-buffered saline [20 mM phosphate buffer (pH 7.4), 150 mM NaCl, 0.02% NaN$_3$] solution. After rinsing with 0.1 M citrate phosphate buffer, pH 3.5, the chromatogram was overlaid with a solution of jack bean β-galactosidase (3.25 units/ml in 0.1 M citrate–phosphate buffer, pH 3.5) (or with the buffer alone as a control), and incubated at 37° for 17 hr in a humidified chamber.

In other experiments (C.-T. Yuen, as cited in Rosenstein et al.[42]) the purified neoglycolipids derived from the milk oligosaccharide lacto-N-neotetraose was subjected to sequential exoglycosidase treatments in solution: The neoglycolipid (30 µg of hexose) was taken up in distilled water (30 µl) containing sodium deoxycholate (40 µg) and sonicated for 5 min at 20° in a sonic water bath before adding jack bean β-galactosidase (Oxford Glycosystems, Abingdon, Oxfordshire, England), 250 mU in 10 µl of 0.4 M citrate phosphate buffer, pH 3.5. The reaction mixture was further sonicated for 20 sec, and incubated at 37° for 18 hr in the presence of toluene. The enzyme was inactivated by heating the reaction mixture in a boiling water bath (2 min) and the neoglycolipid extracted four times with 500 µl of solvent C. The enzyme digestion/extraction procedure was repeated once more. An aliquot of galactosidase-treated neoglycolipid (15 µg of hexose) was taken up in deoxycholate solution as above and further digested with jack bean β-N-acetylhexosaminidase (Oxford Glycosystems), 250 mU in 10 µl of 0.4 M citrate–phosphate buffer, pH 4.5, at 37° for 18 hr, heated in a boiling water bath, and extracted as above.

For removing fucose residues by mild acid hydrolysis, the DHPE derivative of a fucooligosaccharide was prepared and isolated by preparative TLC.[23] Unlike DPPE, which tends to lose the acyl groups on acid treatment, DHPE is unaffected[22] (also C.-T. Yuen and M. S. Stoll, unpublished observations, 1992). The neoglycolipid was heated in 0.02 M sulfuric acid for 30 min at 100°, neutralized with 0.2 M ammonium hydroxide, and dried under nitrogen. Removal of fucose residues was monitored by TLC-LSIMS analysis of the neoglycolipid.

Carbohydrate Recognition Studies

We have studied the binding to neoglycolipids of antibodies,[2,7,15,16,43] various soluble carbohydrate-binding proteins,[16,23a,34,38,41,44–46] whole bac-

[42] I. J. Rosenstein, C.-T. Yuen, M. S. Stoll, and T. Feizi, Infect. Immun. **60**, 5078 (1992).
[43] M. Larkin, W. Knapp, M. S. Stoll, H. Mehmet, and T. Feizi, Clin. Exp. Immunol. **85**, 536 (1991).
[44] R. A. Childs, K. Drickamer, T. Kawasaki, S. Thiel, T. Mizuochi, and T. Feizi, Biochem. J. **262**, 131 (1989).
[45] R. A. Childs, T. Feizi, C.-T. Yuen, K. Drickamer, and M. S. Quesenberry, J. Biol. Chem. **265**, 20770 (1990).
[46] R. A. Childs, J. R. Wright, G. F. Ross, C.-T. Yuen, A. M. Lawson, W. Chai, K. Drickamer, and T. Feizi, J. Biol. Chem. **267**, 9972 (1992).

teria,[42,47] and of Chinese hamster ovary (CHO) cells transfected to express the full-length E-selectin molecule.[23,39] The techniques used, that is, chromatogram overlay, plastic microwell overlay, and inhibition assays with neoglycolipids incorporated into liposomes, are adaptations of existing methods for natural glycolipids.[8–14,48,49] The artificial nature of the lipid moiety is a disadvantage in recognition systems that involve specific portions of the ceramides in native glycolipids. However, an advantage of neoglycolipids is that the homogeneity of the lipid moiety allows oligosaccharides to be studied without effects of lipid heterogeneity. Moreover, by radiolabeling neoglycolipids (A. Herraez, M. S. Stoll, and T. Feizi, procedure under development) it is possible to monitor conveniently the amounts of the lipid-linked oligosaccharides absorbed onto plastic microwells in quantitative binding assay procedures.

The power of the neoglycolipid technology for detecting novel ligands among a heterogeneous array of unknown oligosaccharides released from a mucin-type glycoprotein derived from a human ovarian cystadenoma protein is illustrated in Fig. 2A.[23] Oligosaccharides were released from protein in the presence of 70% (w/v) ethylamine; neoglycolipids derived from four acidic oligosaccharide fractions A1 to A4 were resolved on TLC and tested for E-selectin binding, using as indicator a CHO cell line transfected to express the full-length E-selectin molecule and labeled with [³H]thymidine. Binding was observed to selected components in each of the heterogeneous fractions (Fig. 2A, lanes 1–4). No binding was detected when nontransfected CHO cells were used (not shown). Fraction A4, which gave the fastest migrating neoglycolipid band with E-selectin binding, was investigated further. By TLC-LSIMS analysis (see Characterization of Neoglycolipids by Mass Spectrometry, below) this component was identified as a sulfated tetrasaccharide with a branched sequence, hexose-(deoxyhexose)N-acetylhexosamine-hexose [abbreviated to Hex-(dHex)HexNAc-Hex]. From oligosaccharide fraction A4 the component (A4E) with E-selectin-binding activity was isolated by HPLC, and subjected to methylation analysis before and after mild acid treatment.[23] A4E was thus identified as an equimolar mixture of 3'-sulfated Le^a- and SSEA-1/Le^x-type, β-eliminated peeled products:

[47] I. J. Rosenstein, M. S. Stoll, T. Mizuochi, R. A. Childs, E. F. Hounsell, and T. Feizi, *Lancet* **2**, 1327 (1988).

[48] K.-A. Karlsson and N. Strömberg, *in* "Methods in Enzymology" (V. Ginsburg, ed.), Vol. 138, p. 220. Academic Press, Orlando, FL, 1987.

[49] P. Swank-Hill, L. K. Needham, and R. L. Schnaar, *Anal. Biochem.* **163**, 27 (1987).

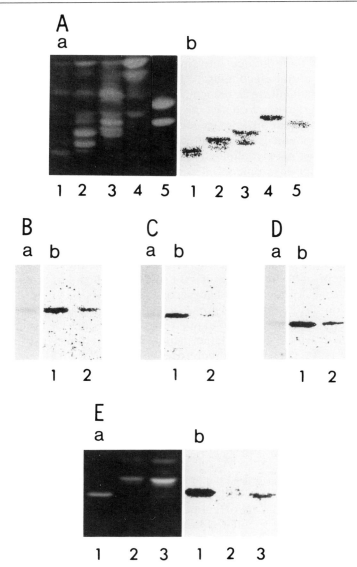

FIG. 2. Chromatogram overlay assays showing *E*-selectin binding to neoglycolipids derived from the O-linked oligosaccharides of a human ovarian cystadenoma glycoprotein and comparisons with binding to already established *E*-selectin ligands.[23] (A) Oligosaccharides were released by a nonreductive β elimination in the presence of 70% (w/v) ethylamine desalted with Bio-Rad AG 50W (H$^+$), deproteinated by passing over a C$_{18}$ Sep-Pak cartridge, and fractionated into acidic and neutral fractions by DEAE-Sephadex A-25 chromatography. The acidic fraction was rechromatographed on DEAE-Sephadex A-25 and a peak eluting

$$HSO_3\text{-}3Gal1\text{-}3GlcNAc1\text{-}3Gal$$

$$\big|\ 1,4$$

$$Fuc$$

$$HSO_3\text{-}3Gal1\text{-}4GlcNAc1\text{-}3Gal$$

$$\big|\ 1,3$$

$$Fuc$$

Further studies[23] have shown that (1) these sulfated oligosaccharides are the most potent ligands so far identified for the *L*-selectin molecules; (2) the presence of the fucose residue is required for *E*-selectin but not for *L*-selectin binding: The neoglycolipid A4E after treatment with acid to release the fucose residue was no longer bound by CHO-*E*-selectin cells (Fig. 2E) but it showed substantial binding by a soluble form of *L*-selectin; (3) *E*-selectin binding to A4E was substantially stronger than to SSEA-1/Le^a- and Le^x-type sequences; compare, for example, the binding intensities of A4E and Le^a-active neoglycolipid in lanes 4 and 5 (Fig. 2A) relative to the intensity of chemical staining of the two neoglycolipids; and (4) the *E*-selectin binding to A4E is equal if not superior to that of the chemically synthesized 3′-sialyl-Le^x/SSEA-1 glycolipid analog with a trisaccharide backbone: NeuAcα2-3Galβ1-4(Fucα1-3)GlcNAcβ1-3Gal-Cer (3′-sialyl-LNFT-Cer). The intensities of binding of these two lipid-linked oligosaccharides were comparable overall, but near the limits of detectable binding; for example, at 5 and 10 pmol the intensities of binding

---

at 0.1 *M* ammonium acetate buffer (pH 5.0) was chromatographed on a BioGel P-4 column, separated into fractions A1–A4 corresponding to 11–15, 8–11, 5–8, and 2–5 hexose units, respectively, and conjugated to DPPE. The four reaction mixtures (lanes 1–4) and the neoglycolipid standards lacto-*N*-tetraose and lacto-*N*-fucopentaose II, 1 nmol of each (lane 5), were subjected to TLC in duplicate in solvent D. One plate (a) was stained with primulin to identify neoglycolipid areas; in a duplicate plate, this area (b) was overlaid with ^{3}H-labeled Chinese hamster ovary cells transfected to express the full-length *E*-selectin molecule, and binding was detected by fluorography. (B–D) Parts (a) contained the purified neoglyco-lipid (DHPE) of the *E*-selectin-binding oligosaccharide (A4E) isolated from fraction A4 (A), 3′-sialyl-LNFT-Cer (B), and 3′-sialyl-LNFP-III-Cer (C) at 110, 110, and 75 pmol, respectively, stained with orcinol; (b) *E*-selectin binding to 10 pmol (lane 1) and 5 pmol (lane 2) of the three lipid-linked oligosaccharides, respectively; these had been chromatographed in solvent D and overlaid with ^{3}H-labeled *E*-selectin cells and the binding intensities compared by fluorography. (E) (a) Primulin staining and (b) *E*-selectin binding; lane 1 contained DHPE derivative of A4E, 100 pmol; lanes 2 and 3 contained acid-treated (defucosylated) neoglycolipid A4E (100 and 300 pmol, respectively); the presence of a trace amount of the original fucosylated neoglycolipid remaining after acid treatment is indicated by *E*-selectin binding in (b), lanes 2 and 3. (Taken from Yuen *et al.*[23] with permission.)

to the A4E neoglycolipids were greater than to the 3′-sialyl analog (see Fig. 2B and C), and they were close to those observed with a chemically synthesized sialyl analog that has the tetrasaccharide backbone NeuAcα2-3Galβ1-4(Fucα1-3)GlcNAcβ1-3Galβ1-4Glc-Cer (3′-sialyl-LNFP-III-Cer) (see Fig. 2D). In separate experiments[23] it was observed that neoglycolipids and corresponding glycosylceramides give equivalent binding intensities, and the lipid-linked oligosaccharides with tetrasaccharide backbones give stronger binding than with a trisaccharide backbone. Thus, equivalent binding intensities were given by neoglycolipids derived from 3′-sialyl-LNFP-II: NeuAcα2-3Galβ1-3(Fucα1-4)GlcNAcβ1-3Galβ1-4Glc, the chemically synthesized glycolipid 3′-sialyl-LNFP-III-Cer, and the 3′-sialyl-LNFP-III neoglycolipid with tetrasaccharide backbones, whereas binding to the 3′-sialyl-LNFT-Cer is weaker.

Characterization of Neoglycolipids by Mass Spectrometry

Neoglycolipids may be analyzed by LSIMS on the conventional target probe after removal of excess reagents, or directly from the surface of the silica gel plates following TLC. The advantage of the latter approach is that purification of the neoglycolipids from the reaction mixture is not required; second, mixtures of neoglycolipids can be analyzed; and third, the ligand-binding experiments may be performed in conjunction with preliminary characterization of components bound with respect to molecular weight and monosaccharide sequence of the oligosaccharides.

Sample Preparation for TLC-LSIMS

Mark the position of primulin-stained neoglycolipid bands on the silica gel surface and aluminum backing of the TLC plate, and set the plate face down on a clean glass surface. With a scalpel blade and metal ruler excise the bands (typically 1.5×5 mm) for mass spectrometry. Attach the TLC strips with the aluminum backing to the LSIMS target probe tip by a $1:6$ (v/v) mixture of water-soluble glue and glycerol. To the sample on silica add 2 μl of solvent F, followed by the liquid matrix (2 μl), consisting of diethanolamine (DEA), tetramethylurea (TMU), and m-nitrobenzyl alcohol (NBA), $2:2:1$, by volume, respectively.

Sample Preparation for Conventional LSIMS

Coat the LSIMS stainless steel target probe with 1 μl of the DEA–TMU–NBA matrix. Take up the purified neoglycolipid in chloroform–methanol–water ($25:25:8$, v/v) and apply onto the probe.

Mass Spectrometry

Practical details of techniques of mass spectrometry for carbohydrate analysis are given elsewhere in this volume.[50] The spectra shown in Figs. 4–8 are obtained directly from the silica surface, using a VG Analytical ZAB2-E mass spectrometer fitted with a cesium ion gun operated at 25 keV and an emission current of 0.5 μA. A conventional secondary electron multiplier is used as detector. Single scans at 30 sec decade^{-1} are acquired with the VG Analytical 11-250J data system in the continuum mode at resolutions up to 2500 resolving power. Mass calibration is carried out with cesium iodide as reference standard.

Mass spectra of neoglycolipids may also be acquired from a lane of the TLC plate ($\sim$75 $\times$ 5 mm) containing all the neoglycolipid bands in a sample, by using a TLC plate drive to pass it through the ionizing region of the ion source while continuously acquiring mass spectra.[51] This approach, which has been applied to natural glycolipids,[52] is less sensitive than when using the optimized conditions for analysis of discrete sample components cut from the TLC plate.[53]

Interpretation of Mass Spectra of Neoglycolipids from Reducing Oligosaccharides

The mass spectra of neoglycolipids derived from reducing oligosaccharides are readily interpretable by virtue of the charge localization effect of the lipid, which not only gives an abundant quasimolecular ion, but dominates fragmentation and gives rise only to ions containing the lipid-linked end.[24] Because of this and the occurrence of fragmentation at all glycosidic bonds irrespective of monosaccharide composition, determination of the oligosaccharide sequence is greatly simplified and unambiguous. Uniquely for oligosaccharide derivatives, the LSIMS fragmentation of the neoglycolipids is the same in positive and negative ion spectra because of the presence of both a protonation and deprotonation site in the lipid moiety. However, negative ion spectra are generally preferred because of their low chemical background in the mass region containing the informative sequence ions and the absence of quasimolecular adduct ions with alkali salts where an acidic group is present. The follow-

[50] A. Dell, A. J. Reason, K.-H. Khoo, M. Panico, R. A. McDowell, and H. R. Morris, this volume [8].

[51] A. M. Lawson, W. Chai, M. Stoll, R. H. Bateman, J. Curtis, E. F. Hounsell, and T. Feizi, *Proc. Am. Soc. Mass Spectrom. Conf. Mass Spectrom. Allied Top., 37th,* Miami Beach, FL, p. 790 (1989).

[52] K.-A. Karlsson, B. Lanne, W. Pimlott, and S. Teneberg, *Carbohydr. Res.* **221,** 49 (1991).

[53] W. Chai, G. C. Cashmore, R. A. Carruthers, M. S. Stoll, and A. M. Lawson, *Biol. Mass Spectrom.* **20,** 169 (1991).

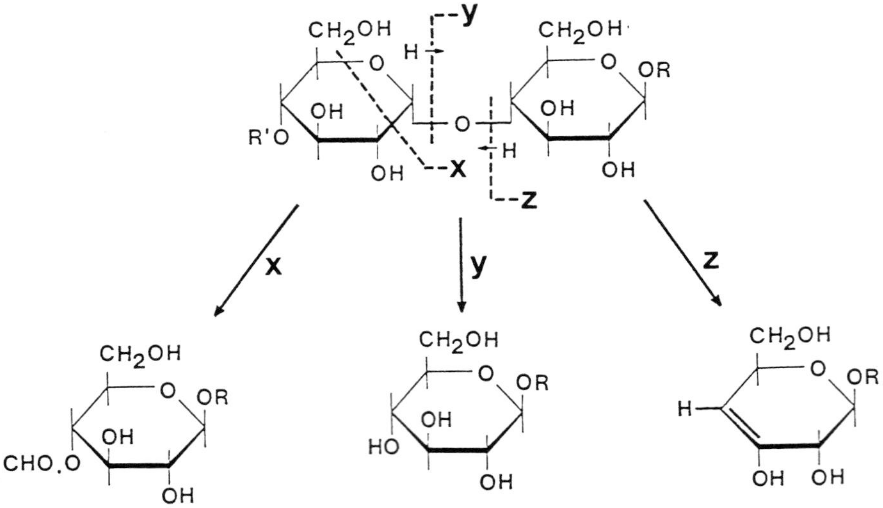

FIG. 3. Principal cleavages that provide monosaccharide sequence information in the spectra of neoglycolipids. These fragmentations occur equally for dHex, Hex, and HexNAc residues. Designations X, Y, and Z are according to Ref. 54. R, Reducing end of molecule, including the DPPE moiety; R', hydrogen or nonreducing end of molecule. (Adapted from Lawson et al.[24])

ing general features are apparent in negative ion spectra: (1) intense $[M - H]^-$ ions from which monosaccharide composition can be deduced in terms of dHex, Hex, HexNAc, NeuAc, and so on, (2) a less abundant series of glycosidic and related fragmentations providing information on monosaccharide sequence, and (3) gaps in the fragmentation pattern indicating branching points in the monosaccharide sequence. The mechanisms of glycosidic cleavage are depicted in Fig. 3, and the mass spectra (Figs. 4–6) of neoglycolipids derived from the following reducing oligosaccharides illustrate the salient features: maltopentaose, $Glc\alpha1$-$4Glc\alpha1$-$4Glc\alpha1$-$4Glc\alpha1$-$4Glc$; the sialyl-Le^a-active oligosaccharide, 3'-sialyl-lacto-N-fucopentaose II (3'-S-LNFP-II),

$$Gal\beta1\text{-}3\,GlcNAc\beta1\text{-}3Gal\beta1\text{-}4Glc$$
$$|\,2,3 \qquad\quad |\;1,4$$
$$NeuAc\alpha \;\; Fuc\alpha$$

the linear Le^a- and Le^x-related oligosaccharide difucosyl-p-lacto-N-hexaose (DFpLNH),

$$Gal\beta1\text{-}3GlcNAc\beta1\text{-}3Gal\beta1\text{-}4GlcNAc\beta1\text{-}3Gal\beta1\text{-}4Glc$$
$$|\;1,4 \qquad\qquad\qquad\quad |\;1,4$$
$$Fuc\alpha \qquad\qquad\qquad\qquad Fuc\alpha$$

and the branched Le^a- and Le^x-related oligosaccharide, difucosyllacto-N-hexaose (DFLNH),

$$
\begin{array}{c}
\text{Fuc}\alpha \\
|\ 1,3 \\
\text{Gal}\beta1\text{-4GlcNAc}\beta1 \\
\end{array}
$$

Fucα
| 1,3
Galβ1-4GlcNAcβ1
                                                  \
                                                   $^6_3$Galβ1-4Glc
                                                  /
Galβ1-3GlcNAcβ1
| 1,4
Fucα

Monosaccharide composition of the neoglycolipid is deduced from the m/z value of the abundant $[M - H]^-$ ion by computing combinations of the masses of component sugars (dHex, 146 Da; Hex, 162 Da; HexNAc, 203 Da; NeuAc, 291 Da), together with the hydrogen at the terminal nonreducing end and ionized DPPE or DHPE at the modified end (692 and 664 Da, respectively). The $[M - H]^-$ of the DPPE neoglycolipid of maltopentaose, for example, has an m/z value of $(162 \times 5) + 692 = 1502$ and of DFpLNH is $(2 \times 146) + (4 \times 162) + (2 \times 203) + 692 = 2038$. The major fragment ions arise by glycosidic cleavage Y (according to the nomenclature of Domon and Costello[54]) at each residue irrespective of composition, together with two other ion series X and Z, as shown in Fig. 3. These cleavages are analogous to those observed in LSIMS analysis of oligosaccharides,[55] and their derivatives formed by reductive amination (see Ref. 56 and references therein). This series of ion triplets, which are characteristic of fragmentation, help to avoid confusion of the Y fragments with molecular ions of lower oligomers of equivalent composition. The regular 162-mass unit differences between the Y fragment ions in the spectrum of the maltopentoase derivative (i.e., m/z 854, 1016, 1178, and 1340; Fig. 4) correspond to the linear sequence of glucose residues. Fucose at the nonreducing terminus of a linear oligosaccharide gives the expected $X, Y,$ and Z fragmentations. However, a fucose branch on a linear chain results in loss of both substituents at the branching residue together with their glycosidic oxygens. Thus in the spectrum of 3′-sialyl-LNFP-II neoglycolipid (Fig. 5) the ion at m/z 1185 occurring as a singlet, and not the usual fragmentation triplet, is diagnostic of the fucose branching point. In the spectrum of the lipid-linked difucosyl oligosaccharide DFpLNH

[54] B. Domon and C. E. Costello, *Glycoconjugate J.* **5,** 397 (1988).
[55] A. Dell, *Adv. Carbohydr. Chem. Biochem.* **45,** 19 (1987).
[56] L. Poulter and A. L. Burlingame, *in* "Methods in Enzymology" (J. McCloskey, ed.), Vol. 193, p. 661. Academic Press, San Diego, 1990.

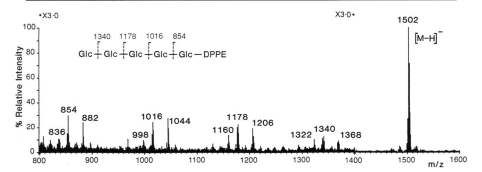

FIG. 4. The negative ion mass spectrum of maltopentaose–DPPE. The monosaccharide sequence information is obtained from the Y fragments (m/z 854, 1016, 1178, and 1340), Z fragments (m/z 836, 998, 1160, and 1322), and X fragments (m/z 882, 1044, 1206, and 1368) (where X, Y, and Z are the fragmentations depicted in Fig. 3). (Adapted from Lawson *et al.*[24])

(Fig. 6A) there are two ions diagnostic of fucose branching (i.e., m/z 1185 and 1696). Neoglycolipids with branching points not involving fucose show distinctive gaps in the sequence-information ions. For example, in the mass spectrum of the DFLNH neoglycolipid (Fig. 6B), there is a gap in the sequence ions from m/z 1527 down to m/z 854; the latter ion arises from the loss of branching galactose and the attached outer chains. The special feature of spectra from N-acetylneuraminic-acid-containing neo-glycolipids is the presence of intense Y cleavage ions ([M − H − 291]⁻, for example, m/z 1527 in Fig. 5) and often, a sodium adduct ion of the molecular species due to the presence of salt ([M − 2H + Na]⁻, m/z

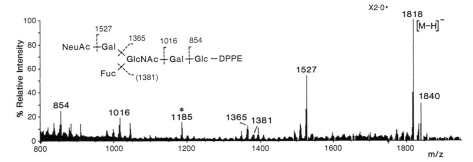

FIG. 5. The negative ion mass spectrum from the sialyl-Leᵃ-active neoglycolipid 3'-S-LNFP-II-DPPE. The major Y fragments only are indicated: m/z 854, 1016, 1365, 1381, and 1527 and the diagnostic fucose branching ion (m/z 1185) (asterisked). Loss of NeuAc, m/z 1527, from the molecular species gives an abundant ion relative to the other fragments. The ion at m/z 1840 arises from sodium adduction to the molecule [M − 2H + Na]⁻.

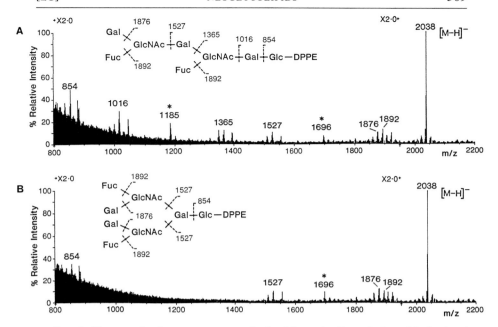

FIG. 6. The negative ion mass spectra obtained from the linear Le^a- and Le^x-related difucosyl neoglycolipid DFpLNH-DPPE (A) and the branched Le^a- and Le^x-related neoglycolipid DFLNH-DPPE (B). The Y fragment ions and their origins are shown and also the ions at m/z 1185 and 1696, diagnostic of fucose branching (asterisked).

1840 as in Fig. 5). The N-acetylneuraminic acid fragmentation from [M − H]$^-$ predominates over cleavages at branch points, and concomitant loss of the N-acetylneuraminic acid and branch residues can occur (e.g., m/z 1381; Fig. 5).

Information on the specific linkage positions of individual monosaccharides is not available from conventional scan spectra; however, intensity differences of fragment ions related to linkage position are apparent in product ion spectra recorded by B/E-linked field scanning where B and E are the magnetic field and electric sector voltage, respectively, or by collision-induced dissociation tandem mass spectrometry,[57] and these spectra will be useful once a database for lipid-linked oligosaccharides is assembled.

The presence of a dehydration product giving an [M − H − 18]$^-$ ion is frequently observed with neoglycolipids derived from oligosaccharides containing 4-linked N-acetylglucosamine at their reducing ends.[24,34] As

[57] W. Chai, G. C. Cashmore, M. S. Stoll, S. J. Gaskell, R. S. Orkiszewski, and A. M. Lawson, *Biol. Mass Spectrom.* **20**, 313 (1991).

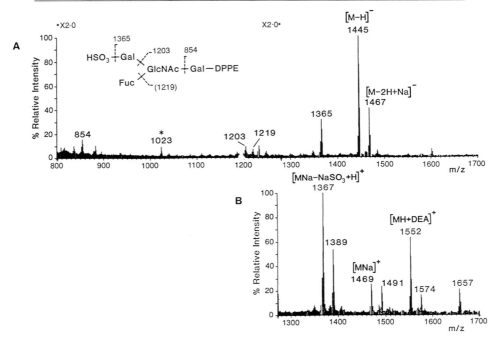

FIG. 7. The negative ion spectrum (A) and the high-mass region of the positive ion spectrum (B) of the DPPE neoglycolipids of the equimolar mixture of the Le^a- and Le^x-type tetrasaccharides (fraction A4E) derived from a human ovarian cystadenoma glycoprotein. In (A), the sequence can be deduced from the Y fragment ion series in the negative ion spectrum: m/z 854, 1203, and 1219, and the diagnostic fucose branch ion at m/z 1023 (asterisked). In the positive ion spectrum the sodium-cationized quasimolecular ions ([MNa]$^+$, m/z 1469, and [MNa$_2$-H]$^+$, m/z 1491) are accompanied by desulfated fragment ions: m/z 1367 and 1389. The ions at m/z 1552, 1574, and 1657 are matrix adduct ions [MH + DEA]$^+$, [MNa + DEA]$^+$, and [MH + 2DEA]$^+$, respectively, where DEA denotes a matrix component diethanolamine. (Taken from Yuen et al.,[23] with permission.)

the elements of water are lost from the terminal N-acetylglucosamine,[57] a different set of sequence ions is formed (each ion showing − 18 units). Although mass spectra from the normal and dehydrated forms of neoglycolipids can still be interpreted, reaction conditions should be adjusted to maximize one or other product in order to maintain the best sensitivity (see Preparation of Neoglycolipids, above).

Experience with TLC-LSIMS of the neoglycolipids derived from sulfated oligosaccharides is illustrated in Fig. 7. This is the sulfated Le^a- and SSEA-1/Le^x-type tetrasaccharide fraction A4E with potent E-selectin binding activity (see Carbohydrate Recognition Studies, above). The ions at m/z 1445 and 1467 in the negative ion spectrum (Fig. 7A) could be assigned as the [M − H]⁻ and [M − 2H + Na]⁻ ions of a lipid-linked oligosaccharide with the composition dHex-Hex₂-HexNAc, with an additional 80 Da equivalent to either a sulfate or phosphate group. From the negative and positive ion spectra of the neoglycolipid (Fig. 7A and B, respectively), and a study of neoglycolipids from sulfated and phosphorylated model oligosaccharides, the presence of sulfate in the tetrasaccharide was deduced. The key feature is the lability of the sulfate linkage relative to the phosphate linkages.[23] In positive mode in particular the desulfation [MNa − NaSO₃ + H]⁺ ion is dominant (Fig. 7B). This contrasts with the results from a phosphorylated neoglycolipid, which gave prominent quasimolecular ions, MH⁺ and MNa⁺, whereas the dephosphorylation fragment ion [MH − HPO₃]⁺ was of low intensity (not shown).

Interpretation of Mass Spectra of Neoglycolipids Derived from Reduced Oligosaccharides

With O-linked oligosaccharides released by alkaline borohydride degradation, the neoglycolipids obtained from the products of mild oxidative cleavage of C-4–C-5 bond of the core N-acetylgalactosaminitol give spectra with the same general features as those from the neoglycolipids of reducing oligosaccharides. The special feature is that the sequence ions indicate the point of attachment of the sugar chains to the core monosaccharide, in particular the ions m/z 835 and 734 show the linkages to C-3 and C-6, respectively, of the GalNAcol. This allows the core type and branching pattern of isolated or simple mixtures of O-linked oligosaccharides to be determined. The mild periodate oxidation reaction is specific for the C-4–C-5 bond of 3- or 6-monosubstituted or 3,6-disubstituted core N-acetylgalactosaminitol,[27,28] and no evidence has been found for cleavage products of any cyclic diol bond of the monosaccharides in an oligosaccharide from LSIMS analyses of a large range of reduced oligosaccharides. The neoglycolipid bands derived from the two oxidation products of the hexasaccharide alditol[58]

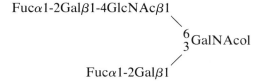

[58] W. Chai, E. F. Hounsell, G. C. Cashmore, J. R. Rosankiewicz, C. J. Bauer, J. Feeney, T. Feizi, and A. M. Lawson, *Eur. J. Biochem.* **203,** 257 (1992).

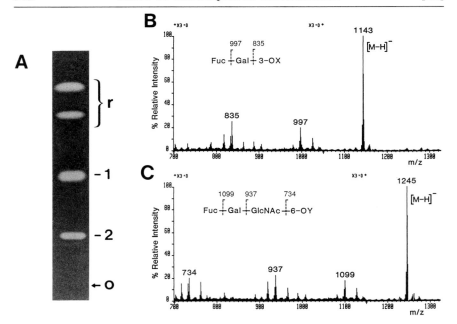

FIG. 8. Primulin-stained TLC of two neoglycolipid bands 1 and 2 (A) and their negative ion mass spectra (B and C). These neoglycolipids, Fuc-Gal-3-OX (B) and Fuc-Gal-GlcNAc-6-OY (C), were produced from the oligosaccharide alditol, Fucα1-2Galβ1-3(Fucα1-2Galβ1-3GlcNAcβ1-6)GalNAcol, by controlled periodate oxidation and conjugation to DPPE. The sequence information ions, m/z 835 and 997 (B) and m/z 734, 937, and 1099 (C), are indicated. OX and OY are fragments of the core GalNAcol linked to DPPE, where OX = —OCH[CH(NHCOCH₃)CH₂OH]CH₂-DPPE and OY = —OCH₂CH₂-DPPE. r, Reagent bands; o, sample origin (A).

and their mass spectra are shown in Fig. 8. The glycosidic fragment ions m/z 997 and 835 (Fig. 8B) are produced from the dHex-Hex sequence in band 1 (Fig. 8A) and the fragment ions m/z 1099, 937, and 734 (Fig. 8C) are produced from the dHex-Hex-HexNAc sequence in band 2 (Fig. 8A).

Sensitivity of LSIMS Analysis of Neoglycolipids

When purified neoglycolipids are added directly to the LSIMS stainless steel target, exceptionally high sensitivity of detection is possible. For example, from a complete mass range scan a 100-fmol sample of maltopentaose as its DPPE neoglycolipid gave a signal-to-noise (S/N) ratio of 12 : 1 for [M − H]⁻.[24] At this concentration, fragment ions were not visible above the background. However, the sequence-determining ions were clearly apparent from 1 pmol of material, and a 5-pmol sample was the optimal amount for analysis.[24]

When analyzed directly from the surface of the silica gel plate, a 50-pmol aliquot of maltopentaose–DPPE produced a spectrum with an $[M - H]^-$ ion having an S/N ratio of 20:1 in addition to fragment ions, whereas only the $[M - H]^-$ ion was detected at the 1-pmol level (amounts quoted here are aliquots from larger samples). When starting with small sample sizes and including all analytical steps, that is, DPPE conjugation, TLC separation, and subjecting the total products to *in situ* LSIMS, the overall sensitivity is again high. Clear $[M - H]^-$ ions above background were obtained from a 3-pmol sample of maltopentaose, whereas $[M - H]^-$ and sequence ions are present in spectra from 100–1000 pmol.[24]

Molecular mass and surface activity influence ionization, and the detection sensitivity differs for individual oligosaccharides. The sensitivity of direct TLC-LSIMS also depends on the extent to which the sample can be brought into solution at the surface of the matrix. For example, with the DPPE-neoglycolipid of the biantennary oligosaccharide

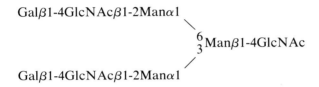

derived from the urine of a patient with G_{M1}-gangliosidosis, 50 and 200 pmol of derivative were required to obtain both $[M - H]^-$ and sequence-determining fragment ions from the sample applied onto the target or analyzed directly from a TLC plate, respectively.[24]

Applications of Neoglycolipids in Biological Systems

On the basis of the initial results of immunochemical experiments using neoglycolipids,[2,7,15] considerable potential was foreseen for these probes[59]: (1) for establishing whether antibodies to glycoproteins are directed at their carbohydrate moieties, (2) for identifying new oligosaccharide epitopes on glycoprotein oligosaccharides, (3) for detecting and selecting for characterization minor antigenic oligosaccharides within mixtures, (4) for immunosequencing with panels of well-characterized monoclonal antibodies, and (5) for use as immunogens. However, the greatest potential for these probes was foreseen[59,60] in the broader aspects of biological chemistry, for identifying new oligosaccharide recognition systems: endogenous lec-

[59] T. Feizi, *Biochem. Soc. Trans.* **16**, 930 (1988).
[60] T. Feizi, *Ciba Found. Symp.* **145**, 62 (1989).

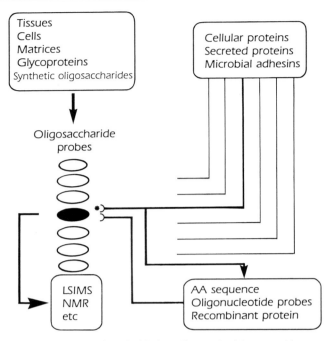

FIG. 9. A general strategy for elucidating oligosaccharide recognition systems. Whole tissues, isolated cells, extracellular matrices or individual glycoproteins, in addition to desired synthetic oligosaccharides, are used to generate neoglycolipids that are used in overlay experiments with radiolabeled proteins. Individual cellular or secreted proteins or microbial adhesins (asterisked) with carbohydrate recognition properties are singled out and character-ized at the level of protein and gene structure. The oligosaccharides recognized are character-ized by state-of-the-art methods of structural analysis such as LSIMS, methylation analysis, and NMR. (From Feizi[60], with permission.)

tins and adhesins of microbial agents (Fig. 9). It was hoped that roles would also be found for common oligosaccharide structures, and that it would be possible to detect and decode the information content of unique or unusual oligosaccharide structures on glycoproteins and proteoglycans, and to make comparisons with natural glycolipids. It was also suggested[59] that neoglycolipids might find uses as substrates in studies of oligosaccha-ride recognition by glycosyltransferases and glycosidases. Their potential value was also considered[59] in the "quality control" of recombinant glyco-proteins, particularly with respect to the presence of immunogenic oligo-saccharides or for monitoring the presence or absence of specific oligosac-charide sequences, using lectins and sequence-specific antibodies. We comment below on some developments in these directions.

Immunochemical Studies

In reviewing novel structure/antigenicity assignments to oligosaccharides made by using neoglycolipids, it should be recalled that the immunogenic potential of oligosaccharides conjugated to lipids was predicted by Wiegandt and Ziegler[18] and was proved by Wood and Kabat,[19] who generated hybridoma-derived monoclonal antibodies to oligosaccharide fragments of dextran after these were conjugated to stearylamine and incorporated into liposomes. Neoglycolipids derived from capsular polysaccharides of *Streptococcus pneumoniae* have also been prepared as potential semisynthetic vaccines.[61] More recently, neoglycolipids derived by conjugation of mannooligosaccharide mixtures derived from phosphomannans of the cell wall of the yeast *Candida albicans* and conjugated to the lipid 4-hexadecylaniline were found to be as immunogenic in mice and rats as the natural bacterial polysaccharides[62] and to be useful in monitoring antibody levels in the sera of infected patients.[63]

By using a sequence-specific monoclonal antibody, an unusual repetitive blood group A sequence was detected[16] as an unsuspected minor component in a blood group A-active oligosaccharide preparation from the human fetal gastrointestinal tract. Binding studies with neoglycolipids derived from a series of structurally defined oligosaccharides served to identify[43] epitopes recognized by a monoclonal antibody, VIB-E3, to the human B lymphocyte antigen CD24: NeuAcα2-6Gal and NeuAcα2-6GalNAc. Two hybridoma antibodies designated L3 and L4 raised to the neural cell adhesion molecule L1 have been shown[63a] to bind to neoglycolipids derived from high-mannose type oligosaccharides released from this glycoprotein.

Studies of Microbial Adhesion

The adhesive specificity of type 1 fimbriated (mannose-sensitive) *Escherichia coli* for high mannose-type oligosaccharides derived from N-glycosylated proteins was directly visualized by using neoglycolipids resolved on chromatograms,[47] in accordance with predictions from earlier inhibition of binding studies. In addition, by using a series of neoglycolipids derived from structurally characterized oligosaccharides from human

---

[61] H. Snippe, J. E. G. van Dam, A. J. van Houte, J. M. N. Willers, J. P. Kamerling, and J. F. G. Vliegenthart, *Infect. Immun.* **42**, 842 (1983).

[62] C. Faille, J. C. Michalski, G. Strecker, D. W. R. Mackenzie, D. Camus, and D. Poulain, *Infect. Immun.* **58**, 3537 (1990).

[63] M. P. Hayette, G. Strecker, C. Faille, D. Dive, D. Camus, D. W. R. Mackenzie, and D. Poulain, *J. Clin. Microbiol.* **30**, 411 (1992).

[63a] B. Schmitz, J. Peter-Katalinic, H. Egge, and M. Schachner, *Glycobiology* (in press).

milk, an adhesive specificity for lactose-related sequences linked to lipids was revealed for *E. coli* (unrelated to the presence of fimbriae).[47] By similar approaches, binding specificities related to lipid-linked lactose or type 1 (Galβ1-3GlcNAc) or type 2 (Galβ1-4GlcNAc) disaccharide units have also been shown for *Pseudomonas aeruginosa* isolated from patients with cystic fibrosis.[23,64] An adhesive specificity for *Fusobacterium nucleatum* (a bacterial component of human gingival plaque in patients with periodontal diseases) was described toward galactose termini on N-linked chains derived from a human protein-rich salivary glycoprotein.[65]

A binding specificity of human serum mannan-binding protein toward high mannose-type N-linked oligosaccharides of the envelope glycoprotein gp120 of human immunodeficiency virus was shown,[37] using neoglycolipid derived from both natural, virus-derived, and recombinant glycoprotein produced in Chinese hamster ovary cells.

Studies of the Combining Specificities of Endogenous Carbohydrate Proteins

In our laboratory, neoglycolipids have been extensively used in studies of the adhesive specificities of mammalian carbohydrate-binding proteins, including members of the C-type lectin, S-type lectin, and pentraxin families, and several different mechanisms have been revealed by which biological specificities may arise through carbohydrate recognition. A carbohydrate-binding specificity has also been assigned[65a] by Schmitz and colleagues to a member of the immunoglobulin superfamily as discussed below.

C-Type Lectin Family. Neoglycolipid libraries derived from various N-glycosylated proteins, and from free oligosaccharides isolated from human milk, urine, and other sources, have allowed close comparisons of the binding specificities of the serum proteins conglutinin and mannan-binding proteins.[34,41,44,45] Both protein types show multiple binding specificities toward N-linked oligosaccharides terminating in mannose or *N*-acetylglucosamine, and toward fucooligosaccharides with fucose 4-linked to *N*-acetylglucosamine (Le^a^-type). And yet the proteins have differing biological activities in the complement system: only conglutinin shows carbohydrate-mediated binding to the complement glycoprotein

[64] R. Ramphal, C. Carnoy, S. Fievre, J.-C. Michalski, N. Houdret, G. Lamblin, G. Strecker, and P. Roussel, *Infect. Immun.* **59**, 700 (1991).

[65] B. L. Gillece-Castro, A. Prakobphol, A. L. Burlingame, H. Leffler, and S. J. Fisher, *J. Biol. Chem.* **266**, 17358 (1991).

[65a] R. Horstkorte, M. Schachner, J. P. Magyar, T. Vorherr, and B. Schmitz, *J. Cell Biol.* **121**, 1409 (1993).

iC3b. This glycoprotein contains high mannose-type oligosaccharides, and once released from protein these oligosaccharides are bound both by conglutinin and human mannan-binding protein (D. Solis, W. Loveless, C.-T. Yuen, M. S. Stoll, A. M. Lawson, and T. Feizi, unpublished observations, 1993). There are subtle differences in the combining specificities of the two proteins: (1) conglutinin (but not mannan-binding protein) shows preferential binding to the high mannose-type oligosaccharides, $Man_7GlcNAc_2$ and $Man_8GlcNAc_2$, with multiple terminal 2-linked mannose residues[34]; and (2) mannan-binding protein binds, in addition, to the SSEA/Le^x-type fucooligosaccharide (lacto-N-fucopentaose III) with fucose 3-linked to N-acetylglucosamine.[44] It seems likely that the different biological activities with respect to iC3b binding arise from the subtle differences in combining specificities, and differential recognition of the high mannose-type oligosaccharides in the context of the carrier protein. In the case of pulmonary surfactant protein A, there is evidence for carbohydrate recognition in the context of particular carrier lipids.[46] For example, the strength of binding to galactosylceramides and to dihexosyl ceramides is influenced by the structures of their lipid moieties, and the corresponding neoglycolipids are weakly bound. The importance of cooperativity for high-avidity carbohydrate binding is apparent from the following observations with natural mannan-binding protein and pulmonary surfactant protein A (both are soluble, oligomeric proteins), the recombinant combining domain of mannan-binding protein (this oligomerizes spontaneously) and of surfactant protein A (exists as a monomer), and the recombinant full-length (transmembrane, monomeric) form of the human endothelial cell adhesion molecule E-selectin[39,44-46]: (1) Only with the oligomeric forms of the soluble proteins could carbohydrate binding be detected in solid-phase binding assays using lipid-linked oligosaccharides immobilized on chromatograms or plastic microwells; (2) the monomeric combining domain of the surfactant protein could inhibit the binding of the natural oligomeric protein; (3) the binding activity of the cell membrane-associated E-selectin toward sialyl-SSEA-1/Le^x and sialyl-Le^a-type oligosaccharides was highly dependent on its density of expression at the cell surface; moreover, cells with the highest density of E-selectin expression showed, in addition, binding to non-sialyloligosaccharide analogs of SSEA-1/Le^x, Le^a, and Le^b types. Collectively these results highlight ways in which biological activities may be (1) maintained for soluble proteins by oligomeric state, and the appropriate presentation of their oligosaccharide ligands on proteins or lipids, or (2) regulated for membrane-associated lectins by variation in levels of expression (degrees of operational oligomerization) and by the degree of clustering of their oligosaccharide ligands.

The neoglycolipid technology has the potential to detect unique or hitherto unsuspected oligosaccharide ligands from among mixtures of unknown oligosaccharides derived from glycoproteins, as in the case of the selectins described in previous sections of this chapter.

S-Type Lectin. Studies of the adhesive specificity of the widely distributed soluble 14-kDa (β-galactoside-binding) lectin, using immobilized neoglycolipids and natural glycolipids, have shown that in lipid-linked oligosaccharides, the disaccharide ligands Galβ1-4Glc/GlcNAc need to be presented as part of tetra- or longer oligosaccharides.[38] This contrasts with the ability of the lectin to bind to these oligosaccharides coupled onto agarose via divinyl sulfone.

Pentraxin. A further example of multispecificity among endogenous lectins is illustrated by amyloid P protein, a component of all forms of amyloid deposits. A binding specificity toward mannose-6-phosphorylated oligosaccharides (as found on lysosomal hydrolases) was detected by the neoglycolipid technology, and toward several sulfated glycolipids, in addition to the known binding of glycosaminoglycans heparin, heparan sulfate, and dermatan sulfate.[66] It was suggested that these properties may be clues to a role for amyloid P in the formation of hard-to-dissolve multicomponent complexes characteristic of amyloid deposits, and the mechanism of the generation of fibril-forming peptides by proteolysis of amyloidogenic proteins.

Immunoglobulin-Like. Neoglycolipids derived from high-mannose type and other types of oligosaccharides have been instrumental in showing that the neural cell adhesion molecule NCAM binds high-mannose type oligosaccharides.[65a] Moreover, evidence has been obtained that oligosaccharides of this type are the recognition structures on a second cell adhesion molecule L1 to which NCAM binds via its fourth immunoglobulin-like domain.[65a]

Studies of Glycosyltransferases

In glycosyltransferase studies, it is highly desirable to have acceptor substrates that are conveniently prepared and readily assayed. Provided it is taken into account that activities of these enzymes may be influenced by the aglycon moieties, the neoglycolipid technology is potentially a versatile means of substrate generation and product analysis. This was reported by Pohlentz *et al.*[22] in a study of sialyltransferases from rat liver Golgi vesicles, using lactose–DHPE and *N*-acetyllactosamine–DHPE as acceptor substrates.

---

[66] R. W. Loveless, G. Floyd-O'Sullivan, J. G. Raynes, C.-T. Yuen, and T. Feizi, *EMBO J.* **11**, 813 (1992).

Studies of Oligosaccharide-Mediated Cell Signaling

In functional studies of glycoconjugates, neoglycolipids show remarkable promise. Early studies of Wiegandt and associates[18] showed that the sialooligosaccharide, $Gal\beta1$-3GalNAc$\beta1$-4(NeuAcα2-3)Gal$\beta1$-4Glc, derived from G_{M1}-ganglioside and conjugated to stearylamine, is bound by cholera toxin, as is the natural glycolipid. Pacuszka *et al.*[67] and Pacuszka and Fishman[68] have now examined the properties of this oligosaccharide linked to DPPE with respect to uptake and metabolism by cultured rat glioma C6 cells that are deficient in G_{M1}-ganglioside. As with the natural glycolipid, the neoglycolipid G_{M1}–DPPE is not only taken up and inserted into the plasma membranes of these cells, where it serves as an effective receptor for cholera toxin, but it is internalized and metabolized. In particular, there is evidence that the internalized neoglycolipid is sorted both to the Golgi compartment (further sialylated) and to lysosomes (degraded to varying degrees). Of special interest is the observation[67] that despite the presence of an artificial lipid moiety, the exogenously added neoglycolipid, like the native glycolipid, serves as a functional receptor that effectively mediates a rise in cAMP levels when the cells are exposed to cholera toxin. As suggested by these authors,[67] neoglycolipids with different lipid moieties may be useful for exploring the functions of gangliosides in other systems. On a more general note, neoglycolipids with different oligosaccharide and lipid moieties hold promise in the unraveling of elements in the cell signaling pathways of the type proposed[69,70] on the basis of indirect evidence.

Addendum

In the course of preparing Le^a or Le^x type neoglycolipids, which have fucose 3- or 4-linked, respectively, to *N*-acetylglucosamine at the reducing end, we have noted that side reactions occur; these include dehydration and peeling. These side reactions are largely eliminated if 150 m*M* potassium trifluoroacetate and 150 m*M* potassium borate adjusted to pH 5 are included in the conjugation reaction mixture. The mechanism of this effect is still under investigation but the procedure may be used to prevent dehydration of neoglycolipids during conjugation of HexNAc terminating oligosaccharides. Excess salt should be removed after conjugation (see Desalting of Reaction Mixture).

[67] T. Pacuszka, R. M. Bradley, and P. H. Fishman, *Biochemistry* **30**, 2563 (1991).
[68] T. Pacuszka and P. H. Fishman, *Biochim. Biophys. Acta* **1083**, 153 (1991).
[69] T. Feizi and R. A. Childs, *Trends Biochem. Sci.* **10**, 24 (1985).
[70] T. Feizi and R. A. Childs, *Biochem. J.* **245**, 1 (1987).

Author Index

Numbers in parentheses are footnote reference numbers and indicate that an authors work is referred to although the name is not cited in the text.

H

Q

R

Subject Index

β-galactosidase (E.coli)

α-mannosidase (jack bean) pg 441, 284

Endoglycosidase F, N-glycosidase free < high mannose / biant complex + hybrid / no tri or tetra structures

O-glycosidase, BSA-free (some O-linked)

N-Glycosidase F, Recombinant < high mannose / hybrid / complex